Stem Cell Research on Cardiology

Stem Cell Research on Cardiology

Editor

Robert David

MDPI • Basel • Beijing • Wuhan • Barcelona • Belgrade • Manchester • Tokyo • Cluj • Tianjin

Editor
Robert David
University of Rostock
Germany

Editorial Office
MDPI
St. Alban-Anlage 66
4052 Basel, Switzerland

This is a reprint of articles from the Special Issue published online in the open access journal *Cells* (ISSN 2073-4409) (available at: https://www.mdpi.com/journal/cells/special_issues/stem_cell_cardiology).

For citation purposes, cite each article independently as indicated on the article page online and as indicated below:

LastName, A.A.; LastName, B.B.; LastName, C.C. Article Title. *Journal Name* **Year**, *Article Number*, Page Range.

ISBN 978-3-03943-192-2 (Pbk)
ISBN 978-3-03943-193-9 (PDF)

Cover image courtesy of Paula Müller and Heiko Lemcke.

Contents

About the Editor . . . . **vii**

Preface to "Stem Cell Research on Cardiology" . . . . **ix**

Mostafa Samak and Rabea Hinkel
Stem Cells in Cardiovascular Medicine: Historical Overview and Future Prospects
Reprinted from: *Cells* **2019**, *8*, 1530, doi:10.3390/cells8121530 . . . . **1**

Sophie Kussauer, Robert David and Heiko Lemcke
hiPSCs Derived Cardiac Cells for Drug and Toxicity Screening and Disease Modeling:
What Micro- Electrode-Array Analyses Can Tell Us
Reprinted from: *Cells* **2019**, *8*, 1331, doi:10.3390/cells8111331 . . . . **29**

**András Horváth, Torsten Christ, Jussi T. Koivumäki, Maksymilian Prondzynski, Antonia
T. L. Zech, Michael Spohn, Umber Saleem, Ingra Mannhardt, Bärbel Ulmer, Evaldas
Girdauskas, Christian Meyer, Arne Hansen, Thomas Eschenhagen and Marc D. Lemoine**
Case Report on: Very Early Afterdepolarizations in HiPSC-Cardiomyocytes—An Artifact by
Big Conductance Calcium Activated Potassium Current ($I_{bk,Ca}$)
Reprinted from: *Cells* **2020**, *9*, 253, doi:10.3390/cells9010253 . . . . **59**

**Haval Sadraddin, Ralf Gaebel, Anna Skorska, Cornelia Aquilina Lux, Sarah Sasse,
Beschan Ahmad, Praveen Vasudevan, Gustav Steinhoff and Robert David**
CD271$^+$ Human Mesenchymal Stem Cells Show Antiarrhythmic Effects in a Novel Murine
Infarction Model
Reprinted from: *Cells* **2019**, *8*, 1474, doi:10.3390/cells8121474 . . . . **73**

**Ruben Daum, Dmitri Visser, Constanze Wild, Larysa Kutuzova, Maria Schneider,
Günter Lorenz, Martin Weiss, Svenja Hinderer, Ulrich A. Stock, Martina Seifert
and Katja Schenke-Layland**
Fibronectin Adsorption on Electrospun Synthetic Vascular Grafts Attracts Endothelial
Progenitor Cells and Promotes Endothelialization in Dynamic In Vitro Culture
Reprinted from: *Cells* **2020**, *9*, 778, doi:10.3390/cells9030778 . . . . **87**

Sara Barreto, Leonie Hamel, Teresa Schiatti, Ying Yang and Vinoj George
Cardiac Progenitor Cells from Stem Cells: Learning from Genetics and Biomaterials
Reprinted from: *Cells* **2019**, *8*, 1536, doi:10.3390/cells8121536 . . . . **117**

Yehuda Wexler and Udi Nussinovitch
The Diagnostic Value of Mir-133a in ST Elevation and Non-ST Elevation Myocardial Infarction:
A Meta-Analysis
Reprinted from: *Cells* **2020**, *9*, 793, doi:10.3390/cells9040793 . . . . **173**

**Marcus-André Deutsch, Stefan Brunner, Ulrich Grabmaier, Robert David, Ilka Ott
and Bruno C. Huber**
Cardioprotective Potential of Human Endothelial-Colony Forming Cells from Diabetic and
Nondiabetic Donors
Reprinted from: *Cells* **2020**, *9*, 588, doi:10.3390/cells9030588 . . . . **187**

**Sarah Sasse, Anna Skorska, Cornelia Aquilina Lux, Gustav Steinhoff, Robert David
and Ralf Gaebel**
Angiogenic Potential of Bone Marrow Derived CD133$^+$ and CD271$^+$ Intramyocardial Stem Cell
Trans- Plantation Post MI
Reprinted from: *Cells* **2020**, *9*, 78, doi:10.3390/cells9010078 203

Feixiang Ge, Zetian Wang and Jianzhong Jeff Xi
Engineered Maturation Approaches of Human Pluripotent Stem Cell-Derived Ventricular
Cardiomyocytes
Reprinted from: *Cells* **2020**, *9*, 9, doi:10.3390/cells9010009 219

**Paula Mueller, Markus Wolfien, Katharina Ekat, Cajetan Immanuel Lang, Dirk Koczan, Olaf
Wolkenhauer, Olga Hahn, Kirsten Peters, Hermann Lang, Robert David and Heiko Lemcke**
RNA-Based Strategies for Cardiac Reprogramming of Human Mesenchymal Stromal Cells
Reprinted from: *Cells* **2020**, *9*, 504, doi:10.3390/cells9020504 229

**Hui-Yung Song, Chian-Shiu Chien, Aliaksandr A. Yarmishyn, Shih-Jie Chou, Yi-Ping Yang,
Mong-Lien Wang, Chien-Ying Wang, Hsin-Bang Leu, Wen-Chung Yu, Yuh-Lih Chang and
Shih-Hwa Chiou**
Generation of *GLA*-Knockout Human Embryonic Stem Cell Lines to Model
Autophagic Dysfunction and Exosome Secretion in Fabry Disease-Associated
Hypertrophic Cardiomyopathy
Reprinted from: *Cells* **2019**, *8*, 327, doi:10.3390/cells8040327 249

**Ludwig T. Weckbach, Andreas Uhl, Felicitas Boehm, Valentina Seitelberger, Bruno C. Huber,
Gabriela Kania, Stefan Brunner and Ulrich Grabmaier**
Blocking LFA-1 Aggravates Cardiac Inflammation in Experimental Autoimmune Myocarditis
Reprinted from: *Cells* **2019**, *8*, 1267, doi:10.3390/cells8101267 267

**Praveen Vasudevan, Ralf Gaebel, Piet Doering, Paula Mueller, Heiko Lemcke, Jan Stenzel,
Tobias Lindner, Jens Kurth, Gustav Steinhoff, Brigitte Vollmar, Bernd Joachim Krause,
Hueseyin Ince, Robert David and Cajetan Immanuel Lang**
18F-FDG PET-Based Imaging of Myocardial Inflammation Predicts a Functional Outcome
Following Transplantation of mESC-Derived Cardiac Induced Cells in a Mouse Model of
Myocardial Infarction
Reprinted from: *Cells* **2019**, *8*, 1613, doi:10.3390/cells8121613 275

**Chiara Gardin, Letizia Ferroni, Christian Latremouille, Juan Carlos Chachques,
Dinko Mitrečić and Barbara Zavan**
Recent Applications of Three Dimensional Printing in Cardiovascular Medicine
Reprinted from: *Cells* **2020**, *9*, 742, doi:10.3390/cells9030742 293

Dominik Schüttler, Sebastian Clauss, Ludwig T. Weckbach and Stefan Brunner
Molecular Mechanisms of Cardiac Remodeling and Regeneration in Physical Exercise
Reprinted from: *Cells* **2019**, *8*, 1128, doi:10.3390/cells8101128 327

About the Editor

Robert David

BIRTHDATE: 1967-05-11

CURRENT INSTITUTE: Clinic for Heart Surgery, University Medical Center Rostock

ACADEMIC BACKGROUND

From-Until: Institution, Area of Specialization, Degree

Since 2016: Clinic for Heart Surgery, University Medical Center Rostock, Cardiovascular repair strategies, Univ.-Prof. permanent

Since 2012: Reference and Translation Center for Cardiac Stem Cell Therapy, University Medical Center Rostock, Cardiovascular repair strategies, Univ.-Prof. Tenure track

2001–2012: Medizinische Klinik und Poliklinik I, Klinikum Großhadern, Stem cell based cardiovascular repair, Priv.-Doz.

1998–2001: University of Ulm, Xenopus embryology, Postdoc

1994–1998: Max-Planck-Institute for Developmental Biology, Tübingen, Xenopus embryology, Ph.D.

1987–1993: Technical University of Munich, Cell Biology, Dipl. Biol.

Major research interests:

Cardiovascular programming and reprogramming

Targeted cardiovascular gene transfer

Novel imaging technologies.

Preface to "Stem Cell Research on Cardiology"

At present, stem cell research in cardiology ranges from fundamental research over translational preclinical experiments up to promising clinical trials, to investigate a putative new medicinal product. The aim of all these efforts is to improve treatment opportunities for cardiac patients and to prevent severe disease courses.

In this context, new technologies are constantly emerging, and offer new impressive possibilities to analyze data and to generate testing capacities suitable for clinical application.

This Special Issue of Cells comprises highly relevant original articles and brief communications, as well as informative reviews analyzing the current situation of cardiac stem cell research. Here, also new elementary aspects of several myocardial diseases and cardiac remodeling, with respect to physiological and pathological events, e.g. arrhythmia and inflammation, are lit up. Furthermore, programming approaches, maturation strategies and the characterization of cardiac cells for translational research are addressed, and the reader will be informed about prospective approaches in 3D cell aggregate application in the field.

We expect this collection of articles to provide valuable input for future research and further discussions in the wide field of cardiac regenerative medicine.

We sincerely wish to thank all authors who have contributed to this Special Issue.

Robert David
Editor

Review

Stem Cells in Cardiovascular Medicine: Historical Overview and Future Prospects

Mostafa Samak [1,2] **and Rabea Hinkel** [1,2,*]

[1] Department of Laboratory Animal Science, Leibnitz-Institut für Primatenforschung, Deutsches Primatenzentrum GmbH, Kellnerweg 4, 37077 Göttingen, Germany; MSamak@dpz.eu
[2] DZHK (German Centre for Cardiovascular Research), Partner Site Göttingen, 37075 Göttingen, Germany
* Correspondence: rhinkel@dpz.eu

Received: 22 October 2019; Accepted: 23 November 2019; Published: 27 November 2019

Abstract: Cardiovascular diseases remain the leading cause of death in the developed world, accounting for more than 30% of all deaths. In a large proportion of these patients, acute myocardial infarction is usually the first manifestation, which might further progress to heart failure. In addition, the human heart displays a low regenerative capacity, leading to a loss of cardiomyocytes and persistent tissue scaring, which entails a morbid pathologic sequela. Novel therapeutic approaches are urgently needed. Stem cells, such as induced pluripotent stem cells or embryonic stem cells, exhibit great potential for cell-replacement therapy and an excellent tool for disease modeling, as well as pharmaceutical screening of novel drugs and their cardiac side effects. This review article covers not only the origin of stem cells but tries to summarize their translational potential, as well as potential risks and clinical translation.

Keywords: iPSC; PSC; ESC; cardiovascular disease; regeneration

---

1. Introduction

Cardiovascular diseases (CVDs) remain a plight to modern-day humans, accounting for over one-third of all deaths worldwide, according to recent World Health Organization (WHO) estimates [1]. In the US alone, one person dies of CVD-related complications every 40 s, mostly ischemic attacks [2]. To this day, catheter-based or surgical interventions, e.g., coronary bypass and implantation of assist devices, are by far the most widely applied clinical measures—albeit with several complications [3,4]. Despite great improvements, most surgical interventions available are mere preservatives, i.e., attempts to sustain the functionally intact heart tissue for as long as possible without structural compensation. Howbeit, due to the progressive nature of CVDs, heart failure (HF) is, in most cases, inevitable [5]. Regardless of etiology and severity, many end-stage HF patients will eventually need cardiac transplantation [6]. With very few treatment options, not to mention the paucity of available donor hearts, the need for alternative therapeutic measures is indispensable.

In recent decades, stem cell (SC) technologies have emerged with a great promise that could be envisaged for almost all human ailments, most importantly for noncommunicable diseases characterized by organ dysfunction and/or degeneration. In this regard, CVDs are certainly the most attractive target for SC-based therapeutic approaches [7–10]. From a mere improvement of cardiac microenvironment, to partial regeneration and/or compensation of lost functional tissue, and ending with a complete fabrication of a surrogate heart, SCs have set the hopes high. Moreover, SC-based technologies have enabled great in-depth understanding of the pathogenesis of CVD entities and served as a platform to test novel therapeutic approaches at minimal risk of adverse events to patients and much lower costs. This article aims at reviewing the available knowledge on SCs and their applications for cardiovascular research, highlighting milestone achievements in both basic and translational research, and expanding in particular on pluripotent SCs.

2. Adult Stem Cells

The body's regenerative capacity is a well-ingrained piece of knowledge from ancient times. Modern science attributes this phenomenon to the presence of resident SC niches in different organs and tissue, i.e., adult SCs. These cells are undifferentiated, but they are capable of self-renewal and differentiation to one or more cell type, which sets them apart on a potency spectrum, e.g., multipotent SCs. Adult SCs' regenerative potential becomes even more conspicuous in organs and/or tissues with high turnover rates, but, more importantly, as a response to tissue injury. A wealth of knowledge is now available on different adult SC populations, and efforts have been made to reap the benefits of these cells to treat CVDs. We highlight below a few examples of adult SCs, which declared themselves as powerful research targets for cardiovascular medicine and made their way to the clinic.

2.1. Skeletal Myoblasts

Intuitively, due to embryonic and morphologic commonalities between skeletal and cardiac muscle tissues, skeletal myoblasts have been among the early attractive research targets for cardiac regeneration. Skeletal myoblasts (SM) constitute a group of satellite cell-derivatives residing within skeletal muscle fibers, which are activated upon injurious insults to migrate, proliferate, and differentiate, forming new muscle fibers, i.e., myogenesis [11]. Facilitated by their being readily accessible from autologous muscle biopsies, rapid in vitro expansion, ischemic tolerability, and low risk of tumorigenicity, the cardiac regenerative potential of SMs has been the subject of several preclinical investigations in both small and large animal models of CVDs [12–17]. Indeed, results from these studies have demonstrated positive outcomes by reducing infarct size, as well as myocardial fibrosis, thwarting ventricular remodeling and improving overall cardiac function. Consequently, several clinical trials were initiated to verify their efficacy [18–22]. Despite initially reported improvements in cardiac parameters of patients transplanted with SMs, many have experienced ventricular arrhythmias, which were later attributed to the lack of electromechanical coupling between the transplanted SM-derived myotubules and resident cardiomyocytes where they failed to form gap junctions [23–26]. Furthermore, larger randomized, placebo-controlled, double-blinded clinical studies not only failed to show any therapeutic benefits of SMs in patients with severe ischemic heart disease at both short- and/or long-term follow-up, but also reported postoperative arrhythmic events even upon prophylactic pharmacological treatment [27–31]. As a result, SMs have lost their popularity as SCs for cardiac applications.

2.2. Bone-Marrow-Derived SCs

Since the mid-20th century, the BM has long been praised for its SC abundance. BM transplantation has been a clinical practice since the mid to late 1960s, intended for correction hematologic, as well as immune disorders. However, reports from the late 1990s first demonstrating the ability of BM-derived cells to migrate to injured tissues and support regeneration have instigated a wave of research on their therapeutic potentials for CVDs [32,33]. Indeed, early studies in animal models of MI corroborated the aforementioned expectations. The first tentative clinical translation of this finding was reported in 2001 in Düsseldorf, Germany, where a MI patient received autologous BM-derived nucleated cells upon catheter angioplasty and reported positive outcomes [34]. This was followed by several controlled clinical studies, albeit with inconsistent findings [35,36].

Generally speaking, BM-derived SCs can be sub-grouped into two large cell populations; hematopoietic (HSCs) and nonhematopoietic SCs. HSCs give rise to all blood-cell types and include a subpopulation of pro-vasculogenic endothelial progenitor cells (EPCs), which can be found in the circulating blood among others [37]. Of the nonhematopoietic BM-derived SCs, mesenchymal stromal/stem cells (MSCs) are the most studied, due to their greater multipotency, manifested in their ability to differentiate into osteoblasts, adipocyte, and chondrocytes under defined in vitro conditions, adding to their reported immune-modulatory and anti-inflammatory properties [38]. With better characterization of these cells based on surface-marker expression, studies were led, examining the therapeutic potential of each BM-derived SC type.

For example, BM-derived CD133- and/or CD34-positive HSCs were utilized for phase I and II clinical trials, where patients of MI received intramyocardial transplantation or intracoronary injections of these cells. Despite short-term follow-ups showing positive outcomes, characterized by enhanced left-ventricular ejection fraction (LVEF) along enhanced myocardial perfusion, these studies failed to show any long-term benefits [39,40]. Most recently, results from randomized, placebo-controlled, double-blinded phase III clinical trials also showed a congruent trajectory [41].

On the other hand, MSCs (CD73-, CD105-, and CD90-positive) have been a subject of greater scrutiny in both basic and translational research. Adding to their paracrine- and exosome-mediated immunosuppressive properties, MSCs are unique in their ability to evade the immune system [42]. This is largely due to their moderate levels of HLA class I expression, while lacking the expression of HLA class II, B7, and CD40 ligand conferring privilege to the immune system of their host, thus enabling allogenic transplantation without the need of concomitant immunosuppression [42,43]. Indeed, studies in large animals have shown improvements in LVEF upon MSC therapy in the setting of myocardial ischemia. Nevertheless, results from translational attempts of these findings in clinical studies fall into a wide spectrum of significance with regard to their benefits, notwithstanding their mode of transplantation (i.e., autologous vs. allogenic) [44]. Despite some showing significant improvements in patients with acute MI, other randomized controlled studies concluded no significant differences [45–48]. Nonetheless, two randomized pilot studies were conducted in 2012 and 2017 in patients with ischemic cardiomyopathy (ICM) and nonischemic dilated cardiomyopathy (NIDCM), respectively, comparing autologous to allogenic MSC therapy [49,50]. Results from these studies, also known as POSEIDON, alluded to the efficacy of MSC therapy in these patient cohorts, with superiority given to allogenic transplantation. However, these studies were limited to the small sample size and lack of a placebo control group.

2.3. Cardiac Progenitor Cells and Stem Cell Niches

Indeed, the heart's endogenous regenerative capacity has been an area of extensive research over the past decades. Contrary to the long-held dogma of being a postmitotic organ, studies have challenged this notion, claiming that the mammalian heart is indeed capable of self-regeneration, albeit exiguously. Studies using mitotic index, as well as DNA labeling, have conveyed the finding that cardiomyocytes can self-renew during adulthood. However, debates have flared as to what extent this self-renewal takes place, and even to the reliability of the methods used to quantify it. Herein, nuclear labeling is not reliable, due to the characteristic polyploidy that human CMs undergo during growth or disease [51–54]. Radiocarbon (^{14}C) dating, on the other hand, has provided more accurate estimates of cardiomyocyte turnover in the adult heart [55]. Interestingly, studies have shown a significant increase in cardiomyocyte count and/or ploidy in neonatal and preadolescent life in both rodents and humans, which contributed to heart growth [56,57].

Furthermore, the existence of SC niches harboring cardiac progenitor cells (CPCs) has also been reported and highlighted by research as evidence of the heart's regenerative capacity, notwithstanding another yet-unresolved debate [51,58]. CPCs are multipotent as was shown by their ability to differentiate to cardiac cell lineages, including cardiomyocytes; they were claimed to confer cardiac tissue repair and regeneration. As a heterogeneous population of cells, they are each identified by expression of distinct markers. Of these cells, c-kit-, Isl1-, or epicardial Tbx18-positive (also WT1-positive) cells are three heavily studied cell populations due to their cardiomyogenic differentiation potential attributed during development, neonatal life, and even in adult hearts.

The c-kit-expressing cells are the most studied CPCs, however, with contradicting reports regarding their significance for cardiac cell repair the in adult postinjury [59–61]. Despite their demonstrated contribution to cardiac regeneration in the neonatal hearts, c-kit-positive CPCs' role in the adult setting of myocardial injury is largely debated [60,62,63]. A recent report alluded to the role of c-kit-positive cells in cardiac adaptation to injury, where c-kit was shown to be upregulated in response to pathological stress [64]. Furthermore, a RNA-sequencing study recently showed that c-kit-positive cells transiently adopt a cardiomyocyte-like pattern of gene expression upon myocardial infarction in vivo [65]. Contrary to these findings, more recent studies by Li and colleagues refuted the myogenic potential of these cells in the adult by using a new genetic-lineage tracing system [66]. Furthermore, the same group has shown that early segregation of myocytes and nonmyocytes during embryonic development (E10.5 to E11.5) is the cut-off line beyond which no contribution to new cardiomyocyte formation occurs, even during neonatal life [67]. Moreover, a study published earlier this year by Elhelaly and colleagues argued that c-kit-positive cells do not contribute to cardiomyogenesis, even during neonatal life [68]. Howbeit, the commonly agreed-upon consensus in the field is that CPCs are remnant SCs from developmental stages whose role in the adult heart, if any, confines to maintaining cardiac tissue homeostasis, and their cardiomyogenic potential in the context of injury is inexistent [69,70]. Importantly, however, the repercussions of the aforementioned findings instigated a wave of research endeavors to exploit the heart's endogenous regenerative capacity for novel therapeutic interventions. In summary, the field of cardiac progenitor cells is controversy discussed, and the regenerative potential (and existence) of the cells in the adult human heart need further investigations.

3. Pluripotent Stem Cells

Despite the efforts that have been made with adult SCs, none of these cells could meet the expectations as a reliable treatment for CVDs. That is because not even the most potent adult SC could provide an appreciable source for myocardial tissue regeneration and/or functional compensation for the lost contractile element of the heart, e.g., as a result of infarction, let alone cardiomyopathies or congenital heart disease [71]. In this regard, the pursuit after functional CMs calls for a different type of SCs, i.e., pluripotent SCs (PSCs).

3.1. Embryonic Stem Cells

The ability of a cell to give rise of all three germ layers of the developing embryo, i.e., pluripotency, is the most vivid and sought-after character of SCs, not only in the context of regenerative medicine, but also for basic research purposes. Pluripotency of embryonic blastocyst inner mass cells was first shown in the mouse as early as 1981 by Evans et al. [72]. In 1998, Thomson et al. first reported the generation of pluripotent embryonic stem cells (ESCs, Figure 1) from human blastocysts, which are capable of self-renewal and differentiation to all three germ layers [73]. Nevertheless, ethical considerations have long hovered over human ESCs (hESCs), as their derivation entails destruction of an embryo. This has prompted legislative issues, in that many countries have imposed bans on their use and/or research funding [74,75]. To add insult to injury, ESCs' ability to form teratomas (tumors of mixed germ layers) when transplanted undifferentiated has further flared the argument against their clinical application, despite efforts to enhance differentiations and purifications protocols [76].

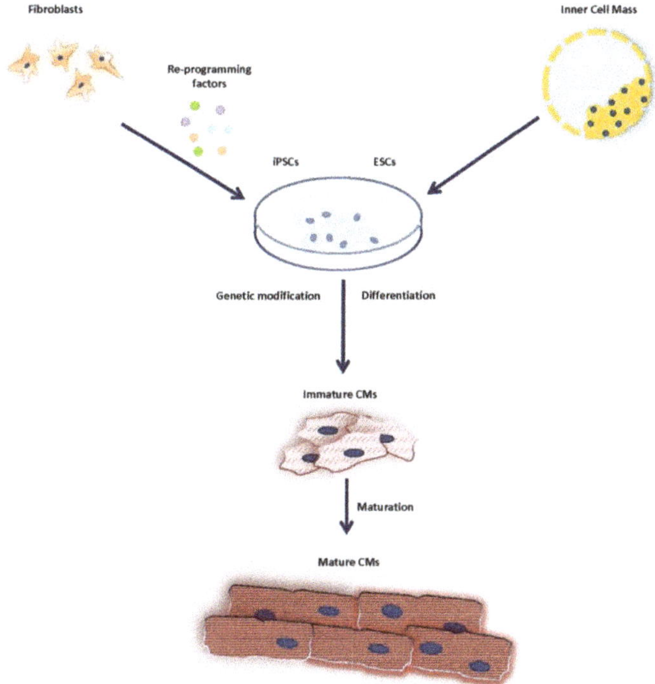

Figure 1. Differentiation of cardiomyocytes derived from pluripotent stem cells.

3.2. Induced Pluripotent Stem Cells

It was not long until the not-so-bright picture of SC research changed. Inspired by the preexisting knowledge of master regulator genes capable of imparting cellular identities, Takahashi and Yamanaka developed the first technique of somatic cell reprogramming in 2006 [77,78]. In their Nobel Prize experiment, induced pluripotent stem cells (iPSCs, Figure 1) could be generated from somatic cells, such as skin fibroblasts, by expression of four transcription factors that were found to be crucial for cellular reprograming to ESC-like inner mass cells, namely Oct3/4, Sox2, c-Myc, and Klf4 [78]. Ever since, scientist have raced to improve the reprogramming efficiencies of iPSCs by manipulating the transcription-factor cocktail and selecting for expression of other transcription factors, such as Nanog and Lin28 [79–81]. Generation of viable and tumor-free whole organisms with iPSCs that were capable of germ-line transmission was also made possible [82]. Unsurprisingly, human iPSCs (hiPSCs) were generated as soon as one year after their first generation in a mouse, and by the same pioneering group of scientists, as well as others [83,84].

3.3. Embryonic Stem Cells Versus Induced Pluripotent Stem Cells

The primary intended purpose of reprogramming of somatic cells and generation of iPSCs was to wipe the initial cellular identity and drive them back to the embryonic inner mass state, and hence serve as a surrogate for embryonically derived cells, i.e., ESCs. Indeed, iPSCs greatly resemble conventional ESCs in terms of growth characteristics, gene-expression profiles, epigenetic status, and developmental potential, which were shown in earlier studies by Yamanaka and colleagues, as well as others [79,84–86]. However, upon comparison of various undifferentiated cell lines, reports argued that iPSCs may not be perfectly identical to conventional ESCs. This is largely attributed to the unique epigenetic signatures of their parent somatic cells. Despite previous studies showing that somatic cells undergo epigenetic remodeling upon reprogramming, studies have shown that iPSCs indeed retain epigenetic patterns of their donor cells, e.g., CpG island methylation [86–90]. Furthermore, gene and miRNA expression signature were also shown to trail along with iPSCs (reviewed in [91]). Upon differentiation to CMs, further comparison of mature CMs differentiated from ESCs and iPSCs can be insightful. In this regard, CMs of either origin were reported to display similar ultrastructural phenotypes, upon electron microscopic examination [92]. In line with these findings, a study by Gupta et al. revealed that global transcriptional profiles of mature CMs derived from either human iPSCs or ESCs are highly similar [93]. However, iPSC-CMs were more likely to share some somatic cell signature with their undifferentiated iPSC-parents. Thus, identification of these variations between iPSC- and ESC-CMs, as well as the interline variability of either type of PSCs, is essential before they are utilized for disease modeling or clinical application.

Unlike ESCs, iPSCs derivation does not involve destruction of embryos, and hence does not fall into the same ethical pitfalls. However, other ethical considerations arose with hiPSCs, especially with regard to the possibility of reproductive cloning, the risk of generating genetically engineered human embryos, and, more extremely, human–animal chimeras [94]. Furthermore, and like ESCs, iPSCs are subject to safety concerns due to their ability to form tumors, even with rigorous protocols of differentiation and selection [95].

In recent years, substantial developments in stem cell technology in terms of reprogramming efficiency and enhancing their clinical applicability have prompted scientist to utilize pluripotent stem cells (PSCs), not only to regenerate, but also to model the human heart for basic research purposes. Furthermore, some countries have tentatively started to loosen their tight regulations, especially on hESCs; a step that coincided with the establishment of stem-cell registries in the US and Europe [96–98]. This has led to several initiatives on stem cell therapy for many disease conditions, including CVDs [99]. As promising as this may sound, several challenges, however, preclude the full realization of PSC-based therapy. In the following, we shall focus on PSCs by addressing efforts made over the past decades to optimize their generation, differentiation, and maturation for CVD research, as well as efficient delivery methods for late clinical and/or translational purposes.

4. Cardiac Stem-Ness

Embryology is the fundament for generation of cardiac cells from PSCs in the laboratory. The heart is the first organ to develop and function during embryogenesis [100]. In the lateral mesoderm, cardiac specification takes place, a process initiated by two T-box transcription factors, Eomesodermin and Brachyury(T), which have been shown to induce the expression of yet another critical factor, namely mesoderm posterior 1(MesP1) [101,102]. MesP1 is a basic helix-loop-helix (bHLH) transcription factor considered to be the master regulator orchestrating the differentiation and commitment of cardiac precursors [101,102]. Cardiac precursors then assume a crescent-shaped structure known as the cardiac crescent, at which cells are irreversibly committed to the cardiac lineage. This is marked by the expression of key transcription factors, namely Nkx2.5, GATA4 and Tbx5 [103]. Two waves of Nkx2.5 expression ensue, depicting the formation of two regions known as the first and second heart fields, which subsequently give rise to different heart chambers, as well as the cardiac outflow tract [76,103].

After all, heart development is a dynamic three-dimensional process governed by an intricate network of signals and gene transcription [104,105]. Howbeit, three major signaling pathways converge to drive the process, from early cardiac tissue specification of mesoderm progenitors to subsequent differentiation into cardiac progenitors, namely BMP (bone morphogenic protein) and Nodal/Activin, both being members of the TGF-β (transforming growth factor beta) cytokine family, and the Wnt/β-catenin [106,107]. Paracrine signals responsible for the fine-tuning of those pathways is crucial for heart development. For example, signals activating the Wnt/β-catenin pathways are essential for early mesoderm induction, whereas inhibitors of the same pathway are subsequently required for precardiac specification [76].

4.1. Generation/Differentiation of Pluripotent Stem Cells

Protocols for in vitro generation of cardiac stem cells (CSCs) from PSCs, either from ESCs or iPSCs, rely primarily on simulating the signaling microenvironment, which induces the aforementioned rudimental pathways, starting by initial epithelial to mesenchymal transition, mesodermal specification, and subsequent cardiogenic differentiation, followed by selection for cardiac markers [108–110]. The initially reported protocols relied simply on serum in culture medium as a source of inducing factors, observing spontaneous formation of aggregates called embryoid bodies (EBs) when cells are plated in suspension [111]. These EBs would later show contractions and positive staining for cardiac markers. This method was first reported in ESCs, however, with very low efficiency [111]. Nevertheless, the EB-based differentiation remained a standard protocol and was also the first differentiation method applied to generate CMs from mouse iPSCs only a couple of years after their first introduction in 2006 [112,113]. The first CMs generated from iPSCs were reported by a team of researchers from Leibniz institute in Germany, with few refinements introduced to the protocol, which led to the differentiation of typical CMs comparable to those generated from ESCs [112]. Interestingly, precisely at the same time and in the same journal issue, the iPSCs-founding team from Kyoto also published a systematic differentiation protocol of mouse iPSCs into cardiac lineages [113]. Nevertheless, and as mentioned before, the efficiency of the EB-based protocols was low, mainly due to the uncontrolled differentiation cues in the supporting media. One of the earliest and most cited protocols to differentiate ESCs to beating CMs was reported by Mummery et al. in 2003, where they delegated the differentiation cues to paracrine signaling of murine visceral endoderm-like cells (END-2) [114]. They compared their generated CMs to primary human fetal CMs, as well as primary human adult CMs, and reported comparable structural and functional properties. Improvements to differentiation protocols by temporal application of cytokines, as well as small molecule inhibitors (e.g., inhibitors of the Wnt pathway) to simulate the developmental processes have also been successful introduced to generate CMs from PSCs [108–110,115–117]. Furthermore, several groups have sought to simplify the differentiation protocols by using chemically defined culture media consisting of only a few components [115–117].

Nevertheless, differentiation of PSCs by using standard protocols usually yields a mixed population [118]. Thus, identification of selection markers is crucial for the purification of cardiomyocyte progenitors. Pioneering studies by Moretti and colleagues have greatly contributed to the refinement of selection protocols for cardiac organogenesis from PSCs [119]. From an embryological standpoint, myocyte progenitors are distinguished from nonmyocytes (vascular progenitors) by consistent expression of Isl1-1 transcription factor, along with Nkx2.5, whereas co-expression of Isl-1 and CD31 is a marker for endothelial progenitors [119]. Among myocytes, cardiomyocytes can be further distinguished from smooth muscle cells (SMCs) by expression of vascular cell adhesion molecule 1 (VCAM 1) and signal regulator protein alpha (SIRPα), both of which were reported to be reliable selection markers in culture conditions, yielding as much as 98% pure-cardiomyocyte populations by antibody-based sorting from PSCs [120,121]. Successful differentiation can be further confirmed by expression of other cardiomyocyte markers, such as cardiac troponins, e.g., TNNI1 [121,122]. Using lentiviral vectors, expression of selection markers, e.g., antibiotic-resistant genes or fluorescent proteins, under control of cardiomyocyte-specific promoter, has also been reported to purify cardiomyocytes [123,124]. Importantly, documented biochemical disparity between CMs and non-CMs in energy metabolism was also exploited for the so-called "metabolic purification" of CMs. In this regard, manipulation of culture conditions by altering the composition of the culture medium (e.g., glucose depletion, lactate, and glutamine supplementation) was found to be crucial for such nongenetic purification of CMs [125].

Finally, studies have pointed out the important role of MicroRNAs (small, noncoding RNAs that regulate gene expression by degradation of messenger RNAs) in CM phenotype differentiation [126–128].

4.2. Maturation of Pluripotent Stem Cells

To be utilized for disease modeling or regenerative medicine, one might expect PSC-CMs to recapitulate the structural and functional characteristics of adult CMs (Figure 1). Nevertheless, CM differentiation of PSCs usually yields immature cells, resembling the embryonic or fetal state [129]. This manifests in their morphology, gene expression, and electrophysiology. More recently, single-cell-transcriptomic analyses have proven to be a powerful tool to understand the transcriptional roadmap of in vitro CM differentiation, and therefore enable a better design of differentiation and maturation protocols [130,131]. The following highlights the major differences between immature PSC-derived CMs and mature and/or adult ones.

Morphologically, PSC-CMs are significantly smaller in size, compared to their adult or matured counterparts. Upon maturation, cells assume an elongated shape, reminiscent of adult CMs [132]. Sarcomeres are much less organized in immature PSC-CMs and become much more organized upon maturation, which usually correlates with isoform switch of sarcomeric proteins. A good example is troponin I, wherein different isoforms distinguish embryonic CMs from adult ones [133,134]. Stoichiometric replacement of the fetal troponin *TNNI1*, encoding slow skeletal troponin I (TnIs), gene with the adult *TNNI3*, encoding adult cardiac troponin I (TnIc), gene was reported in a study by Bedada et al. as a quantifiable marker for maturation in PSC-CMs [135]. Another well-characterized hallmark of mature CMs is the isoform switch of myosin heavy chain (MHC). Two isoforms exist, the alpha isoform (encoded by *MYH6*), also known as the faster isoform, and the beta isoform (encoded by *MYH7*), also known as the slower isoform [136]. Importantly, differences exist between rodents and humans in this regard. In small rodents (mice and rats) with faster heart rates, alpha-MHC isoform predominates and increases upon maturation, whereas, in bovine and human hearts, despite the presence of the alpha-MHC isoform, the beta MHC isoform usually predominates, regardless of the state of development, and increases with age [136,137]. However, most differentiation protocols of human PSC yield CM with both isoforms, but studies have shown that long-term cultures, especially on stiff substrates, lead to a greater shift toward the beta-isoform, reflecting maturation [138]. Titin is another key component of the sarcomere that undergoes isoform switch during maturation. Fetal titin isoforms N2BA 1 and 2 are more compliant, but they switch to the N2B isoform in postnatal and adult cardiomyocytes [139]. Genes encoding structural and force-generating myofibrillar proteins are much poorly expressed in in vitro maturated PSC-CMs when compared to adult- and fetal-heart samples [140]. This might be attributed to the absence of biomechanical stresses in vitro, which are normally present upon heart development in vivo [141].

Electrophysiological, and similar to contractile components, ion-transport related genes, such as those for voltage-gated potassium channels, e.g., *KCNJ2* and Ryanodine receptor *RYR2*, were poorly expressed in immature CMs [142]. The lower expression level of the *KCNJ2*-encoding membrane protein of the inward-rectifier current, as well as genes encoding beta-subunit members of the voltage-gated potassium channels, such as *KCNIP2*, *KCNAB1*, and *KCND3*, all affect both the inward-rectifier (I_{k1}) and the transient-outward (I_{to}) currents, respectively, leading to the characteristic "less negative" resting membrane potential in PSC-CMs (~−60 mV) compared to adult CMs (~−90 mV) [129,132,142,143]. Furthermore, studies have shown that PSC-CMs have few to no T-tubules, which are key components of excitation–contraction coupling (ECC) and a hallmark of mature and/or adult CMs; this is typified by unsynchronized Ca^{2+} transients in immature CMs [144].

Metabolically, immature CMs have few and underdeveloped mitochondria, accounting for a small fraction of the cell volume. Adult CMs, on the other hand, show highly developed, well-distributed, and dense mitochondria, accounting for ~20–40% of the adult myocyte volume. During development, hypoxia is an early trigger for mesoderm cardiac specification [145]. The growing heart, thus, resorts to glycolysis as a major source (80%) of energy. As CMs mature and become terminally differentiated, mitochondrial oxidative capacity increases, with fatty acid β-oxidation (80%) becoming a major source of energy [146]. PSC-CMs recapitulate both mitochondrial structure and glycolytic dependence of embryonic-state CMs [92,147]. Recent studies have shown that tweaking the culture media composition to mimic these metabolic changes, e.g., replacing high-carbohydrate, high-insulin, glucose-based, with low-carb, low-insulin fatty-acid-based media-enhanced maturation [148]. Table 1 summarizes the major differences between human-PSC-derived and adult CM.

Table 1. Major structural, electrophysiological and metabolic differences between human PSC-derived CMs and maturated/adulte CMs.

PSC-Derived CM	Mature/Adult CM
Smaller in size, roundish in shape	Larger in size, elongated in shape
Disorganized sarcomeres	Organized sarcomeres
Slow/skeletal troponin I (TnIs)	Adult cardiac troponin I (TnIc)
Titin N2BA isoform	Titin N2B isoform
Higher αMHC:βMHC	Lower αMHC:βMHC
Poor expression of ion-transport components genes (e.g., KCNJ2, RYR2)	High expression of ion-transport components genes
Less efficient Calcium handling	Improved Calcium handling
Less negative resting membrane potential	More negative resting membrane potential
No or few T-tubules	Abundant T-tubules
Few, underdeveloped mitochondria	Dense, well-distributed and developed mitochondria
Glucose as major energy source	Fatty acids as major energy source

Over the years, efforts have been made to enhance the maturation of PSC-CMs, and these include prolonged cultures, using stiff gel micro-patterned substrates, and application of electrical and/or biochemical stimuli [132,149–152]. The overall goal was to simulate the in vivo environment of the myocardium, where CMs are under constant physical, topographic, and humoral stimuli leading to their structural and functional maturation.

4.3. Engineered Heart Tissue

Importantly, the accumulated knowledge of cardiac stem cell biology and maturation has culminated in the so called "Engineered Heart Tissue" (EHT), a milestone achievement. The nascent EHT is attributed to work done by Zimmermann and Eschenhagen in the early 2000s [153,154]. Ever since, EHT technology has rapidly progressed through refinements in mechanical loading, electrical stimulation, medium supplementation, and miniaturization. The result was a 3D cardiac tissue structure with mature CMs and near-physiological contractile forces [155]. The pioneering work of these scientists has opened the doors for more revolutionary developments, such as 3D bioprinting, organ-on-chip platforms, and laser-cut decellularized myocardium, all with ample opportunities for both basic research and clinical applicability [156–158].

5. Applications of PSCs in Cardiovascular Research

5.1. Pluripotent Stem Cells in Cardiovascular Disease Modeling

The use of PSCs to model cardiac disease in vitro has become highly attractive, especially after the introduction of iPSCs [107]. This is mainly because of inadequacies of other models in terms of sampling, propagation, and maintenance, as for human primary cardiomyocytes, or their ability to fully recapitulate physiological properties of human CMs, as in rodent models. Considering the relative difficulty in cloning and genetically modifying human ESCs, most established models of CVDs are iPSC-based [107]. The feasibility in sampling and propagation of iPSCs, as well as advances in reprogramming protocols, which later adopted nonintegrating genomic approaches to deliver the reprogramming factors, has greatly increased their popularity [159]. Patient-specific iPSC-CMs have enabled the study of genetic variants underlying several CVDs and establish a phenotype–genotype understanding of not only monogenic, but rather complex and difficult-to-model genetic variants (e.g., chromosomal deletions or translocations), and, most important, model congenital heart disease (CHDs) in newborns [160,161]. As a result, several patient-derived iPSCs lines have been developed to model CVDs. The first of such was reported by Carvajal-Vergara et al. in 2010 for LEOPARD syndrome, an autosomal dominant developmental disorder characterized by hypertrophic cardiomyopathy [162]. Ever since, several other cell lines have been reported, mainly modeling cardiac channelopathies, (e.g., long QT syndromes), cardiomyopathies of wide etiology spectrum (e.g., dilated, hypertrophic, arrhythmogenic, Barth syndrome, and Pompe-disease-associated), and infectious myocarditis [163–177].

Despite the previously discussed disparities in structural and electrophysiological characters of iPSC-derived and adult CMs, these studies have shown that patient-specific iPSC-CMs recapitulate their corresponding disease phenotypes. For example, whole-cell patch-clamp analyses of different long QT syndrome (LQTS) patient-derived iPSC-CMs showed typically prolonged APs, decreased rectifier potassium currents I_K, increased late sodium currents I_{NaL}, and impaired voltage-dependent inactivation of the L-type channels (LTC), due to malfunctions in corresponding proteins of potassium (KCNQ1, KCNH2 in LQTS1 and 2), sodium (SCN5A in LQTS3), and calcium (CaV1.2 in LQTS8 or Timothy syndrome) channels, respectively. Moreover, these patient-specific models demonstrated great utility for pharmacological screening of several drugs with disease-modifying abilities, leading to both novel and/or personalized therapeutic strategies (reviewed in [160]).

Finally, patient-specific iPSC-derived non-CMs were also generated, for example, of SMCs or endothelial cells. A more recent example is an elegant publication by Gu et al., utilizing iPSC-derived endothelial cells from patients with autosomal-dominant mutations in *BMPR2* associated with familial pulmonary arterial hypertension (FPAH) [178]. In their study, comparing symptomatic patients with unaffected carriers highlighted important modifiers of the BMP-receptor pathway, as well as differentially expressed genes, which imparted protection against FPAH. Their findings were of great importance as to the identification of multiple genetic factors affecting disease penetrance, which could be therapeutically targeted to modify disease progression and severity.

Importantly, the previous example behooves an important consideration when conducting studies on patient-specific iPSCs for CVD modeling, which pertains to the identification and/or the availability of proper control lines. This is because, even among patient-matched donor cohorts, genetic variability can still confound the analysis of the disease phenotype, especially in the presence of disease modifiers, or when the genotype–phenotype is less conspicuous [169,179]. In such cases, it is possible to rely on more than one control cell line—albeit a laborious approach. Alternatively, the patient's iPSC-CMs can be compared to those from a healthy sibling, thus limiting genetic variability [171]. However, recently developed computational in silico models of iPSC-CMs and their optimization by Paci and colleagues have provided an unprecedented approach to this issue, enabling simulation and calibration of over a thousand diseased or control iPSC-CM models [180–182]. Finally, in case of monogenetic diseases, an isogenic cell line created by correction of the disease-causing mutation in the patient iPSCs by means of gene-editing approaches can serve as the best control cell line (discussed below). An elegant example was reported in a study by Bellin and colleagues, where they used iPSC-CMs from LQTS2 patients with a distinct mutation in potassium channel KCNH2, and compared it to an isogenic control upon correction of the genetic mutation [183]. Furthermore, they reproduced the study model in human ESC-CMs, where they introduced the same mutation, and recapitulated the disease phenotype, thus generating two genetically distinct isogenic pairs of LQTS2 and control lines.

5.2. Pluripotent Stem Cells in Pharmaceutical Screenings

Since their first introduction, iPSC-CMs have become attractive for drug testing, antiquating the hERG test, which utilizes cell lines that stably express the human ether-a-go-go-related gene (hERG) *KCNH2* encoding the I_{Kr} channel involved in cardiac repolarization. Whole-cell patch-clamp screening for compounds that block the I_{Kr} current serves as a good marker of cardiotoxicity, as such blockade leads to the prolongation of the QT interval, i.e., ventricular repolarization, resulting in potentially fatal ventricular tachycardia called Torsade de Pointes [184]. Since the actual risk for cardiac toxicity is not confined to a certain channel and/or mechanism, iPSC-CMs are hence more representative in typifying cardiac toxicity to drugs. Furthermore, recent introduction of automated patch-clamp (APC) devices, all-optical cardiac electrophysiology with novel optogenetic actuation, and video microscopy have all revolutionized drug screening in iPSC-CMs and tissue constructs, enabling high-throughput testing platforms for hundreds of samples and/or drugs, thus creating a wealth of information in short time [185–188]. Furthermore, comprehensive in vitro proarrhythmic Assay (CIPA) has recently emerged as a powerful model to predict cardiac toxicity by integrating the knowledge from both in vitro and recently developed in silico computational models (http://cipaproject.org/about-cipa/) [189]. However, as discussing this is beyond the scope of this review, we refer the reader to the cited work by Paci et al.

5.3. Genetic Modification of Pluripotent Stem Cells

The advent of genome-editing methods has incited great progress in PSC research. Exploiting the cell's inherent DNA-repair mechanisms, such as nonhomologous end-joining (NHEG) or homologous recombination (HR), has long been used to introduce small but disruptive mutations to target genes, either by insertion or deletions of base pairs, also known as "Indels". The discovery and later advances of nucleases that can more specifically target desired sequences, such as zinc-finger nucleases (ZFNs) or transcription activator-like effector nucleases (TALENs), have enabled the study of several disease causing mutations [190–192]. Many PSC-lines have been generated by using this technology for both disease modeling and even clinical applications [193–196]. Vector-mediated delivery of sequence-specific nucleases along with a homologous DNA template to patient-derived iPSCs leads to the excision of targeted locus and, by virtue of cellular homology directed repair (HDR) system, can be corrected by the homologous template with the desired genetic modification. A prominent example is the combination of ZFNs and piggyBac technology which could achieve a biallelic correction of a disease-causing mutation in human iPSCs [197]. In a recent study by Karakikis et al., they reported the use of TALENs to correct gene mutations in patients with hereditary heart failure [198]. These patients harbor an amino acid deletion mutation (R14del) in the coding region of the phospholamban (PLN) gene, which is an important regulator of cardiac calcium cycling in the sarcoplasmic reticulum (SR). They display a phenotype of dilated cardiomyopathy, hypertrophy, episodic ventricular arrhythmia, and overt HF by middle age [199,200]. Skin-derived iPSCs from these patients were isolated, edited, and CM-differentiated, where further analyses showed reversal of the disease's phenotype. Nevertheless, engineering of sequence-specific ZFNs or TALENs, as well as achieving their robust delivery for this purpose, can be laborious and technically challenging, let alone high in cost [192,194,201].

In recent years, CRISPR/Cas9 has emerged as the new horsepower of genome-editing technology, overshadowing ZFNs and TALENs [202]. The system, first described in prokaryotes as part of their adaptive immune system, relies on an RNA-guided endonuclease (Cas9) that localizes to complementary DNA sequences, where it creates double-strand break amenable for correction by the cell's endogenous HR. Provided that a homologous sequence is available, desired gene modifications can be introduced [202]. Indeed, CRISPR/Cas9 has been zealously received by cell biologists as an attractive tool for SC research [203]. In cardiovascular biology, CRISPR/Cas9 was successfully applied to patient-derived iPSC to target disease-causing mutations of CVDs [204–206]. A recent study demonstrated the utility of CRISPR/Cas9 in phenotypic characterization of iPSC-CMs from patients with arrhythmogenic right ventricular dysplasia/cardiomyopathy (ARVD/C) [207]. In this patient cohort, mutations in the *SCN5A* encoding the $Na_V1.5$ sodium channel protein led to the phenotype, which could be reversed in this study upon editing with CRISPR/Cas9.

A study published earlier this year by Seeger and colleagues made use of genome-editing techniques to create isogenic iPSC lines from patients with heterozygous mutations in the myosin-binding protein C3 (*MYBPC3*), which is deemed as the underlying cause of hypertrophic cardiomyopathy (HCM) [208]. Their results refuted previous hypotheses of either MZBPC3 haploinsufficiency or truncated poison peptide as the underlying cause of HCM. However, they were able to provide evidence for chronic activation of the nonsense-mediated decay (NMD) as the initial pathogenic trigger that leads to dysregulated gene expression and aberrant calcium signaling upon MYBPC3 mutations.

The aforementioned examples give a great promise to SC therapy of CVDs. One might also envisage the possibility of autologous cell transplantation of iPSC-derived CMs with rectified mutations to ameliorate or even cure disease conditions. However, great challenges remain as to the validation of these technologies, let alone deciding on a safe and effective clinical setting for PSCs delivery to treat CVDs. The next chapter outlines recent advances in preclinical research on SC-based therapy for CVDs.

6. Translational Potential of PSCs in Cardiovascular Regenerative Therapy

Harnessing the multifaceted potential of SCs for effective therapeutic purposes to treat CVDs is the ultimate goal of the above-introduced laborious efforts of scientists over the past decades. Provided that SC-derived CMs are sufficiently propagated, differentiated, and maturated, their application to the diseased myocardium spans a wide spectrum of delivery methods, from intravenous administration to direct myocardial injection. Nevertheless, several factors are to be considered with regard to engraftment of transplanted cells and integration, as well as functional contribution to host myocardium, electromechanical coupling between graft and host CMs, and long-term survival. The aforementioned limitations have long been challenges to preclinical and translational applications of SC therapy in general, and in cardiac regenerative therapy in particular. The following summarizes advances made in the realm of preclinical and translational research with PSCs over the past decades, in light of examples from small and large animal models and up until the first clinical initiatives.

6.1. Pluripotent Stem Cells in Rodent Models

Earlier studies attempted to engraft human ESC-derived cardiomyocytes in rodent models and reported transient functional improvement in cardiac parameters [209,210]. However, poor engraftment and survival of transplanted cells has been a challenge in these settings. Laflamme et al. utilized pro-survival factor cocktail to limit CM death upon engraftment in infarcted rat heart and reported positive outcomes [108]. To overcome poor engraftment and survival issues, Masumoto and colleagues developed a layered-sheet assembly of three cardiovascular cell populations, namely CMs, endothelial cells, and vascular mural cells, differentiated from mouse ESCs and transplanted into nude-rat model of MI. The transplanted sheets were reported to ameliorate infarct size and improve cardiac function; however, such benefits were shown to be attributed to paracrine-mediated neovascularization and not to actual contribution of transplanted cells [211]. Despite these results, the same group of scientists from Kyoto continued to optimize the stacked-sheet approach, and they recently reported successful long-term survival of engrafted cells through insertion of gelatin hydrogel microspheres between each cardiovascular cell sheet [212]. Analogously, human iPSC-CMs have recently demonstrated favorable therapeutic outcomes when injected in infarcted myocardia of mice. Interestingly, however, the engraftability and survival of those cells depended heavily on the maturation stage [142].

Importantly, electromechanical coupling between the graft and host myocytes is a rather crucial consideration to avoid ventricular arrhythmia. In a Guinea pig model, Shiba et al. reported successful engraftment of human ESC-derived CMs with a 1:1 electrophysiological coupling and improved mechanical function of injured hearts [213].

6.2. Pluripotent Stem Cells in Large-Animal Models

6.2.1. Porcine Models

Swine models were featured in the earliest attempts of cell therapy for heart disease. A number of features make the pig an attractive translational model. These include a heart weight-to-body ratio that is equal to a human's and a similar sinus rate (~90 bpm) [214]. ESC-derived cardiomyocytes have been functionally tested in a swine model of complete atrioventricular block as biologic pacemaker for the treatment of bradycardia [215]. Hereof, Kehat et al. reported survival and functional integration of the transplanted cells, which were able to pace the porcine ventricle with complete heart block [215]. In another porcine model of acute MI, Ye and colleagues used a mixture of cardiovascular cell populations—this time from human iPSCs origin—loaded on a three-dimensional fibrin patch containing IGF-1 (insulin-like growth factor 1) and reported functional integration and significant improvements of several cardiovascular parameters [216]. More recently, Kawamura et al. took a rather unprecedented approach to enhance survival and engraftment of transplanted human iPSC-CMs by combining cell-sheets with pedicle omentum flap as a source of angiogenic factors and reported enhanced engraftment, survival, and therapeutic outcome in a porcine model of ischemic cardiomyopathy [217,218]. The aforementioned tissue sheet technology from Kyoto was also recently applied to a porcine model of MI, where a heterogeneous mixture of cardiovascular cell populations differentiated from human iPSCs and reported functional restoration of the infarcted hearts and attenuated remodeling [219]. Another study from 2018 by Gao and colleagues reported the application of human iPSC-derived fabricated cardiac muscle patches (hCMPs) composed of CMs, smooth muscle and endothelial cells, reprogramed from cardiac fibroblasts and matured in dynamic culture conditions [220]. They transplanted these patches in infarcted pig hearts and demonstrated significant improvements upon histological and functional analyses. Altogether, these results highlight the importance of co-administration of nonmyocyte cardiac cells, which provide paracrine and angiogenic support equally important to both host and graft tissue. Furthermore, they highlight the superiority of modern tissue engineered scaffolds over direct application of cells. Finally, these studies corroborate the utility of swine models for translational cardiovascular research.

6.2.2. Non-Human Primate Models

The utility of non-human primates (NHPs) in regenerative medicine has long been appreciated, especially in transplantation medicine [221]. In this regard, certain macaque species (e.g., *Macaca fascicularis* or Mauritian Cynomolgus macaque) are a valuable preclinical model to study allogenic transplantation of iPSCs [222]. This is because they exhibit limited diversity in their major histocompatibility complex (MHC) genes, which are distributed only among seven haplotypes and are structurally identical to those in humans [223,224]. Indeed, allogenic transplantation of iPSC-CM among MHC-matched Cynomolgus monkeys was shown in a study by Shiba et al. to be immune-tolerable, and improved cardiac contractile function upon MI [224]. Matching MHC antigens between donors and recipients was shown by others to reduce immunogenicity upon allogenic transplantation of iPSC-CM in the Cynomolgus macaque [225].

NHPs continue to provide unmatched insights to PSC-therapy of CVDs in late-translational studies. In a recently published elegant work by Chong et al., human ESC-CMs were utilized in the pigtail macaque (*Macaca nemestrina*) as a model of ischemia-reperfusion injury [226]. They reported significant re-muscularization of the infarcted areas, structural and functional integration of the grafted cells via establishment of adherent junctions, and electromechanical coupling, typified by synchronous calcium transients. Such promising results were slightly relegated by the presence of arrhythmia in the grafted animals—albeit nonfatal.

7. Pluripotent Stem Cells in First Human Trials

The aforementioned successes in late-translational studies with large animal models, as well as the advances made in tissue engineering and grafting techniques, paved the way to the first clinical application of PSC therapy in cardiac settings, which was recently reported in a case study by Menasche et al., using ESC-derived cardiac progenitors [227]. They used the ESC I6 line, which was enriched in vitro by culturing on clinical-grade irradiated human foreskin fibroblasts as feeder cells. Cardiac commitment was then achieved by bone morphogenic protein-2 (BMP-2) and a specific tyrosine kinase inhibitor of the fibroblast growth factor receptor (FGFR), and then confirmed by the expression of the cardiac specific transcription factor Isl-1, as well as the stage-specific embryonic antigen-1 (SSEA-1), which was used for cell purification by immunomagnetic sorting. The cells were embedded in a fibrin scaffold patch and surgically implanted in the infarcted area of a 68-year-old woman patient with severe heart failure. The three-month follow-up showed functional integration of the patch, evident by electrocardiography, and overall symptomatic improvement marked by enhanced left-ventricular ejection fraction (LVEF), with no complications of arrhythmia, tumors, or immunosuppression-related adverse events. The results were encouraging, and the Parisian group conducted a larger-scale study, wherein six patients received cellularized patches of ESC-derived committed cardiac progenitors [228]. Their one-year follow-up demonstrated safety and tolerability of the grafted cells, with no detected tumors. Moreover, they reported modest symptomatic improvements, as well as in different cardiac parameters.

As for iPSCs, a group of scientists from Osaka have reported their granted permission to pursue with their clinical application last year and the results are yet to be reported [229].

8. Conclusions and Remarks

Stem cells are a novel source of cells which might be used as a screening tool for pharmaceutical developments. Here, single cells on iPSC status, as well as differentiated cardiomyocyte progenitors, might be used. In addition, engineered heart tissue displays a second model situation for screening of novel therapeutic options, before applying in animal experiments or clinical trials. Since these systems are based on human sources of cells, testing in these model situations might enhance safety and side effect prediction in novel approaches of cardiovascular therapies. Utilization of stem cells in patients suffering from cardiovascular disease is a second interesting field with great potential. The allogenic transplantation of stem cells requires only a modest immunosuppression and might improve cardiac function and thereby survival, as well as quality of life, in patients suffering from cardiac conditions, e.g., heart failure. However, the clinical potentials, as well as the potential side effects, need to be investigated in clinical trials before establishing stem-cell-based therapy as a standard of care in cardiovascular patients. In summary, stem cells, especially induced pluripotent stem cells, have wide therapeutic potential, but need to be characterized and investigated in more detail in preclinical, as well as clinical, trials to understand in more detail their potentials and risks.

Author Contributions: M.S., writing—original draft and visualization; R.H., writing—review and editing.

Funding: This research received no external funding.

Conflicts of Interest: The authors declare no conflicts of interest.

Abbreviations

CM	cardiomyocytes
CPC	cardiac progenitor cells
CVD	cardiovascular disease
ESC	embryonic stem cells
iPSC	induced pluripotent stem cells
LVEF	left ventricular ejection fraction
MSC	mesenchymal stem cells
PSC	pluripotent stem cells
SM	skeletal myoblast

References

1.   WHO. Cardiovascular Diseases (CVDs). Available online: http://www.who.int/en/news-room/fact-sheets/detail/cardiovascular-diseases- (accessed on 1 June 2019).

2. Benjamin, E.J.; Blaha, M.J.; Chiuve, S.E.; Cushman, M.; Das, S.R.; Deo, R.; de Ferranti, S.D.; Floyd, J.; Fornage, M.; Gillespie, C.; et al. Heart Disease and Stroke Statistics-2017 Update: A Report from the American Heart Association. *Circulation* **2017**, *135*, e146–e603. [CrossRef] [PubMed]

3. Samak, M.; Fatullayev, J.; Sabashnikov, A.; Zeriouh, M.; Schmack, B.; Ruhparwar, A.; Karck, M.; Popov, A.F.; Dohmen, P.M.; Weymann, A. Total Arterial Revascularization: Bypassing Antiquated Notions to Better Alternatives for Coronary Artery Disease. *Med. Sci. Monit. Basic Res.* **2016**, *22*, 107–114. [CrossRef] [PubMed]

4. Fatullayev, J.; Samak, M.; Sabashnikov, A.; Zeriouh, M.; Rahmanian, P.B.; Choi, Y.H.; Schmack, B.; Kallenbach, K.; Ruhparwar, A.; Eghbalzadeh, K.; et al. Continuous-Flow Left Ventricular Assist Device Thrombosis: A Danger Foreseen is a Danger Avoided. *Med. Sci. Monit. Basic Res.* **2015**, *21*, 141–144. [CrossRef] [PubMed]

5. Roger, V.L. Epidemiology of heart failure. *Circ. Res.* **2013**, *113*, 646–659. [CrossRef]

6. Gass, A.L.; Emaminia, A.; Lanier, G.; Aggarwal, C.; Brown, K.A.; Raffa, M.; Kai, M.; Spielvogel, D.; Malekan, R.; Tang, G.; et al. Cardiac Transplantation in the New Era. *Cardiol. Rev.* **2015**, *23*, 182–188. [CrossRef]

7. Gouadon, E.; Moore-Morris, T.; Smit, N.W.; Chatenoud, L.; Coronel, R.; Harding, S.E.; Jourdon, P.; Lambert, V.; Rucker-Martin, C.; Puceat, M. Concise Review: Pluripotent Stem Cell-Derived Cardiac Cells, A Promising Cell Source for Therapy of Heart Failure: Where Do We Stand? *Stem Cells* **2016**, *34*, 34–43. [CrossRef]

8. Sampogna, G.; Guraya, S.Y.; Forgione, A. Regenerative medicine: Historical roots and potential strategies in modern medicine. *J. Microsc. Ultrastruct.* **2015**, *3*, 101–107. [CrossRef]

9. Muller, P.; Lemcke, H.; David, R. Stem Cell Therapy in Heart Diseases-Cell Types, Mechanisms and Improvement Strategies. *Cell. Physiol. Biochem. Int. J. Exp. Cell. Physiol. Biochem. Pharmacol.* **2018**, *48*, 2607–2655. [CrossRef]

10. Viola, J.L.B.; Grad, O. *The Emergence of Tissue Engineering as a Research Field*; National Science Foundation: Alexandria, VA, USA, 14 October 2003.

11. Tajbakhsh, S. Stem cells to tissue: Molecular, cellular and anatomical heterogeneity in skeletal muscle. *Curr. Opin. Genet. Dev.* **2003**, *13*, 413–422. [CrossRef]

12. Al Attar, N.; Carrion, C.; Ghostine, S.; Garcin, I.; Vilquin, J.T.; Hagege, A.A.; Menasche, P. Long-term (1 year) functional and histological results of autologous skeletal muscle cells transplantation in rat. *Cardiovasc. Res.* **2003**, *58*, 142–148. [CrossRef]

13. Suzuki, K.; Murtuza, B.; Suzuki, N.; Smolenski, R.T.; Yacoub, M.H. Intracoronary infusion of skeletal myoblasts improves cardiac function in doxorubicin-induced heart failure. *Circulation* **2001**, *104* (Suppl. 1), I213–I217. [CrossRef] [PubMed]

14. Pouly, J.; Hagege, A.A.; Vilquin, J.T.; Bissery, A.; Rouche, A.; Bruneval, P.; Duboc, D.; Desnos, M.; Fiszman, M.; Fromes, Y.; et al. Does the functional efficacy of skeletal myoblast transplantation extend to nonischemic cardiomyopathy? *Circulation* **2004**, *110*, 1626–1631. [CrossRef] [PubMed]

15. Gavira, J.J.; Perez-Ilzarbe, M.; Abizanda, G.; Garcia-Rodriguez, A.; Orbe, J.; Paramo, J.A.; Belzunce, M.; Rabago, G.; Barba, J.; Herreros, J.; et al. A comparison between percutaneous and surgical transplantation of autologous skeletal myoblasts in a swine model of chronic myocardial infarction. *Cardiovasc. Res.* **2006**, *71*, 744–753. [CrossRef] [PubMed]

16. He, K.L.; Yi, G.H.; Sherman, W.; Zhou, H.; Zhang, G.P.; Gu, A.; Kao, R.; Haimes, H.B.; Harvey, J.; Roos, E.; et al. Autologous skeletal myoblast transplantation improved hemodynamics and left ventricular function in chronic heart failure dogs. *J. Heart Lung Transplant.* **2005**, *24*, 1940–1949. [CrossRef] [PubMed]

17. Gavira, J.J.; Nasarre, E.; Abizanda, G.; Perez-Ilzarbe, M.; de Martino-Rodriguez, A.; Garcia de Jalon, J.A.; Mazo, M.; Macias, A.; Garcia-Bolao, I.; Pelacho, B.; et al. Repeated implantation of skeletal myoblast in a swine model of chronic myocardial infarction. *Eur. Heart J.* **2010**, *31*, 1013–1021. [CrossRef]

18. Menasche, P.; Hagege, A.A.; Vilquin, J.T.; Desnos, M.; Abergel, E.; Pouzet, B.; Bel, A.; Sarateanu, S.; Scorsin, M.; Schwartz, K.; et al. Autologous skeletal myoblast transplantation for severe postinfarction left ventricular dysfunction. *J. Am. Coll. Cardiol.* **2003**, *41*, 1078–1083. [CrossRef]

19. Herreros, J.; Prosper, F.; Perez, A.; Gavira, J.J.; Garcia-Velloso, M.J.; Barba, J.; Sanchez, P.L.; Canizo, C.; Rabago, G.; Marti-Climent, J.M.; et al. Autologous intramyocardial injection of cultured skeletal muscle-derived stem cells in patients with non-acute myocardial infarction. *Eur. Heart J.* **2003**, *24*, 2012–2020. [CrossRef]

20. Siminiak, T.; Fiszer, D.; Jerzykowska, O.; Grygielska, B.; Rozwadowska, N.; Kalmucki, P.; Kurpisz, M. Percutaneous trans-coronary-venous transplantation of autologous skeletal myoblasts in the treatment of post-infarction myocardial contractility impairment: The POZNAN trial. *Eur. Heart J.* **2005**, *26*, 1188–1195. [CrossRef]

21. Smits, P.C.; van Geuns, R.J.; Poldermans, D.; Bountioukos, M.; Onderwater, E.E.; Lee, C.H.; Maat, A.P.; Serruys, P.W. Catheter-based intramyocardial injection of autologous skeletal myoblasts as a primary treatment of ischemic heart failure: Clinical experience with six-month follow-up. *J. Am. Coll. Cardiol.* **2003**, *42*, 2063–2069. [CrossRef]

22. Hagege, A.A.; Marolleau, J.P.; Vilquin, J.T.; Alheritiere, A.; Peyrard, S.; Duboc, D.; Abergel, E.; Messas, E.; Mousseaux, E.; Schwartz, K.; et al. Skeletal myoblast transplantation in ischemic heart failure: Long-term follow-up of the first phase I cohort of patients. *Circulation* **2006**, *114*, I108–I113. [CrossRef]

23. Ince, H.; Petzsch, M.; Rehders, T.C.; Chatterjee, T.; Nienaber, C.A. Transcatheter transplantation of autologous skeletal myoblasts in postinfarction patients with severe left ventricular dysfunction. *J. Endovasc. Ther.* **2004**, *11*, 695–704. [CrossRef]

24. Veltman, C.E.; Soliman, O.I.; Geleijnse, M.L.; Vletter, W.B.; Smits, P.C.; ten Cate, F.J.; Jordaens, L.J.; Balk, A.H.; Serruys, P.W.; Boersma, E.; et al. Four-year follow-up of treatment with intramyocardial skeletal myoblasts injection in patients with ischaemic cardiomyopathy. *Eur. Heart J.* **2008**, *29*, 1386–1396. [CrossRef] [PubMed]

25. Reinecke, H.; Poppa, V.; Murry, C.E. Skeletal muscle stem cells do not transdifferentiate into cardiomyocytes after cardiac grafting. *J. Mol. Cell. Cardiol.* **2002**, *34*, 241–249. [CrossRef] [PubMed]

26. Abraham, M.R.; Henrikson, C.A.; Tung, L.; Chang, M.G.; Aon, M.; Xue, T.; Li, R.A.; B, O.R.; Marban, E. Antiarrhythmic engineering of skeletal myoblasts for cardiac transplantation. *Circ. Res.* **2005**, *97*, 159–167. [CrossRef] [PubMed]

27. Brickwedel, J.; Gulbins, H.; Reichenspurner, H. Long-term follow-up after autologous skeletal myoblast transplantation in ischaemic heart disease. *Interact. Cardiovasc. Thorac. Surg.* **2014**, *18*, 61–66. [CrossRef] [PubMed]

28. Dib, N.; Dinsmore, J.; Lababidi, Z.; White, B.; Moravec, S.; Campbell, A.; Rosenbaum, A.; Seyedmadani, K.; Jaber, W.A.; Rizenhour, C.S.; et al. One-year follow-up of feasibility and safety of the first U.S., randomized, controlled study using 3-dimensional guided catheter-based delivery of autologous skeletal myoblasts for ischemic cardiomyopathy (CAuSMIC study). *JACC Cardiovasc. Interv.* **2009**, *2*, 9–16. [CrossRef]

29. Duckers, H.J.; Houtgraaf, J.; Hehrlein, C.; Schofer, J.; Waltenberger, J.; Gershlick, A.; Bartunek, J.; Nienaber, C.; Macaya, C.; Peters, N.; et al. Final results of a phase IIa, randomised, open-label trial to evaluate the percutaneous intramyocardial transplantation of autologous skeletal myoblasts in congestive heart failure patients: The SEISMIC trial. *Eurointerv. J. Eur. Collab. Work. Group Interv. Cardiol. Eur. Soc. Cardiol.* **2011**, *6*, 805–812. [CrossRef]

30. Menasche, P.; Alfieri, O.; Janssens, S.; McKenna, W.; Reichenspurner, H.; Trinquart, L.; Vilquin, J.T.; Marolleau, J.P.; Seymour, B.; Larghero, J.; et al. The Myoblast Autologous Grafting in Ischemic Cardiomyopathy (MAGIC) trial: First randomized placebo-controlled study of myoblast transplantation. *Circulation* **2008**, *117*, 1189–1200. [CrossRef]

31. Povsic, T.J.; O'Connor, C.M.; Henry, T.; Taussig, A.; Kereiakes, D.J.; Fortuin, F.D.; Niederman, A.; Schatz, R.; Spencer, R.; Owens, D.; et al. A double-blind, randomized, controlled, multicenter study to assess the safety and cardiovascular effects of skeletal myoblast implantation by catheter delivery in patients with chronic heart failure after myocardial infarction. *Am. Heart J.* **2011**, *162*, 654–662. [CrossRef]

32. Ferrari, G.; Cusella-De Angelis, G.; Coletta, M.; Paolucci, E.; Stornaiuolo, A.; Cossu, G.; Mavilio, F. Muscle regeneration by bone marrow-derived myogenic progenitors. *Science* **1998**, *279*, 1528–1530. [CrossRef]

33. Orlic, D.; Kajstura, J.; Chimenti, S.; Jakoniuk, I.; Anderson, S.M.; Li, B.; Pickel, J.; McKay, R.; Nadal-Ginard, B.; Bodine, D.M.; et al. Bone marrow cells regenerate infarcted myocardium. *Nature* **2001**, *410*, 701–705. [CrossRef] [PubMed]

34. Strauer, B.E.; Brehm, M.; Zeus, T.; Gattermann, N.; Hernandez, A.; Sorg, R.V.; Kogler, G.; Wernet, P. Intracoronary, human autologous stem cell transplantation for myocardial regeneration following myocardial infarction. *Deutsche Medizinische Wochenschrift 1946* **2001**, *126*, 932–938. [CrossRef]

35. Wen, Y.; Chen, B.; Wang, C.; Ma, X.; Gao, Q. Bone marrow-derived mononuclear cell therapy for patients with ischemic heart disease and ischemic heart failure. *Expert Opin. Biol. Ther.* **2012**, *12*, 1563–1573. [CrossRef] [PubMed]

36. De Jong, R.; Houtgraaf, J.H.; Samiei, S.; Boersma, E.; Duckers, H.J. Intracoronary stem cell infusion after acute myocardial infarction: A meta-analysis and update on clinical trials. *Circ. Cardiovasc. Interv.* **2014**, *7*, 156–167. [CrossRef] [PubMed]

37. Malliaras, K.; Marban, E. Cardiac cell therapy: Where we've been, where we are, and where we should be headed. *Br. Med. Bull.* **2011**, *98*, 161–185. [CrossRef]

38. Miao, C.; Lei, M.; Hu, W.; Han, S.; Wang, Q. A brief review: The therapeutic potential of bone marrow mesenchymal stem cells in myocardial infarction. *Stem Cell Res. Ther.* **2017**, *8*, 242. [CrossRef]

39. Erbs, S.; Linke, A.; Schachinger, V.; Assmus, B.; Thiele, H.; Diederich, K.W.; Hoffmann, C.; Dimmeler, S.; Tonn, T.; Hambrecht, R.; et al. Restoration of microvascular function in the infarct-related artery by intracoronary transplantation of bone marrow progenitor cells in patients with acute myocardial infarction: The Doppler Substudy of the Reinfusion of Enriched Progenitor Cells and Infarct Remodeling in Acute Myocardial Infarction (REPAIR-AMI) trial. *Circulation* **2007**, *116*, 366–374. [CrossRef]

40. Asahara, T.; Kawamoto, A.; Masuda, H. Concise review: Circulating endothelial progenitor cells for vascular medicine. *Stem Cells* **2011**, *29*, 1650–1655. [CrossRef]

41. Steinhoff, G.; Nesteruk, J.; Wolfien, M.; Kundt, G.; Borgermann, J.; David, R.; Garbade, J.; Grosse, J.; Haverich, A.; Hennig, H.; et al. Cardiac Function Improvement and Bone Marrow Response–: Outcome Analysis of the Randomized PERFECT Phase III Clinical Trial of Intramyocardial CD133(+) Application After Myocardial Infarction. *EBioMedicine* **2017**, *22*, 208–224. [CrossRef]

42. Karantalis, V.; Schulman, I.H.; Balkan, W.; Hare, J.M. Allogeneic cell therapy: A new paradigm in therapeutics. *Circ. Res.* **2015**, *116*, 12–15. [CrossRef]

43. Dimmeler, S.; Zeiher, A.M. Cell therapy of acute myocardial infarction: Open questions. *Cardiology* **2009**, *113*, 155–160. [CrossRef] [PubMed]

44. Kanelidis, A.J.; Premer, C.; Lopez, J.; Balkan, W.; Hare, J.M. Route of Delivery Modulates the Efficacy of Mesenchymal Stem Cell Therapy for Myocardial Infarction: A Meta-Analysis of Preclinical Studies and Clinical Trials. *Circ. Res.* **2017**, *120*, 1139–1150. [CrossRef] [PubMed]

45. Chen, S.L.; Fang, W.W.; Ye, F.; Liu, Y.H.; Qian, J.; Shan, S.J.; Zhang, J.J.; Chunhua, R.Z.; Liao, L.M.; Lin, S.; et al. Effect on left ventricular function of intracoronary transplantation of autologous bone marrow mesenchymal stem cell in patients with acute myocardial infarction. *Am. J. Cardiol.* **2004**, *94*, 92–95. [CrossRef] [PubMed]

46. Lee, J.W.; Lee, S.H.; Youn, Y.J.; Ahn, M.S.; Kim, J.Y.; Yoo, B.S.; Yoon, J.; Kwon, W.; Hong, I.S.; Lee, K.; et al. A randomized, open-label, multicenter trial for the safety and efficacy of adult mesenchymal stem cells after acute myocardial infarction. *J. Korean Med Sci.* **2014**, *29*, 23–31. [CrossRef]

47. Chullikana, A.; Majumdar, A.S.; Gottipamula, S.; Krishnamurthy, S.; Kumar, A.S.; Prakash, V.S.; Gupta, P.K. Randomized, double-blind, phase I/II study of intravenous allogeneic mesenchymal stromal cells in acute myocardial infarction. *Cytotherapy* **2015**, *17*, 250–261. [CrossRef]

48. Gao, L.R.; Pei, X.T.; Ding, Q.A.; Chen, Y.; Zhang, N.K.; Chen, H.Y.; Wang, Z.G.; Wang, Y.F.; Zhu, Z.M.; Li, T.C.; et al. A critical challenge: Dosage-related efficacy and acute complication intracoronary injection of autologous bone marrow mesenchymal stem cells in acute myocardial infarction. *Int. J. Cardiol.* **2013**, *168*, 3191–3199. [CrossRef]

49. Hare, J.M.; Fishman, J.E.; Gerstenblith, G.; DiFede Velazquez, D.L.; Zambrano, J.P.; Suncion, V.Y.; Tracy, M.; Ghersin, E.; Johnston, P.V.; Brinker, J.A.; et al. Comparison of allogeneic vs. autologous bone marrow-derived mesenchymal stem cells delivered by transendocardial injection in patients with ischemic cardiomyopathy: The POSEIDON randomized trial. *JAMA* **2012**, *308*, 2369–2379. [CrossRef]

50. Hare, J.M.; DiFede, D.L.; Rieger, A.C.; Florea, V.; Landin, A.M.; El-Khorazaty, J.; Khan, A.; Mushtaq, M.; Lowery, M.H.; Byrnes, J.J.; et al. Randomized Comparison of Allogeneic Versus Autologous Mesenchymal Stem Cells for Nonischemic Dilated Cardiomyopathy: POSEIDON-DCM Trial. *J. Am. Coll. Cardiol.* **2017**, *69*, 526–537. [CrossRef]

51. Bergmann, O.; Jovinge, S. Cardiac regeneration in vivo: Mending the heart from within? *Stem Cell Res.* **2014**, *13*, 523–531. [CrossRef]

52. Bergmann, O.; Zdunek, S.; Alkass, K.; Druid, H.; Bernard, S.; Frisen, J. Identification of cardiomyocyte nuclei and assessment of ploidy for the analysis of cell turnover. *Exp. Cell Res.* **2011**, *317*, 188–194. [CrossRef]

53. Laflamme, M.A.; Murry, C.E. Heart regeneration. *Nature* **2011**, *473*, 326–335. [CrossRef] [PubMed]

54. Ang, K.L.; Shenje, L.T.; Reuter, S.; Soonpaa, M.H.; Rubart, M.; Field, L.J.; Galinanes, M. Limitations of conventional approaches to identify myocyte nuclei in histologic sections of the heart. *Am. J. Physiol. Cell Physiol.* **2010**, *298*, C1603–C1609. [CrossRef] [PubMed]

55. Bergmann, O.; Bhardwaj, R.D.; Bernard, S.; Zdunek, S.; Barnabe-Heider, F.; Walsh, S.; Zupicich, J.; Alkass, K.; Buchholz, B.A.; Druid, H.; et al. Evidence for cardiomyocyte renewal in humans. *Science* **2009**, *324*, 98–102. [CrossRef] [PubMed]

56. Mollova, M.; Bersell, K.; Walsh, S.; Savla, J.; Das, L.T.; Park, S.Y.; Silberstein, L.E.; Dos Remedios, C.G.; Graham, D.; Colan, S.; et al. Cardiomyocyte proliferation contributes to heart growth in young humans. *Proc. Natl. Acad. Sci. USA* **2013**, *110*, 1446–1451. [CrossRef] [PubMed]

57. Naqvi, N.; Li, M.; Calvert, J.W.; Tejada, T.; Lambert, J.P.; Wu, J.; Kesteven, S.H.; Holman, S.R.; Matsuda, T.; Lovelock, J.D.; et al. A proliferative burst during preadolescence establishes the final cardiomyocyte number. *Cell* **2014**, *157*, 795–807. [CrossRef] [PubMed]

58. Beltrami, A.P.; Barlucchi, L.; Torella, D.; Baker, M.; Limana, F.; Chimenti, S.; Kasahara, H.; Rota, M.; Musso, E.; Urbanek, K.; et al. Adult cardiac stem cells are multipotent and support myocardial regeneration. *Cell* **2003**, *114*, 763–776. [CrossRef]

59. Aguilar-Sanchez, C.; Michael, M.; Pennings, S. Cardiac Stem Cells in the Postnatal Heart: Lessons from Development. *Stem Cells Int.* **2018**, *2018*, 1247857. [CrossRef]

60. Hesse, M.; Fleischmann, B.K.; Kotlikoff, M.I. Concise review: The role of C-kit expressing cells in heart repair at the neonatal and adult stage. *Stem Cells* **2014**, *32*, 1701–1712. [CrossRef]

61. Nigro, P.; Perrucci, G.L.; Gowran, A.; Zanobini, M.; Capogrossi, M.C.; Pompilio, G. c-kit(+) cells: The tell-tale heart of cardiac regeneration? *Cell. Mol. Life Sci.* **2015**, *72*, 1725–1740. [CrossRef]

62. Jesty, S.A.; Steffey, M.A.; Lee, F.K.; Breitbach, M.; Hesse, M.; Reining, S.; Lee, J.C.; Doran, R.M.; Nikitin, A.Y.; Fleischmann, B.K.; et al. c-kit+ precursors support postinfarction myogenesis in the neonatal, but not adult, heart. *Proc. Natl. Acad. Sci. USA* **2012**, *109*, 13380–13385. [CrossRef]

63. Nadal-Ginard, B.; Ellison, G.M.; Torella, D. The cardiac stem cell compartment is indispensable for myocardial cell homeostasis, repair and regeneration in the adult. *Stem Cell Res.* **2014**, *13*, 615–630. [CrossRef] [PubMed]

64. Gude, N.A.; Firouzi, F.; Broughton, K.M.; Ilves, K.; Nguyen, K.P.; Payne, C.R.; Sacchi, V.; Monsanto, M.M.; Casillas, A.R.; Khalafalla, F.G.; et al. Cardiac c-Kit Biology Revealed by Inducible Transgenesis. *Circ. Res.* **2018**, *123*, 57–72. [CrossRef] [PubMed]

65. Hodgkinson, C.P.; Gomez, J.A.; Baksh, S.S.; Payne, A.; Schmeckpeper, J.; Pratt, R.E.; Dzau, V.J. Insights from molecular signature of in vivo cardiac c-Kit(+) cells following cardiac injury and beta-catenin inhibition. *J. Mol. Cell. Cardiol.* **2018**, *123*, 64–74. [CrossRef] [PubMed]

66. Li, Y.; He, L.; Huang, X.; Bhaloo, S.I.; Zhao, H.; Zhang, S.; Pu, W.; Tian, X.; Li, Y.; Liu, Q.; et al. Genetic Lineage Tracing of Nonmyocyte Population by Dual Recombinases. *Circulation* **2018**, *138*, 793–805. [CrossRef] [PubMed]

67. Li, Y.; Lv, Z.; He, L.; Huang, X.; Zhang, S.; Zhao, H.; Pu, W.; Li, Y.; Yu, W.; Zhang, L.; et al. Genetic Tracing Identifies Early Segregation of the Cardiomyocyte and Nonmyocyte Lineages. *Circ. Res.* **2019**, *125*, 343–355. [CrossRef] [PubMed]

68. Elhelaly, W.M.; Cardoso, A.C.; Pereira, A.H.M.; Elnawasany, A.; Ebrahimi, S.; Nakada, Y.; Sadek, H.A. C-Kit Cells Do Not Significantly Contribute to Cardiomyogenesis During Neonatal Heart Regeneration. *Circulation* **2019**, *139*, 559–561. [CrossRef] [PubMed]

69. Tzahor, E.; Poss, K.D. Cardiac regeneration strategies: Staying young at heart. *Science* **2017**, *356*, 1035–1039. [CrossRef]

70. Van Berlo, J.H.; Kanisicak, O.; Maillet, M.; Vagnozzi, R.J.; Karch, J.; Lin, S.C.; Middleton, R.C.; Marban, E.; Molkentin, J.D. c-kit+ cells minimally contribute cardiomyocytes to the heart. *Nature* **2014**, *509*, 337–341. [CrossRef]

71. Sussman, M.A.; Murry, C.E. Bones of contention: Marrow-derived cells in myocardial regeneration. *J. Mol. Cell. Cardiol.* **2008**, *44*, 950–953. [CrossRef]

72. Evans, M.J.; Kaufman, M.H. Establishment in culture of pluripotential cells from mouse embryos. *Nature* **1981**, *292*, 154–156. [CrossRef]

73. Thomson, J.A.; Itskovitz-Eldor, J.; Shapiro, S.S.; Waknitz, M.A.; Swiergiel, J.J.; Marshall, V.S.; Jones, J.M. Embryonic stem cell lines derived from human blastocysts. *Science* **1998**, *282*, 1145–1147. [CrossRef] [PubMed]

74. De Trizio, E.; Brennan, C.S. The business of human embryonic stem cell research and an international analysis of relevant laws. *J. Biolaw Bus.* **2004**, *7*, 14–22. [PubMed]

75. Jain, K.K. Ethical and regulatory aspects of embryonic stem cell research. *Expert Opin. Biol. Ther.* **2005**, *5*, 153–162. [CrossRef] [PubMed]

76. Murry, C.E.; Keller, G. Differentiation of embryonic stem cells to clinically relevant populations: Lessons from embryonic development. *Cell* **2008**, *132*, 661–680. [CrossRef] [PubMed]

77. Davis, R.L.; Weintraub, H.; Lassar, A.B. Expression of a single transfected cDNA converts fibroblasts to myoblasts. *Cell* **1987**, *51*, 987–1000. [CrossRef]

78. Takahashi, K.; Yamanaka, S. Induction of pluripotent stem cells from mouse embryonic and adult fibroblast cultures by defined factors. *Cell* **2006**, *126*, 663–676. [CrossRef]

79. Okita, K.; Ichisaka, T.; Yamanaka, S. Generation of germline-competent induced pluripotent stem cells. *Nature* **2007**, *448*, 313–317. [CrossRef]

80. Masip, M.; Veiga, A.; Izpisua Belmonte, J.C.; Simon, C. Reprogramming with defined factors: From induced pluripotency to induced transdifferentiation. *Mol. Hum. Reprod.* **2010**, *16*, 856–868. [CrossRef]

81. Zhao, Y.; Yin, X.; Qin, H.; Zhu, F.; Liu, H.; Yang, W.; Zhang, Q.; Xiang, C.; Hou, P.; Song, Z.; et al. Two supporting factors greatly improve the efficiency of human iPSC generation. *Cell Stem Cell* **2008**, *3*, 475–479. [CrossRef]

82. Han, J.; Yuan, P.; Yang, H.; Zhang, J.; Soh, B.S.; Li, P.; Lim, S.L.; Cao, S.; Tay, J.; Orlov, Y.L.; et al. Tbx3 improves the germ-line competency of induced pluripotent stem cells. *Nature* **2010**, *463*, 1096–1100. [CrossRef]

83. Takahashi, K.; Tanabe, K.; Ohnuki, M.; Narita, M.; Ichisaka, T.; Tomoda, K.; Yamanaka, S. Induction of pluripotent stem cells from adult human fibroblasts by defined factors. *Cell* **2007**, *131*, 861–872. [CrossRef] [PubMed]

84. Yu, J.; Vodyanik, M.A.; Smuga-Otto, K.; Antosiewicz-Bourget, J.; Frane, J.L.; Tian, S.; Nie, J.; Jonsdottir, G.A.; Ruotti, V.; Stewart, R.; et al. Induced pluripotent stem cell lines derived from human somatic cells. *Science* **2007**, *318*, 1917–1920. [CrossRef] [PubMed]

85. Wernig, M.; Meissner, A.; Foreman, R.; Brambrink, T.; Ku, M.; Hochedlinger, K.; Bernstein, B.E.; Jaenisch, R. In vitro reprogramming of fibroblasts into a pluripotent ES-cell-like state. *Nature* **2007**, *448*, 318–324. [CrossRef] [PubMed]

86. Maherali, N.; Sridharan, R.; Xie, W.; Utikal, J.; Eminli, S.; Arnold, K.; Stadtfeld, M.; Yachechko, R.; Tchieu, J.; Jaenisch, R.; et al. Directly reprogrammed fibroblasts show global epigenetic remodeling and widespread tissue contribution. *Cell Stem Cell* **2007**, *1*, 55–70. [CrossRef] [PubMed]

87. Chin, M.H.; Mason, M.J.; Xie, W.; Volinia, S.; Singer, M.; Peterson, C.; Ambartsumyan, G.; Aimiuwu, O.; Richter, L.; Zhang, J.; et al. Induced pluripotent stem cells and embryonic stem cells are distinguished by gene expression signatures. *Cell Stem Cell* **2009**, *5*, 111–123. [CrossRef]

88. Doi, A.; Park, I.H.; Wen, B.; Murakami, P.; Aryee, M.J.; Irizarry, R.; Herb, B.; Ladd-Acosta, C.; Rho, J.; Loewer, S.; et al. Differential methylation of tissue- and cancer-specific CpG island shores distinguishes human induced pluripotent stem cells, embryonic stem cells and fibroblasts. *Nat. Genet.* **2009**, *41*, 1350–1353. [CrossRef]

89. Marchetto, M.C.; Yeo, G.W.; Kainohana, O.; Marsala, M.; Gage, F.H.; Muotri, A.R. Transcriptional signature and memory retention of human-induced pluripotent stem cells. *PLoS ONE* **2009**, *4*, e7076. [CrossRef]

90. Kim, K.; Doi, A.; Wen, B.; Ng, K.; Zhao, R.; Cahan, P.; Kim, J.; Aryee, M.J.; Ji, H.; Ehrlich, L.I.; et al. Epigenetic memory in induced pluripotent stem cells. *Nature* **2010**, *467*, 285–290. [CrossRef]

91. Hewitt, K.J.; Garlick, J.A. Cellular reprogramming to reset epigenetic signatures. *Mol. Asp. Med.* **2013**, *34*, 841–848. [CrossRef]

92. Gherghiceanu, M.; Barad, L.; Novak, A.; Reiter, I.; Itskovitz-Eldor, J.; Binah, O.; Popescu, L.M. Cardiomyocytes derived from human embryonic and induced pluripotent stem cells: Comparative ultrastructure. *J. Cell. Mol. Med.* **2011**, *15*, 2539–2551. [CrossRef]

93. Gupta, M.K.; Illich, D.J.; Gaarz, A.; Matzkies, M.; Nguemo, F.; Pfannkuche, K.; Liang, H.; Classen, S.; Reppel, M.; Schultze, J.L.; et al. Global transcriptional profiles of beating clusters derived from human induced pluripotent stem cells and embryonic stem cells are highly similar. *BMC Dev. Biol.* **2010**, *10*, 98. [CrossRef] [PubMed]

94. Volarevic, V.; Markovic, B.S.; Gazdic, M.; Volarevic, A.; Jovicic, N.; Arsenijevic, N.; Armstrong, L.; Djonov, V.; Lako, M.; Stojkovic, M. Ethical and Safety Issues of Stem Cell-Based Therapy. *Int. J. Med. Sci.* **2018**, *15*, 36–45. [CrossRef] [PubMed]

95. Kiskinis, E.; Eggan, K. Progress toward the clinical application of patient-specific pluripotent stem cells. *J. Clin. Investig.* **2010**, *120*, 51–59. [CrossRef] [PubMed]

96. Kurtz, A.; Stacey, G.; Kidane, L.; Seriola, A.; Stachelscheid, H.; Veiga, A. Regulatory insight into the European human pluripotent stem cell registry. *Stem Cells Dev.* **2014**, *23* (Suppl. 1), 51–55. [CrossRef]

97. Lee, J.Y.; Lee, D.Y.; Choi, Y.S.; Lee, K.J.; Kim, Y.O. Registration of human embryonic stem cell lines: Korea, 2010. *Osong Public Health Res. Perspect.* **2011**, *2*, 141–147. [CrossRef]

98. Obama, B. Executive Order 13505 of March 9, 2009: Removing Barriers to Responsible Scientific Research Involving Human Stem Cells. Government Publishing Office. 2009. Available online: https://www.gpo.gov/fdsys/pkg/FR-2009-03-11/pdf/E9-5441.pdf (accessed on 9 March 2009).

99. Sanganalmath, S.K.; Bolli, R. Cell therapy for heart failure: A comprehensive overview of experimental and clinical studies, current challenges, and future directions. *Circ. Res.* **2013**, *113*, 810–834. [CrossRef]

100. Olson, E.N. Gene regulatory networks in the evolution and development of the heart. *Science* **2006**, *313*, 1922–1927. [CrossRef]

101. Costello, I.; Pimeisl, I.M.; Drager, S.; Bikoff, E.K.; Robertson, E.J.; Arnold, S.J. The T-box transcription factor Eomesodermin acts upstream of Mesp1 to specify cardiac mesoderm during mouse gastrulation. *Nat. Cell Biol.* **2011**, *13*, 1084–1091. [CrossRef]

102. David, R.; Jarsch, V.B.; Schwarz, F.; Nathan, P.; Gegg, M.; Lickert, H.; Franz, W.M. Induction of MesP1 by Brachyury(T) generates the common multipotent cardiovascular stem cell. *Cardiovasc. Res.* **2011**, *92*, 115–122. [CrossRef]

103. Wu, S.M.; Chien, K.R.; Mummery, C. Origins and fates of cardiovascular progenitor cells. *Cell* **2008**, *132*, 537–543. [CrossRef]

104. Sylva, M.; van den Hoff, M.J.; Moorman, A.F. Development of the human heart. *Am. J. Med. Genet. Part A* **2014**, *164A*, 1347–1371. [CrossRef] [PubMed]

105. De Bakker, B.S.; de Jong, K.H.; Hagoort, J.; de Bree, K.; Besselink, C.T.; de Kanter, F.E.; Veldhuis, T.; Bais, B.; Schildmeijer, R.; Ruijter, J.M.; et al. An interactive three-dimensional digital atlas and quantitative database of human development. *Science* **2016**, *354*. [CrossRef] [PubMed]

106. Gadue, P.; Huber, T.L.; Nostro, M.C.; Kattman, S.; Keller, G.M. Germ layer induction from embryonic stem cells. *Exp. Hematol.* **2005**, *33*, 955–964. [CrossRef] [PubMed]

107. Moretti, A.; Laugwitz, K.L.; Dorn, T.; Sinnecker, D.; Mummery, C. Pluripotent stem cell models of human heart disease. *Cold Spring Harb. Perspect. Med.* **2013**, *3*. [CrossRef] [PubMed]

108. Laflamme, M.A.; Chen, K.Y.; Naumova, A.V.; Muskheli, V.; Fugate, J.A.; Dupras, S.K.; Reinecke, H.; Xu, C.; Hassanipour, M.; Police, S.; et al. Cardiomyocytes derived from human embryonic stem cells in pro-survival factors enhance function of infarcted rat hearts. *Nat. Biotechnol.* **2007**, *25*, 1015–1024. [CrossRef] [PubMed]

109. Ren, Y.; Lee, M.Y.; Schliffke, S.; Paavola, J.; Amos, P.J.; Ge, X.; Ye, M.; Zhu, S.; Senyei, G.; Lum, L.; et al. Small molecule Wnt inhibitors enhance the efficiency of BMP-4-directed cardiac differentiation of human pluripotent stem cells. *J. Mol. Cell. Cardiol.* **2011**, *51*, 280–287. [CrossRef]

110. Zeineddine, D.; Papadimou, E.; Mery, A.; Menard, C.; Puceat, M. Cardiac commitment of embryonic stem cells for myocardial repair. *Methods Mol. Med.* **2005**, *112*, 175–182.

111. Kehat, I.; Kenyagin-Karsenti, D.; Snir, M.; Segev, H.; Amit, M.; Gepstein, A.; Livne, E.; Binah, O.; Itskovitz-Eldor, J.; Gepstein, L. Human embryonic stem cells can differentiate into myocytes with structural and functional properties of cardiomyocytes. *J. Clin. Investig.* **2001**, *108*, 407–414. [CrossRef]

112. Mauritz, C.; Schwanke, K.; Reppel, M.; Neef, S.; Katsirntaki, K.; Maier, L.S.; Nguemo, F.; Menke, S.; Haustein, M.; Hescheler, J.; et al. Generation of functional murine cardiac myocytes from induced pluripotent stem cells. *Circulation* **2008**, *118*, 507–517. [CrossRef]

113. Narazaki, G.; Uosaki, H.; Teranishi, M.; Okita, K.; Kim, B.; Matsuoka, S.; Yamanaka, S.; Yamashita, J.K. Directed and systematic differentiation of cardiovascular cells from mouse induced pluripotent stem cells. *Circulation* **2008**, *118*, 498–506. [CrossRef]

114. Mummery, C.; Ward-van Oostwaard, D.; Doevendans, P.; Spijker, R.; van den Brink, S.; Hassink, R.; van der Heyden, M.; Opthof, T.; Pera, M.; de la Riviere, A.B.; et al. Differentiation of human embryonic stem cells to cardiomyocytes: Role of coculture with visceral endoderm-like cells. *Circulation* **2003**, *107*, 2733–2740. [CrossRef] [PubMed]

115. Burridge, P.W.; Matsa, E.; Shukla, P.; Lin, Z.C.; Churko, J.M.; Ebert, A.D.; Lan, F.; Diecke, S.; Huber, B.; Mordwinkin, N.M.; et al. Chemically defined generation of human cardiomyocytes. *Nat. Methods* **2014**, *11*, 855–860. [CrossRef] [PubMed]

116. Lian, X.; Zhang, J.; Azarin, S.M.; Zhu, K.; Hazeltine, L.B.; Bao, X.; Hsiao, C.; Kamp, T.J.; Palecek, S.P. Directed cardiomyocyte differentiation from human pluripotent stem cells by modulating Wnt/beta-catenin signaling under fully defined conditions. *Nat. Protoc.* **2013**, *8*, 162–175. [CrossRef] [PubMed]

117. Lian, X.; Bao, X.; Zilberter, M.; Westman, M.; Fisahn, A.; Hsiao, C.; Hazeltine, L.B.; Dunn, K.K.; Kamp, T.J.; Palecek, S.P. Chemically defined, albumin-free human cardiomyocyte generation. *Nat. Methods* **2015**, *12*, 595–596. [CrossRef] [PubMed]

118. Yang, L.; Soonpaa, M.H.; Adler, E.D.; Roepke, T.K.; Kattman, S.J.; Kennedy, M.; Henckaerts, E.; Bonham, K.; Abbott, G.W.; Linden, R.M.; et al. Human cardiovascular progenitor cells develop from a KDR+ embryonic-stem-cell-derived population. *Nature* **2008**, *453*, 524–528. [CrossRef] [PubMed]

119. Moretti, A.; Caron, L.; Nakano, A.; Lam, J.T.; Bernshausen, A.; Chen, Y.; Qyang, Y.; Bu, L.; Sasaki, M.; Martin-Puig, S.; et al. Multipotent embryonic isl1+ progenitor cells lead to cardiac, smooth muscle, and endothelial cell diversification. *Cell* **2006**, *127*, 1151–1165. [CrossRef]

120. Elliott, D.A.; Braam, S.R.; Koutsis, K.; Ng, E.S.; Jenny, R.; Lagerqvist, E.L.; Biben, C.; Hatzistavrou, T.; Hirst, C.E.; Yu, Q.C.; et al. NKX2-5(eGFP/w) hESCs for isolation of human cardiac progenitors and cardiomyocytes. *Nat. Methods* **2011**, *8*, 1037–1040. [CrossRef]

121. Dubois, N.C.; Craft, A.M.; Sharma, P.; Elliott, D.A.; Stanley, E.G.; Elefanty, A.G.; Gramolini, A.; Keller, G. SIRPA is a specific cell-surface marker for isolating cardiomyocytes derived from human pluripotent stem cells. *Nat. Biotechnol.* **2011**, *29*, 1011–1018. [CrossRef]

122. Uosaki, H.; Fukushima, H.; Takeuchi, A.; Matsuoka, S.; Nakatsuji, N.; Yamanaka, S.; Yamashita, J.K. Efficient and scalable purification of cardiomyocytes from human embryonic and induced pluripotent stem cells by VCAM1 surface expression. *PLoS ONE* **2011**, *6*, e23657. [CrossRef]

123. Gallo, P.; Grimaldi, S.; Latronico, M.V.; Bonci, D.; Pagliuca, A.; Gallo, P.; Ausoni, S.; Peschle, C.; Condorelli, G. A lentiviral vector with a short troponin-I promoter for tracking cardiomyocyte differentiation of human embryonic stem cells. *Gene Ther.* **2008**, *15*, 161–170. [CrossRef]

124. Kita-Matsuo, H.; Barcova, M.; Prigozhina, N.; Salomonis, N.; Wei, K.; Jacot, J.G.; Nelson, B.; Spiering, S.; Haverslag, R.; Kim, C.; et al. Lentiviral vectors and protocols for creation of stable hESC lines for fluorescent tracking and drug resistance selection of cardiomyocytes. *PLoS ONE* **2009**, *4*, e5046. [CrossRef] [PubMed]

125. Tohyama, S.; Hattori, F.; Sano, M.; Hishiki, T.; Nagahata, Y.; Matsuura, T.; Hashimoto, H.; Suzuki, T.; Yamashita, H.; Satoh, Y.; et al. Distinct metabolic flow enables large-scale purification of mouse and human pluripotent stem cell-derived cardiomyocytes. *Cell Stem Cell* **2013**, *12*, 127–137. [CrossRef] [PubMed]

126. Zhao, Y.; Ransom, J.F.; Li, A.; Vedantham, V.; von Drehle, M.; Muth, A.N.; Tsuchihashi, T.; McManus, M.T.; Schwartz, R.J.; Srivastava, D. Dysregulation of cardiogenesis, cardiac conduction, and cell cycle in mice lacking miRNA-1-2. *Cell* **2007**, *129*, 303–317. [CrossRef] [PubMed]

127. Fu, J.D.; Rushing, S.N.; Lieu, D.K.; Chan, C.W.; Kong, C.W.; Geng, L.; Wilson, K.D.; Chiamvimonvat, N.; Boheler, K.R.; Wu, J.C.; et al. Distinct roles of microRNA-1 and -499 in ventricular specification and functional maturation of human embryonic stem cell-derived cardiomyocytes. *PLoS ONE* **2011**, *6*, e27417. [CrossRef] [PubMed]

128. Hodgkinson, C.P.; Kang, M.H.; Dal-Pra, S.; Mirotsou, M.; Dzau, V.J. MicroRNAs and Cardiac Regeneration. *Circ. Res.* **2015**, *116*, 1700–1711. [CrossRef]

129. Yang, X.; Pabon, L.; Murry, C.E. Engineering adolescence: Maturation of human pluripotent stem cell-derived cardiomyocytes. *Circ. Res.* **2014**, *114*, 511–523. [CrossRef]

130. Friedman, C.E.; Nguyen, Q.; Lukowski, S.W.; Helfer, A.; Chiu, H.S.; Miklas, J.; Levy, S.; Suo, S.; Han, J.J.; Osteil, P.; et al. Single-Cell Transcriptomic Analysis of Cardiac Differentiation from Human PSCs Reveals HOPX-Dependent Cardiomyocyte Maturation. *Cell Stem Cell* **2018**, *23*, 586–598. [CrossRef]

131. Churko, J.M.; Garg, P.; Treutlein, B.; Venkatasubramanian, M.; Wu, H.; Lee, J.; Wessells, Q.N.; Chen, S.Y.; Chen, W.Y.; Chetal, K.; et al. Defining human cardiac transcription factor hierarchies using integrated single-cell heterogeneity analysis. *Nat. Commun.* **2018**, *9*, 4906. [CrossRef]

132. Lundy, S.D.; Zhu, W.Z.; Regnier, M.; Laflamme, M.A. Structural and functional maturation of cardiomyocytes derived from human pluripotent stem cells. *Stem Cells Dev.* **2013**, *22*, 1991–2002. [CrossRef]

133. Bhavsar, P.K.; Dhoot, G.K.; Cumming, D.V.; Butler-Browne, G.S.; Yacoub, M.H.; Barton, P.J. Developmental expression of troponin I isoforms in fetal human heart. *FEBS Lett.* **1991**, *292*, 5–8.

134. Hunkeler, N.M.; Kullman, J.; Murphy, A.M. Troponin I isoform expression in human heart. *Circ. Res.* **1991**, *69*, 1409–1414. [CrossRef] [PubMed]

135. Bedada, F.B.; Chan, S.S.; Metzger, S.K.; Zhang, L.; Zhang, J.; Garry, D.J.; Kamp, T.J.; Kyba, M.; Metzger, J.M. Acquisition of a quantitative, stoichiometrically conserved ratiometric marker of maturation status in stem cell-derived cardiac myocytes. *Stem Cell Rep.* **2014**, *3*, 594–605. [CrossRef] [PubMed]

136. Taegtmeyer, H.; Sen, S.; Vela, D. Return to the fetal gene program: A suggested metabolic link to gene expression in the heart. *Ann. N. Y. Acad. Sci.* **2010**, *1188*, 191–198. [CrossRef] [PubMed]

137. Everett, A.W. Isomyosin expression in human heart in early pre- and post-natal life. *J. Mol. Cell. Cardiol.* **1986**, *18*, 607–615. [CrossRef]

138. Weber, N.; Schwanke, K.; Greten, S.; Wendland, M.; Iorga, B.; Fischer, M.; Geers-Knorr, C.; Hegermann, J.; Wrede, C.; Fiedler, J.; et al. Stiff matrix induces switch to pure beta-cardiac myosin heavy chain expression in human ESC-derived cardiomyocytes. *Basic Res. Cardiol.* **2016**, *111*, 68. [CrossRef]

139. Lahmers, S.; Wu, Y.; Call, D.R.; Labeit, S.; Granzier, H. Developmental control of titin isoform expression and passive stiffness in fetal and neonatal myocardium. *Circ. Res.* **2004**, *94*, 505–513. [CrossRef]

140. Xu, X.Q.; Soo, S.Y.; Sun, W.; Zweigerdt, R. Global expression profile of highly enriched cardiomyocytes derived from human embryonic stem cells. *Stem Cells* **2009**, *27*, 2163–2174. [CrossRef]

141. Blazeski, A.; Zhu, R.; Hunter, D.W.; Weinberg, S.H.; Boheler, K.R.; Zambidis, E.T.; Tung, L. Electrophysiological and contractile function of cardiomyocytes derived from human embryonic stem cells. *Prog. Biophys. Mol. Biol.* **2012**, *110*, 178–195. [CrossRef]

142. Funakoshi, S.; Miki, K.; Takaki, T.; Okubo, C.; Hatani, T.; Chonabayashi, K.; Nishikawa, M.; Takei, I.; Oishi, A.; Narita, M.; et al. Enhanced engraftment, proliferation, and therapeutic potential in heart using optimized human iPSC-derived cardiomyocytes. *Sci. Rep.* **2016**, *6*, 19111. [CrossRef]

143. Synnergren, J.; Ameen, C.; Jansson, A.; Sartipy, P. Global transcriptional profiling reveals similarities and differences between human stem cell-derived cardiomyocyte clusters and heart tissue. *Physiol. Genom.* **2012**, *44*, 245–258. [CrossRef]

144. Lieu, D.K.; Liu, J.; Siu, C.W.; McNerney, G.P.; Tse, H.F.; Abu-Khalil, A.; Huser, T.; Li, R.A. Absence of transverse tubules contributes to non-uniform Ca(2+) wavefronts in mouse and human embryonic stem cell-derived cardiomyocytes. *Stem Cells Dev.* **2009**, *18*, 1493–1500. [CrossRef] [PubMed]

145. Ramirez-Bergeron, D.L.; Runge, A.; Dahl, K.D.; Fehling, H.J.; Keller, G.; Simon, M.C. Hypoxia affects mesoderm and enhances hemangioblast specification during early development. *Development* **2004**, *131*, 4623–4634. [CrossRef] [PubMed]

146. Lopaschuk, G.D.; Jaswal, J.S. Energy metabolic phenotype of the cardiomyocyte during development, differentiation, and postnatal maturation. *J. Cardiovasc. Pharmacol.* **2010**, *56*, 130–140. [CrossRef] [PubMed]

147. Kim, C.; Wong, J.; Wen, J.; Wang, S.; Wang, C.; Spiering, S.; Kan, N.G.; Forcales, S.; Puri, P.L.; Leone, T.C.; et al. Studying arrhythmogenic right ventricular dysplasia with patient-specific iPSCs. *Nature* **2013**, *494*, 105–110. [CrossRef]

148. Mills, R.J.; Titmarsh, D.M.; Koenig, X.; Parker, B.L.; Ryall, J.G.; Quaife-Ryan, G.A.; Voges, H.K.; Hodson, M.P.; Ferguson, C.; Drowley, L.; et al. Functional screening in human cardiac organoids reveals a metabolic mechanism for cardiomyocyte cell cycle arrest. *Proc. Natl. Acad. Sci. USA* **2017**, *114*, E8372–E8381. [CrossRef] [PubMed]

149. Hazeltine, L.B.; Simmons, C.S.; Salick, M.R.; Lian, X.; Badur, M.G.; Han, W.; Delgado, S.M.; Wakatsuki, T.; Crone, W.C.; Pruitt, B.L.; et al. Effects of substrate mechanics on contractility of cardiomyocytes generated from human pluripotent stem cells. *Int. J. Cell Biol.* **2012**, *2012*, 508294. [CrossRef]

150. Kim, D.H.; Lipke, E.A.; Kim, P.; Cheong, R.; Thompson, S.; Delannoy, M.; Suh, K.Y.; Tung, L.; Levchenko, A. Nanoscale cues regulate the structure and function of macroscopic cardiac tissue constructs. *Proc. Natl. Acad. Sci. USA* **2010**, *107*, 565–570. [CrossRef]

151. Lee, Y.K.; Ng, K.M.; Chan, Y.C.; Lai, W.H.; Au, K.W.; Ho, C.Y.; Wong, L.Y.; Lau, C.P.; Tse, H.F.; Siu, C.W. Triiodothyronine promotes cardiac differentiation and maturation of embryonic stem cells via the classical genomic pathway. *Mol. Endocrinol.* **2010**, *24*, 1728–1736. [CrossRef]
152. McDevitt, T.C.; Laflamme, M.A.; Murry, C.E. Proliferation of cardiomyocytes derived from human embryonic stem cells is mediated via the IGF/PI 3-kinase/Akt signaling pathway. *J. Mol. Cell. Cardiol.* **2005**, *39*, 865–873. [CrossRef]
153. Zimmermann, W.H.; Schneiderbanger, K.; Schubert, P.; Didie, M.; Munzel, F.; Heubach, J.F.; Kostin, S.; Neuhuber, W.L.; Eschenhagen, T. Tissue engineering of a differentiated cardiac muscle construct. *Circ. Res.* **2002**, *90*, 223–230. [CrossRef]
154. Zimmermann, W.H.; Melnychenko, I.; Wasmeier, G.; Didie, M.; Naito, H.; Nixdorff, U.; Hess, A.; Budinsky, L.; Brune, K.; Michaelis, B.; et al. Engineered heart tissue grafts improve systolic and diastolic function in infarcted rat hearts. *Nat. Med.* **2006**, *12*, 452–458. [CrossRef] [PubMed]
155. Eschenhagen, T.; Eder, A.; Vollert, I.; Hansen, A. Physiological aspects of cardiac tissue engineering. *Am. J. Physiol. Heart Circ. Physiol.* **2012**, *303*, H133–H143. [CrossRef] [PubMed]
156. Cui, H.; Miao, S.; Esworthy, T.; Zhou, X.; Lee, S.J.; Liu, C.; Yu, Z.X.; Fisher, J.P.; Mohiuddin, M.; Zhang, L.G. 3D bioprinting for cardiovascular regeneration and pharmacology. *Adv. Drug Deliv. Rev.* **2018**, *132*, 252–269. [CrossRef] [PubMed]
157. Sheehy, S.P.; Grosberg, A.; Qin, P.; Behm, D.J.; Ferrier, J.P.; Eagleson, M.A.; Nesmith, A.P.; Krull, D.; Falls, J.G.; Campbell, P.H.; et al. Toward improved myocardial maturity in an organ-on-chip platform with immature cardiac myocytes. *Exp. Biol. Med.* **2017**, *242*, 1643–1656. [CrossRef] [PubMed]
158. Schwan, J.; Kwaczala, A.T.; Ryan, T.J.; Bartulos, O.; Ren, Y.; Sewanan, L.R.; Morris, A.H.; Jacoby, D.L.; Qyang, Y.; Campbell, S.G. Anisotropic engineered heart tissue made from laser-cut decellularized myocardium. *Sci. Rep.* **2016**, *6*, 32068. [CrossRef] [PubMed]
159. Gonzalez, F.; Boue, S.; Izpisua Belmonte, J.C. Methods for making induced pluripotent stem cells: Reprogramming a la carte. *Nat. Rev. Genet.* **2011**, *12*, 231–242. [CrossRef] [PubMed]
160. Yoshida, Y.; Yamanaka, S. Induced Pluripotent Stem Cells 10 Years Later: For Cardiac Applications. *Circ. Res.* **2017**, *120*, 1958–1968. [CrossRef]
161. Bellin, M.; Mummery, C.L. Inherited heart disease—What can we expect from the second decade of human iPS cell research? *FEBS Lett.* **2016**, *590*, 2482–2493. [CrossRef]
162. Carvajal-Vergara, X.; Sevilla, A.; D'Souza, S.L.; Ang, Y.S.; Schaniel, C.; Lee, D.F.; Yang, L.; Kaplan, A.D.; Adler, E.D.; Rozov, R.; et al. Patient-specific induced pluripotent stem-cell-derived models of LEOPARD syndrome. *Nature* **2010**, *465*, 808–812. [CrossRef]
163. Dudek, J.; Cheng, I.F.; Balleininger, M.; Vaz, F.M.; Streckfuss-Bomeke, K.; Hubscher, D.; Vukotic, M.; Wanders, R.J.; Rehling, P.; Guan, K. Cardiolipin deficiency affects respiratory chain function and organization in an induced pluripotent stem cell model of Barth syndrome. *Stem Cell Res.* **2013**, *11*, 806–819. [CrossRef]
164. Fatima, A.; Xu, G.; Shao, K.; Papadopoulos, S.; Lehmann, M.; Arnaiz-Cot, J.J.; Rosa, A.O.; Nguemo, F.; Matzkies, M.; Dittmann, S.; et al. In vitro modeling of ryanodine receptor 2 dysfunction using human induced pluripotent stem cells. *Cell. Physiol. Biochem. Int. J. Exp. Cell. Physiol. Biochem. Pharmacol.* **2011**, *28*, 579–592. [CrossRef] [PubMed]
165. Hinson, J.T.; Chopra, A.; Nafissi, N.; Polacheck, W.J.; Benson, C.C.; Swist, S.; Gorham, J.; Yang, L.; Schafer, S.; Sheng, C.C.; et al. HEART DISEASE. Titin mutations in iPS cells define sarcomere insufficiency as a cause of dilated cardiomyopathy. *Science* **2015**, *349*, 982–986. [CrossRef] [PubMed]
166. Huang, H.P.; Chen, P.H.; Hwu, W.L.; Chuang, C.Y.; Chien, Y.H.; Stone, L.; Chien, C.L.; Li, L.T.; Chiang, S.C.; Chen, H.F.; et al. Human Pompe disease-induced pluripotent stem cells for pathogenesis modeling, drug testing and disease marker identification. *Hum. Mol. Genet.* **2011**, *20*, 4851–4864. [CrossRef] [PubMed]
167. Itzhaki, I.; Maizels, L.; Huber, I.; Zwi-Dantsis, L.; Caspi, O.; Winterstern, A.; Feldman, O.; Gepstein, A.; Arbel, G.; Hammerman, H.; et al. Modelling the long QT syndrome with induced pluripotent stem cells. *Nature* **2011**, *471*, 225–229. [CrossRef]
168. Kujala, K.; Paavola, J.; Lahti, A.; Larsson, K.; Pekkanen-Mattila, M.; Viitasalo, M.; Lahtinen, A.M.; Toivonen, L.; Kontula, K.; Swan, H.; et al. Cell model of catecholaminergic polymorphic ventricular tachycardia reveals early and delayed afterdepolarizations. *PLoS ONE* **2012**, *7*, e44660. [CrossRef]

169. Lahti, A.L.; Kujala, V.J.; Chapman, H.; Koivisto, A.P.; Pekkanen-Mattila, M.; Kerkela, E.; Hyttinen, J.; Kontula, K.; Swan, H.; Conklin, B.R.; et al. Model for long QT syndrome type 2 using human iPS cells demonstrates arrhythmogenic characteristics in cell culture. *Dis. Models Mech.* **2012**, *5*, 220–230. [CrossRef]

170. Ma, D.; Wei, H.; Lu, J.; Ho, S.; Zhang, G.; Sun, X.; Oh, Y.; Tan, S.H.; Ng, M.L.; Shim, W.; et al. Generation of patient-specific induced pluripotent stem cell-derived cardiomyocytes as a cellular model of arrhythmogenic right ventricular cardiomyopathy. *Eur. Heart J.* **2013**, *34*, 1122–1133. [CrossRef]

171. Ma, D.; Wei, H.; Zhao, Y.; Lu, J.; Li, G.; Sahib, N.B.; Tan, T.H.; Wong, K.Y.; Shim, W.; Wong, P.; et al. Modeling type 3 long QT syndrome with cardiomyocytes derived from patient-specific induced pluripotent stem cells. *Int. J. Cardiol.* **2013**, *168*, 5277–5286. [CrossRef]

172. Moretti, A.; Bellin, M.; Welling, A.; Jung, C.B.; Lam, J.T.; Bott-Flugel, L.; Dorn, T.; Goedel, A.; Hohnke, C.; Hofmann, F.; et al. Patient-specific induced pluripotent stem-cell models for long-QT syndrome. *N. Engl. J. Med.* **2010**, *363*, 1397–1409. [CrossRef]

173. Yazawa, M.; Hsueh, B.; Jia, X.; Pasca, A.M.; Bernstein, J.A.; Hallmayer, J.; Dolmetsch, R.E. Using induced pluripotent stem cells to investigate cardiac phenotypes in Timothy syndrome. *Nature* **2011**, *471*, 230–234. [CrossRef]

174. Ojala, M.; Prajapati, C.; Polonen, R.P.; Rajala, K.; Pekkanen-Mattila, M.; Rasku, J.; Larsson, K.; Aalto-Setala, K. Mutation-Specific Phenotypes in hiPSC-Derived Cardiomyocytes Carrying Either Myosin-Binding Protein C Or alpha-Tropomyosin Mutation for Hypertrophic Cardiomyopathy. *Stem Cells Int.* **2016**, *2016*, 1684792. [CrossRef] [PubMed]

175. Wang, G.; McCain, M.L.; Yang, L.; He, A.; Pasqualini, F.S.; Agarwal, A.; Yuan, H.; Jiang, D.; Zhang, D.; Zangi, L.; et al. Modeling the mitochondrial cardiomyopathy of Barth syndrome with induced pluripotent stem cell and heart-on-chip technologies. *Nat. Med.* **2014**, *20*, 616–623. [CrossRef] [PubMed]

176. Sun, N.; Yazawa, M.; Liu, J.; Han, L.; Sanchez-Freire, V.; Abilez, O.J.; Navarrete, E.G.; Hu, S.; Wang, L.; Lee, A.; et al. Patient-specific induced pluripotent stem cells as a model for familial dilated cardiomyopathy. *Sci. Transl. Med.* **2012**, *4*, 130ra47. [CrossRef] [PubMed]

177. Sharma, A.; Marceau, C.; Hamaguchi, R.; Burridge, P.W.; Rajarajan, K.; Churko, J.M.; Wu, H.; Sallam, K.I.; Matsa, E.; Sturzu, A.C.; et al. Human induced pluripotent stem cell-derived cardiomyocytes as an in vitro model for coxsackievirus B3-induced myocarditis and antiviral drug screening platform. *Circ. Res.* **2014**, *115*, 556–566. [CrossRef]

178. Gu, M.; Shao, N.Y.; Sa, S.; Li, D.; Termglinchan, V.; Ameen, M.; Karakikes, I.; Sosa, G.; Grubert, F.; Lee, J.; et al. Patient-Specific iPSC-Derived Endothelial Cells Uncover Pathways that Protect against Pulmonary Hypertension in BMPR2 Mutation Carriers. *Cell Stem Cell* **2017**, *20*, 490–504. [CrossRef]

179. Fatima, A.; Kaifeng, S.; Dittmann, S.; Xu, G.; Gupta, M.K.; Linke, M.; Zechner, U.; Nguemo, F.; Milting, H.; Farr, M.; et al. The disease-specific phenotype in cardiomyocytes derived from induced pluripotent stem cells of two long QT syndrome type 3 patients. *PLoS ONE* **2013**, *8*, e83005. [CrossRef]

180. Paci, M.; Hyttinen, J.; Aalto-Setala, K.; Severi, S. Computational models of ventricular- and atrial-like human induced pluripotent stem cell derived cardiomyocytes. *Ann. Biomed. Eng.* **2013**, *41*, 2334–2348. [CrossRef]

181. Paci, M.; Passini, E.; Severi, S.; Hyttinen, J.; Rodriguez, B. Phenotypic variability in LQT3 human induced pluripotent stem cell-derived cardiomyocytes and their response to antiarrhythmic pharmacologic therapy: An in silico approach. *Heart Rhythm* **2017**, *14*, 1704–1712. [CrossRef]

182. Paci, M.; Polonen, R.P.; Cori, D.; Penttinen, K.; Aalto-Setala, K.; Severi, S.; Hyttinen, J. Automatic Optimization of an in Silico Model of Human iPSC Derived Cardiomyocytes Recapitulating Calcium Handling Abnormalities. *Front. Physiol.* **2018**, *9*, 709. [CrossRef]

183. Bellin, M.; Casini, S.; Davis, R.P.; D'Aniello, C.; Haas, J.; Ward-van Oostwaard, D.; Tertoolen, L.G.; Jung, C.B.; Elliott, D.A.; Welling, A.; et al. Isogenic human pluripotent stem cell pairs reveal the role of a KCNH2 mutation in long-QT syndrome. *EMBO J.* **2013**, *32*, 3161–3175. [CrossRef]

184. Rampe, D.; Brown, A.M. A history of the role of the hERG channel in cardiac risk assessment. *J. Pharmacol. Toxicol. Methods* **2013**, *68*, 13–22. [CrossRef] [PubMed]

185. Ahola, A.; Polonen, R.P.; Aalto-Setala, K.; Hyttinen, J. Simultaneous Measurement of Contraction and Calcium Transients in Stem Cell Derived Cardiomyocytes. *Ann. Biomed. Eng.* **2018**, *46*, 148–158. [CrossRef] [PubMed]

186. Bjork, S.; Ojala, E.A.; Nordstrom, T.; Ahola, A.; Liljestrom, M.; Hyttinen, J.; Kankuri, E.; Mervaala, E. Evaluation of Optogenetic Electrophysiology Tools in Human Stem Cell-Derived Cardiomyocytes. *Front. Physiol.* **2017**, *8*, 884. [CrossRef] [PubMed]

187. Klimas, A.; Ambrosi, C.M.; Yu, J.; Williams, J.C.; Bien, H.; Entcheva, E. OptoDyCE as an automated system for high-throughput all-optical dynamic cardiac electrophysiology. *Nat. Commun.* **2016**, *7*, 11542. [CrossRef]

188. Rajamohan, D.; Kalra, S.; Duc Hoang, M.; George, V.; Staniforth, A.; Russell, H.; Yang, X.; Denning, C. Automated Electrophysiological and Pharmacological Evaluation of Human Pluripotent Stem Cell-Derived Cardiomyocytes. *Stem Cells Dev.* **2016**, *25*, 439–452. [CrossRef]

189. Chi, K.R. Revolution dawning in cardiotoxicity testing. *Nat. Rev. Drug Discov.* **2013**, *12*, 565–567. [CrossRef]

190. Miller, J.C.; Tan, S.; Qiao, G.; Barlow, K.A.; Wang, J.; Xia, D.F.; Meng, X.; Paschon, D.E.; Leung, E.; Hinkley, S.J.; et al. A TALE nuclease architecture for efficient genome editing. *Nat. Biotechnol.* **2011**, *29*, 143–148. [CrossRef]

191. Urnov, F.D.; Rebar, E.J.; Holmes, M.C.; Zhang, H.S.; Gregory, P.D. Genome editing with engineered zinc finger nucleases. *Nat. Rev. Genet.* **2010**, *11*, 636–646. [CrossRef]

192. Kim, H.; Kim, J.S. A guide to genome engineering with programmable nucleases. *Nat. Rev. Genet.* **2014**, *15*, 321–334. [CrossRef]

193. Collin, J.; Lako, M. Concise review: Putting a finger on stem cell biology: Zinc finger nuclease-driven targeted genetic editing in human pluripotent stem cells. *Stem Cells* **2011**, *29*, 1021–1033. [CrossRef]

194. Zou, J.; Maeder, M.L.; Mali, P.; Pruett-Miller, S.M.; Thibodeau-Beganny, S.; Chou, B.K.; Chen, G.; Ye, Z.; Park, I.H.; Daley, G.Q.; et al. Gene targeting of a disease-related gene in human induced pluripotent stem and embryonic stem cells. *Cell Stem Cell* **2009**, *5*, 97–110. [CrossRef] [PubMed]

195. Hockemeyer, D.; Soldner, F.; Beard, C.; Gao, Q.; Mitalipova, M.; DeKelver, R.C.; Katibah, G.E.; Amora, R.; Boydston, E.A.; Zeitler, B.; et al. Efficient targeting of expressed and silent genes in human ESCs and iPSCs using zinc-finger nucleases. *Nat. Biotechnol.* **2009**, *27*, 851–857. [CrossRef] [PubMed]

196. Hockemeyer, D.; Wang, H.; Kiani, S.; Lai, C.S.; Gao, Q.; Cassady, J.P.; Cost, G.J.; Zhang, L.; Santiago, Y.; Miller, J.C.; et al. Genetic engineering of human pluripotent cells using TALE nucleases. *Nat. Biotechnol.* **2011**, *29*, 731–734. [CrossRef]

197. Yusa, K.; Rashid, S.T.; Strick-Marchand, H.; Varela, I.; Liu, P.Q.; Paschon, D.E.; Miranda, E.; Ordonez, A.; Hannan, N.R.; Rouhani, F.J.; et al. Targeted gene correction of alpha1-antitrypsin deficiency in induced pluripotent stem cells. *Nature* **2011**, *478*, 391–394. [CrossRef] [PubMed]

198. Karakikes, I.; Stillitano, F.; Nonnenmacher, M.; Tzimas, C.; Sanoudou, D.; Termglinchan, V.; Kong, C.W.; Rushing, S.; Hansen, J.; Ceholski, D.; et al. Correction of human phospholamban R14del mutation associated with cardiomyopathy using targeted nucleases and combination therapy. *Nat. Commun.* **2015**, *6*, 6955. [CrossRef]

199. Haghighi, K.; Kolokathis, F.; Gramolini, A.O.; Waggoner, J.R.; Pater, L.; Lynch, R.A.; Fan, G.C.; Tsiapras, D.; Parekh, R.R.; Dorn, G.W., 2nd; et al. A mutation in the human phospholamban gene, deleting arginine 14, results in lethal, hereditary cardiomyopathy. *Proc. Natl. Acad. Sci. USA* **2006**, *103*, 1388–1393. [CrossRef]

200. Van der Zwaag, P.A.; van Rijsingen, I.A.; Asimaki, A.; Jongbloed, J.D.; van Veldhuisen, D.J.; Wiesfeld, A.C.; Cox, M.G.; van Lochem, L.T.; de Boer, R.A.; Hofstra, R.M.; et al. Phospholamban R14del mutation in patients diagnosed with dilated cardiomyopathy or arrhythmogenic right ventricular cardiomyopathy: Evidence supporting the concept of arrhythmogenic cardiomyopathy. *Eur. J. Heart Fail.* **2012**, *14*, 1199–1207. [CrossRef]

201. Ovando-Roche, P.; Georgiadis, A.; Smith, A.J.; Pearson, R.A.; Ali, R.R. Harnessing the Potential of Human Pluripotent Stem Cells and Gene Editing for the Treatment of Retinal Degeneration. *Curr. Stem Cell Rep.* **2017**, *3*, 112–123. [CrossRef]

202. Hsu, P.D.; Lander, E.S.; Zhang, F. Development and applications of CRISPR-Cas9 for genome engineering. *Cell* **2014**, *157*, 1262–1278. [CrossRef]

203. Chaterji, S.; Ahn, E.H.; Kim, D.H. CRISPR Genome Engineering for Human Pluripotent Stem Cell Research. *Theranostics* **2017**, *7*, 4445–4469. [CrossRef]

204. Limpitikul, W.B.; Dick, I.E.; Tester, D.J.; Boczek, N.J.; Limphong, P.; Yang, W.; Choi, M.H.; Babich, J.; DiSilvestre, D.; Kanter, R.J.; et al. A Precision Medicine Approach to the Rescue of Function on Malignant Calmodulinopathic Long-QT Syndrome. *Circ. Res.* **2017**, *120*, 39–48. [CrossRef]

205. Yamamoto, Y.; Makiyama, T.; Harita, T.; Sasaki, K.; Wuriyanghai, Y.; Hayano, M.; Nishiuchi, S.; Kohjitani, H.; Hirose, S.; Chen, J.; et al. Allele-specific ablation rescues electrophysiological abnormalities in a human iPS cell model of long-QT syndrome with a CALM2 mutation. *Hum. Mol. Genet.* **2017**, *26*, 1670–1677. [CrossRef] [PubMed]

206. Brodehl, A.; Ebbinghaus, H.; Deutsch, M.A.; Gummert, J.; Gartner, A.; Ratnavadivel, S.; Milting, H. Human Induced Pluripotent Stem-Cell-Derived Cardiomyocytes as Models for Genetic Cardiomyopathies. *Int. J. Mol. Sci.* **2019**, *20*, 4381. [CrossRef] [PubMed]

207. Te Riele, A.S.; Agullo-Pascual, E.; James, C.A.; Leo-Macias, A.; Cerrone, M.; Zhang, M.; Lin, X.; Lin, B.; Sobreira, N.L.; Amat-Alarcon, N.; et al. Multilevel analyses of SCN5A mutations in arrhythmogenic right ventricular dysplasia/cardiomyopathy suggest non-canonical mechanisms for disease pathogenesis. *Cardiovasc. Res.* **2017**, *113*, 102–111. [CrossRef] [PubMed]

208. Seeger, T.; Shrestha, R.; Lam, C.K.; Chen, C.; McKeithan, W.L.; Lau, E.; Wnorowski, A.; McMullen, G.; Greenhaw, M.; Lee, J.; et al. A Premature Termination Codon Mutation in MYBPC3 Causes Hypertrophic Cardiomyopathy via Chronic Activation of Nonsense-Mediated Decay. *Circulation* **2019**, *139*, 799–811. [CrossRef] [PubMed]

209. Van Laake, L.W.; Passier, R.; Monshouwer-Kloots, J.; Verkleij, A.J.; Lips, D.J.; Freund, C.; den Ouden, K.; Ward-van Oostwaard, D.; Korving, J.; Tertoolen, L.G.; et al. Human embryonic stem cell-derived cardiomyocytes survive and mature in the mouse heart and transiently improve function after myocardial infarction. *Stem Cell Res.* **2007**, *1*, 9–24. [CrossRef]

210. Caspi, O.; Huber, I.; Kehat, I.; Habib, M.; Arbel, G.; Gepstein, A.; Yankelson, L.; Aronson, D.; Beyar, R.; Gepstein, L. Transplantation of human embryonic stem cell-derived cardiomyocytes improves myocardial performance in infarcted rat hearts. *J. Am. Coll. Cardiol.* **2007**, *50*, 1884–1893. [CrossRef]

211. Masumoto, H.; Matsuo, T.; Yamamizu, K.; Uosaki, H.; Narazaki, G.; Katayama, S.; Marui, A.; Shimizu, T.; Ikeda, T.; Okano, T.; et al. Pluripotent stem cell-engineered cell sheets reassembled with defined cardiovascular populations ameliorate reduction in infarct heart function through cardiomyocyte-mediated neovascularization. *Stem Cells* **2012**, *30*, 1196–1205. [CrossRef]

212. Matsuo, T.; Masumoto, H.; Tajima, S.; Ikuno, T.; Katayama, S.; Minakata, K.; Ikeda, T.; Yamamizu, K.; Tabata, Y.; Sakata, R.; et al. Efficient long-term survival of cell grafts after myocardial infarction with thick viable cardiac tissue entirely from pluripotent stem cells. *Sci. Rep.* **2015**, *5*, 16842. [CrossRef]

213. Shiba, Y.; Fernandes, S.; Zhu, W.Z.; Filice, D.; Muskheli, V.; Kim, J.; Palpant, N.J.; Gantz, J.; Moyes, K.W.; Reinecke, H.; et al. Human ES-cell-derived cardiomyocytes electrically couple and suppress arrhythmias in injured hearts. *Nature* **2012**, *489*, 322–325. [CrossRef]

214. Lelovas, P.P.; Kostomitsopoulos, N.G.; Xanthos, T.T. A comparative anatomic and physiologic overview of the porcine heart. *J. Am. Assoc. Lab. Anim. Sci.* **2014**, *53*, 432–438. [PubMed]

215. Kehat, I.; Khimovich, L.; Caspi, O.; Gepstein, A.; Shofti, R.; Arbel, G.; Huber, I.; Satin, J.; Itskovitz-Eldor, J.; Gepstein, L. Electromechanical integration of cardiomyocytes derived from human embryonic stem cells. *Nat. Biotechnol.* **2004**, *22*, 1282–1289. [CrossRef] [PubMed]

216. Ye, L.; Chang, Y.H.; Xiong, Q.; Zhang, P.; Zhang, L.; Somasundaram, P.; Lepley, M.; Swingen, C.; Su, L.; Wendel, J.S.; et al. Cardiac repair in a porcine model of acute myocardial infarction with human induced pluripotent stem cell-derived cardiovascular cells. *Cell Stem Cell* **2014**, *15*, 750–761. [CrossRef] [PubMed]

217. Kawamura, M.; Miyagawa, S.; Fukushima, S.; Saito, A.; Miki, K.; Ito, E.; Sougawa, N.; Kawamura, T.; Daimon, T.; Shimizu, T.; et al. Enhanced survival of transplanted human induced pluripotent stem cell-derived cardiomyocytes by the combination of cell sheets with the pedicled omental flap technique in a porcine heart. *Circulation* **2013**, *128*, S87–S94. [CrossRef]

218. Kawamura, M.; Miyagawa, S.; Fukushima, S.; Saito, A.; Miki, K.; Funakoshi, S.; Yoshida, Y.; Yamanaka, S.; Shimizu, T.; Okano, T.; et al. Enhanced Therapeutic Effects of Human iPS Cell Derived-Cardiomyocyte by Combined Cell-Sheets with Omental Flap Technique in Porcine Ischemic Cardiomyopathy Model. *Sci. Rep.* **2017**, *7*, 8824. [CrossRef]

219. Ishigami, M.; Masumoto, H.; Ikuno, T.; Aoki, T.; Kawatou, M.; Minakata, K.; Ikeda, T.; Sakata, R.; Yamashita, J.K.; Minatoya, K. Human iPS cell-derived cardiac tissue sheets for functional restoration of infarcted porcine hearts. *PLoS ONE* **2018**, *13*, e0201650. [CrossRef]

220. Gao, L.; Gregorich, Z.R.; Zhu, W.; Mattapally, S.; Oduk, Y.; Lou, X.; Kannappan, R.; Borovjagin, A.V.; Walcott, G.P.; Pollard, A.E.; et al. Large Cardiac Muscle Patches Engineered from Human Induced-Pluripotent Stem Cell-Derived Cardiac Cells Improve Recovery from Myocardial Infarction in Swine. *Circulation* **2018**, *137*, 1712–1730. [CrossRef]
221. Bontrop, R.E. Non-human primates: Essential partners in biomedical research. *Immunol. Rev.* **2001**, *183*, 5–9. [CrossRef]
222. Ishigaki, H.; Shiina, T.; Ogasawara, K. MHC-identical and transgenic cynomolgus macaques for preclinical studies. *Inflamm. Regen.* **2018**, *38*, 30. [CrossRef]
223. Deleidi, M.; Hargus, G.; Hallett, P.; Osborn, T.; Isacson, O. Development of histocompatible primate-induced pluripotent stem cells for neural transplantation. *Stem Cells* **2011**, *29*, 1052–1063. [CrossRef]
224. Shiba, Y.; Gomibuchi, T.; Seto, T.; Wada, Y.; Ichimura, H.; Tanaka, Y.; Ogasawara, T.; Okada, K.; Shiba, N.; Sakamoto, K.; et al. Allogeneic transplantation of iPS cell-derived cardiomyocytes regenerates primate hearts. *Nature* **2016**, *538*, 388–391. [CrossRef] [PubMed]
225. Kawamura, T.; Miyagawa, S.; Fukushima, S.; Maeda, A.; Kashiyama, N.; Kawamura, A.; Miki, K.; Okita, K.; Yoshida, Y.; Shiina, T.; et al. Cardiomyocytes Derived from MHC-Homozygous Induced Pluripotent Stem Cells Exhibit Reduced Allogeneic Immunogenicity in MHC-Matched Non-human Primates. *Stem Cell Rep.* **2016**, *6*, 312–320. [CrossRef] [PubMed]
226. Chong, J.J.; Yang, X.; Don, C.W.; Minami, E.; Liu, Y.W.; Weyers, J.J.; Mahoney, W.M.; Van Biber, B.; Cook, S.M.; Palpant, N.J.; et al. Human embryonic-stem-cell-derived cardiomyocytes regenerate non-human primate hearts. *Nature* **2014**, *510*, 273–277. [CrossRef] [PubMed]
227. Menasche, P.; Vanneaux, V.; Hagege, A.; Bel, A.; Cholley, B.; Cacciapuoti, I.; Parouchev, A.; Benhamouda, N.; Tachdjian, G.; Tosca, L.; et al. Human embryonic stem cell-derived cardiac progenitors for severe heart failure treatment: First clinical case report. *Eur. Heart J.* **2015**, *36*, 2011–2017. [CrossRef] [PubMed]
228. Menasche, P.; Vanneaux, V.; Hagege, A.; Bel, A.; Cholley, B.; Parouchev, A.; Cacciapuoti, I.; Al-Daccak, R.; Benhamouda, N.; Blons, H.; et al. Transplantation of Human Embryonic Stem Cell-Derived Cardiovascular Progenitors for Severe Ischemic Left Ventricular Dysfunction. *J. Am. Coll. Cardiol.* **2018**, *71*, 429–438. [CrossRef] [PubMed]
229. Faculty of Medicine, Osaka University. iPS Cell-Based Therapy for Heart Disease: Clinical Application iPS Cell-Derived Cardiomyocytes. Available online: http://www.med.osaka-u.ac.jp/eng/archives/2777 (accessed on 1 June 2019).

Review

hiPSCs Derived Cardiac Cells for Drug and Toxicity Screening and Disease Modeling: What Micro-Electrode-Array Analyses Can Tell Us

Sophie Kussauer, Robert David * and Heiko Lemcke

Department Cardiac Surgery, Medical Center, University of Rostock, 18057 Rostock, Germany;
Sophie.Kussauer@med.uni-rostock.de (S.K.); Heiko.Lemcke@med.uni-rostock.de (H.L.)
* Correspondence: Robert.David@med.uni-rostock.de; Tel.: +49-381-4988973

Received: 25 September 2019; Accepted: 23 October 2019; Published: 28 October 2019

Abstract: Human induced pluripotent stem cell (iPSC)-derived cardiomyocytes (CM) have been intensively used in drug development and disease modeling. Since iPSC-cardiomyocyte (CM) was first generated, their characterization has become a major focus of research. Multi-/micro-electrode array (MEA) systems provide a non-invasive user-friendly platform for detailed electrophysiological analysis of iPSC cardiomyocytes including drug testing to identify potential targets and the assessment of proarrhythmic risk. Here, we provide a systematical overview about the physiological and technical background of micro-electrode array measurements of iPSC-CM. We introduce the similarities and differences between action- and field potential and the advantages and drawbacks of MEA technology. In addition, we present current studies focusing on proarrhythmic side effects of novel and established compounds combining MEA systems and iPSC-CM. MEA technology will help to open a new gateway for novel therapies in cardiovascular diseases while reducing animal experiments at the same time.

Keywords: cardiomyocytes; multi-electrode-array; micro-electrode-array; MEA; drug/toxicity screening; field potential

---

1. Introduction

The first generation of induced pluripotent stem cells (iPSCs) by Yamanka and co-workers in 2006 was a milestone for stem cell research as it allows the in vitro production of human cells without ethical concerns. Like embryonic stem cells, iPSCs have the capability to differentiate into any cell type, including cardiomyocytes, therefore providing an easy accessible cellular source for the generation of cardiac organoids and tissue structures [1–3].

One possible application for iPSC-derived cardiomyocytes (CMs) is their use in cell therapy replacing damaged tissue by in vitro generated CMs. As cardiovascular diseases are the major cause of death worldwide such regenerative approaches are needed for the development of novel treatment options. The potential and feasibility of iPSC-CM transplantation has been investigated in small and large animal models [4–7].

Thereby, an important future option of iPSC-CMs will be their generation from patient specific tissue enabling the implementation of autologous cell transplantation strategies. In this respect, iPSC-CMs can be used for the development of personalized drug screening approaches and clinically relevant diseases models. Therefore, iPSCs enable cost-effective methods to identify potential drug targets, even more accurately than animal models or other in vitro cell systems. Successful pre-clinical application of iPSC-derived cardiomyocytes for drug screening assays has been lately demonstrated by the CiPA initiative, which was initiated to assess the proarrhythmic risk of novel cardio therapeutics. A myriad of studies investigated in vitro drug effects on different ion channels of iPSC-CMs [8–11], reflecting the importance of electrophysiological measurements using stem cell derived cardiac cells.

However, the maturation of iPSC derived CMs is still a critical point for their application in cardiovascular research as well as for clinical applications. Besides metabolic and structural maturation, proper ion channel composition is crucial for the development of a mature cardiac phenotype. During the last decade, extensive analyses have been performed on the electrophysiological properties of iPSC-CMs [12–15]. Several ion channels and ion currents have been found to be present in iPSC-CMs, including sodium (I_{Na}), potassium (I_{K1} and I_{Kr}), L-type and T-type calcium channels, etc. Although multiple differentiation protocols have been developed, researchers failed to generate fully mature cardiomyocytes in vitro, possessing identical electrophysiological properties as their native adult counterparts [16–18]. Moreover, it has largely been shown that iPSC-CMs represent a heterogeneous population of electrophysiological phenotypes, i.e., atrial, ventricular and nodal-like cells [19], each characterized by a specific electrical profile. Therefore, it is important to obtain electrophysiological data for detailed characterization of iPSC-CMs, in particular when differentiation into a certain cardiac subtype is desired [20,21].

Typical approaches to investigate the electrophysiological properties of stem cell derived CMs will be discussed in the following paragraph.

2. Methods for Electrophysiological Characterization of iPSC-CMs

Several different techniques exist to study the electrophysiological properties of cardiac cells, including patch clamp analysis, MEA measurement and fluorescence dye-based assessment of the membrane potential. Each of these techniques has its own advantages and limitations, which are described in detail in the following.

2.1. Patch Clamping

Patch clamping is the gold standard technique for the acquisition of ion current data and detailed measurement of action potential (AP) properties in individual cells. The basic principle of patch clamp relies on a blunt ended glass pipette that is sealed onto the cellular membrane to obtain a so-called gigaseal [22].

In the "current patch clamp" mode the membrane potential is recorded while the current applied by the patch pipette is controlled by the operator [23]. The current patch clamp technique allows detection of APs that occur spontaneously or after stimulation with a current change induced by the recording pipette. Considering the fact that iPSC-CMs also contain non-beating populations, current patch clamp methodology allows the detection of AP patterns in these quiescent cells [23,24]. Moreover, detailed AP features, such as AP duration, amplitude, beating rate and mean diastolic potential can reliably be acquired with current patch clamping [25].

When precise characterization of ion channel subtypes is desired, "voltage patch clamp" is performed to measure individual ion currents. Unlike in the current patch clamp mode, the operator keeps the membrane potential at a certain value that enables detection of the net membrane current. In hiPSC-CMs, voltage patch clamp has been successfully applied to obtain data about ion channel density, voltage dependency and activation/deactivation characteristics [23].

However, these manual patch clamp methods are complex, technically challenging procedures that require high operator skills as well as a biophysical background for data interpretation. Another limitation is the low throughput since measurements are usually performed on the single cell level. Therefore, automated patch clamp devices have been developed to overcome the aforementioned drawbacks of manual patch clamp approaches [26]. Automatic platforms profoundly increase the efficiency of electrophysiological data recording by assessment of 10–700 cells at the same time [27]. However, while automated systems are capable to analyze hundreds of cells under variable experimental conditions, the accuracy of obtained data is reduced if compared to manual patch clamping [28,29]. High-throughput analysis is realized by analysis of single cell suspensions, in contrast to manual patch clamping where cells are usually processed in an adherent state. Recently developed systems are equipped with temperature control, optical stimulation and internal perfusion systems to ensure high data quality and reproducibility [27,30]. Data consistency and robustness is further determined by the homogeneity and density of the applied single cell suspension—a point that is particularly important for iPSC-CMs that are sensitive to dissociation as it can affect membrane proteins and electrical physiology of the cell, including ion channel expression [31,32]. Automatic techniques also do not provide the possibility of selective cell capturing. Hence, the system demands highly purified cell populations, which could be challenging when working with CMs differentiated from iPSCs that commonly represent a mixture of different cardiac subtypes [29].

2.2. Optical Recordings of the Membrane Potential

An indirect technique to assess electrophysiological data of iPSC-CMs is the application of voltage-sensitive dyes that change fluorescence intensity or emission spectra upon alteration of the membrane potential. Considering fluorescence microscopy as one of the most commonly used methods in cell research, utilizing voltage-sensitive dyes is operationally simple and does not require special instrumentation. Moreover, it is less invasive and it enables monitoring of voltage dynamics over thousands of cells with very high temporal resolution [33]. Several studies have proven feasibility of voltage sensor probes for drug screening experiments in iPSC-CMs [34–36]. Recently, Takaki and colleagues [37] applied voltage sensitive dyes for the identification of distinct cardiac subtypes in an iPSC-CM population. Further, the authors were able to detect differences in the AP pattern in iPSC-CMs obtained from patients suffering from the long QT syndrome, compared to control cells.

Alternatively, voltage sensitive probes can be engineered as fluorescent proteins that are stably expressed in target cells. Compared to voltage-sensitive dyes, these proteins possess lower phototoxicity, thus, facilitating long-term measurements. These genetically encoded probes are designed by conjugating a voltage-sensing domain to a single fluorescent protein, a fluorescence resonance energy transfer (FRET) pair or rhodopsin proteins [38,39]. Changes of the membrane potential induce conformational rearrangement of the voltage-sensor, which in turn modulates the emission spectra of the attached fluorescent protein. The latest generation of genetically encoded voltage sensors, such as ArcLight, Archer1 or QuasAr1, show large fluorescence alteration upon depolarizing events (40%–80% for a 100 mV depolarization) associated with faster on/off kinetics (1–10 ms). Shaheen et al. generated ArcLight expressing human iPSC-CMs to establish a 2D cardiac tissue platform for optical mapping and pharmacological studies [40]. Former data confirmed the suitability of genetically encoded voltage sensors for iPSC-CM drug screening applications and disease modeling attributed to altered AP phenotypes [41–44]. However, there are certain limitations of this technology. Like fluorescence dyes, genetically encoded voltage indicators provide only relative, not absolute values for the membrane potential [45]. The lower on/off kinetics increase the probability of losing high frequency AP elements [41,45]. Furthermore, introduction of voltage-sensitive proteins like ArcLight could affect the electrophysiological properties of iPSC-CMs. This needs to be carefully addressed by the operator as well as proper folding and membrane integration of the voltage sensitive probe.

2.3. MEA-Based Analysis of Cell Behavior

A common MEA system is composed of dot-like electrodes arranged in two-dimensional grids that measure the fluctuations in the extracellular field potential (FP) of an attached cell layer in respect to a reference electrode placed outside the grid (Figure 1). MEA is a non-invasive, label free methodology that has been initially applied to investigate neuronal activity [46]. However, in recent years an increasing number of studies have taken advantage of MEAs to particularly analyze compound-induced cardiac toxicity in iPSC-CMs [47–49]. Like optical recordings of the membrane potential, MEA systems allow non-invasive and cost-effective measurements at high throughput scale, and long-term observations [46,50]. On the other hand, Rynnännen et al. published data of a custom-made MEA platform for FP detection based on a single cell analysis [51]. In contrast to conventional MEA systems, this optimized device demonstrated a modified layout of larger electrodes, most suitable for observation of single iPSC-CMs. Similarly, agarose micro-chambers printed on MEA have been found to facilitate single cell detection of FPs in stem cell-derived CMs [52]. Moreover, electrophysiological assessment using MEAs is not only restricted to cell culture but can also be performed on the tissue level to better simulate in vivo conditions, as shown for murine and human heart tissue slices [53–55].

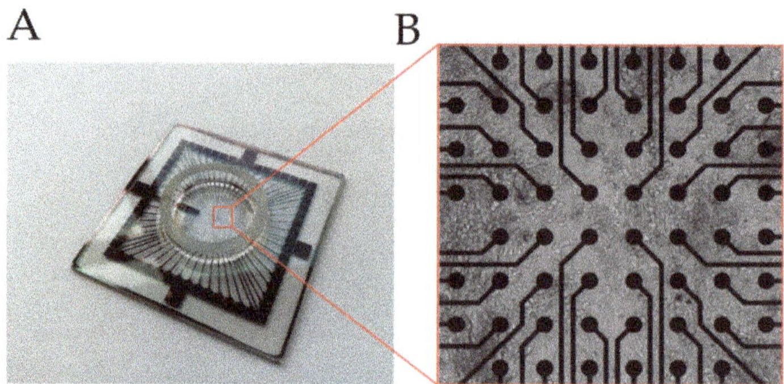

Figure 1. (**A**) Glass multi-/micro-electrode array (MEA) chip used to detect field potential (FP) of cells. (**B**) Cells seeded on an MEA surface, grown on top of the electrodes (black dots), Video S1.

An advantage of the MEA technology is its high flexibility as it can be combined with other detection methodologies to multiply the number of parameters describing cellular functions. The main parameter assessed is the FP of spontaneously beating CMs that can be correlated with certain elements of the AP pattern. Additionally, newly developed platforms provide the possibility to detect impedance of the attached cell layer [56,57]. Unlike the FP that reflects the electrical activity, impedance corresponds to the mechanical movement of the cell on the electrode. It is influenced by cell density, cell number and the extent of cell adhesion. Thus, measuring impedance helps to acquire valuable information about beating behavior, proliferation, cell death and viability [22,58].

A relationship between contraction parameters and electrophysiological activity has also been investigated by combination of MEA and high-speed video microscopy, followed by motion based image analysis of beating cells [59]. Likewise, fluorescence microscopy was used to correlate FP measurements with subcellular information [60]. However, the combined setup of MEA platforms with optical techniques requires certain structural features to achieve optimal visualization of target cells, such as transparent electrodes [51,60,61].

In another study, Siemenov et al. developed a combined scanning ion conductance microscopy–MEA system for simultaneous detection of cell surface morphology and FP in cardiomyocytes [62]. The platform reveals morpho-dynamic parameters, including maximum displacement and cell volume changes in a time-dependent manner. Together with the FP data obtained from MEA measurements, the authors were able to reconstruct 3-dimensional motion of the cell surface over a complete contraction-relaxation cycle [62].

In order to obtain reproducible and reliable experimental data, a number of points need to be considered when working with MEA systems that are particularly important for drug screening assays. Since individual iPSC-CMs show variations in AP waveforms [63], confluent monolayer cell sheets are preferred to reduce the variability of the acquired FP patterns. In this regard, cell density needs to be carefully addressed by the operator as it was found to influence electrical remodeling of CMs derived from human iPSCs [64].

3. Action Potential vs. Field Potential

Both, AP and FP are parameters describing the membrane potential of cardiomyocytes or any other cell type that is electrically active. They are generated by ion currents between the extra- and intracellular space, tightly regulated by several different membrane-located ion channels [65]. In drug development, pharmacological compounds are classified in respect to their cardiotoxic effects based on the AP or/and FP pattern in iPSC-derived CMs [34]. In addition, electrophysiology is used to identify and characterize the different cardiac subtypes in iPSC derived CM populations, which is crucial for specific cell programming strategies [20,66,67].

3.1. Action Potential in Native Cardiac Cells and iPSC Derived CMs

The AP represents the time-dependent alterations of the membrane potential in CMs that occur during the contraction of heart tissue. This requires a well-defined orchestration of numerous ion channels. Since the human heart comprises different cardiomyocyte subtypes, the AP pattern varies significantly, depending on the regions of heart (e.g., atrium, sinus node and ventricle) [68].

Despite this electrical heterogeneity, each subtype specific pattern consists of five different phases, reflecting the activity of certain ion channels (Figure 2A). Based on an incoming depolarization stimulus, opening of voltage-gated Na^+ channels induces sodium influx into the cytoplasm, resulting in a rapid depolarization of the membrane potential up to +20 to +40 mV (Phase 0). Subsequently, phase 1 is determined by time and voltage-dependent opening and closing of various ion transporters permeable to Na^+, Ca^{2+} and K^+, leading to a slight, transient hyperpolarization of the membrane potential (−10 to −30 mV). The following phase 2 is characterized by a relatively high capacity of the cell membrane and it is primarily driven by depolarization-dependent L-type Ca^{2+} channels. Due to a balanced interplay between inward currents of calcium and efflux of potassium phase 2 demonstrates a plateau period, which is particularly prominent in ventricular muscle cells [68,69]. During the plateau phase, Ca^{2+} channel conductance decreases while the outward current of K^+ inclines. This in turn promotes further repolarization (phase 3) leading to a resting potential of ~−85 mV (phase 4).

As stated above, AP patterns are unique for each cardiac subtype resulting from different ion channel composition within the cellular membrane (Figure 2A). Nodal cells, found in sinoatrial and an atrioventricular AV node or His-bundles, are capable to generate their own AP without an additional depolarizing stimulus. Compared to atrial and ventricular cells, the resting potential in nodal cells is unstable, begins at ~−60 mV (vs. ~−85 mV in atrial and ventricular cells) and gradually increases towards a threshold. This "pacemaker potential" is generated by K^+ channels that open slowly upon depolarization and deactivates with time. Concurrently, depolarization of about −60 mV activates a nodal specific Na^+ channel, known as the "funny channel", causing an increase of the intracellular Na^+ level. Once a threshold is reached, opening of voltage-gated Ca^{2+} channels induce a strong upstroke. This is in contrast to atrial and ventricular cells where a Na^+ influx mainly contributes to the rapid depolarization in phase 0.

Differences in the AP pattern are also distinct between atrial and ventricular regions of the heart. Atrial CMs undergo a more rapid early repolarization and demonstrate a less profound plateau phase, followed by slow phase 3 repolarization (Figure 2A). These differences emanate from specific K^+ channels, expressed in atrial cells, but not found in ventricular tissue [68].

Analysis and classification of AP patterns of iPSC-CMs is challenging as they demonstrate a large amount of variability, which supports the notion that CMs derived from iPSCs are a mixture of different cardiac subtypes of distinct maturation level [16,19,70]. Although multiple differentiation protocols have been established, researchers failed to generate fully mature cardiomyocytes in vitro possessing identical electrophysiological properties as their native adult counterparts [16,18]. For example, Ronaldson-Bouchard et al. have shown structural and metabolic maturity and adult-state like gene-expression of three iPSC cell lines after cultivation as cardiac tissues for four weeks, it remains to be seen whether this approach can be generally transferred to any iPSC cell line [71]. In addition, it is difficult to compare APs among different studies because of various experimental conditions used. Nevertheless, a common feature of iPSC-CMs compared to native CMs is their ability to generate APs without the need of an external, depolarizing signal, indicating a relative similarity with nodal tissue cells. Indeed, iPSC-CMs were found to express funny channels to drive spontaneous activity. Another nodal-cell characteristic of iPSC-CMs is their relatively positive resting potential of −60 to −70 mV that mainly relies on a lower or absent expression of the K$^+$ channel I$_{K1}$ [16,72], allowing the depolarizing funny current to trigger the AP. This low level of I$_{K1}$ further evokes a slower upstroke velocity in phase 0. Computational simulations revealed that an increasing expression of I$_{K1}$ in iPSC-CMs would induce a more negative and more stable resting potential [16,72–74]. These data also indicate an immature state of iPSC derived-CMs and suggests possible limitations for cardiovascular research and clinical applications.

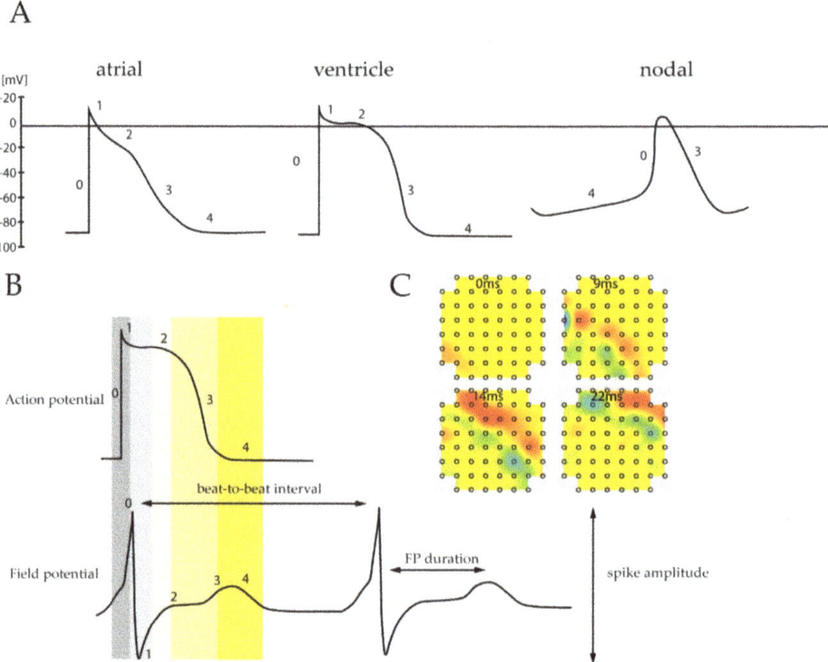

Figure 2. (**A**) Subtype specific pattern of the cardiac action potential. Ventricular, atrial and nodal cells are characterized by unique depolarization and repolarization processes leading to different action potential (AP) waveforms. Numbers correspond to the different phases that reflect the activity of involved ion channels. (**B**) Comparison of the different phases between recorded action potential and field potential. As field potential measurements allow reconstruction of the corresponding action potential it provides important physiological parameters of electrically active cells, including spike amplitude, FP interval, etc. (**C**) Moreover, MEA analysis can be applied to obtain data about prolongation velocity and direction of the field potential spreading throughout the cell layer.

3.2. Field Potential

Classically, the cardiac action potential of single cells is analyzed using patch clamp devices, which allow detection of each individual ion current contributing to the AP pattern [69,75]. In contrast, MEA does not directly measure the AP but rather record cardiac FP instead, shown in Figure 2B. The FP encompasses the spatiotemporal electrical activity of cell clusters attached to the electrode, thus, it is the superposition of all ionic processes, ranging from fast action potentials to slowest fluctuations [76,77]. The measured FP arises from spreading of the cardiac AP throughout the cell monolayer relative to the recording electrodes. Therefore, it is comparable to the clinical electrocardiogram signal that represents voltage change over time due to electrical activity of the heart [47].

Since the biophysical processes underlying the generation of FPs are well known, it is possible to reconstruct the corresponding AP pattern and to extract important physiological parameters [76].

Figure 2B depicts the different phases of a typical ventricular AP pattern and the corresponding FP measured by MEA. The FP waveform contains a strong transient spike attributed to the Na^+ influx and associated membrane depolarization, followed by a gentle incline based on the intracellular increase of Ca^{2+} level and ending with repolarization associated with K^+ efflux. In addition, a comparative analysis of patch clamp data and MEA recordings revealed that duration of the FPs correlate well with the length of the QT interval of APs [78]. Similar results were obtained by Asakura et al., showing that MEA-based FP detection can be applied to determine the prolongation of the QT interval following drug administration in iPSC-derived CMs [49]. In addition to the QT interval and K^+/Ca^+ flux, the FP pattern provides valuable information about the beating frequency as well as AP duration (Table 1, Figure 2B). Moreover, since MEA measurements are commonly performed on cell monolayers, propagation and direction of the AP can be determined (Figure 2C, Video S2).

Table 1. Functional parameters acquired by FP measurements using MEA Systems.

FP Morphology	Physiological Parameter
Spatiotemporal Assessment	Propagation velocity, direction origin of AP spread
FP Duration	QT interval of AP
FPs Over Time	Beating frequency
Spike Amplitude	Na^+ current
Spike Plateau	Ca^{2+}/K^+ current
Beat-to-Beat Interval	AP duration

For a more precise comparison of FPs and patch clamp measurements, MEA platforms have been developed that allow the detection of APs. These local extracellular AP assays, utilize electrodes capable to apply electrical stimulation in order to induce small pores in the cellular membrane for the acquisition of stable AP patterns over longer timescales [79,80]. In addition, the use of 3-dimensional electrodes can facilitate the coupling intensity and decrease the membrane resistance of individual cells required for intracellular recordings [81,82].

4. Application of MEAs for Cardiotoxic Risk Assessment

In 2014 the US department of health and human services estimated that nearly 1 million patients show adverse drug reactions each year—among these drug induced arrhythmias are the leading cause [83].

The comprehensive in vitro proarrhythmia assay (CiPA) initiative was originated for drug proarrhythmic potential assessment in order to analyze several known drugs and substances on their potential to affect the cardiac system. Thus, a list of 28 relevant drugs with a potential effect was published. The list reaches from vandatanib, clarithromycin, droperidol over metoprolol to tamoxifen and verapamil to name only a few. The drugs were categorized into high risk, intermediate risk and no or very low risk for torsade-de-pointes-tachycardia (TdP; Table 2). TdP is characterized by polymorphic ventricular tachyarrhythmias, which can follow drug induced delayed ventricular repolarization (OT interval prolongation) [84,85]. For the categorization the initiative recommends and describes assays that are mechanistically based in vitro assays and are composed of four different steps that in total should give a comprehensive overview of the possible proarrhythmic potential:

Table 2. List of CiPA compounds defined by CiPA initiative * (May 2016) [86].

High TdP Risk	Intermediate TdP Risk	No or Very Low TdP Risk
	Astemizole	
	Chlorpromazine	Diltiazem
Azimilide	Cisapride	Loratadine
Bepridil	Clarithromycin	Metoprolol
Dofetilide	Clozapine	Mexiletine
Ibutilide	Domperidone	Nifedipine
Quinidine	Droperidol	Nitrendipine
Vandetanib	Terfenadine	Ranolazine
Disopyramide	Pimozide	Tamoxifen
D,l Sotalol	Risperidone	Verapamil
	Ondansetron	

First, the effect of potential drugs and substances on several cardiac ion currents, which are defined as a core set of ion channel types needs to be analyzed. Second, the electrophysiological properties are simulated in in silico models. Since ventricular cardiomyocytes can be generated from human stem cells they represent a promising platform for drug testing, consequently the drug effects are measured in this in vitro setting as the third step. Finally, the expected and unexpected effects on the entire human organism need to be clinically evaluated [84].

Since the inception of the CiPA initiative in 2013, the analysis of the listed drugs was set into focus of research by the research community. Several cell lines (including self-generated and commercially available cell lines like iCell Cardiomyocytes (Fuji), Pluricytes (Pluriomics), Cor4u (Ncardia), Axol Bioscience, i-HCm (Cell applications), ASC (Applied Stem Cell, ix Cells Biotechnologies), CDI (Cellular Dynamics International), Cellartis (Clontech, Takara), ReproCardio (ReproCELL) and ACCEGEN (immortalized from patients or transdifferentiated from hSC) have been used to analyze the impact of compound administration on cardiomyocyte electrophysiology. The methodology used for the measurements range from (automated) patch clamp over MEA to optical measurement.

There is a tendency noticeable promoting the assessment of not only hERG inhibition or QT prolongation but also analysis of Nav1.5 (voltage gated Na^+ channel), Cav1.2 (voltage gated Ca^{2+} channel) or of index of cardiac electrophysiological balance (assesses balance between OT interval and QRS duration).

The Consortium for Safety Assessment using Human iPS cells (CSAHI) was established in 2013 by the Japan Pharmaceutical Manufacturers Association in order to "give recommendations for the usage of human iPS-cell derived cardiomyocytes, hepatocytes and neurons in drug testing evaluation" [87]. The CSAHI study is also aiming at the analysis of potential unknown effects on the cardiac system and tries to overcome/decrease proarrhythmic-risk market withdrawal. There have been several substances being tested that are not listed on the CiPA list. CSAHI provides a generalizable platform with the promising method for prediction of cardiotoxicity [88]. There are different parameters that are analyzed for the named prediction, such as QT prolongation, arrhythmia, but not only using the hERG assay. The latter is known to be an inaccurate predictor, because it only focuses on the inhibition of this particular ion channel—yet for cardiac adverse effects, mostly a broader range of (different) ion channels is affected [89].

Using animal hearts as a model, (which is also done for drug testing approaches) or the guinea pig papillary muscle action potential assay (qpAPD), is also not sufficient due to interspecies differences in electrophysiological properties and different responding behavior to drugs [87,90].

The tested substances are modulating a range of cardiac ion currents and consequently can have multiple arrhythmogenic effects. Due to this fact, multiple parameters are required to be evaluated, especially when using MEA technology. The electrophysiological response to drugs can be analyzed using the heart rate, field potential duration (FPD) and the corrected FPD (cFPD), all indicating arrhythmia-like waveforms [85]. Moreover, further analytical parameters are available using impedance measurement, deformation analysis or high content imaging [87].

The usage of human iPSC-CMs for drug testing is a promising tool due to their large-scale production circumventing the lack of a source for human adult cells [85]. However, they are displaying/carrying the disadvantage of not being identical to isolated primary adult cardiomyocytes and showing indicators for immature state, iPSC-CM are comparable in expression of cardiomyocyte marker. Especially the expression of relevant cardiac ion channels such as I_{Na}, I_{CaL}, I_f, I_{to}, I_{K1}, I_{Kr} and I_{Ks} has been analyzed [91,92]. Compared to previously used single ionic current model approaches they have shown a higher sensitivity and specificity [85]. Techniques such as qpAPD evaluated several false negative results [87], leading to possibly high risk in drug administration on patients. Many attempts have been carried out to analyze the response of human iPSC-CM to the administration of not only cardiogenic drugs and to further compare it to human adult cardiomyocytes. Yet, limitations must be kept in mind when transferring results from cell models to the human organism.

Since comparability within CiPA associated data generation is crucial Kanda et al. aimed to develop a standardized protocol for the experimental data generation including experimental conditions and calibration compounds providing it to a big community [63].

In order to validate the reliability and comparability of iPSC-CM based drug testing CiPA associated studies have been examined using a batch of known (formerly analyzed) drugs, various commercial cell lines, different electrophysiological platforms and multiple experimental sites [85,93]. Differences between the various analyzed combinations could be seen but also representative effects on depolarization, confirming the utility of the CiPA paradigm. Promoting the concept of CiPA, the CSAHI study from Japan HEART TEAM could not detect any inter-facility variability [90] and is providing new insights from their large scale drug testing combining electrophysiological data (from the MEA platform) with gene expression profiles [87,88]. A comprehensive overview of tested substances on their proarrhythmic risk/cardiac side effects using the combination of MEA technology with iPSC-CM is given in Tables 3 and 4. Table 3 summarizes drugs with a primarily non-cardiac medical indication such as antibiotic or antipsychotic drugs. Anti-arrhythmic drugs and cardiac ion-channel blocker are included in Table 4, containing cardiogenic substances.

To consider the influence of serum containing medium during administration and measurement of potential cardiotoxic drugs Schocken et al. compared serum containing and serum free medium in pro-arrhythmia risk assessment. The solubility of a drug connected with the precise drug concentration as well as cardiomyocyte electrophysiology may be affected by the serum. Mostly the precise serum composition is unknown. Using a high-throughput MEA 25 substances have been analyzed, showing differences in drug availability and the tendency of serum to influence the FPD in an increasing or decreasing manner for several drugs [94].

To further improve and expand the system of iPSC-CM drug testing, Zeng et al. addressed the diversity of iPSC-CM models from different gender and ethnical origin with known pharmaceuticals, detecting possible inter-sex differences [95]. Therefore, they prefer/vote for generalized pre-set acceptance criteria for iPSC-CMs. Burnett et al. recently published a study using not only a population-based CM model, generated from cells of 43 individuals (both gender and diverse ancestry) to defeat the drawback of inter-individual variability but also tested a large scale of substances of pharmaceuticals, environment and food. Both for control and substrate administration they found inter-individual variability, increasing the requirement of population based-models (to reproduce a whole population) [96].

Table 3. MEA based safety testing of drugs without cardiac indication using human induced pluripotent stem cell cardiomyocyte (hiPSC-CM).

Substance	(Site of) Action	Effect	Min. Effective Conc.	Cell Type/ Subtype	Differentiation Protocol	Age/ Maturation State	Platform	Reference
Alfuzosin	Treatment of benign prostatic enlargement, a hERG-channel blocker	Clinical QT prolongation	30 nM	hiPSC-CM iCell™ (mixture of ventricular, atrial, nodal cells)	n/a	32 days of differentiation +15–26 days	MEA	[87,88]
Astemizole	Antihistaminergic drug, H1 receptor antagonist, multi-channel block	Repolarization prolongation/arrhythmogenic effects, hERG channel blockade	3–10 nM	hiPSC-CM iCell™/iCell²™/Cor4U® (mixture of ventricular, atrial, nodal cells)	n/a	32 days of differentiation (+15–26 days) n/a	MEA	[87,88,93]
BaCl$_2$	Digitalis like activity, stimulation tonic contraction in muscle, used as contrast agent	Chronotropic effect K$^+$ and Ca^{2+} modulation	-	hiPSC-CM iCell™ (mixture of ventricular, atrial, nodal cells)	n/a	32 days of differentiation +15–26 days	MEA	[87]
Blebbistatin	Myosin II ATPase inhibitor	Increase in beating frequency, beating arrest (30 μM)	1–30 μM	hiPSC-CM iCell™ (mixture of ventricular, atrial, nodal cells)	n/a	Min. 32 days of differentiation	MEA	[88]
Carbachol	Parasympathomimetic drug cholinergic agonist K$_{Ach-}$ channel, glaucoma treatment	Negative chronotropic effects, FPDc prolongation, decrease in beating frequency	10 μM	Double reporter cell line, subtypes: ventricular, atrial, nodal, TBX5 Nkx2.5/hiPSC-CM (iCell™ mixture of ventricular, atrial, nodal cells)	2D n/a	35-40 day of differentiation 32 days of differentiation	Patch clamp, MEA	[21,88]
Chlorpromazine	Anti-psychotic drug, multi-channel block	Early afterdepolarization, beating arrest	10 μM	hiPSC-CM iCell™ mixture of ventricular, atrial, nodal cells)	n/a	32 days of differentiation	MEA	[88]
Chromanol 293B	IKv7.1 channel Blocker	Prolong FPD in control cells,		LQTS cells and control (patient- derived cells) n/a	3D	30-60 days of differentiation +50 days	MEA	[97]
Cisapride	Prokinetic gastrointestinal drug, multi-channel block	Prolongation of FPD, Repolarization delays/arrhythmogenic effects Prolongation of QT from patients with long QT syndrome	100 nM	hiPSC-CM Cor4U® and iCell™ mixture of ventricular, atrial, nodal cells Wt iPSC, and from patient with LQTS (n/a)	n/a 3D	10 days after differentiation, min. 32 days of differentiation n/a	MEA, automated patch clamp	[26,87,88,93, 98,99]

Table 3. *Cont.*

Substance	(Site of) Action	Effect	Min. Effective Conc.	Cell Type/ Subtype	Differentiation Protocol	Age/ Maturation State	Platform	Reference
Clarithromycin	Antibiotic drug	Repolarization prolongation, arrhythmogenic effects	-	hiPSC-CM iCell^2™/Cor4U® (mixture of ventricular, atrial, nodal cells)	n/a	32 days of differentiation n/a	MEA/VSO	[93]
Clozapine	Anti-psychotic drug, multi-channel block	Shortening of FPDc, increase in beat frequency	0.3–1 µM	hiPSC-CM iCell™ mixture of ventricular, atrial, nodal cells)	n/a	32 days of differentiation	MEA	[88]
Domperidone	Dopamine-antagonist, anti-nausea drug hERG- channel blocker	Repolarization prolongation, arrhythmogenic effects	10 nM	hiPSC-CM iCell™/iCell^2™/Cor4U® mixture of ventricular, atrial, nodal cells)	n/a	32 days of differentiation n/a	MEA/VSO	[88,93]
Doxorubicin	anthracycline chemotherapy agent	Decrease in FPD, beat frequency and spike amplitude	1 µM	hiPSC-CMs iCell™ 50% ventricular, 10% atrial cells patient derived cells (n/a)	n/a 2D	32 days of differentiation 20–30 days of differentiation	MEA	[100,101]
Droperidol	Neuroleptic drug	Repolarization prolongation, arrhythmogenic effects	-	hiPSC-CM iCell^2™/Cor4U® (mixture of ventricular, atrial, nodal cells)	n/a	32 days of differentiation n/a	MEA/VSO	[93]
Fluoxetine	Anti-depressant drug	Clinical QT prolongation	-	hiPSC-CM iCell™ (mixture of ventricular, atrial, nodal cells)	n/a	32 days of differentiation +15–26 days	MEA	[87]
Isoproterenol	Bronchodilator	Chronotropic effect, K^+ and Ca^{2+} Modulation, FPDc shortening, increasing beating frequency	3–100 nM	hiPSC-CM iCell™ (mixture of ventricular, atrial, nodal cells)	n/a	32 days of differentiation +15–26 days	MEA	[87,88]
Loratadine	Anti- histaminergic drug, H1 receptor block, multi-channel block	Increase in beating frequency	0.1–3 µM	hiPSC-CM iCell™ (mixture of ventricular, atrial, nodal cells)	n/a	32 days of differentiation	MEA	[88]
Moxifloxacin	Anti-biotic drug, multi-channel block	Repolarization delay	10 µM	hiPSC-CM iCell™/Cor4U® (mixture of ventricular, atrial, nodal cells) GE Healthcare (Cytiva™), Stanford Cardiac Institute	n/a	32 days of differentiation +14–24 days n/a	MEA	[85]

Table 3. *Cont.*

Substance	(Site of) Action	Effect	Min. Effective Conc.	Cell Type/ Subtype	Differentiation Protocol	Age/ Maturation State	Platform	Reference
Ondansetron	Antiemetic drug, serotonin-receptor block	Repolarization prolongation, arrhythmogenic effects	30 nM	hiPSC-CM iCell^2 ™/Cor4U® (mixture of ventricular, atrial, nodal cells)	n/a	32 days of differentiation n/a	MEA/VSO	[93]
Pimozide	Anti-psychotic drug, multi-channel block	Repolarization prolongation/arrhythmogenic effects	3–10 nM	hiPSC-CM iCell™/iCell^2 ™/Cor4U® (mixture of ventricular, atrial, nodal cells)	n/a	32 days of differentiation +15–26 days n/a	MEA	[87,93]
Risperidon	Anti-psychotic drug, serotonin-receptor block	Repolarization prolongation	3–30 nM	hiPSC-CM iCell^2™/Cor4U® (mixture of ventricular, atrial, nodal cells)	n/a	32 days of differentiation n/a	MEA	[93]
Sunitinib	Anti-cancer drug, tyrosine kinase inhibitor	FPDc prolongation, early afterdepolarization	0.3–10 μM	hiPSC-CM iCell™ (mixture of ventricular, atrial, nodal cells)	n/a	32 days of differentiation	MEA	[88]
Terfenadine	Anti-histaminergic drug, H1 receptor block	FPDc prolongation, decrease in spike amplitude, repolarization prolongation	100–1000 nM	hiPSC-CM iCell™/iCell^2™/Cor4U® (mixture of ventricular, atrial, nodal cells)	n/a	32 days of differentiation n/a	MEA	[93,98]
Tetrodotoxin (TTX)	Neurotoxic drug, (voltage sensitive) Na$_v$ (1.1, 1.7, 1.5)- channel block	Decrease in slope, depolarization potential and action potential duration	10 μM	hiPSC-CM Cor4U® mixture of ventricular, atrial, nodal cells	n/a	10 days after differentiation	Automated patch clamp	[26]
Thioridazine	Sedative, anti- psychotic drug, multi-channel block	Repolarization delays/arrhythmogenic effects	100 nM	hiPSC-CM iCell™ (mixture of ventricular, atrial, nodal cells)	n/a	32 days of differentiation +15–26 days	MEA	[87,88]
Tolterodine	Treatment of urinary incontinence, muscarinic receptor antagonist	clinical QT prolongation, early afterdepolarization	100–300 nM	hiPSC-CM iCell™ (mixture of ventricular, atrial, nodal cells)	n/a	32 days of differentiation +15–26 days	MEA	[87]
Vanoxerine	Serotonin-dopamine reuptake inhibitor	Clinical QT prolongation, multiple ion-channel effects, early afterdepolarizations	100 nM	hiPSC-CM iCell™ (mixture of ventricular, atrial, nodal cells)	n/a	32 days of differentiation +15–26 days	MEA	[87]
Vandetanib	Anti-cancer drug for thyroid gland, kinase inhibitor	Repolarization prolongation, arrhythmia like events	0.1–1 μM	hiPSC-CM iCell^2™/Cor4U® (mixture of ventricular, atrial, nodal cells)	n/a	32 days of differentiation n/a	MEA/VSO	[93]

Table 4. MEA based safety testing of drugs with cardiac indication using hiPSC-CM.

Substance	(Side of) Action	Effect	Min. Effective Conc.	Cell Type	Differentiation Protocol	Age/Maturation State	Platform	Reference
Amiodarone	Class III anti-arrhythmic drug, multi-channel block	Clinical QT prolongation	0.1–1 μM	hiPSC-CM iCell™ (mixture of ventricular, atrial, nodal cells)	n/a	32 days of differentiation +15–26 days	MEA	[87,88]
Azimilide	Class III anti-arrhythmic drug	FPDc prolongation, decrease in beating frequency, early after depolarization	0.3–1 μM	hiPSC-CM iCell™ (mixture of ventricular, atrial, nodal cells)	n/a	32 days of differentiation	MEA	[88]
Bay K 8644	Agonist of voltage sensitive dihydropyridine (DHP; L-Typ) Calcium channel	FPDc prolongation, decrease in beat frequency, positive inotropic	0.3–3 nM	hiPSC-CM iCell™ (mixture of ventricular, atrial, nodal cells)	n/a	32 days of differentiation +15–26 days	MEA	[87,88]
Bepridil	Class IV anti-arrhythmic drug, multi-channel block	Repolarization delays/arrhythmogenic effects	0.1–1 μM	hiPSC-CM iCell™/iCell2™/Cor4U® (mixture of ventricular, atrial, nodal cells)	n/a	32 days of differentiation +15–26 days n/a	MEA	[87,88,93]
Dofetilide	Class III anti-arrhythmic drug, multi-channel block	Increase in FPD, TdP arrhythmias	3–100 nM	hiPSC-CM iCell™/iCell2™/Cor4U® (mixture of ventricular, atrial, nodal cells) Pluricytes™	n/a	32 days of differentiation n/a n/a	MEA	[88,93,102]
E-4031	Class III anti-arrhythmic drug, hERG- channel block	prolonged FPD, severe arrhythmia in LQTS iPSC-CM	30–100 nM	hiPSC-CM iCell™/Cor4U® (mixture of ventricular, atrial, nodal cells) GE Healthcare (Cytiva™), Stanford Cardiac Institute	n/a3D	32 days of differentiation +14–24 days n/a 30–60 days of differentiation +50 days	MEA	[85,97,98]
Flecainide	Class Ic anti-arrhythmic drug, multi-channel block	Decrease in spike amplitude, FPDc prolongation	1 μM	LQTS cells and control hiPSC-CM iCell™/Cor4U® (mixture of ventricular, atrial, nodal cells) GE Healthcare (Cytiva™), Stanford Cardiac InstituteCPVT cells and control	n/a 2D/3D	32 days of differentiation +14–24 days n/a 20–30 days of beating	MEA Patch clamp	[85,98,103]

Table 4. *Cont.*

Substance	(Side of) Action	Effect	Min. Effective Conc.	Cell Type	Differentiation Protocol	Age/Maturation State	Platform	Reference
Ibutilide	Class III Anti-arrhythmic drug, multi-channel block	Arrhythmia like events, early after depolarizations	1–100 nM	hiPSC-CM iCell™/iCell^2™/Cor4U® (mixture of ventricular, atrial, nodal cells)	n/a	32 days of differentiation n/a	MEA/VSO	[88,93]
Ivabradin	Treatment of stable angina pectoris, If-channel inhibitor, heart rate reducing drug	Prolongation in APD, decrease in beating frequency	1 μM	Double reporter cell line, subtypes: ventricular, atrial, nodal, TBX5 Nkx2.5/hiPSC-CM	2D	35–40 days of differentiation	Patch clamp	[21]
JNJ303	IKv7.1- channel inhibitor	Small prolongation of FPDc	300 nM	hiPSC-CM iCell™/Cor4U® (mixture of ventricular, atrial, nodal cells) GE Healthcare (Cytiva™), Stanford Cardiac Institute	n/a	32 days of differentiation n/a	MEA	[85]
Levocromakalim	Vasodilating drug, K$_{ATP}$ opener	Membrane hyperpolarization, decrease in FPDc and beating frequency	1–3 μM	hiPSC-CM iCell™ (mixture of ventricular, atrial, nodal cells)	n/a	32 days of differentiation +15–26 days	MEA	[87,88]
Metoprolol	Anti- arrhythmic, anti-hypertonic drug, ß1-adreno receptor block	Induced arrhythmias, hERG block at higher concentrations	100 μM	hIPSC CM iCell^2™/Cor4U® (mixture of ventricular, atrial, nodal cells) CPVT cells and control hIPSC-CM	n/a 2D/3D	32 days of differentiation n/a 20–30 days of beating	MEA/VSO Patch clamp	[93,103]
Mexiletine	Class Ib anti-arrhythmic drug, Inhibiting Nav1.5- also hERG block	Reduce spike amplitude, cessation of spontaneous beating (100 μM)	1–10 μM,	iCell™/iCell^2™ Cor4U® (mixture of ventricular, atrial, nodal cells) GE Healthcare (Cytiva™), Stanford Cardiac Institute	n/a	32 days of differentiation +14–24 days n/a	MEA	[85,88,93,98]
Mibefradil	Treatment of angina pectoris and hypertension, multi-channel block	Shortening in FPDc, increase in beat frequency	0.3–1 μM	hiPSC-CM iCell™ (mixture of ventricular, atrial, nodal cells)	n/a	32 days of differentiation +15–26 days	MEA	[87,88]
Nifedipin	Vasodilating drug, I$_{CaL}$ block	Shortening of FPDc, increase in beating rate	0.3–1 μM	hiPSC-CM iCell™/Cor4U® (mixture of ventricular, atrial, nodal cells) GE Healthcare (Cytiva™), Stanford Cardiac Institute	n/a	32 days of differentiation +10 days n/a	MEA, automated patch clamp	[26,85,98]

Table 4. *Cont.*

Substance	(Side of) Action	Effect	Min. Effective Conc.	Cell Type	Differentiation Protocol	Age/Maturation State	Platform	Reference
NS- 1643	hERG-channel activator	Repolarization effect, decrease in FPDc, increase in beating frequency	3 µM	hiPSC-CM iCell™ (mixture of ventricular, atrial, nodal cells)	n/a	32 days of differentiation +15–26 days	MEA	[87,88]
Ouabain	Cardiac glycoside, Na^+-K^+- ATPase inhibitor	Repolarization effects, decrease in FPDc	10–100 nM	hiPSC-CM iCell™ (mixture of ventricular, atrial, nodal cells)	n/a	32 days of differentiation +15–26 days	MEA	[87,88]
Propranolol	Class II anti-arrhythmic drug, beta- receptor block	Early afterdepolarization, decrease in beating frequency	10 µM	hiPSC-CM iCell™ (mixture of ventricular, atrial, nodal cells)	n/a	32 days of differentiation	MEA	[88]
Quinidine	Class Ia anti-arrhythmic drug, multi-channel block (Nav1.5, Cav1.2, hERG)	FPDc prolongation, reduced spike amplitude, repolarization delays/arrhythmogenic effects	0.3–10 µM	hiPSC-CM iCell™/iCell²™ Cor4U® (mixture of ventricular, atrial, nodal cells) GE Healthcare (Cytiva™), Stanford Cardiac Institute Fibroblast-derived iPSC-CM (67% ventricular, 5% nodal, 28% atrial)	n/a 3D	32 days of differentiation +15–26 days 65–95 days after differentiation induction	MEA, Low impedance MEA	[85,87,88,93, 98,104]
Ranolazine	Angina pectoris treatment, multichannel (Na and hERG block)	FPDc prolongation, clinical QT prolongation, repolarization prolongation	0.3 µM, clinical conc. <100 µM	hiPSC-CM iCell iCell²™ Cor4U® (mixture of ventricular, atrial, nodal cells) GE Healthcare (Cytiva™), Stanford Cardiac Institute hiPSC-CMs	32 days of differentiation +15–26 days (14–24 days) n/a	MEA	[85,87,88,93]
Sotalol	Anti-arrhythmic drug, beta adreno receptor block	Repolarization prolongation, arrhythmogenic effects, hERG-channel block	15 µM	iCell²™/Cor4U® (mixture of ventricular, atrial, nodal cells) Fibroblast-derived iPSC-CM (67% ventricular,5% nodal, 28% atrial)	n/a 3D	32 days of differentiation n/a 65–95 days after differentiation induction	Low impedance MEA	[93,104]

Table 4. *Cont.*

Substance	(Side of) Action	Effect	Min. Effective Conc.	Cell Type	Differentiation Protocol	Age/Maturation State	Platform	Reference
Verapamil	Class VI anti-arrhythmic drug, inhibits hERG, I_{Cal}-typ calcium channels, Multi-channel block	Shortening of FPDc, increase in spontaneous beat rate; shortening in APD_{20} and APD_{90}	0.1–0.3 μM; 1 μM	hiPSC-CM iCell™ (mixture of ventricular, atrial, nodal cells) Pluricytes™	n/a	32 days of differentiation n/a	MEA; patch clamp	[98,102]
Vernakalant	Class III anti-arrhythmic drug, used for cardioversion of atrial fibrillation, atrial potassium- channel block	APD prolongation, partly arrhythmogenic effects,	-	Double reporter cell line, different subtypes (ventricular phenotype)	2D	20–30 days post induction of differentiation	Patch clamp	[21]
ZD 7288	Selective hyperpolarization-activated cyclic nucleotide-gated channel blocker, I_{f}- current inhibitor	Negative chronotropic effect, FPDc prolongation	3–30 nM	hiPSC-CM iCell™ (mixture of ventricular, atrial, nodal cells)	n/a	32 days of differentiation +15–26 days	MEA	[87,88]

5. Disease Modeling Using hiPSC-CM

Cardiovascular diseases (CVD) are the number one cause of death globally according to the WHO. This group of disease is not only present in developed countries but also in low income countries. CVD affects both the blood vessels and the heart and range from coronary heart disease, thrombosis, over structural abnormalities of the heart and arrhythmias to stroke and heart failure [83].

In order to understand the pathological mechanisms of the disease iPSC-derived cardiomyocytes generated from patient cells like dermal fibroblasts or blood cells showed high potential as a tool to not only generate immortalized cell lines from healthy donors but also from diseased patients [105]. The generated cell lines can show disease-specific phenotypes [99]. Especially physiological characteristics can be determined when carrying the mutations in the cardiac relevant genes. This suggests that the disease phenotype can be recapitulated in vitro. To further analyze the pathogenic mechanisms causing the disease and the disease specific phenotype patient originated iPSC-CM were analyzed and afterwards genetically fixed. Here, RNA interference can be used for silencing or suppressing mutant genes [106]. To determine and control the effects caused by mutations the initial genetic defect was induced and replicated in hESC derived CM on the contrary. A correction of the defect led to a normalization of the analyzed parameter whereas the induction led to the diseased phenotype [107]. Besides a better understanding of heart disease, the transgenic models should contribute to testing of cardiogenic drugs on their positive therapeutic as well as detrimental side effects. Moreover, the sensitivity of patients for side effects of drugs can be evaluated and, the clinical vulnerability of high-risk groups (population) to drug induced-cardiotoxicity can be set into consideration. This testing of therapies *in vitro* is another promising tool [99,108].

For the analysis and validation of the generated iPSC-CM based disease models patch clamp analyses are still the gold standard, despite the number of MEA based measurements in increasing (Figure 3). In this part we want to give a selective overview about already developed disease models for cardiac disease generated from human iPSCs, where the usage of MEA platform is additionally mentioned.

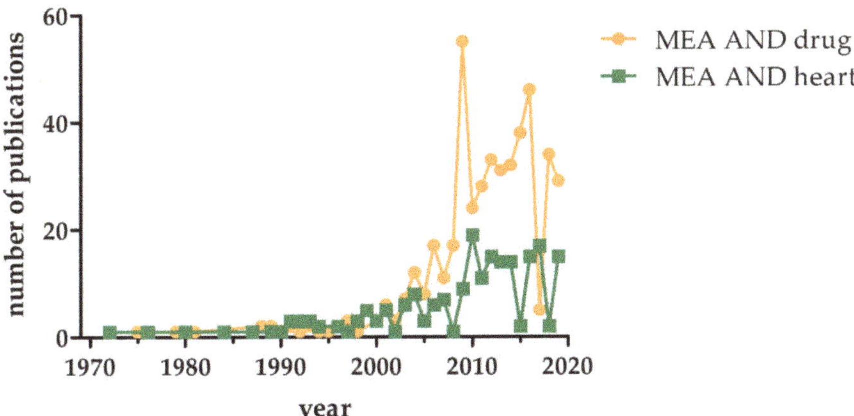

Figure 3. Increase of PubMed listed publications involving MEA based analysis of "heart" or "drugs" over the last five decades. The terms (multielectrode array and drug) or (microelectrode array and drug) and (multielectrode array and heart) or (microelectrode array and heart) were used for the PubMed search (date: Sept 2019).

6. Overview of Developed Disease Models

Ping Liang et al. generated a library of iPSC-derived CM from patients suffering from various hereditary cardiac disorders to show that cardiac drug toxicity differs between different pathophysiological conditions. The iPSC-CM was generated from patients with hereditary long-QT syndrome, familial hypertrophic cardiomyopathy and familial dilated cardiomyopathy. They have shown that patients that already suffer from a heart disease have a higher incidence to show adverse effects arising from their medical treatment. They seem to have a higher sensitivity to cardiotropic drugs and can have a higher risk for arrhythmias, which possibly are leading to death [99].

In 2014 Zhang et al. generated cardiomyocytes derived from iPSC from patients with recessive, life-threatening cardiac arrhythmia of Jarvell and Lang–Nielsen syndrome. They gave new insights into the pathological mechanisms and showed enhanced sensitivity to proarrhythmic drugs in the generated cell-based disease model using MEA technology and patch clamp [109].

Considering the literature of the last years for ion channelopathies, these appear to be in focus of disease modeling, revealing many well-established human iPSC generated disease models. Among these the long QT syndrome is the most common. The first model has been developed by Moretti et al. (2010). Ventricular and atrial cells in contrast to nodal type or healthy control cells, have shown significantly increased APDs [110]. The response of sporadic Long QT1-iPSC-CM to small molecule inhibitors has been analyzed measuring changes in the FPD with the MEA platform [97].

A mutation in the gene of the sodium voltage-gated channel (Nav1.5) alpha subunit 5 (SCN5A) for example is leading to conduction defects, phenotypes of the LQT3 and Brugada syndrome due to a gain and loss of function [24]. A review on modeling long QT syndrome with the aid of iPSC-CM can be found by Sala et al. [111].

Another channelopathy that has been used for the generation of a disease model is the catecholaminergic polymorphic ventricular tachycardia (CPVT). An incorrect and insufficient Ca^{2+} handling (inclusive spontaneous release or sequestration) is leading to this adrenergically mediated polymorphic ventricular tachycardia [112]. Sasaki et al. generated CM from CPVT patient- derived iPSCs and identified S107 as a potential therapeutic agent since a pre-incubation with S107 led to a reduction of isoprenaline induced delayed afterdepolarizations [113]. Acimovic et al. (2018) developed a CPVT model using a novel ryanodine receptor mutation and further analyzed the response to a treatment with flecainide and metoprolol [103].

Furthermore, models of structural myopathies have been developed: among these, the hypertrophic cardiomyopathy (HCM) and the familial dilated cardiomyopathy (DCM) are the most common being analyzed. HCM associated death is mostly caused by ventricular fibrillation developed from ventricular arrhythmias. Despite that not all pathophysiological mechanisms are known yet, some detected mutations have been the basis of developed disease models. HCM iPSC derived cardiomyocytes have shown enhanced sarcomere arrangement [114], deviating electromechanical properties such as delayed after depolarizations and calcium handling [115]. DCM is leading to systolic dysfunction such as decreased ejection fraction due to an expanded size of the left ventricle combined with decreased chamber thickness. Several known mutations have been fundamental for the generation of a DCM model, including *MYH7* [116], *TNNT* [117], *LMNA* (laminin A/c) [118], Desmin [119], Titin [120] and RBM20 RNA-binding motif protein 20 [121,122]. On the cellular level an increase in cell size, abnormal sarcomere structure and organization e.g., sarcomeric α actinin and defective calcium handling (altering calcium machinery) [121] could be seen within these models. For further detailed and completely information we would recommend the review by Giacomelli et al. [123].

Moreover, cardiomyocytes have been generated from patients with Duchenne muscular dystrophy iPSCs showing phenotypical deficiency of dystrophin and increased Ca levels in resting cells. The before mentioned disease caused by a high quantity of mutations is characterized by a knockout of the dystrophin protein resulting in muscle degeneration [124]. For further detailed reading we recommend the systematic review on cardiomyopathy phenotypes in iPSC-CM (HCM and DCM) by Eschenhagen and Carrier [125].

Besides the generation of disease models from patient derived cells, new gene editing technologies find application: de la Roche et al. generated a model for the Brugada syndrome carrying the A735V mutation in the *SCN5A* gene introduced by CRISPR/Cas9 in order to be independent of the patient's genetic background. The generated CM showed electrophysiological alteration such as decreased upstroke velocity and sodium current density [126].

Among the current literature the number of developed models that are not primarily heart diseases (structural caused heart disease) but show heart related symptoms is increasing. For example a model for Chagas disease, a parasite caused infection by *Trypanosoma cruzi* that is associated with cardiomyopathy symptoms since the parasites are replicating in the cardiomyocytes, was developed by Bozzi et al. after infection of iPSC-CM with *Trypanosoma cruzi* [127]. Lee et al. developed a model for autosomal dominant polycystic kidney disease (ADPKD) form patient derived iPSC-CM showing mutations in the *PKD1/PKD2* gene [128], to name only two of them.

Since the generation of disease models with patient derived iPSC-CM was developed and practiced within the last years, the first scientific outcome has been produced in a more detailed understanding of the pathophysiological mechanisms that underlie known symptoms. Several recent studies were performed:

Using patient derived iPSC-CM from Fabry disease combined with gene editing technology Birket et al. analyzed the functional consequences of underlying genetic defects. They have shown the accumulation of GL-3 and alterations in excitability and calcium handling in cardiomyocytes. Moreover LIMP-2, shown to accumulate in the cells, was detected as a new potential biomarker [129].

Caluori et al. developed a system combining MEA platform with cantilever of atomic force microscopy (AFM) in order to analyze the topography, beating force and electric events simultaneously of control and DCM-derived iPSC-CM. Meanwhile substances for the cell characterization and toxicity testing have been administrated [130].

Moreover, microtissues have been generated from the disease model cardiomyocytes for hypertrophic cardiomyopathy HCM [131] and for catecholaminergic polymorphic ventricular tachycardia CPVT [132]. The HCM model was used to analyze contraction force development and calcium transient under mechanical overload revealing that both genetic defects and environmental stress reinforce dysfunctions in contractility [131]. The CPVT tissue construct examined molecular and cellular abnormalities and is confirming the approach of generating tissue like structures for the analysis of arrhythmia, which is mostly generated by the connection of many cells [132].

Another approach for the further use of these disease models and an improved reliability of the results/predictions are large scale simulations in in silico models. In order to enable this approach for the LQT syndrome with regards to the phenotypical variation (due to the large amount of different mutations) this system was developed and showed comparable results to experimental data concerning electrophysiological properties. Additionally, this simulation can facilitate the understanding of further biophysical mechanisms [133].

7. Conclusions and Outlook

Cardiomyocytes generated from human iPSCs are used to study heart development, cardiac function and heart diseases, and to develop novel pharmacological therapeutics [11,21,23,134]. This involves a detailed electrophysiological characterization of these cardiomyocytes under physiological and pathological conditions. Due to its non-invasive, label-free character MEA methodology has become a widely used tool to assess the electrical properties of cells in vitro. Besides iPSC-derived cardiomyocytes, acquisition of the FP by MEA assays can also be applied to any other cell type that shows electrical activity, including adult cardiomyocytes or neuronal cells. However, MEA measurements are not only restricted to cell monolayers, but can also be performed on slices of cardiac and brain tissue to better simulate in vivo conditions. Similarly, the replication of human physiology can be enhanced when MEA technology is combined with in vitro generated organoids. Therefore, MEA platforms can become a valuable tool in organ-on-a-chip engineering, promoting the clinical translation of acquired data [135,136].

As a high-throughput technology, MEA devices are particular important for the identification of novel therapeutics in drug research. The combination of MEA systems and iPSCs has been successfully applied in pharmacological studies and disease modeling, suggesting iPSC-based MEA measurements as a powerful approach for the development of personalized therapies that allow a more specific therapeutic intervention compared to conventional treatment options [137–139]. In this concept, patient-derived iPSCs are obtained by reprogramming of fibroblasts and induced to differentiate into cardiomyocytes. Subsequent analyses, e.g., transcriptomics and proteomics, enable the identification of molecular targets needed for pharmacological therapy. In a second step, the potency and efficiency of tested drugs is evaluated by MEA measurements, which in turn can provide important information to establish personalized drug treatments. Thus, MEA technology will help to open a new gateway for novel therapies in cardiovascular diseases.

Supplementary Materials: The supplementary materials are available online at http://www.mdpi.com/2073-4409/8/11/1331/s1.

Funding: This work was supported by the EU structural Fund (ESF/14-BM-A55-0024/18). In addition, R.D. and H.L. are supported by the DFG (DA1296/6-1), the German Heart Foundation (F/01/12), the FORUN Program of Rostock University Medical Centre (889001 and 889003) and the Josef and Käthe Klinz Foundation (T319/29737/2017). R.D. and S.K. are supported by the DAMP Foundation and the BMBF (VIP+ 00240).

Conflicts of Interest: The authors declare no conflict of interest. The funders had no role in the design of the study; in the collection, analyses, or interpretation of data; in the writing of the manuscript, or in the decision to publish the results.

Abbreviations

2D	Two-dimensional
ADPKD	Autosomal dominant polycystic kidney disease
AFM	Atomic force microscopy
AP	Action potential
APD	Action potential duration
ATM	Atomic force microscopy
AV Node	Atrioventricular node
HCM	Hypertrophic cardiomyopathy
Ca^{2+}	Calcium
CiPA	Comprehensive in vitro proarrhythmia assay
CM	Cardiomyocytes
CPVT	Catecholaminergic polymorphic ventricular tachycardia
CSAHI	Consortium for Safety Assessment using Human iPS cells
CVD	Cardiovascular disease
DCM	Familial dilated cardiomyopathy
FP	Field potential
FPDc	Field potential duration (corrected)
FRET	Fluorescence resonance energy transfer
GL-3	Basic helix-loop-helix proteins, GLABRA3
gpAPD	Guinea pig papillary muscle action potential assay
hERG	Human ether-a-go-go Related gene-voltage sensitive potassium channel
hiPSC	Human induced pluripotent stem cell
K^+	Potassium
LIMP-2	Lysosomal integral membrane protein 2
LMNA	Lamin A/C
MEA	Multi-/micro-electrode array
MYH7	Myosin heavy chain 7
Na^+	Sodium

PKD1/2	Polycystic kidney disease 1
RNA	Ribonucleic acid
SCN5A	Sodium voltage-gated channel alpha subunit 5
TBX5	T-box transcription factor
TdP	Torsade-de-Pointes-Tachycardia
TnnT	TroponinT
VSO	Voltage-sensitive optical system
WT	Wild type

References

1. Takahashi, K.; Yamanaka, S. Induction of Pluripotent Stem Cells from Mouse Embryonic and Adult Fibroblast Cultures by Defined Factors. *Cell* **2006**, *126*, 663–676. [CrossRef] [PubMed]
2. Yoshida, Y.; Yamanaka, S. Induced Pluripotent Stem Cells 10 Years Later. *Circ. Res.* **2017**, *120*, 1958–1968. [CrossRef] [PubMed]
3. Zuppinger, C. 3D Cardiac Cell Culture: A Critical Review of Current Technologies and Applications. *Front. Cardiovasc. Med.* **2019**, *6*, 87. [CrossRef] [PubMed]
4. Rojas, S.V.; Kensah, G.; Rotaermel, A.; Baraki, H.; Kutschka, I.; Zweigerdt, R.; Martin, U.; Haverich, A.; Gruh, I.; Martens, A. Transplantation of purified iPSC-derived cardiomyocytes in myocardial infarction. *PLoS ONE* **2017**, *12*, e0173222. [CrossRef] [PubMed]
5. Gao, L.; Gregorich, Z.R.; Zhu, W.; Mattapally, S.; Oduk, Y.; Lou, X.; Kannappan, R.; Borovjagin, A.V.; Walcott, G.P.; Pollard, A.E.; et al. Large Cardiac Muscle Patches Engineered From Human Induced-Pluripotent Stem Cell–Derived Cardiac Cells Improve Recovery From Myocardial Infarction in Swine. *Circulation* **2018**, *137*, 1712–1730. [CrossRef]
6. Shadrin, I.Y.; Allen, B.W.; Qian, Y.; Jackman, C.P.; Carlson, A.L.; Juhas, M.E.; Bursac, N. Cardiopatch platform enables maturation and scale-up of human pluripotent stem cell-derived engineered heart tissues. *Nat. Commun.* **2017**, *8*, 1825. [CrossRef]
7. Zhao, X.; Chen, H.; Xiao, D.; Yang, H.; Itzhaki, I.; Qin, X.; Chour, T.; Aguirre, A.; Lehmann, K.; Kim, Y.; et al. Comparison of Non-human Primate versus Human Induced Pluripotent Stem Cell-Derived Cardiomyocytes for Treatment of Myocardial Infarction. *Stem Cell Rep.* **2018**, *10*, 422–435. [CrossRef]
8. Pfeiffer-Kaushik, E.R.; Smith, G.L.; Cai, B.; Dempsey, G.T.; Hortigon-Vinagre, M.P.; Zamora, V.; Feng, S.; Ingermanson, R.; Zhu, R.; Hariharan, V.; et al. Electrophysiological characterization of drug response in hSC-derived cardiomyocytes using voltage-sensitive optical platforms. *J. Pharmacol. Toxicol. Methods* **2018**, *99*, 106612. [CrossRef]
9. Huo, J.; Wei, F.; Cai, C.; Lyn-Cook, B.; Pang, L. Sex-Related Differences in Drug-Induced QT Prolongation and Torsades de Pointes: A New Model System with Human iPSC-CMs. *Toxicol. Sci.* **2018**, *167*, 360–374. [CrossRef]
10. Izumi-Nakaseko, H.; Hagiwara-Nagasawa, M.; Naito, A.T.; Goto, A.; Chiba, K.; Sekino, Y.; Kanda, Y.; Sugiyama, A. Application of human induced pluripotent stem cell-derived cardiomyocytes sheets with microelectrode array system to estimate antiarrhythmic properties of multi-ion channel blockers. *J. Pharmacol. Sci.* **2018**, *137*, 372–378. [CrossRef]
11. Edwards, S.L.; Zlochiver, V.; Conrad, D.B.; Vaidyanathan, R.; Valiquette, A.M.; Joshi-Mukherjee, R. A Multiwell Cardiac μGMEA Platform for Action Potential Recordings from Human iPSC-Derived Cardiomyocyte Constructs. *Stem Cell Rep.* **2018**, *11*, 522–536. [CrossRef] [PubMed]
12. Itzhaki, I.; Maizels, L.; Huber, I.; Zwi-Dantsis, L.; Caspi, O.; Winterstern, A.; Feldman, O.; Gepstein, A.; Arbel, G.; Hammerman, H.; et al. Modelling the long QT syndrome with induced pluripotent stem cells. *Nature* **2011**, *471*, 225–229. [CrossRef] [PubMed]
13. Braam, S.R.; Tertoolen, L.; van de Stolpe, A.; Meyer, T.; Passier, R.; Mummery, C.L. Prediction of drug-induced cardiotoxicity using human embryonic stem cell-derived cardiomyocytes. *Stem Cell Res.* **2010**, *4*, 107–116. [CrossRef] [PubMed]
14. Goineau, S.; Castagné, V. Electrophysiological characteristics and pharmacological sensitivity of two lines of human induced pluripotent stem cell derived cardiomyocytes coming from two different suppliers. *J. Pharmacol. Toxicol. Methods* **2018**, *90*, 58–66. [CrossRef] [PubMed]

15. Grimm, F.A.; Blanchette, A.; House, J.S.; Ferguson, K.; Hsieh, N.-H.; Dalaijamts, C.; Wright, A.A.; Anson, B.; Wright, F.A.; Chiu, W.A.; et al. A human population-based organotypic in vitro model for cardiotoxicity screening. *ALTEX* **2018**, *35*, 441–452. [CrossRef] [PubMed]

16. Goversen, B.; van der Heyden, M.A.G.; van Veen, T.A.B.; de Boer, T.P. The immature electrophysiological phenotype of iPSC-CMs still hampers in vitro drug screening: Special focus on IK1. *Pharm. Ther.* **2018**, *183*, 127–136. [CrossRef] [PubMed]

17. Paci, M.; Hyttinen, J.; Rodriguez, B.; Severi, S. Human induced pluripotent stem cell-derived versus adult cardiomyocytes: An\textlessi\textgreaterin silico\textless/i\textgreater electrophysiological study on effects of ionic current block. *Br. J. Pharmacol.* **2015**, *172*, 5147–5160. [CrossRef]

18. Tan, S.H.; Ye, L. Maturation of Pluripotent Stem Cell-Derived Cardiomyocytes: A Critical Step for Drug Development and Cell Therapy. *J. Cardiovasc. Transl. Res.* **2018**, *11*, 375–392. [CrossRef]

19. Karakikes, I.; Ameen, M.; Termglinchan, V.; Wu, J.C. Human Induced Pluripotent Stem Cell-Derived Cardiomyocytes: Insights into Molecular, Cellular, and Functional Phenotypes. *Circ. Res.* **2015**, *117*, 80–88. [CrossRef]

20. Hausburg, F.; Jung, J.J.; David, R. Specific cell (re-)programming: Approaches and perspectives. In *Advances in Biochemical Engineering/Biotechnology*; Springer: Basel, Switzerland, 2017; Volume 163, pp. 71–115.

21. Zhang, J.Z.; Termglinchan, V.; Shao, N.-Y.; Itzhaki, I.; Liu, C.; Ma, N.; Tian, L.; Wang, V.Y.; Chang, A.C.Y.; Guo, H.; et al. A Human iPSC Double-Reporter System Enables Purification of Cardiac Lineage Subpopulations with Distinct Function and Drug Response Profiles. *Cell Stem. Cell* **2019**, *24*, 802–811. [CrossRef]

22. Obergrussberger, A.; Stölzle-Feix, S.; Becker, N.; Brüggemann, A.; Fertig, N.; Möller, C. Novel Screening Techniques For Ion Channel Targeting Drugs. *Channels* **2015**, *9*, 367–375. [CrossRef] [PubMed]

23. Casini, S.; Verkerk, A.O.; Remme, C.A. Human iPSC-Derived Cardiomyocytes for Investigation of Disease Mechanisms and Therapeutic Strategies in Inherited Arrhythmia Syndromes: Strengths and Limitations. *Cardiovasc. Drugs Ther.* **2017**, *31*, 325–344. [CrossRef] [PubMed]

24. Davis, R.P.; Casini, S.; van den Berg, C.W.; Hoekstra, M.; Remme, C.A.; Dambrot, C.; Salvatori, D.; Oostwaard, D.W.; Wilde, A.A.M.; Bezzina, C.R.; et al. Cardiomyocytes derived from pluripotent stem cells recapitulate electrophysiological characteristics of an overlap syndrome of cardiac sodium channel disease. *Circulation* **2012**, *125*, 3079–3091. [CrossRef] [PubMed]

25. Sallam, K.; Li, Y.; Sager, P.T.; Houser, S.R.; Wu, J.C. Finding the Rhythm of Sudden Cardiac Death. *Circ. Res.* **2015**, *116*, 1989–2004. [CrossRef] [PubMed]

26. Scheel, O.; Frech, S.; Amuzescu, B.; Eisfeld, J.; Lin, K.-H.; Knott, T. Action Potential Characterization of Human Induced Pluripotent Stem Cell–Derived Cardiomyocytes Using Automated Patch-Clamp Technology. *Assay Drug Dev. Technol.* **2014**, *12*, 457–469. [CrossRef]

27. Obergrussberger, A.; Goetze, T.A.; Brinkwirth, N.; Becker, N.; Friis, S.; Rapedius, M.; Haarmann, C.; Rinke-Weiß, I.; Stölzle-Feix, S.; Brüggemann, A.; et al. An update on the advancing high-throughput screening techniques for patch clamp-based ion channel screens: Implications for drug discovery. *Expert Opin. Drug Discov.* **2018**, *13*, 269–277. [CrossRef]

28. Franz, D.; Olsen, H.L.; Klink, O.; Gimsa, J. Automated and manual patch clamp data of human induced pluripotent stem cell-derived dopaminergic neurons. *Sci. Data* **2017**, *4*, 170056. [CrossRef]

29. Yajuan, X.; Xin, L.; Zhiyuan, L. A Comparison of the Performance and Application Differences Between Manual and Automated Patch-Clamp Techniques. *Curr. Chem. Geno.* **2012**, *6*, 87. [CrossRef]

30. Bell, D.C.; Dallas, M.L. Using automated patch clamp electrophysiology platforms in pain-related ion channel research: Insights from industry and academia. *Br. J. Pharmacol.* **2018**, *175*, 2312–2321. [CrossRef]

31. Huang, H.-L.; Hsing, H.-W.; Lai, T.-C.; Chen, Y.-W.; Lee, T.-R.; Chan, H.-T.; Lyu, P.-C.; Wu, C.-L.; Lu, Y.-C.; Lin, S.-T.; et al. Trypsin-induced proteome alteration during cell subculture in mammalian cells. *J. Biomed. Sci.* **2010**, *17*, 36. [CrossRef]

32. Rajamohan, D.; Kalra, S.; Duc Hoang, M.; George, V.; Staniforth, A.; Russell, H.; Yang, X.; Denning, C. Automated Electrophysiological and Pharmacological Evaluation of Human Pluripotent Stem Cell-Derived Cardiomyocytes. *Stem Cells Dev.* **2016**, *25*, 439–452. [CrossRef] [PubMed]

33. Storace, D.; Rad, M.S.; Han, Z.; Jin, L.; Cohen, L.B.; Hughes, T.; Baker, B.J.; Sung, U. Genetically encoded protein sensors of membrane potential. In *Membrane Potential Imaging in the Nervous System and Heart*; Springer: Basel, Switzerland, 2015; pp. 493–509.

34. del Álamo, J.C.; Lemons, D.; Serrano, R.; Savchenko, A.; Cerignoli, F.; Bodmer, R.; Mercola, M. High throughput physiological screening of iPSC-derived cardiomyocytes for drug development. *Biochim. Et Biophys. Acta Mol. Cell Res.* **2016**, *1863*, 1717–1727. [CrossRef] [PubMed]

35. Bedut, S.; Seminatore-Nole, C.; Lamamy, V.; Caignard, S.; Boutin, J.A.; Nosjean, O.; Stephan, J.-P.; Coge, F. High-throughput drug profiling with voltage- and calcium-sensitive fluorescent probes in human iPSC-derived cardiomyocytes. *Am. J. Physiol. Heart Circ. Physiol.* **2016**, *311*, 44–53. [CrossRef]

36. Hortigon-Vinagre, M.P.; Zamora, V.; Burton, F.L.; Green, J.; Gintant, G.A.; Smith, G.L. The use of ratiometric fluorescence measurements of the voltage sensitive dye Di-4-ANEPPS to examine action potential characteristics and drug effects on human induced pluripotent stem cell-derived cardiomyocytes. *Toxicol. Sci.* **2016**, *154*, 320–331. [CrossRef] [PubMed]

37. Takaki, T.; Inagaki, A.; Chonabayashi, K.; Inoue, K.; Miki, K.; Ohno, S.; Makiyama, T.; Horie, M.; Yoshida, Y. Optical Recording of Action Potentials in Human Induced Pluripotent Stem Cell-Derived Cardiac Single Cells and Monolayers Generated from Long QT Syndrome Type 1 Patients. *Stem Cells Int.* **2019**, *2019*, 1–12. [CrossRef]

38. St-Pierre, F.; Chavarha, M.; Lin, M.Z. Designs and sensing mechanisms of genetically encoded fluorescent voltage indicators. *Curr. Opin. Chem. Biol.* **2015**, *27*, 31–38. [CrossRef]

39. Han, Z.; Jin, L.; Platisa, J.; Cohen, L.B.; Baker, B.J.; Pieribone, V.A. Fluorescent Protein Voltage Probes Derived from ArcLight that Respond to Membrane Voltage Changes with Fast Kinetics. *PLoS ONE* **2013**, *8*, e81295. [CrossRef]

40. Shaheen, N.; Shiti, A.; Huber, I.; Shinnawi, R.; Arbel, G.; Gepstein, A.; Setter, N.; Goldfracht, I.; Gruber, A.; Chorna, S.V.; et al. Human Induced Pluripotent Stem Cell-Derived Cardiac Cell Sheets Expressing Genetically Encoded Voltage Indicator for Pharmacological and Arrhythmia Studies. *Stem Cell Rep.* **2018**, *10*, 1879–1894. [CrossRef] [PubMed]

41. Leyton-Mange, J.S.; Mills, R.W.; Macri, V.S.; Jang, M.Y.; Butte, F.N.; Ellinor, P.T.; Milan, D.J. Rapid cellular phenotyping of human pluripotent stem cell-derived cardiomyocytes using a genetically encoded fluorescent voltage sensor. *Stem Cell Rep.* **2014**, *2*, 163–170. [CrossRef]

42. Shinnawi, R.; Huber, I.; Maizels, L.; Shaheen, N.; Gepstein, A.; Arbel, G.; Tijsen, A.J.; Gepstein, L. Monitoring human-induced pluripotent stem cell-derived cardiomyocytes with genetically encoded calcium and voltage fluorescent reporters. *Stem Cell Rep.* **2015**, *5*, 582–596. [CrossRef]

43. Song, L.; Awari, D.W.; Han, E.Y.; Uche-Anya, E.; Park, S.-H.E.; Yabe, Y.A.; Chung, W.K.; Yazawa, M. Dual optical recordings for action potentials and calcium handling in induced pluripotent stem cell models of cardiac arrhythmias using genetically encoded fluorescent indicators. *Stem Cells Transl. Med.* **2015**, *4*, 468–475. [CrossRef] [PubMed]

44. Herron, T.J. Calcium and voltage mapping in hiPSC-CM monolayers. *Cell Calcium.* **2016**, *59*, 84–90.

45. Herron, T.J.; Lee, P.; Jalife, J. Optical Imaging of Voltage and Calcium in Cardiac Cells & Tissues. *Circ. Res.* **2012**, *110*, 609–623. [PubMed]

46. Li, X.; Zhang, R.; Zhao, B.; Lossin, C.; Cao, Z. Cardiotoxicity screening: A review of rapid-throughput in vitro approaches. *Arch. Toxicol.* **2016**, *90*, 1803–1816. [CrossRef] [PubMed]

47. Yamamoto, W.; Asakura, K.; Ando, H.; Taniguchi, T.; Ojima, A.; Uda, T.; Osada, T.; Hayashi, S.; Kasai, C.; Miyamoto, N.; et al. Electrophysiological characteristics of human iPSC-derived cardiomyocytes for the assessment of drug-induced proarrhythmic potential. *PLoS ONE* **2016**, *11*, e0167348. [CrossRef]

48. Sala, L.; Ward-van Oostwaard, D.; Tertoolen, L.G.J.; Mummery, C.L.; Bellin, M. Electrophysiological Analysis of human Pluripotent Stem Cell-derived Cardiomyocytes (hPSC-CMs) Using Multi-electrode Arrays (MEAs). *J. Vis. Exp.* **2017**, *123*, e55587. [CrossRef]

49. Asakura, K.; Hayashi, S.; Ojima, A.; Taniguchi, T.; Miyamoto, N.; Nakamori, C.; Nagasawa, C.; Kitamura, T.; Osada, T.; Honda, Y.; et al. Improvement of acquisition and analysis methods in multi-electrode array experiments with iPS cell-derived cardiomyocytes. *J. Pharmacol. Toxicol. Methods* **2015**, *75*, 17–26. [CrossRef]

50. Zhu, H.; Scharnhorst, K.S.; Stieg, A.Z.; Gimzewski, J.K.; Minami, I.; Nakatsuji, N.; Nakano, H.; Nakano, A. Two dimensional electrophysiological characterization of human pluripotent stem cell-derived cardiomyocyte system. *Sci. Rep.* **2017**, *7*, 43120. [CrossRef]

51. Ryynänen, T.; Pekkanen-Mattila, M.; Shah, D.; Kreutzer, J.; Kallio, P.; Lekkala, J.; Aalto-Setälä, K. Microelectrode array for noninvasive analysis of cardiomyocytes at the single-cell level. *Jpn. J. Appl. Phys.* **2018**, *57*, 117001. [CrossRef]

52. Kaneko, T.; Toriumi, H.; Shimada, J.; Nomura, F. Extracellular field potential recording of single cardiomyocytes in agarose microchambers using microelectrode array. *J. Appl. Phys.* **2018**, *57*, 03EB03. [CrossRef]

53. Kang, C.; Qiao, Y.; Li, G.; Baechle, K.; Camelliti, P.; Rentschler, S.; Efimov, I.R. Human Organotypic Cultured Cardiac Slices: New Platform For High Throughput Preclinical Human Trials. *Sci. Rep.* **2016**, *6*, 28798. [CrossRef]

54. Lane, J.D.; Montaigne, D.; Tinker, A. Tissue-Level Cardiac Electrophysiology Studied in Murine Myocardium Using a Microelectrode Array: Autonomic and Thermal Modulation. *J. Membr. Biol.* **2017**, *250*, 471–481. [CrossRef] [PubMed]

55. Chowdhury, R.A.; Tzortzis, K.N.; Dupont, E.; Selvadurai, S.; Perbellini, F.; Cantwell, C.D.; Ng, F.S.; Simon, A.R.; Terracciano, C.M.; Peters, N.S. Concurrent micro-to macro-cardiac electrophysiology in myocyte cultures and human heart slices. *Sci. Rep.* **2018**, *8*, 6947. [CrossRef] [PubMed]

56. Doerr, L.; Thomas, U.; Guinot, D.R.; Bot, C.T.; Stoelzle-Feix, S.; Beckler, M.; George, M.; Fertig, N. New Easy-to-Use Hybrid System for Extracellular Potential and Impedance Recordings. *J. Lab. Autom.* **2015**, *20*, 175–188. [CrossRef] [PubMed]

57. Qian, F.; Huang, C.; Lin, Y.D.; Ivanovskaya, A.N.; O'Hara, T.J.; Booth, R.H.; Creek, C.J.; Enright, H.A.; Soscia, D.A.; Belle, A.M.; et al. Simultaneous electrical recording of cardiac electrophysiology and contraction on chip. *Lab. A Chip* **2017**, *17*, 1732–1739. [CrossRef] [PubMed]

58. Takasuna, K.; Asakura, K.; Araki, S.; Ando, H.; Kazusa, K.; Kitaguchi, T.; Kunimatsu, T.; Suzuki, S.; Miyamoto, N. Comprehensive in Vitro Cardiac Safety Assessment Using Human Stem Cell Technology: Overview of Csahi Heart Initiative. *J. Pharma. Toxico. Meth.* **2017**, *83*, 42–54. [CrossRef]

59. Hayakawa, T.; Kunihiro, T.; Ando, T.; Kobayashi, S.; Matsui, E.; Yada, H.; Kanda, Y.; Kurokawa, J.; Furukawa, T. Image-based evaluation of contraction-relaxation kinetics of human-induced pluripotent stem cell-derived cardiomyocytes: Correlation and complementarity with extracellular electrophysiology. *J. Mol. Cell. Cardiol.* **2014**, *77*, 178–191. [CrossRef]

60. Cools, J.; Jin, Q.; Yoon, E.; Alba Burbano, D.; Luo, Z.; Cuypers, D.; Callewaert, G.; Braeken, D.; Gracias, D.H. A Micropatterned Multielectrode Shell for 3D Spatiotemporal Recording from Live Cells. *Adv. Sci.* **2018**, *5*, 1700731. [CrossRef]

61. Nagarah, J.M.; Stowasser, A.; Parker, R.L.; Asari, H.; Wagenaar, D.A. Optically transparent multi-suction electrode arrays. *Front. Neurosci.* **2015**, *9*, 384. [CrossRef]

62. Simeonov, S.; Schäffer, T.E. Ultrafast imaging of cardiomyocyte contractions by combining scanning ion conductance microscopy with a microelectrode array. *Anal. Chem.* **2019**, *91*, 9648–9655. [CrossRef]

63. Kanda, Y.; Yamazaki, D.; Kurokawa, J.; Inutsuka, T.; Sekino, Y. Points to consider for a validation study of iPS cell-derived cardiomyocytes using a multi-electrode array system. *J. Pharmacol. Toxicol. Methods* **2016**, *81*, 196–200. [CrossRef]

64. Uesugi, M.; Ojima, A.; Taniguchi, T.; Miyamoto, N.; Sawada, K. Low-density plating is sufficient to induce cardiac hypertrophy and electrical remodeling in highly purified human iPS cell-derived cardiomyocytes. *J. Pharmacol. Toxicol. Methods* **2014**, *69*, 177–188. [CrossRef]

65. Amin, A.S.; Tan, H.L.; Wilde, A.A.M. Cardiac ion channels in health and disease. *Heart Rhythm.* **2010**, *7*, 117–126. [CrossRef] [PubMed]

66. Jung, J.J.; Husse, B.; Rimmbach, C.; Krebs, S.; Stieber, J.; Steinhoff, G.; Dendorfer, A.; Franz, W.-M.; David, R. Programming and Isolation of Highly Pure Physiologically and Pharmacologically Functional Sinus-Nodal Bodies from Pluripotent Stem Cells. *Stem Cell Rep.* **2014**, *2*, 592–605. [CrossRef] [PubMed]

67. Protze, S.I.; Liu, J.; Nussinovitch, U.; Ohana, L.; Backx, P.H.; Gepstein, L.; Keller, G.M. Sinoatrial node cardiomyocytes derived from human pluripotent cells function as a biological pacemaker. *Nat. Biotechnol.* **2017**, *35*, 56–68. [CrossRef] [PubMed]

68. Liu, J.; Laksman, Z.; Backx, P.H. The electrophysiological development of cardiomyocytes. *Adv. Drug Deliv. Rev.* **2016**, *96*, 253–273.

69. Feher, J. The Cardiac Action Potential. In *Quantitative Human Physiology*; Elsevier: Amsterdam, The Netherlands, 2017; pp. 528–536.

70. Du, D.T.M.; Hellen, N.; Kane, C.; Terracciano, C.M.N. Action potential morphology of human induced pluripotent stem cell-derived cardiomyocytes does not predict cardiac chamber specificity and is dependent on cell density. *Biophys. J.* **2015**, *108*, 1–4. [CrossRef]

71. Ronaldson-Bouchard, K.; Ma, S.P.; Yeager, K.; Chen, T.; Song, L.; Sirabella, D.; Morikawa, K.; Teles, D.; Yazawa, M.; Vunjak-Novakovic, G. Advanced maturation of human cardiac tissue grown from pluripotent stem cells. *Nature* **2018**, *556*, 239–243. [CrossRef]

72. Vaidyanathan, R.; Markandeya, Y.S.; Kamp, T.J.; Makielski, J.C.; January, C.T.; Eckhardt, L.L. IK1-enhanced human-induced pluripotent stem cell-derived cardiomyocytes: An improved cardiomyocyte model to investigate inherited arrhythmia syndromes. *Am. J. Physiol. Heart Circ. Physiol.* **2016**, *310*, 1611–1621.

73. Goversen, B.; Becker, N.; Stoelzle-Feix, S.; Obergrussberger, A.; Vos, M.A.; van Veen, T.A.B.; Fertig, N.; de Boer, T.P. A Hybrid Model for Safety Pharmacology on an Automated Patch Clamp Platform: Using Dynamic Clamp to Join iPSC-Derived Cardiomyocytes and Simulations of Ik1 Ion Channels in Real-Time. *Front. Physiol.* **2018**, *8*, 1094. [CrossRef]

74. Jonsson, M.K.B.; Vos, M.A.; Mirams, G.R.; Duker, G.; Sartipy, P.; De Boer, T.P.; Van Veen, T.A.B. Application of human stem cell-derived cardiomyocytes in safety pharmacology requires caution beyond hERG. *J. Mol. Cell. Cardiol.* **2012**, *52*, 998–1008. [CrossRef]

75. Tertoolen, L.G.J.; Braam, S.R.; van Meer, B.J.; Passier, R.; Mummery, C.L. Interpretation of field potentials measured on a multi electrode array in pharmacological toxicity screening on primary and human pluripotent stem cell-derived cardiomyocytes. *Biochem. Biophys. Res. Commun.* **2018**, *497*, 1135–1141. [CrossRef]

76. Clements, M. Multielectrode array (MEA) assay for unit 22.4 profiling electrophysiological drug effects in human stem cell-derived cardiomyocytes. *Curr. Protoc. Toxicol.* **2016**, *2016*, 1–32.

77. Buzsáki, G.; Anastassiou, C.A.; Koch, C. The origin of extracellular fields and currents — EEG, ECoG, LFP and spikes. *Nat. Rev. Neurosci.* **2012**, *13*, 407–420. [CrossRef] [PubMed]

78. Halbach, M.D.; Egert, U.; Hescheler, J.; Banach, K. Estimation of action potential changes from field potential recordings in multicellular mouse cardiac myocyte cultures. *Cell. Physiol. Biochem.* **2003**, *13*, 271–284. [CrossRef] [PubMed]

79. Hayes, H.B.; Nicolini, A.M.; Arrowood, C.A.; Chvatal, S.A.; Wolfson, D.W.; Cho, H.C.; Sullivan, D.D.; Chal, J.; Fermini, B.; Clements, M.; et al. Novel method for action potential measurements from intact cardiac monolayers with multiwell microelectrode array technology. *Sci. Rep.* **2019**, *9*, 11893. [CrossRef] [PubMed]

80. Jans, D.; Callewaert, G.; Krylychkina, O.; Hoffman, L.; Gullo, F.; Prodanov, D.; Braeken, D. Action potential-based MEA platform for in vitro screening of drug-induced cardiotoxicity using human iPSCs and rat neonatal myocytes. *J. Pharmacol. Toxicol. Methods* **2017**, *87*, 48–52. [CrossRef] [PubMed]

81. Xie, C.; Lin, Z.; Hanson, L.; Cui, Y.; Cui, B. Intracellular recording of action potentials by nanopillar electroporation. *Nat. Nanotechnol.* **2012**, *7*, 185–190. [CrossRef] [PubMed]

82. Fendyur, A.; Spira, M.E. Toward on-chip, in-cell recordings from cultured cardiomyocytes by arrays of gold mushroom-shaped microelectrodes. *Front. Neuroeng.* **2012**, *5*, 21. [CrossRef] [PubMed]

83. WHO. Cardiovascular diseases. 2019. Available online: https://www.who.int/health-topics/cardiovascular-diseases (accessed on 28 October 2019).

84. Fermini, B.; Hancox, J.C.; Abi-Gerges, N.; Bridgland-Taylor, M.; Chaudhary, K.W.; Colatsky, T.; Correll, K.; Crumb, W.; Damiano, B.; Erdemli, G.; et al. A New Perspective in the Field of Cardiac Safety Testing through the Comprehensive In Vitro Proarrhythmia Assay Paradigm. *J. Biomol. Screen* **2016**, *21*, 1–11. [CrossRef]

85. Millard, D.; Dang, Q.; Shi, H.; Zhang, X.; Strock, C.; Kraushaar, U.; Zeng, H.; Levesque, P.; Lu, H.-R.; Guillon, J.-M.; et al. Cross-Site Reliability of Human Induced Pluripotent stem cell-derived Cardiomyocyte Based Safety Assays Using Microelectrode Arrays: Results from a Blinded CiPA Pilot Study. *Toxicol. Sci.* **2018**, *164*, 550–562. [CrossRef]

86. CiPA Initiative. 2019. Available online: https://cipaproject.org (accessed on 28 October 2019).

87. Kitaguchi, T.; Moriyama, Y.; Taniguchi, T.; Maeda, S.; Ando, H.; Uda, T.; Otabe, K.; Oguchi, M.; Shimizu, S.; Saito, H.; et al. CSAHi study: Detection of drug-induced ion channel/receptor responses, QT prolongation, and arrhythmia using multi-electrode arrays in combination with human induced pluripotent stem cell-derived cardiomyocytes. *J. Pharmacol. Toxicol. Methods* **2017**, *85*, 73–81. [CrossRef]

88. Nozaki, Y.; Honda, Y.; Watanabe, H.; Saiki, S.; Koyabu, K.; Itoh, T.; Nagasawa, C.; Nakamori, C.; Nakayama, C.; Iwasaki, H.; et al. CSAHi study-2: Validation of multi-electrode array systems (MEA60/2100) for prediction of drug-induced proarrhythmia using human iPS cell-derived cardiomyocytes: Assessment of reference compounds and comparison with non-clinical studies and clinical information. *Regul. Toxicol. Pharmacol.* **2017**, *88*, 238–251.

89. Sager, P.T.; Gintant, G.; Turner, J.R.; Pettit, S.; Stockbridge, N. Rechanneling the cardiac proarrhythmia safety paradigm: A meeting report from the Cardiac Safety Research Consortium. *Am. Heart J.* **2014**, *167*, 292–300. [CrossRef] [PubMed]

90. Kitaguchi, T.; Moriyama, Y.; Taniguchi, T.; Ojima, A.; Ando, H.; Uda, T.; Otabe, K.; Oguchi, M.; Shimizu, S.; Saito, H.; et al. CSAHi study: Evaluation of multi-electrode array in combination with human iPS cell-derived cardiomyocytes to predict drug-induced QT prolongation and arrhythmia — Effects of 7 reference compounds at 10 facilities. *J. Pharmacol. Toxicol. Methods* **2016**, *78*, 93–102. [CrossRef] [PubMed]

91. Ma, J.; Guo, L.; Fiene, S.J.; Anson, B.D.; Thomson, J.A.; Kamp, T.J.; Kolaja, K.L.; Swanson, B.J.; January, C.T. High purity human-induced pluripotent stem cell-derived cardiomyocytes: Electrophysiological properties of action potentials and ionic currents. *Am. J. Physiol. Heart Circ. Physiol.* **2011**, *301*, 2006–2017. [CrossRef] [PubMed]

92. van den Heuvel, N.H.L.; van Veen, T.A.B.; Lim, B.; Jonsson, M.K.B. Lessons from the heart: Mirroring electrophysiological characteristics during cardiac development to in vitro differentiation of stem cell derived cardiomyocytes. *J. Mol. Cell. Cardiol.* **2014**, *67*, 12–25. [CrossRef] [PubMed]

93. Blinova, K.; Dang, Q.; Millard, D.; Smith, G.; Pierson, J.; Guo, L.; Brock, M.; Lu, H.R.; Kraushaar, U.; Zeng, H.; et al. International Multisite Study of Human-Induced Pluripotent Stem Cell-Derived Cardiomyocytes for Drug Proarrhythmic Potential Assessment. *Cell Rep.* **2018**, *24*, 3582–3592. [CrossRef]

94. Schocken, D.; Stohlman, J.; Vicente, J.; Chan, D.; Patel, D.; Matta, M.K.; Patel, V.; Brock, M.; Millard, D.; Ross, J.; et al. Comparative analysis of media effects on human induced pluripotent stem cell-derived cardiomyocytes in proarrhythmia risk assessment. *J. Pharmacol. Toxicol. Methods* **2018**, *90*, 39–47. [CrossRef]

95. Zeng, H.; Wang, J.; Clouse, H.; Lagrutta, A.; Sannajust, F. HiPSC-CMs from different sex and ethnic origin donors exhibit qualitatively different responses to several classes of pharmacological challenges. *J. Pharmacol. Toxicol. Methods* **2019**, *99*, 106598. [CrossRef]

96. Burnett, S.D.; Blanchette, A.D.; Grimm, F.A.; House, J.S.; Reif, D.M.; Wright, F.A.; Chiu, W.A.; Rusyn, I. Population-based toxicity screening in human induced pluripotent stem cell-derived cardiomyocytes. *Toxicol. Appl. Pharmacol.* **2019**, *381*, 114711. [CrossRef]

97. Egashira, T.; Yuasa, S.; Suzuki, T.; Aizawa, Y.; Yamakawa, H.; Matsuhashi, T.; Ohno, Y.; Tohyama, S.; Okata, S.; Seki, T.; et al. Disease characterization using LQTS-specific induced pluripotent stem cells. *Cardiovasc. Res.* **2012**, *95*, 419–429. [CrossRef]

98. Harris, K.; Aylott, M.; Cui, Y.; Louttit, J.B.; McMahon, N.C.; Sridhar, A. Comparison of Electrophysiological Data From Human-Induced Pluripotent Stem Cell–Derived Cardiomyocytes to Functional Preclinical Safety Assays. *Toxicol. Sci.* **2013**, *134*, 412–426. [CrossRef]

99. Liang, P.; Lan, F.; Lee, A.S.; Gong, T.; Sanchez-Freire, V.; Wang, Y.; Diecke, S.; Sallam, K.; Knowles, J.W.; Wang, P.J.; et al. Drug Screening Using a Library of Human Induced Pluripotent Stem Cell–Derived Cardiomyocytes Reveals Disease-Specific Patterns of Cardiotoxicity. *Circulation* **2013**, *127*, 1677–1691. [CrossRef] [PubMed]

100. Maillet, A.; Tan, K.; Chai, X.; Sadananda, S.N.; Mehta, A.; Ooi, J.; Hayden, M.R.; Pouladi, M.A.; Ghosh, S.; Shim, W.; et al. Modeling Doxorubicin-Induced Cardiotoxicity in Human Pluripotent Stem Cell Derived-Cardiomyocytes. *Sci. Rep.* **2016**, *6*, 25333. [CrossRef]

101. Burridge, P.W.; Li, Y.F.; Matsa, E.; Wu, H.; Ong, S.-G.; Sharma, A.; Holmström, A.; Chang, A.C.; Coronado, M.J.; Ebert, A.D.; et al. Human induced pluripotent stem cell-derived cardiomyocytes recapitulate the predilection of breast cancer patients to doxorubicin-induced cardiotoxicity. *Nat. Med.* **2016**, *22*, 547–556. [CrossRef] [PubMed]

102. Mulder, P.; de Korte, T.; Dragicevic, E.; Kraushaar, U.; Printemps, R.; Vlaming, M.L.H.; Braam, S.R.; Valentin, J.-P. Predicting cardiac safety using human induced pluripotent stem cell-derived cardiomyocytes combined with multi-electrode array (MEA) technology: A conference report. *J. Pharmacol. Toxicol. Methods* **2018**, *91*, 36–42. [CrossRef] [PubMed]

103. Acimovic, I.; Refaat, M.; Moreau, A.; Salykin, A.; Reiken, S.; Sleiman, Y.; Souidi, M.; Přibyl, J.; Kajava, A.; Richard, S.; et al. Post-Translational Modifications and Diastolic Calcium Leak Associated to the Novel RyR2-D3638A Mutation Lead to CPVT in Patient-Specific hiPSC-Derived Cardiomyocytes. *JCM* **2018**, *7*, 423. [CrossRef]

104. Navarrete, E.G.; Liang, P.; Lan, F.; Sanchez-Freire, V.; Simmons, C.; Gong, T.; Sharma, A.; Burridge, P.W.; Patlolla, B.; Lee, A.S.; et al. Screening Drug-Induced Arrhythmia Using Human Induced Pluripotent Stem Cell-Derived Cardiomyocytes and Low-Impedance Microelectrode Arrays. *Circulation* **2013**, *128*, 3–13. [CrossRef] [PubMed]

105. Ebert, A.D.; Svendsen, C.N. Human stem cells and drug screening: Opportunities and challenges. *Nat. Rev. Drug Discov.* **2010**, *9*, 367–372. [CrossRef] [PubMed]

106. Wallace, E.; Howard, L.; Liu, M.; O'Brien, T.; Ward, D.; Shen, S.; Prendiville, T. Long QT Syndrome: Genetics and Future Perspective. *Pediatr Cardiol* **2019**, *40*, 1–12. [CrossRef]

107. Bellin, M.; Casini, S.; Davis, R.P.; D'Aniello, C.; Haas, J.; Ward-van Oostwaard, D.; Tertoolen, L.G.J.; Jung, C.B.; Elliott, D.A.; Welling, A.; et al. Isogenic human pluripotent stem cell pairs reveal the role of a KCNH2 mutation in long-QT syndrome: Isogenic pairs of LQT2 pluripotent stem cells. *Embo J.* **2013**, *32*, 3161–3175. [CrossRef]

108. Brandão, K.O.; Tabel, V.A.; Atsma, D.E.; Mummery, C.L.; Davis, R.P. Human pluripotent stem cell models of cardiac disease: From mechanisms to therapies. *Dis. Model. Mech.* **2017**, *10*, 1039–1059. [CrossRef]

109. Zhang, M.; D'Aniello, C.; Verkerk, A.O.; Wrobel, E.; Frank, S.; Ward-van Oostwaard, D.; Piccini, I.; Freund, C.; Rao, J.; Seebohm, G.; et al. Recessive cardiac phenotypes in induced pluripotent stem cell models of Jervell and Lange-Nielsen syndrome: Disease mechanisms and pharmacological rescue. *Proc. Natl. Acad. Sci. USA* **2014**, *111*, 5383–5392. [CrossRef] [PubMed]

110. Moretti, A.; Bellin, M.; Welling, A.; Jung, C.B.; Lam, J.T.; Bott-Flügel, L.; Dorn, T.; Goedel, A.; Höhnke, C.; Hofmann, F.; et al. Patient-Specific Induced Pluripotent Stem-Cell Models for Long-QT Syndrome. *N. Engl. J. Med.* **2010**, *363*, 1397–1409. [CrossRef] [PubMed]

111. Sala, L.; Gnecchi, M.; Schwartz, P.J. Derived from Human-induced Pluripotent Stem Cells. *Arrhythmia Electrophysiol. Rev.* **2019**, *8*, 105. [CrossRef] [PubMed]

112. Liu, N.; Ruan, Y.; Priori, S.G. Catecholaminergic Polymorphic Ventricular Tachycardia. *Prog. Cardiovasc. Dis.* **2008**, *51*, 23–30. [CrossRef]

113. Sasaki, K.; Makiyama, T.; Yoshida, Y.; Wuriyanghai, Y.; Kamakura, T.; Nishiuchi, S.; Hayano, M.; Harita, T.; Yamamoto, Y.; Kohjitani, H.; et al. Patient-Specific Human Induced Pluripotent Stem Cell Model Assessed with Electrical Pacing Validates S107 as a Potential Therapeutic Agent for Catecholaminergic Polymorphic Ventricular Tachycardia. *PLoS ONE* **2016**, *11*, e0164795. [CrossRef]

114. Carvajal-Vergara, X.; Sevilla, A.; D'Souza, S.L.; Ang, Y.-S.; Schaniel, C.; Lee, D.-F.; Yang, L.; Kaplan, A.D.; Adler, E.D.; Rozov, R.; et al. Patient-specific induced pluripotent stem-cell-derived models of LEOPARD syndrome. *Nature* **2010**, *465*, 808–812. [CrossRef]

115. Lan, F.; Lee, A.S.; Liang, P.; Sanchez-Freire, V.; Nguyen, P.K.; Wang, L.; Han, L.; Yen, M.; Wang, Y.; Sun, N.; et al. Abnormal Calcium Handling Properties Underlie Familial Hypertrophic Cardiomyopathy Pathology in Patient-Specific Induced Pluripotent Stem Cells. *Cell Stem Cell* **2013**, *12*, 101–113. [CrossRef]

116. Yang, K.-C.; Breitbart, A.; De Lange, W.J.; Hofsteen, P.; Futakuchi-Tsuchida, A.; Xu, J.; Schopf, C.; Razumova, M.V.; Jiao, A.; Boucek, R.; et al. Novel Adult-Onset Systolic Cardiomyopathy Due to MYH7 E848G Mutation in Patient-Derived Induced Pluripotent Stem Cells. *JACC Basic Transl. Sci.* **2018**, *3*, 728–740. [CrossRef]

117. Sun, N.; Yazawa, M.; Liu, J.; Han, L.; Sanchez-Freire, V.; Abilez, O.J.; Navarrete, E.G.; Hu, S.; Wang, L.; Lee, A.; et al. Patient-Specific Induced Pluripotent Stem Cells as a Model for Familial Dilated Cardiomyopathy. *Sci. Transl. Med.* **2012**, *4*, 47–130. [CrossRef]

118. Siu, C.-W.; Lee, Y.-K.; Ho, J.C.-Y.; Lai, W.-H.; Chan, Y.-C.; Ng, K.-M.; Wong, L.-Y.; Au, K.-W.; Lau, Y.-M.; Zhang, J.; et al. Modeling of lamin A/C mutation premature cardiac aging using patient-specific induced pluripotent stem cells. *Aging* **2012**, *4*, 803–822. [CrossRef]

119. Tse, H.-F.; Ho, J.C.Y.; Choi, S.-W.; Lee, Y.-K.; Butler, A.W.; Ng, K.-M.; Siu, C.-W.; Simpson, M.A.; Lai, W.-H.; Chan, Y.-C.; et al. Patient-specific induced-pluripotent stem cells-derived cardiomyocytes recapitulate the pathogenic phenotypes of dilated cardiomyopathy due to a novel DES mutation identified by whole exome sequencing. *Hum. Mol. Genet.* **2013**, *22*, 1395–1403. [CrossRef] [PubMed]

120. Hinson, J.T.; Chopra, A.; Nafissi, N.; Polacheck, W.J.; Benson, C.C.; Swist, S.; Gorham, J.; Yang, L.; Schafer, S.; Sheng, C.C.; et al. Titin mutations in iPS cells define sarcomere insufficiency as a cause of dilated cardiomyopathy. *Science* **2015**, *349*, 982–986. [CrossRef] [PubMed]

121. Streckfuss-Bömeke, K.; Tiburcy, M.; Fomin, A.; Luo, X.; Li, W.; Fischer, C.; Özcelik, C.; Perrot, A.; Sossalla, S.; Haas, J.; et al. Severe DCM phenotype of patient harboring RBM20 mutation S635A can be modeled by patient-specific induced pluripotent stem cell-derived cardiomyocytes. *J. Mol. Cell. Cardiol.* **2017**, *113*, 9–21. [CrossRef] [PubMed]

122. Wyles, S.P.; Li, X.; Hrstka, S.C.; Reyes, S.; Oommen, S.; Beraldi, R.; Edwards, J.; Terzic, A.; Olson, T.M.; Nelson, T.J. Modeling structural and functional deficiencies of RBM20 familial dilated cardiomyopathy using human induced pluripotent stem cells. *Hum. Mol. Genet.* **2016**, *25*, 254–265. [CrossRef]

123. Giacomelli, E.; Mummery, C.L.; Bellin, M. Human heart disease: Lessons from human pluripotent stem cell-derived cardiomyocytes. *Cell. Mol. Life Sci.* **2017**, *74*, 3711–3739. [CrossRef]

124. Lin, B.; Li, Y.; Han, L.; Kaplan, A.D.; Ao, Y.; Kalra, S.; Bett, G.C.L.; Rasmusson, R.L.; Denning, C.; Yang, L. Modeling and study of the mechanism of dilated cardiomyopathy using induced pluripotent stem cells derived from individuals with Duchenne muscular dystrophy. *Dis. Models Mech.* **2015**, *8*, 457–466. [CrossRef]

125. Eschenhagen, T.; Carrier, L. Cardiomyopathy phenotypes in human-induced pluripotent stem cell-derived cardiomyocytes—a systematic review. *Pflug. Arch. Eur. J. Physiol.* **2019**, *471*, 755–768. [CrossRef]

126. de la Roche, J.; Angsutararux, P.; Kempf, H.; Janan, M.; Bolesani, E.; Thiemann, S.; Wojciechowski, D.; Coffee, M.; Franke, A.; Schwanke, K.; et al. Comparing human iPSC-cardiomyocytes versus HEK293T cells unveils disease-causing effects of Brugada mutation A735V of NaV1.5 sodium channels. *Sci. Rep.* **2019**, *9*, 11173. [CrossRef]

127. Bozzi, A.; Sayed, N.; Matsa, E.; Sass, G.; Neofytou, E.; Clemons, K.V.; Correa-Oliveira, R.; Stevens, D.A.; Wu, J.C. Using Human Induced Pluripotent Stem Cell-Derived Cardiomyocytes as a Model to Study Trypanosoma cruzi Infection. *Stem Cell Rep.* **2019**, *12*, 1232–1241. [CrossRef]

128. Lee, J.-J.; Cheng, S.-J.; Huang, C.-Y.; Chen, C.-Y.; Feng, L.; Hwang, D.-Y.; Kamp, T.J.; Chen, H.-C.; Hsieh, P.C.H. Primary cardiac manifestation of autosomal dominant polycystic kidney disease revealed by patient induced pluripotent stem cell-derived cardiomyocytes. *EBioMedicine* **2019**, *40*, 675–684. [CrossRef]

129. Birket, M.J.; Raibaud, S.; Lettieri, M.; Adamson, A.D.; Letang, V.; Cervello, P.; Redon, N.; Ret, G.; Viale, S.; Wang, B.; et al. A Human Stem Cell Model of Fabry Disease Implicates LIMP-2 Accumulation in Cardiomyocyte Pathology. *Stem Cell Rep.* **2019**, *13*, 380–393. [CrossRef] [PubMed]

130. Caluori, G.; Pribyl, J.; Pesl, M.; Jelinkova, S.; Rotrekl, V.; Skladal, P.; Raiteri, R. Non-invasive electromechanical cell-based biosensors for improved investigation of 3D cardiac models. *Biosens. Bioelectron.* **2019**, *124*, 129–135. [CrossRef] [PubMed]

131. Ma, Z.; Huebsch, N.; Koo, S.; Mandegar, M.A.; Siemons, B.; Boggess, S.; Conklin, B.R.; Grigoropoulos, C.P.; Healy, K.E. Contractile deficits in engineered cardiac microtissues as a result of MYBPC3 deficiency and mechanical overload. *Nat. Biomed. Eng.* **2018**, *2*, 955–967. [CrossRef] [PubMed]

132. Park, S.-J.; Zhang, D.; Qi, Y.; Li, Y.; Lee, K.Y.; Bezzerides, V.J.; Yang, P.; Xia, S.; Kim, S.L.; Liu, X.; et al. Insights Into the Pathogenesis of Catecholaminergic Polymorphic Ventricular Tachycardia From Engineered Human Heart Tissue. *Circulation* **2019**, *140*, 390–404. [CrossRef]

133. Paci, M.; Casini, S.; Bellin, M.; Hyttinen, J.; Severi, S. Large-Scale Simulation of the Phenotypical Variability Induced by Loss-of-Function Long QT Mutations in Human Induced Pluripotent Stem Cell Cardiomyocytes. *Int. J. Mol. Sci.* **2018**, *19*, 3583. [CrossRef]

134. Satsuka, A.; Kanda, Y. Cardiotoxicity assessment of drugs using human iPS cell-derived cardiomyocytes: From proarrhythmia risk to cardiooncology. *Curr. Pharm. Biotechnol.* **2019**. [CrossRef]

135. Zhang, B.; Korolj, A.; Lai, B.F.L.; Radisic, M. Advances in organ-on-a-chip engineering. *Nat. Rev. Mater.* **2018**, *3*, 257–278. [CrossRef]

136. Maoz, B.M.; Herland, A.; Henry, O.Y.F.; Leineweber, W.D.; Yadid, M.; Doyle, J.; Mannix, R.; Kujala, V.J.; FitzGerald, E.A.; Parker, K.K.; et al. Organs-on-Chips with combined multi-electrode array and transepithelial electrical resistance measurement capabilities. *Lab. Chip* **2017**, *17*, 2294–2302. [CrossRef]

137. Di Sanzo, M.; Cipolloni, L.; Borro, M.; La Russa, R.; Santurro, A.; Scopetti, M.; Simmaco, M.; Frati, P. Clinical Applications of Personalized Medicine: A New Paradigm and Challenge. *Curr. Pharm. Biotechnol.* **2017**, *18*, 194–203. [CrossRef]

138. Chen, I.Y.; Matsa, E.; Wu, J.C. Induced pluripotent stem cells: At the heart of cardiovascular precision medicine. *Nat. Rev. Cardiol.* **2016**, *13*, 333–349. [CrossRef]

139. Hamazaki, T.; El Rouby, N.; Fredette, N.C.; Santostefano, K.E.; Terada, N. Concise Review: Induced Pluripotent Stem Cell Research in the Era of Precision Medicine. *Stem Cells* **2017**, *35*, 545–550. [CrossRef] [PubMed]

Article

Case Report on: Very Early Afterdepolarizations in HiPSC-Cardiomyocytes—An Artifact by Big Conductance Calcium Activated Potassium Current ($I_{bk,Ca}$)

András Horváth [1,2,3,†,‡], Torsten Christ [1,2,*,†], Jussi T. Koivumäki [4], Maksymilian Prondzynski [1,2,§], Antonia T. L. Zech [1,2], Michael Spohn [5], Umber Saleem [1,2], Ingra Mannhardt [1,2], Bärbel Ulmer [1,2], Evaldas Girdauskas [2,6], Christian Meyer [3,7], Arne Hansen [1,2], Thomas Eschenhagen [1,2] and Marc D. Lemoine [1,2,7,*]

[1] Institute of Experimental Pharmacology and Toxicology, University Medical Center Hamburg-Eppendorf, 20246 Hamburg, Germany; bandi185@yahoo.com (A.H.); mprondzynski@gmail.com (M.P.); a.zech@uke.de (A.T.L.Z.); u.saleem@uke.de (U.S.); i.mannhardt@uke.de (I.M.); b.ulmer@uke.de (B.U.); ar.hansen@uke.de (A.H.); t.eschenhagen@uke.de (T.E.)

[2] DZHK (German Center for Cardiovascular Research), Partner Site Hamburg/Kiel/Lübeck, 20246 Hamburg, Germany; e.girdauskas@uke.de

[3] Department of Pharmacology and Pharmacotherapy, Faculty of Medicine, University of Szeged, 6721 Szeged, Hungary; chr.meyer@uke.de

[4] BioMediTech, Faculty of Medicine and Health Technology, Tampere University, 33520 Tampere, Finland; j.koivumaeki@gmail.com

[5] Bioinformatics Core, University Medical Center Hamburg-Eppendorf, 20246 Hamburg, Germany; m.spohn@uke.de

[6] Department of Cardiovascular Surgery, University Heart Center, 20246 Hamburg, Germany

[7] Department of Cardiology-Electrophysiology, University Heart Center, 20246 Hamburg, Germany

*   Correspondence: t.christ@uke.de (T.C.); m.lemoine@uke.de (M.D.L.); Tel.: +49-40-7410-52180 (T.C.); +49-40-7410-52438 (M.D.L.)

†   These authors contributed equally to this work.

‡   Current Address: Nanion Technologies GmbH, Ganghoferstraße 70a, 80339 München, Germany.

§   Current Address: Department of Cardiology, Boston Children's Hospital, Harvard Medical School, Boston, MA 02115, USA.

Received: 28 October 2019; Accepted: 15 January 2020; Published: 20 January 2020

Abstract: Human induced pluripotent stem cell-derived cardiomyocytes (hiPSC-CMs) represent an unlimited source of human CMs that could be a standard tool in drug research. However, there is concern whether hiPSC-CMs express all cardiac ion channels at physiological level and whether they might express non-cardiac ion channels. In a control hiPSC line, we found large, "noisy" outward K^+ currents, when we measured outward potassium currents in isolated hiPSC-CMs. Currents were sensitive to iberiotoxin, the selective blocker of big conductance Ca^{2+}-activated K^+ current ($I_{BK,Ca}$). Seven of 16 individual differentiation batches showed a strong initial repolarization in the action potentials (AP) recorded from engineered heart tissue (EHT) followed by very early afterdepolarizations, sometimes even with consecutive oscillations. Iberiotoxin stopped oscillations and normalized AP shape, but had no effect in other EHTs without oscillations or in human left ventricular tissue (LV). Expression levels of the alpha-subunit ($K_{Ca}1.1$) of the BK_{Ca} correlated with the presence of oscillations in hiPSC-CMs and was not detectable in LV. Taken together, individual batches of hiPSC-CMs can express sarcolemmal ion channels that are otherwise not found in the human heart, resulting in oscillating afterdepolarizations in the AP. HiPSC-CMs should be screened for expression of non-cardiac ion channels before being applied to drug research.

Keywords: human induced pluripotent stem cell-derived cardiomyocytes (hiPSC-CMs); iPS cells; stem cells; big conductance calcium activated potassium channel (BK); Maxi-K; slo1; $K_{Ca1.1}$; iberiotoxin; long QT syndrome

---

1. Introduction

Human induced pluripotent stem-cell derived cardiomyocytes (hiPSC-CMs) have gained interest as a human model to study heart physiology and pathophysiology [1–4], cardiovascular pharmacology [5] and cardiac repair [6]. In this context, it is important that hiPSC-CMs share properties of human adult CMs. Many reports claimed that characteristics of hiPSC-CMs might differ from adult CMs, a finding frequently interpreted as immaturity. On the one hand, immaturity could be caused by the lack of cardiac ion channels or by differences in expression levels such as the inward rectifier potassium current (I_{K1}) [7–10]. On the other hand, hiPSC-CMs can show ion currents, which are absent in adult human CMs such as the T-type calcium current [11,12]. Based on these findings, there is a need for a detailed electrophysiological characterization of hiPSC-CMs before using them as a model for human CMs. Here we report the coincidently found expression of a non-cardiac ion channel in hiPSC-CMs as a peculiarity of a single control cell line: the big conductance calcium activated potassium current (BK_{Ca}; alternatively used names: Maxi-K, slo1, $K_{Ca1.1}$).

BK_{Ca} is a voltage- and calcium-gated potassium channel ($K_{Ca1.1}$) generating huge conductivity for potassium. BK_{Ca} is widely expressed in the human body, mainly in neural cells, blood vessels, kidney, but not in in cardiomyocytes. Recently, artificial expression of BK_{Ca} was shown to be able to shorten action potentials (APs) in murine CMs [13]. Consequently, expression of the non-cardiac BK_{Ca} current in human CMs was proposed as a genetic therapy for the long QT syndrome. However, the contribution of individual ion channels to repolarization and resulting AP shape differs from murine to human CMs. Putative impact of BK_{Ca} to human cardiac electrophysiology remains unclear.

The goal of this study was to draw attention to the fact that hiPSC-CMs can express non-cardiac ion channels. In addition, we had the chance to elucidate how the expression of BK_{Ca} may influence the membrane potential of human CMs. Based on our findings, we propose a regular assessment for expression of non-cardiac ion channels as a part of quality control when using hiPSC-CMs.

2. Methods

2.1. Generation of hiPSC and Engineered Heart Tissue (EHT)

The hiPSC line C25 (kind gift from Alessandra Moretti, Munich, Germany) was reprogrammed by lentivirus [1] and was expanded in FTDA medium and differentiated in a three step protocol based on growth factors and a small molecule Wnt inhibitor DS07 (kind gift from Dennis Schade, Dortmund, Germany) as previously published [14–16]. The hiPSC line ERC018 were generated in-house from skin fibroblasts of a healthy subject using the CytoTune™-iPS Sendai Reprogramming Kit (Thermo Fisher Scientific, Waltham, MA, USA) and differentiated to cardiomyocytes as described for C25. In brief, confluent undifferentiated cells were dissociated (0.5 mM EDTA; 10 min) and cultivated in spinner flasks (30×10^6 cells/100 mL; 40 rpm) for embryoid body formation overnight. Mesodermal differentiation was initiated in embryoid bodies over three days in suspension culture with growth factors (BMP-4 (R&D Systems, 314-BP), activin-A (R&D Systems, 338-AC) and FGF2 (PeproTech, 100-18B)). Cardiac differentiation was performed either in adhesion or in suspension culture with Wnt-inhibitor DS07. Cells were cultured in a humidified temperature and gas-controlled incubator (37 °C, 5% CO_2, 5% O_2 and 21% O_2 for final cardiac differentiation). At day 14 the spontaneously beating hiPSC-CMs were dissociated with collagenase II (Worthington, LS004176; 200 U/mL, 3.5 h). For quality control, dissociated cells were analyzed by flow cytometry as described before [14,17] with anti-cardiac troponin T-FITC (Miltenyi, clone REA400, 130-112-756). All differentiation runs utilized for this study had 64–96% cardiac Troponin T positive cells (Supplementary Materials Figure S2). Further characterization of the non-CMs within the EHT was evaluated previously by our group [6], showing low expression of vimentin-positive fibroblast-like markers and the virtual absence of endothelial, neuronal, and endodermal markers. ICell and iCell^2 cardiomyocytes are commercially available hiPSC-CM lines purchased from Fujifilm Cellular Dynamics (Madison, Wisconsin, USA) and were included in expression analysis after culture in EHT. For three-dimensional culture EHTs were generated with 1×10^6 hiPSC-CM/100 μL EHT mastermix as previously described [18]. EHTs were cultured in a 37 °C, 7% CO_2 and 40% O_2 humidified cell culture incubator with a medium consisting of DMEM (F0415, Biochrom; Berlin, Germany), 10% heat-inactivated horse serum (Gibco 26050, Thermo Fisher Scientific, Waltham, MA, USA), 1% penicillin/streptomycin (Gibco 15140), insulin (10 μg/mL; Sigma-Aldrich I9278, St. Louis, MO, USA) and aprotinin (33 μg/mL; Sigma-Aldrich A1153). EHTs were cultured for at least 3 weeks to allow maturation. The work with hiPSC was approved by the Ethical Committee of the University Medical Center Hamburg-Eppendorf (Az. PV4798, 28.10.2014). All donors gave written informed consent.

2.2. Human Adult Heart Tissue

This investigation conforms to all principles outlined by the Declaration of Helsinki and the Medical Association of Hamburg. All materials from patients were taken with informed consent of the donors. Left ventricular free wall and left ventricular septum samples were obtained from patients undergoing heart transplantation or from aortic valve surgery.

2.3. Current Recordings

HiPSC-CMs in EHT were isolated with collagenase II for 5 h (200 U/mL, Worthington, LS004176 dissolved in HBSS-Puffer without Mg^{2+} or Ca^{2+}, Gibco 14175-053 and 1 mM HEPES; pH 7,4), and re-plated on gelatin-coated coverslips for 24–48 h in order to maintain adherence under perfusion. Outward K^+ currents were measured at 37 °C, using the whole-cell configuration of the patch clamp technique. Axopatch 200B amplifier (Axon Instruments, Foster City, CA, USA) and ISO2 software were used for data acquisition and analysis (MFK, Niederhausen, Germany). Heat-polished pipettes were pulled from borosilicate filamented glass (Hilgenberg, Malsfeld, Germany). Tip resistances were 2.5–5 MΩ when filled with pipette solution. Seal resistances were 2–4 GΩ. The cells were investigated in a small perfusion chamber placed on the stage of an inverse microscope. Application of drugs was performed by a system for rapid solution changes (Cell Micro Controls, Virginia Beach, VA, USA; ALA Scientific Instruments, Long Island, NY, USA) [19]. The experiments were performed with the following bath solution (in mM): NaCl 120, KCl 5.4, HEPES 10, $CaCl_2$ 2, $MgCl_2$ 1 and glucose 10 (pH 7.4, adjusted with NaOH). Outward currents were elicited by 1000 ms depolarizing test pulses from −80 to +70 mV (0.2 Hz). The pipette solution included (in mM): DL-Aspartate potassium salt 80, KCl 40, NaCl 8, HEPES 10, Mg-ATP 5, Tris-GTP 0.1, EGTA 5 and $CaCl_2$ 4.4, pH 7.4, adjusted with KOH [20].

2.4. AP Recordings

To record APs in intact EHT and in left ventricular trabeculae we used sharp microelectrodes as reported previously [5,9,21]. Microelectrode tip resistances were 20–50 MΩ when filled with 3 mM KCl. APs were elicited by field stimulation at 1 Hz, 0.5 ms stimulus duration and 50% above threshold intensity. The following bath solution was used (in mM): NaCl 125, KCl 5.4, $MgCl_2$ 0.6, $CaCl_2$ 1, NaH_2PO_4 0.4, NaH_2CO_3 22 and glucose 5.5 and was equilibrated with O_2–CO_2 (95:5). The experiments were performed at 37 °C.

2.5. Molecular Biology

Total RNA was extracted from snap frozen LV and EHT using an RNAeasy mini Kit (Qiagen, Valencia, CA, USA). RNA concentration was determined per fluorometric quantitation with Qubit^{TM} (Thermo Fisher Scientific; Waltham, MA, USA) according to the manufacturer's instructions. Of total

RNA 50 ng was used for expression analysis by nanoString nCounter® SPRINT Profiler according to the manufacturer's instructions. Raw data were analyzed with nSolver™ Data Analysis Software including background subtraction using negative controls and normalization to two housekeeping genes (GAPDH and PGK1).

2.6. Mathematical Modelling and Computer Simulations

To simulate the BK_{Ca} current in silico, we extended a previously published formulation [22] to include (1) calcium dependence according to the Lin et al. [23] data, and (2) inactivation kinetics according to Ding et al. [24] data. To demonstrate the contribution of BK_{Ca} current on AP repolarization, we integrated the BK_{Ca} model to a well-established human left ventricular CM model [25].

2.7. Statistical Analysis

Results are presented as mean values ± SEM. Area under the curve was calculated by using the GraphPad Prism Software 5.02 (GraphPad Software, San Diego, CA, USA). Statistical differences were evaluated by using the Student's *t*-test (paired or unpaired) or repeated measures ANOVA, followed by a Bonferroni test, where appropriate. A value of $p < 0.05$ was considered to be statistically significant.

2.8. Drugs

All drugs and chemicals were obtained from Sigma-Aldrich (St. Louis, MI, USA) except for iberiotoxin (IBTX, Tocris Bioscience, Bristol, UK).

3. Results

3.1. Outward Potassium Currents in hiPSC-CMs, Appearance of $I_{BK,Ca}$

Large, transient outward currents were elicited in hiPSC-CMs (Figure 1A) by depolarizing test pulses. We found in several C25-hiPSC-CMs a large, inactivating outward current followed by a late sustained current with an irregular shape during the entire depolarizing test pulse. The irregular-shaped "noisy" currents were similar to BK_{Ca} currents, which were reported previously in mesenchymal stem cells [26]. Similar to that report we used rather high test pulse potentials (+70 mV) from physiological resting membrane potential of −80 mV and increased the free Ca^{2+} concentration of the pipette solution from 2 to 4.4 mM to facilitate the detection of BK_{Ca} [26]. The selective $I_{BK,Ca}$ blocker IBTX (100 nM) was used to identify the $I_{BK,Ca}$ (Figure 1A). Of the hiPSC-CMs 76% (19 out of 25) showed IBTX-sensitive outward currents suggesting the presence of BK_{Ca}; the area under the curve was reduced by IBTX from 30.6 ± 5.1 pAs/pF to 20.2 ± 4.1 pAs/pF ($n = 19$, $p < 0.0001$, paired *t* test, Figure 1B). IBTX inhibited both the peak and late current density (peak from 82.9 ± 11.5 pA/pF to 44.8 ± 7.6 pA/pF, $p < 0.0001$, Figure 1C and late: from 29.2 ± 5.4 pA/pF to 17.8 ± 3.4 pA/pF; $n = 19$, $p = 0.004$, Figure 1D). In IBTX-insensitive hiPSC-CMs, baseline values of outward peak and late currents were smaller compared to IBTX-sensitive hiPSC-CMs. Furthermore, in hiPSC-CMs without the irregular-shaped outward current IBTX did not change peak or late currents (peak: 45.4 ± 6.4 pA/pF baseline vs. 43.4 ± 3.9 pA/pF IBTX, $p = 0.594$, $n = 6$ and late: 8.3 ± 2.6 pA/pF baseline vs. 7.5 ± 1.8 pA/pF IBTX, $n = 6$, $p = 0.363$; paired *t* test). HiPS-CMs from the commercially available iCell cell line did not show irregular-shaped "noisy" currents or IBTX-sensitivity (Supplementary Materials Figure S1).

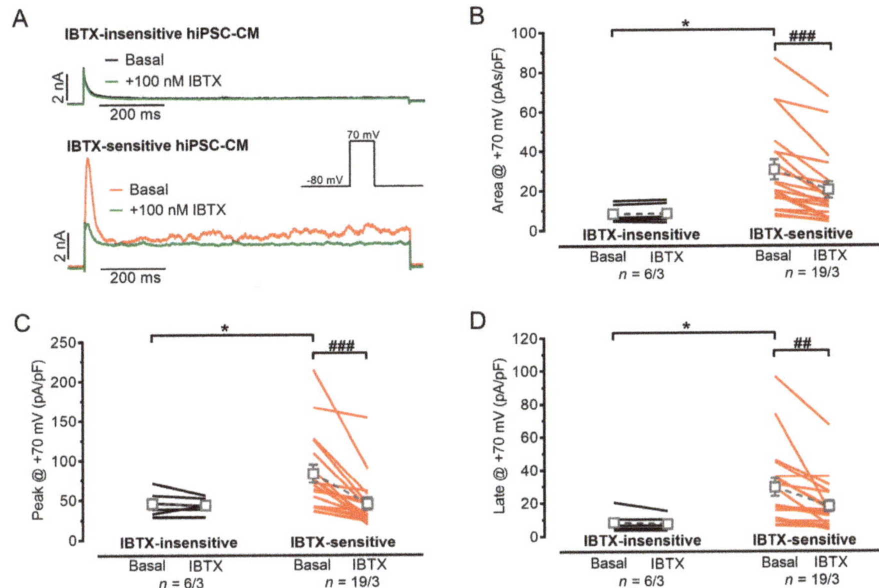

Figure 1. Outward currents in C25 human induced pluripotent stem cell-derived cardiomyocytes (hiPSC-CMs) and the effect of iberiotoxin (IBTX). (**A**) Original outward traces before (black) and after (green) exposure of 100 nM IBTX in insensitive (upper, black directly underlying green curve) and sensitive (lower panel) hiPSC-CMs. (**B–D**). Summary of IBTX (100 nM) effects in insensitive (left panel) and sensitive (right panel) hiPSC-CMs quantified by area under the curve (**B**), peak current (**C**) and current at the end of the test pulse (late current, **D**). Mean values ± SEM. * $p < 0.05$, unpaired Student's *t* test for basal values in insensitive vs. sensitive hiPSC-CMs; ## $p < 0.01$, ### $p < 0.001$; paired Student's *t* test for basal vs. IBTX; n = number of isolated cells/number of individual differentiation batches.

3.2. Action Potentials with Strong Initial Repolarization and Oscillations Are Sensitive to Iberiotoxin

APs recorded in C25-EHTs, exhibiting the IBTX-sensitive irregular-shaped outward current, showed a pronounced initial repolarization ("notch") below the later plateau level of the AP and the initial notch was followed by a (very) early afterdepolarization (Figure 2). In some APs the notch was followed by a peculiar oscillation during plateau phase of the AP. This peculiarity of notch/oscillation was only observed in some of independent differentiation batches of the C25 cell line, but never in any other of three in-house cell lines or five commercial cell lines investigated with sharp microelectrodes as previously described [5,9,27–30]. When the notch/oscillation was detected in one EHT, all impalements showed this peculiarity including all EHTs from this differentiation batch. From all C25 batches investigated with sharp microelectrode, we detected seven with notch/oscillations and 11 without. The passage number of individual differentiation batches was not significantly different with or without notch/oscillations (77.1 ± 5.9, $n = 7$ vs. 67.9 ± 7.9, $n = 9$; $p = 0.391$) as well as the hiPSC-CM differentiation efficiency (81% ± 5% vs. 85% ± 3% TnT + cells, $p = 0.51$). In addition, spontaneous beating frequency was not different between EHTs with or without notch/oscillations (1.09 ± 0.19 Hz, $n = 9$ vs. 0.92 ± 0.07 Hz; $n = 14$; $p = 0.34$). EHTs were treated with 300 nM ivabradine to allow pacing at 1 Hz as previously described [5]. Under these conditions, APD_{90} was slightly longer in EHTs with notch/oscillations than without (279 ± 12 ms, $n = 18/9/6$ vs. 226 ± 6 ms, $n = 30/17/9$; $p = 0.0044$). Take-off potential, upstroke velocity and AP amplitude were not significantly different. The number of oscillations within one AP varied between 1 and 6. In case of several oscillations, there was a rather constant cycle length of oscillations in different EHTs (22.8 ± 0.9 ms, $n = 11$). In case of multiple oscillations, the amplitude decreased with each subsequent oscillation (ranging from 50 to

10 mV, Figure 2B). We used the specific inhibitor IBTX (100 nM) in order to evaluate whether oscillations originate from $I_{BK,Ca}$. To quantify effects of IBTX on early repolarization we analyzed the membrane potential when the initial repolarization reached its lowest point (on average 9.2 ± 0.7 ms after AP upstroke).

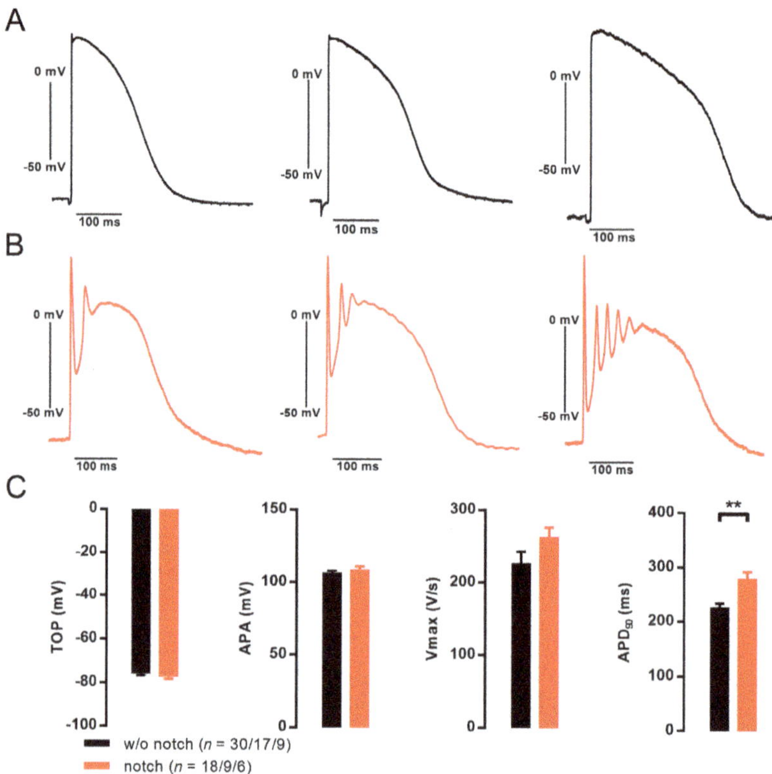

Figure 2. Pronounced notch with oscillation in the plateau phase of action potential (AP) recorded from engineered heart tissue (EHT) derived by C25 human induced pluripotent stem cell-derived cardiomyocytes (hiPSC-CMs). (**A**) Original AP recordings without (**A**) and with (**B**) notch followed by oscillating afterdepolarizations in EHT from cell line C25. (**C**) Summary of the results for take-off potential (TOP), AP amplitude (APA), maximum upstroke velocity (V_{max}), AP duration at 90% repolarization (APD$_{90}$), ** $p < 0.01$, unpaired Student's *t* test; n = number of impalements/number of EHTs/number of individual differentiation batches.

In case of notch/oscillations, IBTX (100 nM) lifted the membrane potential at this time point from −43.5 ± 5.6 to −13.2 ± 8.3 mV (n = 7, p = 0.001, paired *t*-test). APD$_{10}$, APD$_{20}$, APD$_{50}$ and APD$_{70}$ but not APD$_{90}$ was significantly prolonged by IBTX (Figure 3). IBTX did not affect AP duration in EHTs without notch/oscillations. In addition, IBTX did not show any effect on AP in human LV tissue (Figure 3).

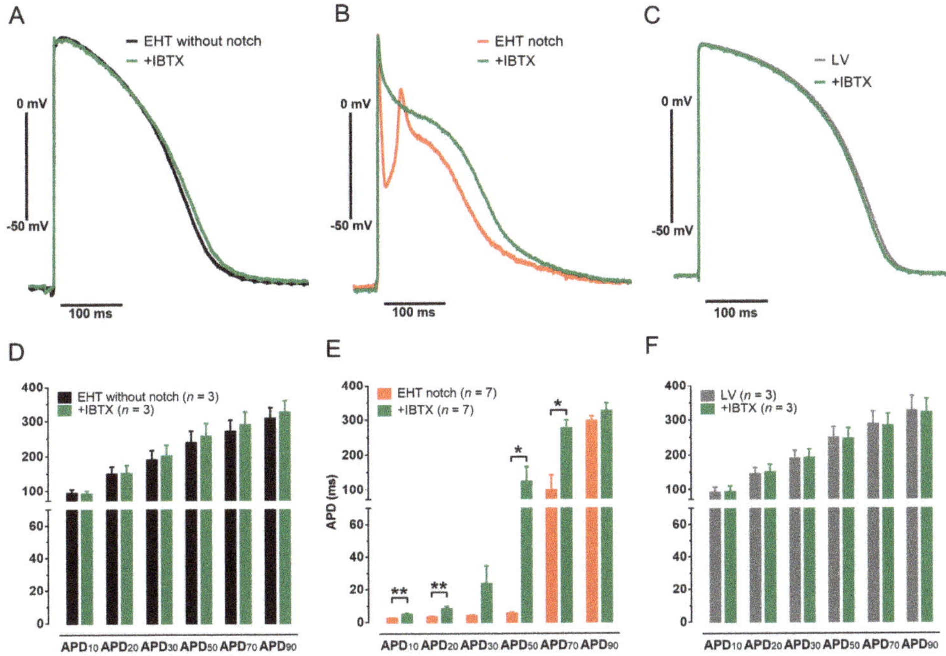

Figure 3. Effects of iberiotoxin (IBTX) on action potentials (APs) in C25 engineered heart tissue (EHT) and in human left ventricular tissue (LV). Original AP recordings before and after exposure to IBTX (100 nM, green) in EHTs with (**A**) and without (**B**) notch/oscillations from cell line C25 and in LV (**C**). Mean values of AP duration at different levels of repolarization (APD$_{10-90}$) before and after exposure to IBTX (100 nM) in EHTs with (**D**) and without (**E**) notch in cell line C25 and in LV (**F**). Mean values ± SEM, * $p < 0.05$, ** $p < 0.01$, paired Student's *t* test.

3.3. EHT with Notch/Oscillations in the AP Showed Large BK$_{Ca}$ Expression

The BK$_{Ca}$ channel is formed as a tetramer of the channel forming alpha-subunit K$_{Ca}$1.1 encoded by the *KCNMA1* gene. We performed retrospectively expression analysis in order to investigate whether there was an association between appearance of oscillations and the expression level of *KCNMA1*. The mRNA level for the channel forming alpha-subunit K$_{Ca}$1.1 showed very low expression in human LV tissue. The same holds true for the in-house control cell line ERC018 and the commercially available cell lines iCell and iCell^2, where notch or oscillations in the AP have never been detected [5]. In contrast, all EHTs from C25 from which we could record notch/oscillations showed a substantial increase in *KCNMA1* mRNA, which is encoding for K$_{Ca}$1.1 (Figure 4). In addition, RNA sequencing showed a 12-fold higher expression level for *KCNMA1* in C25 than in the control cell line ERC001 (Supplementary Figure S4). Other genes related to *KCNMA1* did not show major alterations in expression. Immunofluorescence analysis of hiPSC-CMs revealed the expression of the alpha-1 subunit of the BK$_{Ca}$ channel in the C25 cell line with enhanced signal intensity at cell-cell contacts, whereas no signal was detected in the control cell line [30] (Supplementary Figure S5).

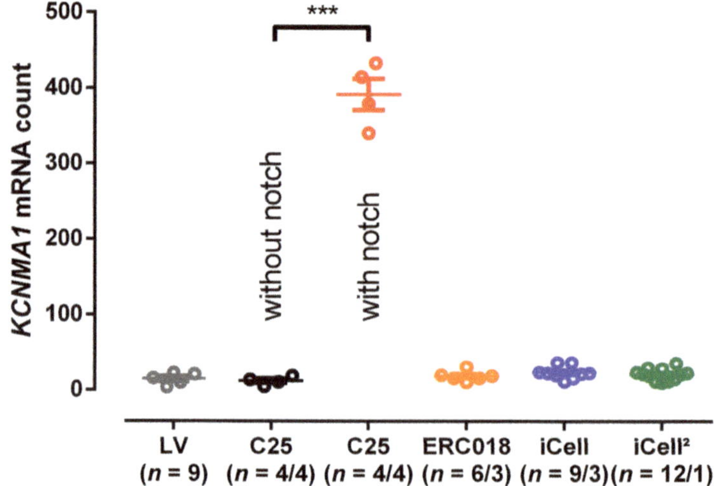

Figure 4. Expression analysis for the alpha subunit (*KCNMA1*) of big conductance calcium activated potassium channel. Individual expression levels of *KCNMA1* in left ventricular tissue (LV) and engineered heart tissue (EHT) from cell line C25 (without and with notch/oscillations), ERC018, iCell and iCell². Mean values ± SEM, *** $p < 0.001$, 1-way ANOVA with multiple comparison, n = number of patients for LV and n = number of EHT/number of individual differentiation batches.

3.4. Computer Simulations Corroborate the Effect of the Non-Cardiac BK_{Ca} to Human AP

To evaluate the impact of BK_{Ca} on repolarization quantitatively, we integrated a BK_{Ca} model into a human ventricular CM model. Simulation results in Figure 5 demonstrated that a reasonably sized BK_{Ca} current can cause a deep notch in the AP. The dynamics of the notch (Figure 5A) match the kinetics of the simulated BK_{Ca} current (Figure 5B). Although inclusion of BK_{Ca} did not cause oscillations of the membrane voltage similar to in vitro observations, the deep notch enhanced drastically L-type Ca^{2+} current ($I_{Ca,L}$, Figure 5C).

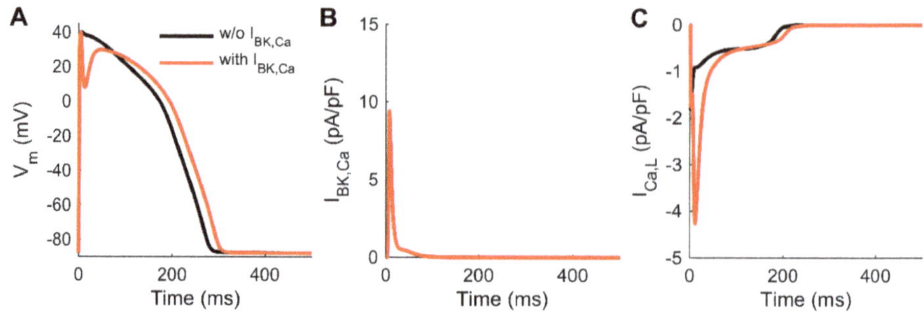

Figure 5. Impact of including BK_{Ca} in a mathematical model of human ventricular cardiomyocytes. BK_{Ca} current (**B**) causes a deep notch in the action potential (**A**), and a substantial increase in the amplitude of L-type Ca^{2+} current (**C**).

4. Discussion

The main findings of this study were:

(1)   HiPSC-CMs could express ion channels otherwise not found in adult cardiomyocytes.

(2)   In human adult ventricular myocardium, BK_{Ca} did not contribute to AP shape.

(3)   In hiPSC-CMs BK_{Ca} could induce irregular AP shapes with oscillations resembling very early after depolarizations.

BK_{Ca} channels are widely expressed in various cell types, including electrically excitable and non-excitable cells [31]. Due to the Ca^{2+}-sensitivity, they provide relevant negative feedback mechanism in the regulation of intracellular Ca^{2+} elevation and membrane potential [32]. Almost every cell type expresses BK_{Ca} in the inner mitochondrial membrane (BK_{mito}) [33]. In CMs, pharmacological opening of BK_{mito} reduces ischemia reperfusion injury [34]. Sarcolemmal expression of BK_{Ca} is typically found in vascular smooth muscle cells, regulating myogenic tone and thereby blood flow. In CMs from rodents and cardiac tissue from humans, sarcolemmal expression of BK_{Ca} is very low or almost non-existent [35–37]. Here we confirm low expression levels of BK_{Ca} in human LV myocardium. Among CMs from other species, IBTX sensitive currents were found only in cultured embryonal chicken CMs [38]. Contribution to repolarization in this species is unclear. In rat ventricular myocardium, IBTX does not affect AP shape [39]. Here, we demonstrated for the first time that human ventricular APs were insensitive to IBTX. In consequence, the detection of $I_{BK,Ca}$ in C25-hiPSC-CMs during patch clamp recordings was unexpected and surprising, since BK_{Ca} is not known to be expressed in the sarcolemma. The effectivity of the selective blocker of $I_{BK,Ca}$ [26,40], IBTX, which only binds from the external side of the channel [41], confirmed the hypothesis that a non-cardiac channel is present and active at transmembrane level in the hiPSC-CMs. These findings were also confirmed by expression analysis in which the existence of the notch/oscillation in the AP correlated with the high expression levels of *KCNMA1*, encoding for the alpha-subunit ($K_{Ca}1.1$) of the BK_{Ca}. In contrast, *KCNMA1* expression was low in human LV tissue and correspondingly IBTX did not show any effect on APs from human LV, which to our knowledge has not been described before.

Recently, overexpression of non-cardiac BK_{Ca} in CMs was proposed as a treatment for LQTS [13], since BK_{Ca} is a hyperpolarizing channel, which might shorten human AP. Support for this idea came from a study describing the electrophysiological function of BK_{Ca} in HL-1 cells by viral overexpression, a cell line derived from a murine atrial tumor. BK_{Ca} overexpression reduced the very short AP of this murine model by 50% (APD_{90} from 30 to 14 ms) and was proposed as a potential genetic therapy to reduce AP duration (APD) of the LQT syndrome [13]. In contrast to the experiments in HL-1 cells, we observed that in hiPSC-CMs, the presence of BK_{Ca} induces oscillations in the early plateau phase and no speed-up in the final repolarization [13].

The pronounced initial repolarization could be confirmed by the in-silico integration of the $I_{BK,Ca}$ in modeling ventricular myocyte AP (Figure 5). However, the inclusion of BK_{Ca} could not resemble oscillations of the membrane voltage. Nevertheless, we would expect that oscillation might be induced by an alternating feedback mechanism [32] of the $I_{BK,Ca}$ and the L-type Ca^{2+} current ($I_{Ca,L}$), since the deep notch drastically enhanced L-type Ca^{2+} current (Figure 5C). The afterdepolarizations following the initial repolarization might be also due to activation of $I_{Ca,T}$ [11] or the sodium calcium exchanger, however, the exact mechanism remains unclear. More accurate and detailed mathematical modeling would require in vitro data on spatial distribution and localization of the BK_{Ca} and Ca^{2+} channels, which is beyond the scope of this study.

Afterdepolarizations might complicate the evaluation of how the BK_{Ca} affects APD. In vitro, there was a slightly longer APD_{90} when BK_{Ca} was present, which could be confirmed in silico. However, the inhibition of $I_{BK,Ca}$ by iberiotoxin in AP with notch/oscillations did not significantly alter APD_{90}, averaged values tended even to longer APD_{90}. The apparent differential contribution of BK_{Ca} to the APD revealed at baseline level or by pharmacological intervention might be due to remodeling of other ion channels downstream to BK expression or due to potential off-target effects of iberiotoxin. Taken together, our results do not support the idea that BK_{Ca} overexpression can cure LQTS in humans, since BK_{Ca} might lead to afterdepolarization and arrhythmia.

Previously, it was shown that the $I_{BK,Ca}$ current contribute to outward currents in the murine sinoatrial node and the selective blocker paxilline decreased beating rate by more than 50% [42].

However, the exact role of $I_{BK,Ca}$ in pacemaking is widely unclear, since substantial decrease in diastolic depolarization by paxilline does not fit to very small contribution of BK_{Ca} to total potassium outward currents activated at positive membrane potentials. Nevertheless, spontaneous beating is a peculiarity of hiPSC-CMs and the autonomic activity of hiPSC-CMs is not fully understood. Therefore, it seems reasonable to speculate that BK_{Ca} may be involved in pacemaking of hiPSC-CMs. However, we would not expect a large impact of $I_{BK,Ca}$ to pacemaking in EHT, since beating rate in EHT with and without expressing BK_{Ca} did not differ.

The reason for the unexpected expression of BK_{Ca} in individual differentiation batches of a single hiPSC-CM cell line (C25) is very difficult to evaluate retrospectively. Cell line C25 did not show chromosomal anomalies at passage number 40 and 92 (Supplementary Figure S3). Since 6 out of 7 individual differentiation batches were done from lower passage numbers than 92, it seems that BK expression was not the consequence of karyotype abnormalities. As notch/oscillation were observed also in preparations with very high differentiation efficiencies (90% and 93%) and hiPSC-CM fraction in EHT was even enriched in comparison to 2D culture (Supplementary Materials Figure S2B), we are confident that BK expression is not due to an extraordinary high fraction of non-cardiac cells within the EHT. Strong batch effects, as either all or none of the EHTs from one preparation depicted signs of BK_{Ca} expression, argue for an upregulation of BK_{Ca} due to events occurring during stem cell culture and cardiac differentiation. There are two factors that might have raised the likelihood for spontaneous mutations in the stem-cell culture. The C25 cell line was reprogrammed by lentivirus and it was passaged to very high number (up to 107 passages). To avoid these factors, we changed to Sendai virus and restricted passage number for future experiments. Since *KCNMA1* related genes did not show major alterations in contrast to *KCNMA1* itself, we would account a genetic alteration more likely than an upregulation due to regulatory pathways. In addition, various reports show that iPS-cells frequently acquire genetic alterations in cell culture. Kilpinen et al. [43] showed that chromosome 10, harboring BK_{Ca}/*KCNMA1*, was among the most susceptible loci to copy number alterations. In addition, in a study searching for variants that provide mutated cells with a growth advantage in culture, *KCNMA1* candidate mosaic variants were identified in two independent hES cell lines [44]. Thus, a genetic alteration leading to a reoccurring overgrowth of BK misexpressing hiPSC is the most likely explanation for our finding. Regulatory expression profiling might reveal this change of BK_{Ca} activity in advance.

5. Conclusions

Our results clearly demonstrated that hiPSC-CMs could possess even non-cardiac ion channels affecting AP waveform causing afterdepolarizations and oscillations in the AP. This might serve as an example that iPS cell culture could lead to genetic alterations with functional consequences. Therefore, we felt that cell culture and differentiation protocols should be standardized as much as possible and that expression of non-cardiac sarcolemmal ion channels should be considered before hiPSC-CMs are used for pharmacological studies. Screening for the expression of the *KCNMA1* gene might be one potential quality parameter.

6. Limitations

We are aware that the description of the BK_{Ca} in a single control cell line of hiPSC-CMs is of limited transferability, especially since we could not uncover conditions or mechanisms of the unexpected BK_{Ca} expression. Nevertheless, presence of BK_{Ca} is a good example that hiPSC-CMs can express non-cardiac proteins with a huge impact on physiological parameters. Although IBTX has been described as selective for $I_{BK,Ca}$ [41], non-specific effects of IBTX cannot be completely excluded. LV tissue was obtained from patients with heart disease; a potential difference to healthy LV tissue is unclear. We implemented the BK in the previously developed hiPSC-CM model [5], but we failed to induce notch/oscillations in that model. Obviously, our present model does not reflect the cellular ultrastructure at the level of detail that would be required for simulating the putative close proximity of the BK and CaL channels.

Supplementary Materials: The following are available online at http://www.mdpi.com/2073-4409/9/1/253/s1; Figure S1: Outward currents in iCell human induced pluripotent stem cell-derived cardiomyocytes (hiPSC-CMs) and the effect of iberiotoxin (IBTX) and Figure S2: Cardiomyocyte content analyzed by flow cytometry. Figure S3: Karyotype of hiPS-cells. Figure S4: RNA sequencing of BK related genes in hiPSC-CMs. Figure S5: Immunoflurorescence analysis of BK in hiPSC-CMs.

Author Contributions: Conceptualization, A.H. (András Horváth), T.C. and M.D.L.; methodology, A.H. (András Horváth), T.C. and M.D.L.; software, J.T.K., M.S.; validation, T.C. and M.D.L.; formal analysis, M.D.L.; investigation, A.H. (András Horváth), M.P., M.S. and M.D.L.; resources, M.P., A.T.L.Z., M.S., U.S., I.M., B.U. and E.G.; data curation, T.C. and M.D.L.; writing—original draft preparation, T.C. and M.D.L.; writing—review and editing, A.H. (András Horváth), T.C., J.T.K., M.P., A.T.L.Z., I.M., B.U., C.M., E.G., A.H. (Arne Hansen), T.E. and M.D.L.; visualization, T.C. and M.D.L.; supervision, T.C., T.E. and M.D.L.; funding acquisition, T.C., A.H. (Arne Hansen), J.T.K., C.M., T.E. and M.D.L. All authors have read and agreed to the published version of the manuscript.

Funding: This work was supported by the German Centre for Cardiovascular Research (DZHK) and the German Ministry of Research Education (BMBF) to M.D.L. & A.H. (Arne Hansen), the German Research Foundation (DFG, 3423/5-1) to A.H. (Arne Hansen), the European Research Council Advanced Grant (IndivuHeart, 340248) to T.E., the European Union's Horizon 2020 research and innovation programme under the Marie Sklodowska-Curie grant (AFib-TrainNet, 675351) to T.E. and T.C.; the Research Promotion Fund of the Faculty of Medicine (Hamburg) to M.P. and M.D.L. ("Clinician Scientist Program" and "Project funding for young scientists"), the Pirkanmaa Regional Fund of the Finnish Cultural Foundation and the Academy of Finland Centre of Excellence in Body-on-Chip Research to J.T.K.

Acknowledgments: The authors thank Alessandra Moretti and Dennis Schade for their kind contribution of materials. The authors gratefully acknowledge expert technical advice and help in providing hiPSC-CMs and EHTs from Anika Knaust, Tessa Werner, Mirja L. Schulze, Marta Lemme, Anna Steenpass, Thomas Schulze, Birgit Klampe, Lisa Krämer, Aya Domke-Shibamiya and Sandra Laufer. FACS analyses were performed at the UKE FACS Sorting Core Unit.

Conflicts of Interest: I.M., A.H. (Arne Hansen) and T.E. are cofounder of EHT Technologies GmbH, Hamburg.

Abbreviations

AP	Action potential
APD	Action potential duration
APD_{90}	Action potential duration at 90% repolarization
BK_{Ca}	Big conductance calcium activated potassium channel, (Maxi-K, slo1, $K_{Ca1.1}$)
CM	Cardiomyocytes
EAD	Early afterdepolarization
EHT	Engineered heart tissue
hiPSC-CMs	Human induced pluripotent stem cell-derived cardiomyocytes
IBTX	Iberiotoxin
$I_{BK,Ca}$	Big-conductance calcium activated potassium current
$I_{Ca,L}$	L-type calcium current
LV	Left ventricle
RMP	Resting membrane potential

References

1. Moretti, A.; Bellin, M.; Welling, A.; Jung, C.B.; Lam, J.T.; Bott-Flügel, L.; Dorn, T.; Goedel, A.; Höhnke, C.; Hofmann, F.; et al. Patient-Specific Induced Pluripotent Stem-Cell Models for Long-QT Syndrome. *N. Engl. J. Med.* **2010**, *363*, 1397–1409. [CrossRef] [PubMed]

2. Itzhaki, I.; Maizels, L.; Huber, I.; Zwi-Dantsis, L.; Caspi, O.; Winterstern, A.; Feldman, O.; Gepstein, A.; Arbel, G.; Hammerman, H.; et al. Modelling the long QT syndrome with induced pluripotent stem cells. *Nature* **2011**, *471*, 225–229. [CrossRef] [PubMed]

3. Carvajal-vergara, X.; Sevilla, A.; Souza, S.L.D.; Ang, Y.; Lee, D.; Yang, L.; Kaplan, A.D.; Adler, E.D.; Rozov, R.; Ge, Y.; et al. Patient-specific induced pluripotent stem-cell-derived models of LEOPARD syndrome. *Nature* **2010**, *465*, 808–812. [CrossRef] [PubMed]

4. Liang, P.; Lan, F.; Lee, A.S.; Gong, T.; Sanchez-Freire, V.; Wang, Y.; Diecke, S.; Sallam, K.; Knowles, J.W.; Wang, P.J.; et al. Drug screening using a library of human induced pluripotent stem cell-derived cardiomyocytes reveals disease-specific patterns of cardiotoxicity. *Circulation* **2013**, *127*, 1677–1691. [CrossRef]

5. Lemoine, M.D.; Krause, T.; Koivumaki, J.T.; Prondzynski, M.; Schulze, M.L.; Girdauskas, E.; Willems, S.; Hansen, A.; Eschenhagen, T.; Christ, T. Human Induced Pluripotent Stem Cell-Derived Engineered Heart Tissue as a Sensitive Test System for QT Prolongation and Arrhythmic Triggers. *Circ. Arrhythmia Electrophysiol.* **2018**, *11*, e006035. [CrossRef]

6. Weinberger, F.; Breckwoldt, K.; Pecha, S.; Kelly, A.; Geertz, B.; Starbatty, J.; Yorgan, T.; Cheng, K.-H.; Lessmann, K.; Stolen, T.; et al. Cardiac repair in guinea pigs with human engineered heart tissue from induced pluripotent stem cells. *Sci. Transl. Med.* **2016**, *8*, ra148–ra363. [CrossRef]

7. Ma, J.; Guo, L.; Fiene, S.J.; Anson, B.D.; Thomson, J.A.; Kamp, T.J.; Kolaja, K.L.; Swanson, B.J.; January, C.T.; Kl, K.; et al. High purity human-induced pluripotent stem cell-derived cardiomyocytes: Electrophysiological properties of action potentials and ionic currents. *Am. J. Physiol. Heart Circ. Physiol.* **2011**, *301*, 2006–2017. [CrossRef]

8. Herron, T.J.; Da Rocha, A.M.; Campbell, K.F.; Ponce-Balbuena, D.; Willis, B.C.; Guerrero-Serna, G.; Liu, Q.; Klos, M.; Musa, H.; Zarzoso, M.; et al. Extracellular matrix-mediated maturation of human pluripotent stem cell-derived cardiac monolayer structure and electrophysiological function. *Circ. Arrhythmia Electrophysiol.* **2016**, *9*, e003638. [CrossRef]

9. Horváth, A.; Lemoine, M.D.; Löser, A.; Mannhardt, I.; Flenner, F.; Uzun, A.U.; Neuber, C.; Breckwoldt, K.; Hansen, A.; Girdauskas, E.; et al. Low Resting Membrane Potential and Low Inward Rectifier Potassium Currents Are Not Inherent Features of hiPSC-Derived Cardiomyocytes. *Stem Cell Rep.* **2018**, *10*, 822–833. [CrossRef]

10. Vaidyanathan, R.; Markandeya, Y.S.; Kamp, T.J.; Makielski, J.C.; Janaury, C.T.; Eckhardt, L.L. IK1-Enhanced Human Induced Pluripotent Stem Cell-Derived Cardiomyocytes: An Improved Cardiomyocyte Model to Investigate Inherited Arrhythmia Syndromes. *Am. J. Physiol. Heart Circ. Physiol.* **2016**, *310*, H1611–H1621. [CrossRef]

11. Uzun, A.U.; Mannhardt, I.; Breckwoldt, K.; Horváth, A.; Johannsen, S.S.; Hansen, A.; Eschenhagen, T.; Christ, T. Ca^{2+}-currents in human induced pluripotent stem cell-derived cardiomyocytes effects of two different culture conditions. *Front. Pharmacol.* **2016**, *7*, 300. [CrossRef] [PubMed]

12. Ivashchenko, C.Y.; Pipes, G.C.; Lozinskaya, I.M.; Lin, Z.; Xiaoping, X.; Needle, S.; Grygielko, E.T.; Hu, E.; Toomey, J.R.; Lepore, J.J.; et al. Human-induced pluripotent stem cell-derived cardiomyocytes exhibit temporal changes in phenotype. *Am. J. Physiol. Heart Circ. Physiol.* **2013**, *305*, H913–H922. [CrossRef] [PubMed]

13. Stimers, J.R.; Song, L.; Rusch, N.J.; Rhee, S.W. Overexpression of the large-conductance, Ca^{2+}-activated K^+(BK) channel shortens action potential duration in HL-1 cardiomyocytes. *PLoS ONE* **2015**, *10*, e0130588. [CrossRef] [PubMed]

14. Mannhardt, I.; Breckwoldt, K.; Letuffe-Brenière, D.; Schaaf, S.; Schulz, H.; Neuber, C.; Benzin, A.; Werner, T.; Eder, A.; Schulze, T.; et al. Human Engineered Heart Tissue: Analysis of Contractile Force. *Stem Cell Rep.* **2016**, *7*, 1–14. [CrossRef] [PubMed]

15. Lanier, M.; Schade, D.; Willems, E.; Tsuda, M.; Spiering, S.; Kalisiak, J.; Mercola, M.; Cashman, J.R. Wnt Inhibition Correlates with Human Embryonic Stem Cell Cardiomyogenesis: A Structure—Activity Relationship Study Based on Inhibitors for the Wnt Response. *J. Med. Chem.* **2012**, *55*, 697–708. [CrossRef] [PubMed]

16. Breckwoldt, K.; Letuffe-Brenière, D.; Mannhardt, I.; Schulze, T.; Ulmer, B.; Werner, T.; Benzin, A.; Klampe, B.; Reinsch, M.C.; Laufer, S.; et al. Differentiation of cardiomyocytes and generation of human engineered heart tissue. *Nat. Protoc.* **2017**, *12*, 1177–1197. [CrossRef]

17. Ulmer, B.M.; Stoehr, A.; Schulze, M.L.; Patel, S.; Gucek, M.; Mannhardt, I.; Funcke, S.; Murphy, E.; Eschenhagen, T.; Hansen, A. Contractile Work Contributes to Maturation of Energy Metabolism in hiPSC-Derived Cardiomyocytes. *Stem Cell Rep.* **2018**, *10*, 834–847. [CrossRef]

18. Schaaf, S.; Eder, A.; Vollert, I.; Stöhr, A.; Hansen, A.; Eschenhagen, T. Generation of Strip-Format Fibrin-Based Engineered Heart Tissue (EHT). In *Cardiac Tissue Engineering: Methods and Protocols*; Radisic, M., Black, D.L., III, Eds.; Springer: New York, NY, USA, 2014; pp. 121–129. ISBN 978-1-4939-1047-2.

19. Christ, T.; Boknik, P.; Wöhrl, S.; Wettwer, E.; Graf, E.M.; Bosch, R.F.; Knaut, M.; Schmitz, W.; Ravens, U.; Dobrev, D. L-type Ca2+ current downregulation in chronic human atrial fibrillation is associated with increased activity of protein phosphatases. *Circulation* **2004**, *110*, 2651–2657. [CrossRef]

20. Dobrev, D.; Friedrich, A.; Voigt, N.; Jost, N.; Wettwer, E.; Christ, T.; Knaut, M.; Ravens, U. The G protein-gated potassium current IK,ACh is constitutively active in patients with chronic atrial fibrillation. *Circulation* **2005**, *112*, 3697–3706. [CrossRef]

21. Lemoine, M.D.; Mannhardt, I.; Breckwoldt, K.; Prondzynski, M.; Flenner, F.; Ulmer, B.; Hirt, M.N.; Neuber, C.; Horváth, A.; Kloth, B.; et al. Human iPSC-derived cardiomyocytes cultured in 3D engineered heart tissue show physiological upstroke velocity and sodium current density. *Sci. Rep.* **2017**, *7*, 5464. [CrossRef]

22. Tabak, J.; Tomaiuolo, M.; Gonzalez-Iglesias, A.E.; Milescu, L.S.; Bertram, R. Fast-Activating Voltage- and Calcium-Dependent Potassium (BK) Conductance Promotes Bursting in Pituitary Cells: A Dynamic Clamp Study. *J. Neurosci.* **2011**, *31*, 16855–16863. [CrossRef] [PubMed]

23. Lin, M.T.; Hessinger, D.A.; Pearce, W.J.; Longo, L.D. Modulation of BK channel calcium affinity by differential phosphorylation in developing ovine basilar artery myocytes. *Am. J. Physiol. Heart Circ. Physiol.* **2006**, *291*, H732–H740. [CrossRef] [PubMed]

24. Ding, J.P.; Li, Z.W.; Lingle, C.J. Inactivating BK Channels in Rat Chromaffin Cells May Arise from Heteromultimeric Assembly of Distinct Inactivation-Competent and Noninactivating Subunits. *Biophys. J.* **1998**, *74*, 268–289. [CrossRef]

25. O'Hara, T.; Virág, L.; Varró, A.; Rudy, Y. Simulation of the undiseased human cardiac ventricular action potential: Model formulation and experimental validation. *PLoS Comput. Biol.* **2011**, *7*, e1002061. [CrossRef]

26. Heubach, J.F.; Graf, E.M.; Leutheuser, J.; Bock, M.; Balana, B.; Zahanich, I.; Christ, T.; Boxberger, S.; Wettwer, E.; Ravens, U. Electrophysiological properties of human mesenchymal stem cells. *J. Physiol.* **2004**, *554*, 659–672. [CrossRef]

27. Schulze, M.L.; Lemoine, M.D.; Fischer, A.W.; Scherschel, K.; David, R.; Hansen, A.; Eschenhagen, T.; Ulmer, B.M. Biomaterials Dissecting hiPSC-CM pacemaker function in a cardiac organoid model. *Biomaterials* **2019**, *206*, 133–145. [CrossRef]

28. Lemme, M.; Ulmer, B.M.; Lemoine, M.D.; Zech, A.T.L.; Flenner, F.; Ravens, U.; Reichenspurner, H.; Rol-Garcia, M.; Smith, G.; Hansen, A.; et al. Atrial-like Engineered heart tissue: An In vitro Model of the Human Atrium, Stem Cell Reports. *Stem Cell Rep.* **2018**, *11*, 1–13. [CrossRef]

29. Lemme, M.; Braren, I.; Prondzynski, M.; Ulmer, M.; Schulze, M.L.; Ismaili, D.; Meyer, C.; Hansen, A.; Christ, T.; Lemoine, M.D.; et al. Chronic intermittent tachypacing by an optogenetic approach induces arrhythmia vulnerability in human engineered heart tissue. *Cardiovasc. Res.* **2019**. [CrossRef]

30. Prondzynski, M.; Lemoine, M.D.; Zech, A.T.L.; Horvath, A.; Di Mauro, V.; Koivumaki, J.T.; Kresin, N.; Busch, J.; Krause, T.; Kramer, E.; et al. Disease modeling of a mutation in alpha-actinin 2 guides clinical therapy in hypertrophic cardiomyopathy. *EMBO Mol. Med.* **2019**, *11*, e11115. [CrossRef]

31. Jan, L.Y.; Jan, Y.N. Cloned potassium channels from eukaryotes and prokaryotes. *Annu. Rev. Neurosci.* **1997**, *20*, 91–123. [CrossRef]

32. Cui, J.; Yang, H.; Lee, U.S. Molecular mechanisms of BK channel activation. *Cell. Mol. Life Sci. CMLS* **2009**, *66*, 852–875. [PubMed]

33. Singh, H.; Stefani, E.; Toro, L. Intracellular BK Ca (iBK Ca) channels. *J. Physiol.* **2012**, *590*, 5937–5947. [CrossRef] [PubMed]

34. Shi, Y.; Jiang, M.T.; Su, J.; Hutchins, W.; Konorev, E.; Baker, J.E. Mitochondrial big conductance KCa channel and cardioprotection in infant rabbit heart. *J. Cardiovasc. Pharmacol.* **2007**, *50*, 497–502. [CrossRef] [PubMed]

35. Tseng-crank, J.; Godinot, N.; Johansent, T.E.; Ahringt, P.K.; Strobaek, D.; Mertz, R.; Foster, C.D.; Olesen, S.-P.; Reinhart, P.H. Cloning, expression, and distribution of a Ca^{2+} -activated K^{+} channel ß-subunit from human brain. *PNAS* **1996**, *93*, 9200–9205. [CrossRef] [PubMed]

36. Ko, J.H.; Ibrahim, M.A.; Park, W.S.; Ko, E.A.; Kim, N.; Warda, M.; Lim, I.; Bang, H.; Han, J. Cloning of large-conductance Ca2+-activated K+ channel α-subunits in mouse cardiomyocytes. *Biochem. Biophys. Res. Commun.* **2009**, *389*, 74–79. [CrossRef] [PubMed]

37. Tseng-crank, J.; Foster, C.D.; Krause, J.D.; Mertz, R.; Godinot, N.; Dichiara, T.J.; Reinhart, P.H.; Drive, M.; Carolina, N. Cloning, Expression, and Distribution of Functionally Distinct Ca^{2+} -Activated K-Channel Isoforms from Human Brain. *Neuron* **1994**, *13*, 1315–1330. [CrossRef]

38. Liu, S.; Gao, X.; Wu, X.; Yu, Y.; Yu, Z.; Zhao, S.; Zhao, H. BK channels regulate calcium oscillations in ventricular myocytes on different substrate stiffness. *Life Sci.* **2019**, *235*, 116802.

39. Takamatsu, H.; Nagao, T.; Ichijo, H.; Adachi-Akahane, S. L-type Ca^{2+} channels serve as a sensor of the SR Ca^{2+} for tuning the efficacy of Ca^{2+}-induced Ca^{2+} release in rat ventricular myocytes. *J. Physiol.* **2003**, *552*, 415–424. [CrossRef]

40. Son, Y.K.; Hong, D.H.; Choi, T.-H.; Choi, S.W.; Shin, D.H.; Kim, S.J.; Jung, I.D.; Park, Y.-M.; Jung, W.-K.; Kim, D.-J.; et al. The inhibitory effect of BIM (I) on L-type Ca^{2+} channels in rat ventricular cells. *Biochem. Biophys. Res. Commun.* **2012**, *423*, 110–115. [CrossRef]

41. Candia, S.; Garcia, M.L.; Latorre, R. Mode of action of iberiotoxin, a potent blocker of the large conductance Ca(2+)-activated K+ channel. *Biophys. J.* **1992**, *63*, 583–590. [CrossRef]

42. Lai, M.H.; Wu, Y.; Gao, Z.; Anderson, M.E.; Dalziel, J.E.; Meredith, A.L. BK channels regulate sinoatrial node firing rate and cardiac pacing in vivo. *AJP: Heart Circ. Physiol.* **2014**, *307*, H1327–H1338. [CrossRef] [PubMed]

43. Kilpinen, H.; Goncalves, A.; Leha, A.; Afzal, V.; Alasoo, K.; Ashford, S.; Bala, S.; Bensaddek, D.; Casale, F.P.; Culley, O.J.; et al. Common genetic variation drives molecular heterogeneity in human iPSCs. *Nature* **2017**, *546*, 370–375. [CrossRef] [PubMed]

44. Merkle, F.T.; Ghosh, S.; Kamitaki, N.; Mitchell, J.; Avior, Y.; Mello, C.; Kashin, S.; Mekhoubad, S.; Ilic, D.; Charlton, M.; et al. Human pluripotent stem cells recurrently acquire and expand dominant negative P53 mutations. *Nature* **2017**, *545*, 229–233. [CrossRef] [PubMed]

Article

CD271$^+$ Human Mesenchymal Stem Cells Show Antiarrhythmic Effects in a Novel Murine Infarction Model

Haval Sadraddin [1,†], Ralf Gaebel [1,2,*,†], Anna Skorska [1,2], Cornelia Aquilina Lux [1], Sarah Sasse [1], Beschan Ahmad [1], Praveen Vasudevan [1,2], Gustav Steinhoff [1,2,†] and Robert David [1,2,†]

1   Department of Cardiac Surgery, Rostock University Medical Center, 18059 Rostock, Germany; haval.sadraddin@evkln.de (H.S.); anna.skorska@med.uni-rostock.de (A.S.); cornelia.lux@gmx.de (C.A.L.); sarah-sasse@t-online.de (S.S.); beschan.ahmad@med.uni-goettingen.de (B.A.); praveen.vasudevan@med.uni-rostock.de (P.V.); gustav.steinhoff@med.uni-rostock.de (G.S.); robert.david@med.uni-rostock.de (R.D.)
2   Department Life, Light & Matter (LL&M), University of Rostock, 18057 Rostock, Germany
*   Correspondence: ralf.gaebel@med.uni-rostock.de; Tel.: +49-381-4946105; Fax: +49-381-4946214
†   Authors contributed equally to this work.

Received: 12 September 2019; Accepted: 19 November 2019; Published: 21 November 2019

Abstract: Background: Ventricular arrhythmias (VA) are a common cause of sudden death after myocardial infarction (MI). Therefore, developing new therapeutic methods for the prevention and treatment of VA is of prime importance. Methods: Human bone marrow derived CD271$^+$ mesenchymal stem cells (MSC) were tested for their antiarrhythmic effect. This was done through the development of a novel mouse model using an immunocompromised Rag2$^{-/-}$ γc$^{-/-}$ mouse strain subjected to myocardial "infarction-reinfarction". The mice underwent a first ischemia-reperfusion through the left anterior descending (LAD) artery closure for 45 min with a subsequent second permanent LAD ligation after seven days from the first infarct. Results: This mouse model induced various types of VA detected with continuous electrocardiogram (ECG) monitoring via implanted telemetry device. The immediate intramyocardial delivery of CD271$^+$ MSC after the first MI significantly reduced VA induced after the second MI. Conclusions: In addition to the clinical relevance, more closely reflecting patients who suffer from severe ischemic heart disease and related arrhythmias, our new mouse model bearing reinfarction warrants the time required for stem cell engraftment and for the first time enables us to analyze and verify significant antiarrhythmic effects of human CD271$^+$ stem cells in vivo.

Keywords: arrhythmia; electrocardiography; cardiac regeneration; stem cells

---

1. Introduction

Deaths arising from ventricular arrhythmias (VA) after acute cardiac events, in Germany, have increased by 100% over the last twenty years [1]. Therefore, new therapeutic methods for the prevention of VA following myocardial infarction (MI) need to be developed. There are reports that have indicated that transplanting of various stem cell populations post MI can influence the heart rhythm. Some have been defined as pro-arrhythmogenic cells, i.e., skeletal myoblasts, while others have shown antiarrhythmic effects, such as embryonic cardiomyocytes and mesenchymal stem cells (MSC) [2–5]. Our research group has focused on the role of human bone marrow (BM) derived hematopoiesis such as MSC in regenerative medicine [6–9]. In previous studies we have shown that it is safe to use CD117$^+$ stem cells for myocardial regeneration, as their electrophysiological properties make them very unlikely to produce hazardous action potentials or contribute to arrhythmias [10,11].

However, the availability of a realistic small animal model reflecting the situation in patients is of utmost importance. For testing the rhythmological behavior of human cells, $Rag2^{-/-}\ \gamma c^{-/-}$ has been selected. These mice (strain C, 129S4-$Rag2^{tm1.1Flv}Il2rg^{tm1.1Flv}$/J) have been derived from a V17 embryonic stem cell line (BALB/c × 129 heterozygote) targeted for the Rag2 and Il2rg genes and lack B, T, and natural killer (NK) cells [12]. They have a deletion of the common cytokine receptor γ chain (γc) gene, and thus reduced numbers of peripheral T and B lymphocytes, and absent natural killer cell (NK) activity. A genetic cross with a recombinase activating gene 2 (RAG2) deficient strain produced mice doubly homozygous for the γc and RAG2 null alleles ($Rag2^{-/-}\ \gamma c^{-/-}$) with a stable phenotype characterized by the absence of all T lymphocyte, B lymphocyte, and NK cell function [13]. This phenotype makes the completely a lymphoid strain useful for studies on human tissue xenotransplantation [14].

As previously shown, stem cell antiarrhythmic effects appear after completion of an approximately one-week engraftment period [4]. The major objective of the present study was to develop a mouse model that enables us to reproduce VA several days after an initial MI. In vivo electrophysiological mouse models are well established: Previously, Berul et al. established an open chest epicardial study of conduction properties with the ability to induce second and third degree heart blocks, but no induction of tachyarrhythmias after programmed stimulation [15]. Hagendorff et al. used a rapid transesophageal atrial stimulation with the ability to induce some types of heart block and atrial tachyarrhythmias with induction of VA [16]. Roell et al. used a method relying on burst and extra stimulus pacing mediated induction of VA, leading to induction of ventricular tachycardia (VT) in 38.9% of the non-infarcted control group and 96.4% of the infarcted group [4].

This mouse model can also be considered representative for patients that suffer from severe ischemic heart disease and related arrhythmias. In this respect the application of the second MI at various points in time will be of interest.

2. Materials and Methods

2.1. Bone Marrow Aspiration

Informed donors gave written consent to the aspiration of their BM according to the Declaration of Helsinki. The ethical committee of the University of Rostock approved the presented study (registered as no. A201023) as of 29 April 2010. The BM samples were obtained by sternal aspiration from patients undergoing coronary artery bypass graft surgery at Rostock University, Germany. The donors had no BM or hematological diseases and were not receiving any immune suppressing medication. The amount of the aspirated BM ranged from 60–100 mL. The BM samples were received from four donors. Two donors got more than 200,000 isolated CD271^{+} MSC in their BM samples and each one has been used for 2 mice (100,000 cells/animal). Anticoagulation was achieved by heparinization with 250 i.E./mL sodium heparine (B. Braun Melsungen AG, Melsungen, Germany).

2.2. Cell Isolation

The mononuclear cells (MNC) were isolated by density gradient centrifugation using Lymphocyte Separation Medium (LSM; 1.077g/L, PAA Laboratories GmbH, Pasching, Austria). The MNC were indirectly labeled with CD271 APC and anti-APC micro beads (Miltenyi Biotec, Bergisch Gladbach, Germany) and CD271^{+} cells were enriched by positive magnetic selection using the magnet activated cell sorting (MACS) system (Miltenyi Biotec).

2.3. Flow Cytometric Analysis

Purity and viability of all cell isolations were verified by flow cytometry. Antibodies were all mouse anti-human and appropriate mouse isotype antibodies were applied for validation of our gating strategy. Anti-CD271 allophycocyanine was obtained from Miltenyi Biotec. Anti-CD45 allophycocyanin-H7, anti-CD45 Horizon V500, anti-CD29 allophycocyanin, anti-CD44 peridinin-chlorophyll-protein Cyanine 5.5 tandem dye, anti-CD73 phycoerythrin, as well as 7-aminoactinomycin were from BD Biosciences

(Heidelberg, Germany) and anti-CD105 Alexa Fluor 488 was purchased from AbD Serotec GmbH (Puchheim, Germany). Near-IR live dead stain and 4′,6-diamidino-2-phenylindole (DAPI) were obtained from Thermo Fisher Scientific (Waltham, Massachusetts, USA).

Cells were suspended in MACS buffer containing PBS, 2mM EDTA, and 0.5% bovine serum albumin. To reduce unspecific binding, FcR blocking reagent (Miltenyi Biotec) was added to all samples. Cells were incubated with the antibodies for 10 min in the dark at 4 °C. RBC lysis buffer (red blood cell, eBioscience GmbH, Frankfurt/Main, Germany) was added for 10 min on ice. Samples were analyzed by BD LSRII flow cytometer and data were analyzed using FACSDiva software, version 6.1.2 (both Becton Dickinson, Franklin Lakes, New Jersey, USA). For optimal multicolor setting and correction of the spectral overlap, single stained controls were utilized, and gating strategy was performed with matched isotype/fluorescence minus one control. The $CD271^+$ cells with a low granularity (SSC^{low}) were included for further flow cytometric analysis and positivity for mesenchymal markers such as CD29, CD44, CD73, CD105, and CD271 was evaluated based on viable $CD45^-$ cells in an established 6-fold staining, as previously described [17]. After performing antibody staining, as described above, 15 µM DAPI was added; cells were incubated for 2 min and then immediately acquired.

2.4. Animals

All animal procedures were performed according to the guidelines from Directive 2010/63/EU of the European Parliament on the protection of animals used for scientific purposes. The federal animal care committee of LALLF Mecklenburg-Vorpommern (Germany) approved the study protocol (approval number LALLF M-V/TSD/7221.3-1-020/14). The $Rag2^{-/-}\gamma c^{-/-}$ mice (strain C, $129S4\text{-}Rag2^{tm1.1Flv}Il2rg^{tm1.1Flv}/J$) were purchased from the Jackson Laboratory (USA).

2.5. Ambulatory ECG Monitoring and Ligation of the Left Anterior Descending Artery

For ambulatory electrocardiogram (ECG) monitoring, twelve-to-fourteen-week-old female $Rag2^{-/-}\gamma c^{-/-}$ mice ($n = 22$) were implanted with a telemetric device (Data Sciences International, DSI, New Brighton, Minnesota, USA) 7 days before performing the myocardial intervention. The mice were anesthetized with pentobarbital (50 mg/kg, intraperitoneal). An incision was made on the left thorax along the ribs to insert a telemetric transmitter (TA11ETA-F10 Implant; DSI) into a subcutaneous pocket with paired wire electrodes placed over the thorax with leads tunneled to the right upper and left lower thorax. The heart rate, PR interval and QRS complex as the entities of an ECG were recorded using Ponemah Physiology Platform (DSI) until 48 h post infarction. Telemetric ECG signal was qualitatively analyzed, and VA were detected and counted with ECG auto 1.5.11.26 software (EMKA Technologies, Paris, France).

Mice were randomly assigned into three groups: (untreated reinfarction, URI; stem cell treated reinfarction, SRI; and MI control, MIC). First, animals of all three groups underwent thoracotomy, as well as the ligation of the left anterior descending (LAD) artery. After 45 min, each mouse received an intramyocardial application of 20 µL BD Matrigel™ Matrix (BD Biosciences, San Jose, CA, USA) and MACS Buffer in a ratio of 1/2. For stem cell treatment (SRI group) a total of 100,000 $CD271^+$ MSC were injected at 4 sites along the border of the blanched myocardium (25,000 cells per injection point). Subsequently, the node was reopened (here, ischemia-reperfusion) leaving the ligature in the heart looped around the LAD with their ends kept inserted in a soft flexible tube placed subcutaneously (Figure 1A).

After an observation period of 7 days, every mouse underwent a second thoracotomy and, subsequently, animals of URI, as well as the SRI group, underwent permanent LAD ligation at the same site and using the suture from the first ligation (here, reinfarction). The rubber tube has facilitated finding of the suture material which was easily exempted from the surrounding tissue. MIC operated mice underwent an identical second surgical procedure without LAD ligation.

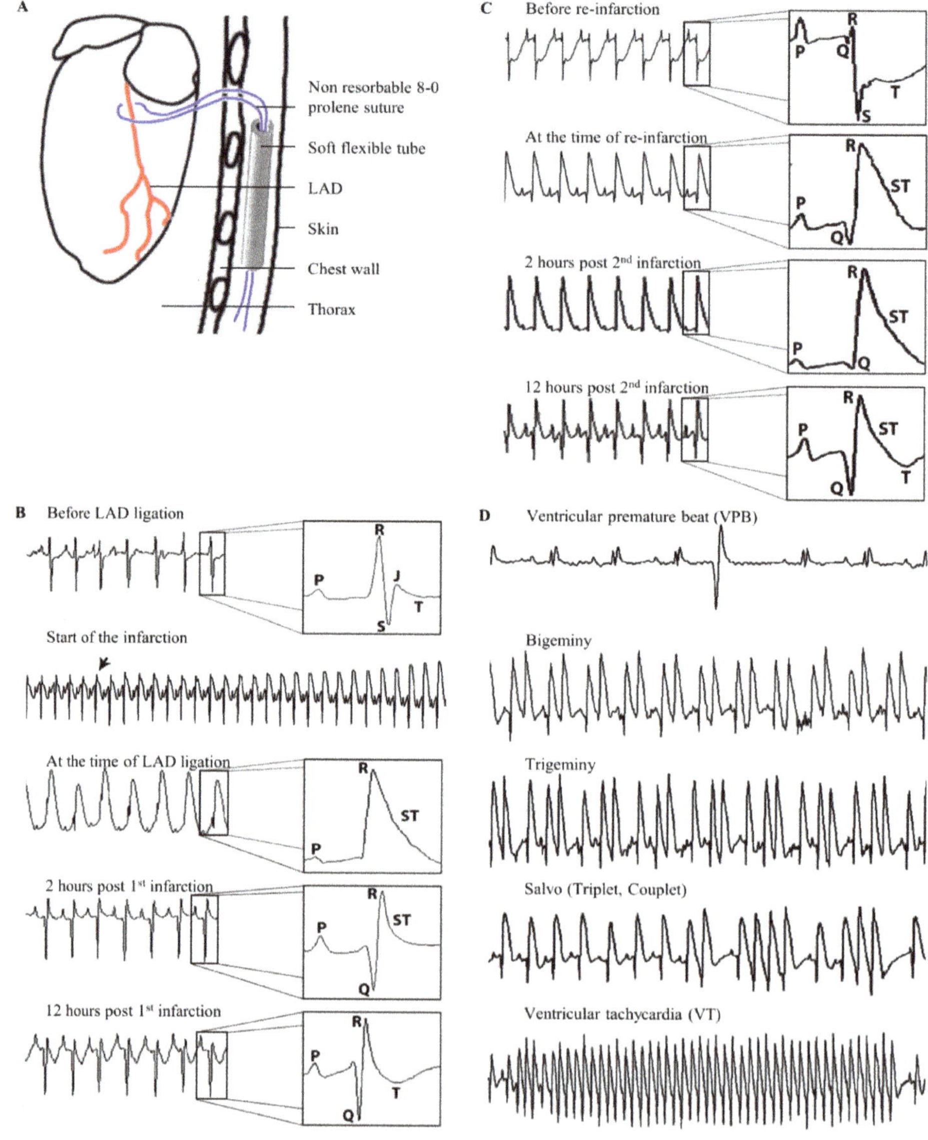

Figure 1. Induction of ventricular arrhythmias (VA). Schematic drawing of the loose left anterior descending (LAD) ligature positioning (**A**). Mouse ECG changes in relation to the time of the firs LAD ligation during ischemia-reperfusion infarction (**B**). Mouse ECG changes in relation to the time of the permanent second LAD-ligation (reinfarction and URI group **C**). Mouse ECG strips showing different types of observed VA (**D**).

2.6. Basic ECG Parameters

To evaluate the antiarrhythmic mechanism of the CD271$^+$ MSC the basic ECG parameters of the mouse groups were measured at different time points. This was performed automatically with the use of ecgAUTO v3.3.0.28 software after manual definition of the start of P wave, the start of Q wave, R wave, end of S wave, and end of T wave. Then, the analysis was done for many subsequent beats at a specific time point. We measured the RR-interval, QRS duration, and QT-interval in milliseconds. The corrected QT-intervals (QTc) were measured using the formula $QTc = QT/(RR/100)^{1/2}$, as previously described [18]. The basic ECG parameters were measured at the time points (1) 48 h after the first LAD ligation, (2) immediately before, and (3) 48 h after the second intervention.

2.7. Definitions of Various Ventricular Arrhythmias

The first temporary occlusion of the LAD resulted in a rapid onset of hyper acute T waves along with ST-segment elevation, as observed in the ECG record. Through reopening of the LAD 45 min post ligation, and subsequent myocardial reperfusion, ST-segment returned back to the isoelectric line and a development of Q waves could be observed. Figure 1B shows the recorded ECG before LAD occlusion, at the time of LAD occlusion, immediately after reopening of the LAD and reperfusion, as well as 12 h after the induction of the MI with clear Q waves and recognizable T waves.

Figure 1C shows the onset of the second infarction after permanently ligating the suture already placed at the time of the first ischemia one week earlier. The ECG immediately prior to performing reinfarction through the second LAD ligation shows deep S waves with deeply depressed ST-segment 7 days after the first operation (ischemia-reperfusion). The induction of the second MI with permanent ligation of the LAD led into an immediate ST-segment elevation in the ECG with subsequent development of deep Q waves. Two hours after reinfarction the ECG shows, a still elevated ST-segment reflecting the permanent LAD ligation. At 12 h post the second MI, Q waves developed with still elevated ST-segment with reappearance of T waves.

Both, the first and second infarctions resulted in the development of different types of VA, namely ventricular premature beats (VPB), ventricular bigeminies and trigeminies (BG/TG), ventricular salvos, as well as ventricular tachycardia (VT) (Figure 1D).

We have analyzed, named, and defined the observed VA in this mouse model in accordance with a standard that fits the previously established guidelines for labeling and describing each type of VA [19] as follows:

2.7.1. Ventricular premature beats

We have defined VPB, as a ventricular electrical complex which varies in voltage (i.e., height) or duration (i.e., width) from the preceding non-VPB ventricular complex and occurs prematurely in relation to it. This means that VPB are not preceded by a P wave if they appear during the early cardiac cycle or they have shorter PR-intervals than those of non-VPB if they appear later in the cardiac cycle.

2.7.2. Bigeminies and trigeminies

A minimum sequence of VPB, normal sinus beat, and VPB repeated at least three times has been defined as bigeminy. The number of repetitions has been defined as three in our model. Additionally, we have defined a sequence of VPB, two normal sinus beats, and a consecutive VPB with a repetition of three times as trigeminy.

2.7.3. Salvos

The observation of two or three consecutive VPB has been defined as salvo [19].

2.7.4. Ventricular tachycardia

VT is defined as a sequence of four or more consecutive VPB [19,20]. This varies between three or more consecutive VPB to ten or more consecutive VPB [21,22]. A detailed description of the morphological and durational subtypes and classifications of VT has not been performed in these mouse models.

2.8. Organ Harvesting

Every mouse underwent euthanization, 48 h after the second intervention by cervical dislocation. Each heart was removed, embedded in O.C.T.™ Compound (Sakura Finetek, Alphen aan den Rijn, Netherlands) and snap frozen in liquid nitrogen. For further histological examination of the infarction area the heart tissue was divided into four horizontal levels from the apex to the base and cut into 5 μm thick slices.

2.9. Human Cells Detection

For detection of human cells on mouse heart cryosections, monoclonal anti-human nuclei primary antibody (Chemicon, Temecula, CA, USA) was used in combination with M.O.M.™ Mouse Ig Blocking Reagent and M.O.M.™ Protein Concentrate following the instructions of Vector® M.O.M.™ Immunodetection Kit (Linaris, Dossenheim, Germany). Donkey anti-mouse Alexa Fluor® 594 served as conjugated secondary antibody followed by counterstaining with DAPI (both from Thermo Fisher Scientific).

2.10. Infarction Size and Leukocytes Infiltration Area Analysis

Randomly chosen histological heart sections of four horizontal infarct levels were stained with Fast Green FCF (Sigma-Aldrich, Saint Louis, Missouri, USA) and Sirius Red (Chroma Waldeck GmbH & Co. KG, Münster, Germany) assessing tissue localization and distribution of connective fibers. Two contiguous levels of the heart (n = 6 for each group) which represent the major infarction ratio were quantitatively estimated using computer aided image analysis (AxioVision LE Rel.4.5 software, Carl Zeiss AG, Oberkochen, Germany). To evaluate leukocytes infiltration area 48 h after the second ligation, the two contiguous levels of the heart which represent the major infarct ratio were stained with Hematoxylin (Merck, Darmstadt, Germany) and Eosin (Thermo Shandon Ltd., Runcorn, UK) and representative images were taken using computerized planimetry (AxioVision LE Rel.4.5 software).

2.11. Statistical Analysis

Statistical analysis was performed using SigmaStat (Version 3.5, Systat Software, San Jose, CA, USA) and IBM SPSS Statistics for Windows (Version 22.0., IBM Corp., Armonk, NY, USA). The comparisons of two experimental groups were performed using Mann–Whitney U test. p values ≤ 0.05 were considered as statistically significant.

3. Results

3.1. Immunophenotypic Analysis of CD271$^+$ Mesenchymal Stem Cells

We isolated CD271$^+$ stem cells according to our established protocol which yields all mesenchymal colony forming progenitors [17,23]. Again, flow cytometric analysis confirmed a mesenchymal phenotype, reflected by significant overexpression of CD73 and CD105 markers in the isolated CD271$^+$ cell fraction as compared with the entire BM-MNC fraction (9.2% ± 2.8% vs. 0.6% ± 0.2% for CD73 and 8.3% ± 2.8% vs. 0.5% ± 0.2% for CD105, respectively, Figure 2). Moreover, only a subfraction of CD271$^+$ cells coexpressed all of the mesenchymal markers while being CD45$^-$ (1.2% ± 0.3% CD271$^+$CD44$^+$CD73$^+$CD105$^+$ of CD271$^+$ cells).

3.2. Induced Ventricular Arrhythmias

Aiming to develop a murine in vivo model to study human stem cell transplantation, we found that VA are only detectable in the first day after induction of MI. To study engraftment and any alteration in the development of transplanted human cells in a murine model it is necessary to use an immunodeficient mouse strain. The trial of testing the potential antiarrhythmic effects of CD271$^+$ MSC through catheter-based transjugular intracardial burst stimulation for induction of VA, a described method by Roell et al. [4], was not promising. We tested six immune compromised mice in which three mice underwent MI and three were healthy animals. All mice of the infarction group and the healthy group developed VA. This observation did not allow for the creation of a proper control group for comparison. Therefore, burst stimulation does not appear to be a suitable approach when using this mouse strain. Consequently, we developed a new mouse model in which we re-induce VA one week after a first ischemia reperfusion by performing a second permanent ligation. This served to simulate the situation in patients where the affected coronary vessel tends to reclose after successful initial recanalization.

In total, one animal (URI group) out of 22 mice died one-hour post induction of the second infarction, due to development of sinus, and, subsequently, third degree heart block. Consequently, this specimen was excluded from the statistical analysis. No other complications were observed after subcutaneous implantation of the telemeter and first LAD ligation, as well as the second LAD ligation.

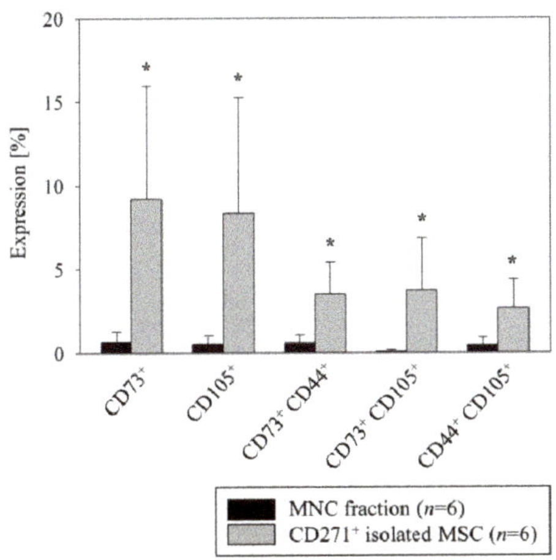

Figure 2. Flow cytometric analysis of MACS-isolated human BM CD271+ stem cells. The freshly isolated BM derived CD271+ stem cells showed a mesenchymal identity by a predominant expression of CD73 and CD105 MSC markers as compared with the entire MNC fraction. Mean ± SD, * $p \leq 0.015$ (Mann–Whitney U Test).

3.3. Induction of Ventricular Arrhythmias after the Reinfarction

To simplify the study of the mechanisms underlying the developed VA, the ECG analyses were subdivided into an acute phase until 15 min and 15 to 45 min post intervention, as well as a delayed phase after 12 h (Table 1), as previously classified by [24]. There were no significant differences in any of the evolved arrhythmias between the groups after the first infarction. Likewise, quantitatively, there was no observed difference in the frequency of the development of VPB, ventricular salvos, and BG/TG 12 h after the first myocardial insult (Figure 3A).

The quantitative assessment of VPB events within 12 h following the second LAD ligation (URI) revealed a significant difference as compared with the control group MIC (1105.0 ± 1146.72 vs. 7.5 ± 8.98, respectively, Figure 3B). This significant difference is also evident in the time frame between 45 min and 12 h after the permanent infarction (URI 1082.2 ± 1127.77 vs. MIC 3.3 ± 2.42, Figure 3C). There was no significant difference in the occurrence of VPB after the first LAD ligation of both groups (URI 60.1 ± 42.19 vs. MIC 259.0 ± 457.69) for the first 12 h and at the time point between 45 min and 12 h (URI 54.5 ± 39.85 vs. MIC 246.8 ± 440.84).

Table 1. Developed ventricular arrhythmias at various time points.

	First LAD Ligation (Mean ± SD)			Second LAD Ligation (Mean ± SD)		
	URI (*n* = 9)	SRI (*n* = 6)	MIC (*n* = 6)	URI (*n* = 9)	SRI (*n* = 6)	MIC (*n* = 6)
VPB						
0–12 h	60.1 ± 42.19	47.5 ± 25.8	259.0 ± 457.69	1105.0 ± 1146.72	178.16 ± 370.12	7.5 ± 8.98
0–15 min	1.8 ± 4.25	2.0 ± 3.63	7.5 ± 14.08	7.1 ± 7.97	4.0 ± 4.04	0.3 ± 0.81
15–45 min	3.6 ± 4.30	3.0 ± 4.69	4.6 ± 5.57	15.6 ± 22.75	0.33 ± 0.51	3.8 ± 7.22
45 min–12 h	54.5 ± 39.85	42.5 ± 27.38	246.8 ± 440.84	1082.2 ± 1127.77	173.8 ± 371.56	3.3 ± 2.42
BG/TG						
0–12 h	1.1 ± 1.45	3.66 ± 4.17	23.6 ± 49.35	113.8 ± 146.02	9.5 ± 18.41	0 ± 0
0–15 min	0 ± 0	0.16 ± 0.40	1.3 ± 3.26	0.5 ± 0.72	0 ± 0	0 ± 0
15–45 min	0.1 ± 0.33	0.16 ± 0.40	0.6 ± 1.21	0.6 ± 1.65	0 ± 0	0 ± 0
45 min–12 h	1.0 ± 1.50	3.33 ± 4.32	21.6 ± 45.89	112.6 ± 144.78	9.5 ± 18.41	0 ± 0
Salvos						
0–12 h	9.0 ± 9.89	3.5 ± 3.27	25.3 ± 29.96	201.7 ± 296.77	1.0 ± 1.67	0.5 ± 0.83
0–15 min	0 ± 0	1.16 ± 1.83	1.0 ± 1.54	0.3 ± 0.50	1.0 ± 1.67	0 ± 0
15–45 min	0.5 ± 0.88	0.5 ± 0.54	12.1 ± 0.64	0 ± 0	0 ± 0	0.5 ± 0.83
45 min–12 h	8.4 ± 10.17	1.83 ± 2.56	12.1 ± 18.17	201.4 ± 296.89	0 ± 0	0 ± 0
VT						
0–12 h	1.3 ± 2.69	0.66 ± 0.51	4.1 ± 3.97	32.6 ± 52.51	1.0 ± 1.26	0 ± 0
0–15 min	0.5 ± 0.83	0.16 ± 0.40	0 ± 0	0 ± 0	0.83 ± 1.16	0 ± 0
15–45 min	0.3 ± 1.00	0.16 ± 0.40	2.0 ± 4.42	0 ± 0	0.16 ± 0.40	0 ± 0
45 min–12 h	1.0 ± 2.64	0.33 ± 0.51	1.6 ± 1.86	32.6 ± 52.51	0 ± 0	0 ± 0

Interestingly, while a significant occurrence of VT after performing the second LAD ligation was identified for URI (32.6 ± 52.5), one week after the onset of the first ligation, none of that was observed in the MIC. This occurred during the first 12 h after the second infarction (Figure 3B) and also in the time frame between 45 min and 12 h after the second infarction (Figure 3C). There was no significant difference in terms of the development of VT after the first infarction (URI 1 ± 2.6 vs. MIC 1.6 ± 1.8).

3.4. Antiarrhythmic Effects of CD271$^+$ MSC Engraftment

The intramyocardial implantation of human CD271$^+$ MSC did not lead to a significant difference in the number of VA early after the first MI (Figure 3A), but showed antiarrhythmic effects by significantly reducing the quantitatively measured VPB which occurred after a reinfarction during the time frame between LAD ligation until 12 h, thereafter (URI 1105.0 ± 1146.72 vs. SRI 178.16 ± 370.12, Figure 3B). Such significant antiarrhythmic behavior was also reflected in a reduction of the number of VT 45 min after the second LAD ligation until 12 h post reinfarction (URI 32.6 ± 52.51 vs. SRI 0 ± 0, Figure 3C). Moreover, there was no significant difference in BG/TG and salvos occurrence between the two groups.

In order to further assess the mechanism of antiarrhythmic effect, ECG monitoring was used. Accordingly, we measured the mean QRS duration and the corrected QT-interval (Figure 3D). Immediately prior to the second LAD ligation (seven days after the first intervention), the stem cell treated animals had a significantly shorter QRS duration (URI 20.77 ± 1.98 vs. SRI 14.66 ± 0.61 and MIC 19.83 ± 0.54 milliseconds, Figure 3E). The significant shorter QTc-interval in SRI (URI 64.25 ± 3 vs. SRI 53.54 ± 3.52 and MIC 67.25 ± 4.14 milliseconds, Figure 3F) could be secondary to the shorter QRS duration as the QRS duration is an integral part of QT duration. There was no statistically significant difference in the mean QRS durations and QTc between URI and MIC immediately prior to the second intervention, as well as between any of the groups 48 h after the first LAD ligation.

After nine days, the engrafted human cells were successfully detected on SRI mouse heart cryosections predominantly in the peri-infarct area by immunofluorescent staining (Figure 3G).

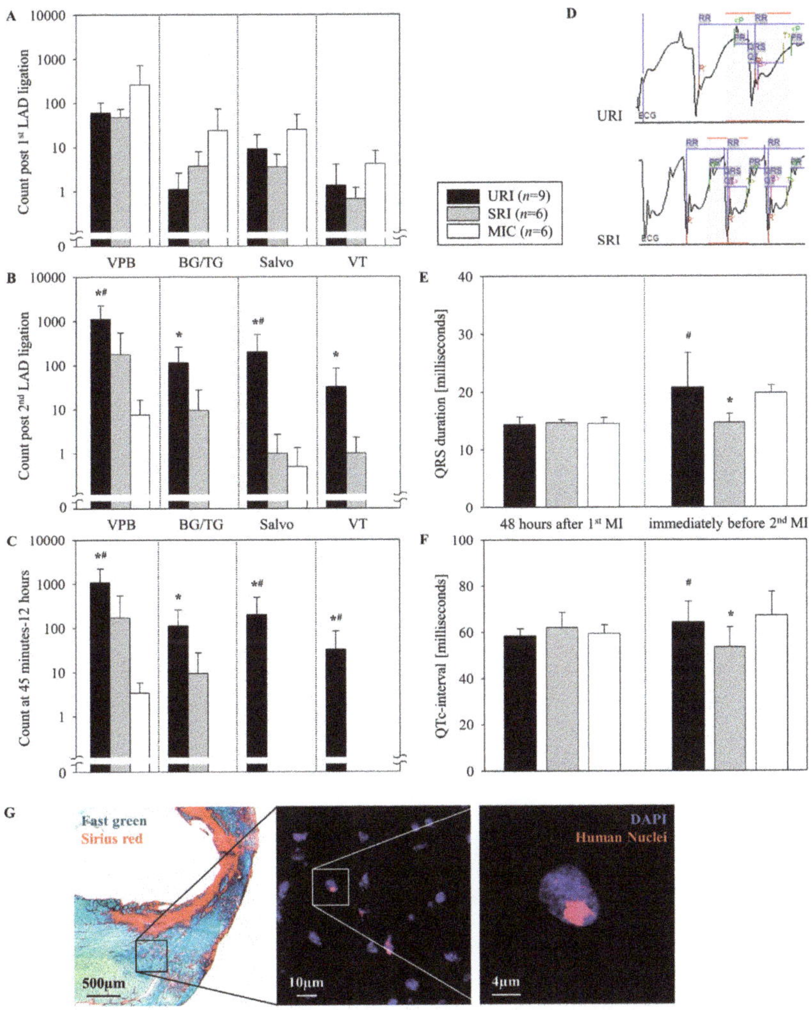

Figure 3. Comparison of developed VA. Until 12 h post LAD ligation (**A,B**). At the time period 45 min to 12 h post the second LAD ligation (**C**). ECG monitoring immediately prior to the second LAD ligation (**D**). QRS duration and QTc-interval 48 h post the first infarction and immediately prior to the second intervention (**E,F**). Mean ± SD, * $p \leq 0.05$ as compared with MIC, # $p \leq 0.05$ as compared with SRI (Mann–Whitney *U* Test). Representative images illustrate the engrafted human stem cells 9 days post transplantation performed for the SRI experimental group which the remaining human MSC were found predominantly in the peri-infarct area as with the Fast Green and Sirius Red staining method (the left picture) confirmed (**G**).

3.5. Alterations of the Infarct Scar

The first ligation of the LAD and opening of the node after 45 min (ischemia-reperfusion) consistently resulted in a transmural MI with its typical histologic changes including the thinning of the left ventricular free wall (Fast Green) and extensive collagen deposition (Sirius Red) nine days post intervention (Figure 4A). After stem cell injection we observed no significant reduction in infarct scar formation (SRI, 14.78% ± 5.85%) in contrast to untreated infarction (URI 17.28% ± 5.10% URI), as well as to control group (MIC, 19.42% ± 3.66%, Figure 4C).

The first ligation of the LAD followed by a permanent second LAD ligation resulted in leukocyte infiltration within the infarction area as depicted in the images of Hematoxylin and Eosin stained slices from hearts 48 h post infarction (Figure 4B). This cellular intervention consequently led to a significant enlargement of the myocardial scar area in URI (38.54% ± 14.28%), as well as SRI (29.36% ± 8.63%) in comparison to MIC (20.1% ± 4.04%, Figure 4C).

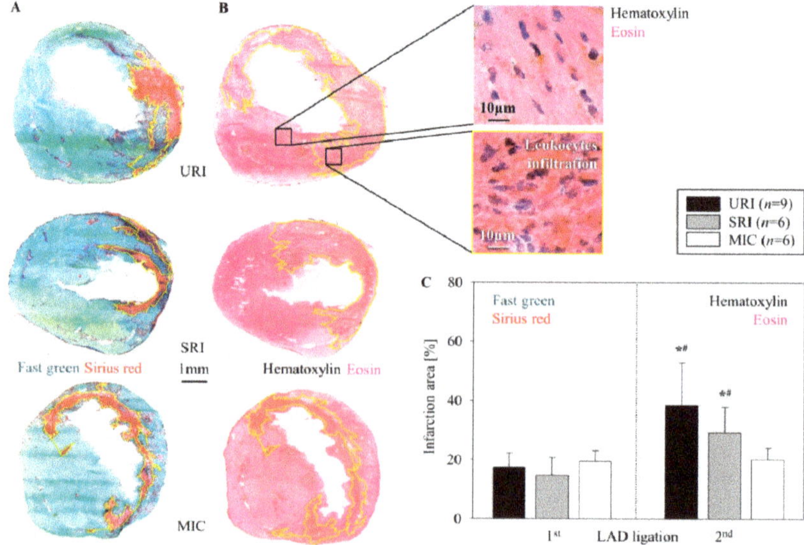

Figure 4. Alterations in MI size. Representative images show the infarction area (enclosed within the yellow border) for URI, SRI, and MIC 9 days after the first LAD ligation (Fast Green and Sirius Red staining) (**A**), as well as 48 h after the second LAD ligation (Hematoxylin and Eosin staining) (**B**). Significant increase of the infarction size after the second LAD ligation, mean ± SD, * $p \leq 0.009$ as compared with the first LAD ligation, # $p \leq 0.041$ in contrast to MIC (Mann–Whitney U test) (**C**).

4. Discussion

The purpose of this study was to develop a mouse model to reproduce VA that typically occurs in patients several days after an initial MI. To study engraftment and development of human cells in a murine model it was necessary to use an immunodeficient mouse strain. During development of an in vivo model of immunodeficient mice for the study of human stem cell transplantation, we found that VA are mostly seen only on the first day after induced MI. Our results supported the hypothesis, that a therapeutic stem cell treatment could have no such antiarrhythmic effect early after their transplantation, as previously shown for other cells [4]. For testing potential antiarrhythmic effects of intramyocardially transplanted cells, Roell et al. induced VA in vivo through transjugular venous catheter mediated burst stimulation of the myocardium after inducing MI [4]. However, this conventional approach could not be utilized in immunodeficient mice, because of the sensitivity of these animals as we observed. For this reason, there was an urgent need for the development of a new mouse model that used immunodeficient mice for in vivo study of antiarrhythmic effects exerted by various human-derived stem cells. On the basis of this, we aimed to develop a new mouse model with re-induced VA by performing a permanent second LAD ligation seven days after re-perfused of the first MI. Indeed, performing the second ligation, one week after the first infarction in Rag2$^{-/-}$γc$^{-/-}$ mice, reproduced cardiac arrhythmias. Other previously described small animal models in this field have utilized immune competent mice or rats subjected to higher stress after performing recurrent transient ischemia following an operatively instrumented mice with the capability of subsequent recurrent transient closure of the LAD [25,26].

The previously described development of VPB in the control group after anesthesia and surgery without performing LAD ligation [27] has also been seen in our mouse model in the time period between 15 and 45 min after re-thoracotomy (without the second LAD ligation) in MIC group. These developed VPB were statistically not significant as compared with the SRI and URI groups.

Our stem cell therapy resulted in shorter QRS duration, shorter QTc-intervals, and decreased occurrence of VT in the early period after the second MI. The observed shortening of QRS indicates the ability of MSC to improve cardiac electric conduction, which is in line with findings from Boink et al. who showed shortened QRS duration in a canine model of MI after MSC transplantation. [28]. However, antiarrhythmic behaviors of human MSC have also been shown in a clinical trial after their intravenous injection following a re-perfused MI [29]. The positive effect may be due to the possible coupling of the MSC with cardiomyocytes through connexin 43 (Cx43) bridges, as it has been found in an in vitro study by our research group [17]. This may be especially relevant as Cx43 plays a major role in post ischemia and post infarction cardiac arrhythmias in the time period between 45 min and 12 h after MI in different animal models [24].

As there is a close relationship between the infarct size and the incidence of VA [30], the significant reduction of the developed VA in the SRI group could not be attributed to the infarct size because there was no significant difference in the infarction area between the SRI and the URI group, early after the second infarction. The Hematoxylin and Eosin staining of the heart sections 48 h after the second intervention (second LAD ligation in SRI and URI groups, re-thoracotomy in MIC group) shows no infarct area expansion and new inflammatory cell infiltration in the MIC group. This indicates that the stained inflammatory cells in SRI and URI groups cannot be remaining cells of the initial MI and their infiltration is the early consequence of the second infarction. Additionally, the lack of intensive leukocyte infiltration in the MIC group after re-thoracotomy (no second LAD ligation) seen with Hematoxylin and Eosin staining of the heart sections nine days after the first operation (LAD ligation) in our mouse strain supports the finding of Yang et al. [31]. The described initial neutrophil infiltration in the infarct border one to two days after LAD ligation in immunocompetent C57BL/6J mice used by Yang and associates has been clearly seen in our SRI and URI groups 48 h after the second infarction. However, the described later lymphocytic infiltration, seven to 14 days after myocardial injury could not be noticeable in our complete alymphoid mouse strain, Rag2$^{-/-}$γc$^{-/-}$ nine days after induced MI.

In this study, we introduce our novel mouse model for testing arrhythmias in the Rag2$^{-/-}$γc$^{-/-}$ mouse. Importantly, CD271$^+$ MSC transplanted into the infarcted area were retained and did not bear any proarrhythmic properties but rather antiarrhythmic effects. We have successfully utilized the model to test potential antiarrhythmic effects of human BM derived CD271$^+$ stem cells in vivo and showed their safety and efficacy. Therefore, we conclude, that human BM derived MSC, i.e., CD271$^+$ are suitable cell types to prevent arrhythmias after MI. This finding supports the previous observations on the safety of intramyocardial transplantation of BM derived stem cells and can be further evaluated for future clinical implication of cell transplantation in the field of electrophysiology [32]. Overall, our novel mouse model offers a new option for testing potential antiarrhythmic effects of cell transplantation therapies with high clinical relevance.

Author Contributions: H.S., R.G., G.S., and R.D. designed and supervised the research study; H.S., R.G., B.A., A.S., C.A.L., S.S., and P.V. conducted the experiments and acquired the data; H.S. and R.G. contributed to the analysis and interpretation of the results; H.S., R.G., G.S., and R.D. were involved in writing and correction of the manuscript, which all authors reviewed and approved.

Funding: This work was supported by the Federal Ministry of Education and Research Germany (FKZ 0312138A and FKZ 316159), the State Mecklenburg-Western Pomerania with EU Structural Funds (ESF/IVWM-B34-0030/10 and ESF/IVBM-B35-0010/12) and the DFG (DA1296/6-1) and the German Heart Foundation (F/01/12). In addition, R.D. is supported by the EU Structural Fund (ESF/14-BM-A55-0024/18), the FORUN Program of Rostock University

Medical Centre (889001 and 889003), the Josef and Käthe Klinz Foundation (T319/29737/2017), the DAMP Foundation, and the BMBF (VIP+ 00240).

Conflicts of Interest: The authors declare that they have no conflict of interest.

References

1. Herzstiftung eV D. 28. Deutscher Herzbericht. Deutsche Herzstiftung eV. 2016. Available online: www.herzstiftung.de/herzbericht (accessed on 18 August 2019).
2. Smits, P.C.; van Geuns, R.J.M.; Poldermans, D.; Bountioukos, M.; Onderwater, E.E.M.; Lee, C.H.; Maat, A.P.W.M.; Serruys, P.W. Catheter-based intramyocardial injection of autologous skeletal myoblasts as a primary treatment of ischemic heart failure: Clinical experiencewith six-month follow-up. *J. Am. Coll. Cardiol.* **2003**, *42*, 2063–2069. [CrossRef] [PubMed]
3. Hagège, A.A.; Marolleau, J.P.; Vilquin, J.T.; Alhéritière, A.; Peyrard, S.; Duboc, D.; Abergel, E.; Messas, E.; Mousseaux, E.; Schwartz, K.; et al. Skeletal myoblast transplantation in ischemic heart failure: Long-term follow-up of the first phase I cohort of patients. *Circulation* **2006**, *114*, 108–113. [CrossRef]
4. Roell, W.; Lewalter, T.; Sasse, P.; Tallini, Y.N.; Choi, B.R.; Breitbach, M.; Doran, R.; Becher, U.M.; Hwang, S.-M.; Bostani, T.; et al. Engraftment of connexin 43-expressing cells prevents post-infarct arrhythmia. *Nature* **2007**, *450*, 819. [CrossRef] [PubMed]
5. Hwang, H.J.; Chang, W.; Song, B.-W.; Song, H.; Cha, M.J.; Kim, I.K.; Lim, S.; Choi, E.J.; Ham, O.; Lee, S.Y.; et al. Antiarrhythmic potential of mesenchymal stem cell is modulated by hypoxic environment. *J. Am. Coll. Cardiol.* **2012**, *60*, 1698–1706. [CrossRef] [PubMed]
6. Steinhoff, G.; Nesteruk, J.; Wolfien, M.; Kundt, G.; Börgermann, J.; David, R.; Garbade, J.; Große, J.; Haverich, A.; Hennig, H.; et al. Cardiac function improvement and bone marrow response outcome analysis of the randomized perfect phase III clinical trial of intramyocardial CD133$^+$ application after myocardial infarction. *EBioMedicine* **2017**, *22*, 208–224. [CrossRef] [PubMed]
7. Mueller, P.; Gaebel, R.; Lemcke, H.; Wiekhorst, F.; Hausburg, F.; Lang, C.; Zarniko, N.; Westphal, B.; Steinhoff, G.; David, R. Intramyocardial fate and effect of iron nanoparticles co-injected with MACS® purified stem cell products. *Biomaterials* **2017**, *135*, 74–84. [CrossRef] [PubMed]
8. Skorska, A.; Müller, P.; Gaebel, R.; Große, J.; Lemcke, H.; Lux, C.A.; Bastian, M.; Hausburg, F.; Zarniko, N.; Bubritzki, S.; et al. GMP-conform on-site manufacturing of a CD133$^+$ stem cell product for cardiovascular regeneration. *Stem Cell Res. Ther.* **2017**, *8*, 33. [CrossRef]
9. Gaebel, R.; Furlani, D.; Sorg, H.; Polchow, B.; Frank, J.; Bieback, K.; Wang, W.; Klopsch, C.; Ong, L.-L.; Li, W.; et al. Cell origin of human mesenchymal stem cells determines a different healing performance in cardiac regeneration. *PLoS ONE* **2011**, *6*, e15652. [CrossRef]
10. Ludwig, M.; Skorska, A.; Tölk, A.; Hopp, H.-H.; Patejdl, R.; Li, J.; Steinhoff, G.; Noack, T. Characterization of ion currents of murine CD117$^+$ stem cells in vitro and their modulation under AT2R stimulation. *Acta Physiol.* **2013**, *208*, 274–287. [CrossRef]
11. Ludwig, M.; Tölk, A.; Skorska, A.; Maschmeier, C.; Gaebel, R.; Lux, C.A.; Steinhoff, G.; David, R. Exploiting AT2R to improve CD117 stem cell function in vitro and in vivo-perspectives for cardiac stem cell therapy. *Cell Physiol. Biochem.* **2015**, *37*, 77–93. [CrossRef]
12. Song, J.; Willinger, T.; Rongvaux, A.; Eynon, E.E.; Stevens, S.; Manz, M.G.; Flavell, R.A.; Galán, J.E. A mouse model for the human pathogen Salmonella typhi. *Cell Host Microbe* **2010**, *8*, 369–376. [CrossRef] [PubMed]
13. Goldman, J.P.; Blundell, M.P.; Lopes, L.; Kinnon, C.; Di Santo, J.P.; Thrasher, A.J. Enhanced human cell engraftment in mice deficient in RAG2 and the common cytokine receptor gamma chain. *Br. J. Haematol.* **1998**, *103*, 335–342. [CrossRef] [PubMed]
14. Mazurier, F.; Fontanellas, A.; Salesse, S.; Taine, L.; Landriau, S.; Moreau-Gaudry, F.; Reiffers, J.; Peault, B.; Di Santo, J.P.; De Verneuil, H. A novel immunodeficient mouse model–RAG2 x common cytokine receptor gamma chain double mutants-requiring exogenous cytokine administration for human hematopoietic stem cell engraftment. *J. Interf. Cytokine Res.* **1999**, *19*, 533–541. [CrossRef] [PubMed]
15. Berul, C.I.; Aronovitz, M.J.; Wang, P.J.; Mendelsohn, M.E. In vivo cardiacelectrophysiology studies in the mouse. *Circulation* **1996**, *94*, 2641–2648. [CrossRef] [PubMed]

16. Hagendorff, A.; Schumacher, B.; Kirchhoff, S.; Lüderitz, B.; Willecke, K. Conductiondisturbances and increased atrialvulnerability in connexin40-deficientmice analysed by transoesophageal stimulation. *Circulation* **1999**, *99*, 1508–1515. [CrossRef]

17. Lemcke, H.; Gaebel, R.; Skorska, A.; Voronina, N.; Lux, C.A.; Petters, J.; Sasse, S.; Zarniko, N.; Steinhoff, G.; David, R. Mechanisms of stem cell based cardiac repair-gap junctional signaling promotes the cardiac lineage specifcation of mesenchymal stem cells. *Sci. Rep.* **2017**, *7*, 9755. [CrossRef]

18. Mitchell, G.F.; Jeron, A.; Koren, G. Measurement of heart rate and Q-T interval in the con-scious mouse. *Am. J. Physiol.* **1998**, *274*, 747–751. [CrossRef]

19. Curtis, M.J.; Hancox, J.C.; Farkas, A.; Wainwright, C.L.; Stables, C.L.; Saint, D.A.; Clements-Jeweryg, H.; Lambiaseh, P.D.; Billmani, G.E.; Janse, M.J.; et al. The Lambeth Conventions (II): Guidelines for the study of animal and human ventricular and supraVA. *Pharmacol. Ther.* **2013**, *139*, 213–248. [CrossRef]

20. Walker, M.J.A.; Curtis, M.J.; Hearse, D.J.; Campbell, R.W.F.; Janse, M.J.; Yellon, D.M.; Cobbe, S.M.; Coker, S.J.; Harness, J.B.; Harron, D.W.G.; et al. The Lambeth Conventions: Guidelines for the study of arrhythmias in ischaemia, infarction, and reperfusion. *Cardiovasc. Res.* **1988**, *22*, 447–455. [CrossRef]

21. Zipes, D.P.; Camm, A.J.; Borggrefe, M.; Buxton, A.E.; Chaitman, B.; Fromer, M. ACC/AHA/ESC 2006 guidelines for management of patients with VA and the prevention of sudden cardiac death–executive summary. *Eur. Heart J.* **2006**, *27*, 2099–2140. [CrossRef]

22. Fiedler, V.B. Reduction of myocardial infarction and dysrhythmic activity by nafazatrom in the conscious rat. *Eur. J. Pharmacol.* **1983**, *88*, 263–267. [CrossRef]

23. Quirici, N.; Soligo, D.; Bossolasco, P.; Servida, F.; Lumini, C.; Deliliers, G.L. Isolation of bone marrow mesenchymal stem cells by anti-nerve growth factor receptor antibodies. *Exp. Hematol.* **2002**, *30*, 783–791. [CrossRef]

24. Wit, A.L.; Peters, N.S. The role of gap junctions in the arrhythmias of ischemia and infarction. *Heart Rhythm* **2012**, *9*, 308–311. [CrossRef] [PubMed]

25. Lujan, H.L.; DiCarlo, S.E. Reperfusion-induced sustained ventricular tachycardia, leading to ventricular fibrillation, in chronically instrumented, intact, conscious mice. *Physiol. Rep.* **2014**, *2*, e12057. [CrossRef]

26. Leprán, I.; Koltai, M.; Siegmund, W.; Szekeres, L. Coronary artery ligation, early arrhythmias, and determination of the ischemic area in conscious rats. *J. Pharmacol. Methods* **1983**, *9*, 219–230. [CrossRef]

27. Betsuyaku, T.; Kanno, S.; Lerner, D.L.; Schuessler, R.B.; Saffitz, J.E.; Yamada, K.A. Spontaneous and inducible ventricular arrhythmias after myocardial infarction in mice. *Cardiovasc. Pathol.* **2004**, *13*, 156–164. [CrossRef]

28. Boink, G.J.J.; Lu, J.; Driessen, H.E.; Duan, L.; Sosunov, E.A.; Anyukhovsky, E.P.; Shlapakova, I.N.; Lau, D.H.; Rosen, T.S.; Danilo, P.; et al. Effect of skeletal muscle Na^+ channel delivered via a cell platform on cardiac conduction and arrhythmia induction. *Circ. Arrhyth. Electrophysiol.* **2012**, *5*, 831–840. [CrossRef]

29. Hare, J.M.; Traverse, J.H.; Henry, T.D.; Dib, N.; Strumpf, R.K.; Schulman, S.P.; Gerstenblith, G.; De Maria, A.N.; Denktas, A.E.; Gammon, R.S.; et al. A Randomized, Double-Blind, Placebo-Controlled, Dose-Escalation Study of Intravenous Adult Human Mesenchymal Stem Cells (Prochymal) After Acute Myocardial Infarction. *J. Am. Coll. Cardiol.* **2009**, *54*, 2277–2286. [CrossRef]

30. Bhar-Amato, J.; Davies, W.; Agarwal, S. Ventricular Arrhythmia after Acute Myocardial Infarc-tion: 'The Perfect Storm'. *Arrhyth. Electrophysiol. Rev.* **2017**, *6*, 134–139. [CrossRef]

31. Yang, F.; Liu, Y.H.; Yang, X.P.; Xu, J.; Kapke, A.; Carretero, O.A. Myocardial Infarction and Cardiac Remodelling in Mice. *Exp. Physiol.* **2002**, *87*, 547–555. [CrossRef]

32. Beeres, S.L.; Zeppenfeld, K.; Bax, J.J.; Dibbets-Schneider, P.; Stokkel, M.P.; Fibbe, W.E.; van der Wall, E.E.; Atsma, D.E.; Schalij, M.J. Electrophysiological and arrhythmogenic effects of intramyocardial bone marrow cell injection in patients with chronic ischemic heart disease. *Heart Rhythm* **2007**, *4*, 257–265. [CrossRef] [PubMed]

Article

Fibronectin Adsorption on Electrospun Synthetic Vascular Grafts Attracts Endothelial Progenitor Cells and Promotes Endothelialization in Dynamic In Vitro Culture

Ruben Daum [1], Dmitri Visser [1], Constanze Wild [2], Larysa Kutuzova [3], Maria Schneider [2], Günter Lorenz [3], Martin Weiss [1,4], Svenja Hinderer [1], Ulrich A. Stock [5], Martina Seifert [2] and Katja Schenke-Layland [1,4,6,7,*]

[1]   NMI Natural and Medical Sciences Institute at the University of Tübingen, 72770 Reutlingen, Germany; ruben.daum@nmi.de (R.D.); dmitri.visser@nmi.de (D.V.); martin.weiss@med.uni-tuebingen.de (M.W.); hinderer@polymedics.de (S.H.)
[2]   Institute of Medical Immunology and BIH Center for Regenerative Therapies (BCRT), Charité-Universitätsmedizin Berlin, Corporate Member of Freie Universität Berlin, Humboldt-Universität zu Berlin, and Berlin Institute of Health, 13353 Berlin, Germany; constanze.wild@charite.de (C.W.); maria.schneider@charite.de (M.S.); martina.seifert@charite.de (M.S.)
[3]   Applied Chemistry, University of Reutlingen, 72762 Reutlingen, Germany; larysa.kutuzova@reutlingen-univeristy.de (L.K.); guenter.lorenz@reutlingen-university.de (G.L.)
[4]   Department of Women's Health, Research Institute for Women's Health, Eberhard-Karls-University Tübingen, 72076 Tübingen, Germany
[5]   Department of Cardiothoracic Surgery, Royal Brompton and Harefield Foundation Trust, Harefield Hospital Hill End Rd, Harefiled UB9 6JH, UK; u.stock@rbht.nhs.uk
[6]   Cluster of Excellence iFIT (EXC 2180) "Image-Guided and Functionally Instructed Tumor Therapies", Eberhard-Karls-University Tübingen, 72076 Tübingen, Germany
[7]   Department of Medicine/Cardiology, Cardiovascular Research Laboratories, David Geffen School of Medicine at UCLA, Los Angeles, CA 90095, USA
*   Correspondence: katja.schenke-layland@med.uni-tuebingen.de; Tel.: +49-707-1298-5205

Received: 25 February 2020; Accepted: 19 March 2020; Published: 23 March 2020

Abstract: Appropriate mechanical properties and fast endothelialization of synthetic grafts are key to ensure long-term functionality of implants. We used a newly developed biostable polyurethane elastomer (TPCU) to engineer electrospun vascular scaffolds with promising mechanical properties (E-modulus: 4.8 ± 0.6 MPa, burst pressure: 3326 ± 78 mmHg), which were biofunctionalized with fibronectin (FN) and decorin (DCN). Neither uncoated nor biofunctionalized TPCU scaffolds induced major adverse immune responses except for minor signs of polymorph nuclear cell activation. The in vivo endothelial progenitor cell homing potential of the biofunctionalized scaffolds was simulated in vitro by attracting endothelial colony-forming cells (ECFCs). Although DCN coating did attract ECFCs in combination with FN (FN + DCN), DCN-coated TPCU scaffolds showed a cell-repellent effect in the absence of FN. In a tissue-engineering approach, the electrospun and biofunctionalized tubular grafts were cultured with primary-isolated vascular endothelial cells in a custom-made bioreactor under dynamic conditions with the aim to engineer an advanced therapy medicinal product. Both FN and FN + DCN functionalization supported the formation of a confluent and functional endothelial layer.

Keywords: vascular graft; endothelialization; tissue engineering; decorin; fibronectin; electrospinning; endothelial progenitor cells; bioreactor; biostable polyurethane

1. Introduction

Atherosclerotic cardiovascular disease is one of the leading causes of death worldwide [1,2]. It includes all medical conditions, where blood flow to organs and limbs is reduced due to plaque deposition. Surgical intervention is required to reopen or replace the defective vessel. The use of autografts, like the saphenous vein or mammary artery, are still the standard clinical approach for the replacement of small diameter blood vessels [3]. However, mechanical or size mismatches, and mainly the scarce availability make alternative grafts necessary [4,5]. In this context, two strategies have emerged in recent years: synthetic substitutes and biological grafts [4]. Although large-diameter synthetic substitutes (>6 mm) are successfully used, small diameter grafts (<6 mm) show low patency rates due to their tendency to elicit thrombosis and the formation of intimal hyperplasia [6–8]. Appropriate mechanical properties and biocompatibility of the synthetic graft as well as a fast endothelialization after implantation are key properties to ensure a long-term functional implant. In addition, the graft should evoke a balanced immune reaction. On the one hand, a moderate immune response is beneficial in order to promote tissue regeneration. On the other hand, chronic immune responses can lead to inflammation, fibrosis, or calcification and should be avoided to ensure long-term function of the vascular graft [9].

Electrospinning has proven to be a suitable method for the fabrication of fibrous scaffolds and vascular constructs as it mimics the highly porous structure and physical properties of the extracellular matrix (ECM) of the native tissue. Due to their high porosity, pore interconnectivity, and large surface area, the fibrous scaffolds are able to promote cell adhesion, cell alignment, and cell proliferation [10–13]. In addition, in order to elicit in situ endothelialization in the body, the material surface can be functionalized with bioactive molecules. A central challenge in this context is the attraction, adhesion, and proliferation of endothelial progenitor cells (EPCs) or endothelial cells (ECs) to form a complete endothelium. Several strategies to address this issue have been described: immobilization of antibodies targeting markers for EPCs such as vascular endothelial growth factor receptor 2 (VEGFR2) and platelet endothelial cell adhesion molecule (PECAM-1) [14,15]; modification of the surface with peptides such as the Arg-Gly-Asp (RGD) or Cys-Ala-Gly (CAG) sequence [16,17]; immobilization of growth factors such as the vascular endothelial growth factor (VEGF) or stromal cell-derived factor-1 (SDF-1) [18,19]; immobilization of oligonucleotides and aptamers [20,21]; and surface modification with oligosaccharides and phospholipids [22,23]. However, it is necessary to develop surfaces with improved biocompatible, bioactive, targeted, and stable biofunctionalization [24].

A recent study described the attraction of EPCs by immobilized recombinant human decorin (DCN) [25]. The small leucine-rich proteoglycan plays a pivotal role in the ECM [26]. It is named after its first known function as a modulator of collagen fibrillogenesis [27]. In recent years, it has been shown that DCN influences a variety of biological processes in addition to its structural function. It is involved in cell attachment [28–30], proliferation [31,32], and migration [28,29,31,33]. Furthermore, it has been described that DCN inhibits the proliferation and migration of vascular smooth muscle cells but does not affect ECs [28,31]. With a proportion of 22% of all proteoglycans in the vessel wall, it also influences many biological processes in vascular homeostasis and angiogenesis [34–36]. Depending on the molecular environment, it can act pro-angiogenic or antiangiogenic [26,34]. For instance, DCN was shown to interact antagonistically with the mesenchymal epithelial transition factor (c-MET) and the VEGFR2, which significantly influences angiogenesis [26,34,37,38]. In addition, DCN binds to the transforming growth factor β (TGF-β), which in turn has an inhibiting effect on the endothelial-mesenchymal transition and fibrosis [26,39,40]. These properties make the protein a promising candidate for improving the endothelialization of a vascular graft. Another highly relevant ECM protein is fibronectin (FN). Since FN interacts with cells via the integrins $\alpha_5\beta_1$ or $\alpha_v\beta_3$, it is a suitable protein for bioactivating a material surface [41–44]. It is of interest with regard to endothelialization, as it plays a pivotal role in wound healing [45,46]. Several studies described the coating of FN in combination with collagens type I [47] and type IV [48], with fibrinogen and tropoelastin [49], hepatocyte growth factor [50], heparin, and VEGF [51] and with SDF-1α [19] to improve reendothelialization. However, it has never been used in combination with DCN before.

Tissue engineering can be used as an alternative strategy to obtain a functional endothelium in a synthetic graft utilizing a patient's own cells [52]. After implantation, the tissue-engineered vascular graft (TEVG) is replaced by the host's cells and ECM and is thereby degraded [4]. However, the loss of mechanical properties due to a too rapid degradation and unfavorable biological reactions to the degradation products remain a major challenge [1,53]. A recent study addressed this problem by producing a TEGV that consists of a combination of a biodegradable and biostable polymer [54].

In our study, a newly developed biostable polyurethane elastomer was used to develop an electrospun scaffold with mechanical properties that are comparable to native vascular tissues, and a bioactive surface that attracts endothelial progenitor cells or promotes endothelialization [55]. For this purpose, planar and tubular electrospun scaffolds (Figure 1a) were biofunctionalized with FN, DCN, or FN and DCN in combination (FN + DCN; Figure 1b,c). The influence of the FN- and DCN-coated scaffolds on human immune cell features was examined (Figure 1d). Subsequently, the functionality of the electrospun scaffolds was further investigated. First, endothelial progenitor cell homing was simulated in vitro by attracting endothelial colony forming cells (ECFCs) with a potent angiogenic capacity and the capability to support vascular repair (Figure 1e,f). Secondly, in a classical TEVG approach primary-isolated vascular endothelial cells (vECs) were cultured in a custom-made bioreactor to create an advanced therapy medicinal product (ATMP) (Figure 1g).

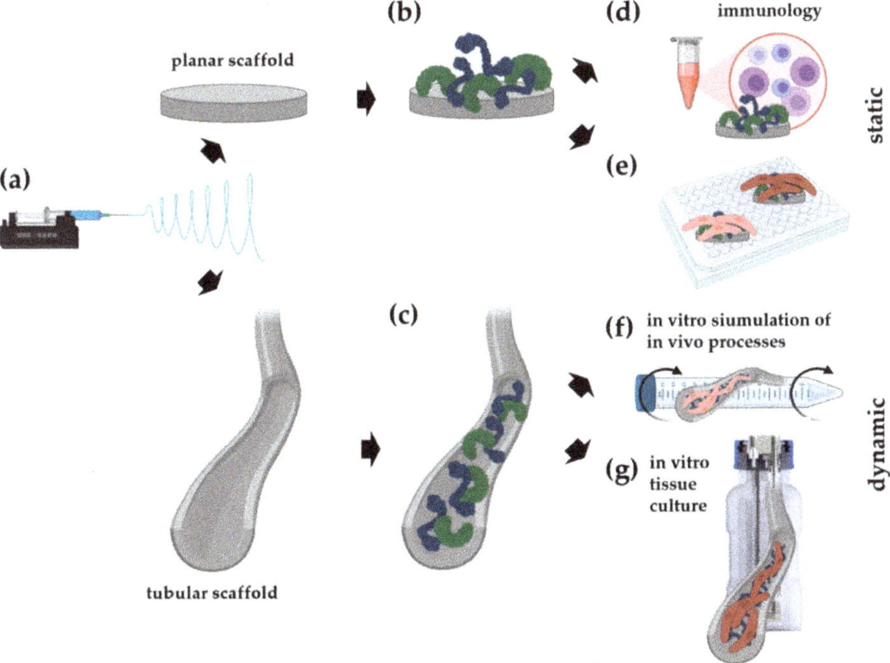

Figure 1. A newly developed polyurethane is used to produce planar and tubular electrospun scaffolds (**a**), which are biofunctionalized with either fibronectin (FN) or decorin (DCN) or with both extracellular matrix (ECM) proteins in combination (**b,c**). Besides investigating the immunology (**d**) and endothelial colony forming cell (ECFC) behavior on either planar (**e**) or in tubular scaffolds (**f**), the tubular scaffolds were also cultured with primary-isolated vascular endothelial cells (vECs) in an tissue-engineered vascular graft (TEVG) approach (**g**) in order to assess an ECM protein-improved endothelialization.

2. Materials and Methods

2.1. Electrospun Scaffold Fabrication

Planar and tubular scaffolds were produced by electrospinning of soft thermoplastic polycarbonate-urethane (TPCU). This elastomeric material was synthesized in our laboratory for special medical applications using the multistep one-pot approach [56], which gives good control of the polymer architecture in catalyst-free systems. In more detail, a long-chain aliphatic polycarbonate with more than 72% (*w/w*) in the TPCU formulation provides an additional crystallization of the soft segment, which enhances biostability of the implantable material as well as improves its mechanical properties. In vitro biostability of the TPCU was studied previously from a mechanical point of view under long-term oxidative treatment [55]. Cytocompatibility of the TPCU material was also demonstrated [57]. By adjusting the respective parameters to achieve a stable process and appropriate mechanical properties of the scaffold (Figure S1a), 0.1 g/mL of the polymer was dissolved in 1,1,1,3,3,3 hexafluoro-2-propanol (804515, Merck, Darmstadt, Germany) and electrospun with the process conditions summarized in Table 1. The electrospinning process was carried out in a temperature- and humidity-controlled electrospinning apparatus (EC-CLI, IME Technologies, Eindhoven, Netherlands).

Table 1. Process conditions for electrospinning planar and tubular scaffolds.

Description	Value
Distance	25 cm
Needle i.d.	0.4 mm
Voltage	18 kV/−0.2 kV (needle/collector)
Temperature	23 °C
Humidity	40%
Mandrel diameter [1]	6 mm
Mandrel rotation speed [1]	2000 rpm
Needle translation distance [1]	80 mm
Volume	6 mL
Flow rate	4 mL/h

i.d.= inner diameter; [1] tubular scaffolds.

2.2. Biofunctionalization of the Scaffolds

Before biofunctionalization, the appropriate disinfection method was investigated. Since ethanol did not affect the scaffold in terms of its mechanical properties (Figure S1b), the constructs were disinfected with 70% ethanol for 20 min and afterwards washed three times for 10 min with phosphate-buffered saline (PBS). Microbiological studies were carried out on the scaffolds to investigate the effectiveness of the disinfection method (Figure S3). The scaffolds were functionalized by protein adsorption. They were incubated for 2 h at 37 °C with 20 µg/mL human plasma FN (F1056, Sigma-Aldrich, St. Louis, USA) or 20 µg/mL recombinant full-length human DCN [25], individually or in combination. Excess protein was removed by washing the scaffolds with PBS.

2.3. Morphological and Mechanical Characterization of the Electrospun Scaffolds

For the morphological characterization, punches from the electrospun scaffolds were examined by scanning electron microscopy (SU8030, Hitachi, Tokyo, Japan) followed by the analysis using ImageJ and the DiameterJ package [58] to assess the pore and fiber sizes. For the investigation of the mechanical properties, a ring tensile test was performed based on the methods described by Laterreur et al. [59] in order to determine the circumferential tensile strength and burst pressure. Briefly, the tubular scaffolds were cut into pieces with the length $L_0 = 7$ mm, clamped into a uniaxial tensile testing device (Zwick Roell, Ulm, Germany), and stretched over a distance s with a velocity of 50 mm/min until rupture.

On the basis of the stress–strain curves (Figure S1c), the burst pressure P_b was then calculated by relating the registered force at rupture F_b to the elongation s_b as follows:

$$P_b = \frac{F_b \pi}{L_0 d_{pin}(\pi + 2) + 2L_0 s_b} \tag{1}$$

where d_{pin} represents the diameter of the pins that were used in the ring tensile test. A derivation of Equation (1) is provided by Lattereur et al. [59]. Using an OCA40 (DataPhysics Instruments GmbH, Filderstadt, Germany), the wettability of the scaffolds was analyzed as previously described [60]. A waterdrop with a volume of 2 µL was placed onto the scaffold and measured using the SCA20 software (DataPhysics Instruments, Filderstadt, Germany). The water absorption ability was determined by weighing the specimens in their dry and wet states after submerging the specimens in water for 1 h. The relative weight increase is referred to as the swelling ratio.

2.4. Immune Cell/Scaffold Co-Culture Assays

Polymorph nuclear cells (PMNs) were isolated from freshly donated human blood and peripheral blood mononuclear cells (PBMCs) from buffy coats according to the ethical approval by the local ethics committee at the Charité Berlin (EA2/139/10 approved on 10th December 2010; EA1/226/14 approved on 24th July 2014) and as recently described [61]. Monocytes were magnetically sorted via CD14 beads (130-050-201, Miltenyi Biotec, Bergisch Gladbach, Germany) from PBMCs as previously described [62]. Monocytes were differentiated into M0 macrophages by adding 50 ng/mL of macrophage colony-stimulating factor (M-CSF) (130-096-491, Miltenyi Biotec) to the culture medium for 7 days. All immune cell co-cultures were performed in Roswell Park Memorial Institute (RPMI) 1640 medium (F1415, Biochrom GmbH, Berlin, Germany) with 10% human serum type AB (H4522, Sigma-Aldrich), 1% L-glutamine (25030-024, Thermo Fisher Scientific, Waltham, MA, USA), and 1% penicillin/streptomycin (15140-122, Thermo Fisher Scientific).

First, the scaffold punches were incubated with 100 µg/mL of recombinant full-length human DCN [25] or 20 µg/mL of FN (F1056, Sigma-Aldrich) at 37 °C for 4 h. Next, punches were washed with PBS (L1825, Biochrom GmbH), placed into a well of a 48-well plate, and kept in place with a silicon ring (Ismatec, Wertheim, Germany). Thereafter, the different immune cell types were applied as follows:

Human PMNs were cultured on the uncoated, DCN- or FN-coated scaffolds; 0.2×10^6 PMNs in 200 µL of complete RPMI were seeded directly on the scaffold punches. Unstimulated cells were used as a negative control, and PMNs that were stimulated with 500 ng/mL of lipopolysaccharide (LPS; 297-473-0, Sigma-Aldrich) served as a positive control. LPS is a component of the bacterial cell membrane that triggers the activation of immune cells. After 4 h of culture, cells were harvested only by careful resuspension, stained with human-specific antibodies for CD11b (1:100; 557701, BD Bioscience, San Jose, CA, USA) and CD66b (1:200; 305107, BioLegend, Fell, Germany), and measured by flow cytometry (CytoFLEX LX, Beckman Coulter, Inc., Brea, CA, USA) as described recently [61]. The determined mean fluorescence intensities (MFIs) of marker expression were normalized to the MFI of unstimulated PMNs directly after isolation.

Human monocytes or M0 macrophages were cultured on the uncoated, DCN- or FN-coated scaffolds; 0.2×10^6 cells in 350 µL of complete RPMI were seeded directly on the scaffold punches. Monocytes that were stimulated with 100 ng/mL of LPS served as a positive control, and unstimulated monocytes served as a negative control. Macrophages cultured without any stimulus were used as negative control. To induce the polarization into the M1 phenotype, 20 ng/mL of IFNγ (130-096-486, Miltenyi Biotec) and 100 ng/mL of LPS were added to the medium of M0 macrophages. After two days of culture, monocytes/macrophages were harvested, stained with human-specific antibodies for CD80 (1:20; 305208, BioLegend) and human leukocyte antigen DR isotype (HLA-DR) (1:200; 307616, BioLegend), and measured by flow cytometry. Cells were detached by adding 100 µL of Accutase (A11105-01, Thermo Fisher Scientific) and incubating the cells at 37 °C for 30 min. The determined MFIs of the marker expression were normalized to the MFI of the unstimulated cells.

PBMCs were cultured on the uncoated, DCN- or FN-coated scaffolds; 0.3×10^6 cells were seeded in 400 µL of complete RPMI directly on the scaffold punches. Unstimulated PBMCs served as a negative control. For the positive controls, PBMCs were stimulated with anti-CD28 (556620, BD Bioscience)/anti-CD3 (OKT3, Janssen-Cilag, Neuss, Germany) antibodies. After three days of culture, PBMCs were harvested, stained with human-specific antibodies for CD69 (1:50; 310926 BioLegend), CD25 (1:50; 302605, BioLegend) and HLA-DR (1:100; 307640, BioLegend), and measured by flow cytometry. PBMCs were detached by adding 100 µL of Accutase and by incubating the cells at 37 °C for 30 min. After gating for single and living cells the CD14− and CD14+ populations were defined. For CD3+ cells, the MFI of the activation markers CD25, CD69, and HLA-DR was determined. The determined MFIs of the marker expression were normalized to the MFI of unstimulated PBMCs.

Co-culture supernatants of monocytes and macrophages were collected and the tumor necrosis factor alpha (TNFα) concentration was analyzed by ELISA (430205, BioLegend) according to the manufacturer´s instructions.

2.5. Cell Culture of Primary Endothelial Cells and Endothelial Colony Forming Cells

Human primary-isolated vECs were isolated from foreskin biopsies under the ethics approval no 495/2018BO2 by enzymatic digestion with dispase and trypsin as previously described [63]. The vECs were cultured in endothelial cell growth medium and SupplementMix (C-22020, PromoCell, Heidelberg, Germany), supplemented with 1% penicillin-streptomycin (15140122, Thermo Fisher Scientific).

Human ECFCs (00189423, Lonza, Basel, Switzerland) were cultured in endothelial cell growth medium-2 with supplements (CC-3162, Lonza). Instead of the supplied fetal bovine serum, 5% of human serum (H4522, Sigma-Aldrich) was used. In addition, 1% L-Glutamine (21051024, Thermo Fisher Scientific) and 1% penicillin-streptomycin (15140122, Thermo Fisher Scientific) were added to the cell culture medium.

Both cell types were cultured at 37 °C and 5% CO_2 and passaged at approximately 80% confluence. The vECs were used for the experiment after 2–4 passages.

2.6. Cell Seeding and Culture on Planar Scaffolds

Prior to cell culture experiments, general biocompatibility of the electrospun scaffolds was examined with a cytotoxicity test based on EN ISO 10993-5 [64]. Briefly, the scaffolds were incubated for 72 h at 37 °C and 5% CO_2 in 1 mL endothelial cell growth medium supplemented with 1% penicillin-streptomycin at an extraction ratio of 0.1 mg/mL; 2×10^4 vECs seeded in a 96-well plate were then exposed for 24 h to the extracts supplied with the cell culture medium supplements. Endothelial cell growth medium without the scaffolds served as a negative control. Cells exposed to 1% SDS served as positive control. The extraction and control medium were removed, and an MTS (3-(4,5-dimethylthiazol-2-yl)-5-(3-carboxymethoxyphenyl)-2-(4-sulfophenyl)-2H-tetrazolium) assay (CellTiter 96Aqueous One Solution Cell Proliferation Assay, Promega, Madison, WI, USA) was performed according to the manufacturer's protocol; 20 µL MTS solution and 100 µL cell culture medium were added to each well. After 30 min of incubation at 37 °C, the absorbance of each well was measured at 450 nm using a microplate reader (PHERAstar, BMG Labtech, Ortenberg, Germany). Cell viability was determined by the absorbance of the samples relative to the negative control. No toxic effect of the material was observed (Figure S2a). Biofunctionalization of the scaffolds was then carried out as described above. Cells were seeded afterwards onto the biofunctionalized scaffolds with a diameter of 6 mm, which were placed in a 96-well plate. For the vECs, 5×10^3 cells/well and, for the ECFCs, 1×10^4 cells/well were seeded in 150 µL of the appropriate medium. If required, media change was carried out every 3 days.

2.7. Endothelial Colony Forming Cells (ECFC) Seeding Under Dynamic Conditions

The tubular electrospun scaffolds were cut to 6 cm length and biofunctionalized with FN and DCN alone or in combination as described above. A cell suspension of 4×10^5 ECFCs/mL was pipetted into the tubular constructs. Afterwards, the constructs were closed at both ends and put in 15-mL centrifuge tubes filled with the corresponding cell culture medium. Placed on a roller mixer (RM5, CAT, Ballrechten-Dottingen, Germany), the tubes were rotated with 60 rpm for 24 h at 37 °C and 5% CO_2. For cell number analysis, the attached cells were stained with 4′,6-diamidino-2-phenylindole (DAPI) (1:50, 10236276001, Roche Diagnostics, Mannheim, Germany) and counted.

2.8. Development of a Bioreactor System for Tissue-Engineered Vascular Graft (TEVG) Culture

The TEVG approach was performed with a custom-made bioreactor setup. The culture chamber consists of a 250-mL glass bottle (Schott Duran, Wertheim, Germany) and encloses a removable custom-designed graft frame that holds the vascular graft. A computer-aided design (CAD) model for the graft frame was created in Solidworks (Dassault Systèmes, Vélizy-Villacoublay, France) and milled out of polyether ether ketone (PEEK; ADS Kunststofftechnik, Ahaus, Germany) using a 2.5-axis flatbed milling setup (Isel, Eichenzell, Germany) with computer numerical control (CNC). The constructed parts were subjected to the aforementioned cytotoxicity test to ensure no toxic leachables are released into the medium under culture (Figure S2b). The modular design of the culture chamber allows for a toolless assembly of the bioreactor system under a sterile bench.

The graft frame—once inserted into the culture chamber—is connected to medium reservoirs and a bubble trap with flexible silicone tubing. Sterile gas exchange is facilitated by sterile filters connected to the medium reservoirs. The entire setup is driven by a multichannel roller pump (Ismatec) (Figure 2).

The flow rates Q for dynamic culture were determined with a derived formulation of the Hagen–Poiseuille equation for laminar flow in straight circular pipes with internal radius r:

$$\tau = \frac{4\mu Q}{\pi r^3} \tag{2}$$

where μ denotes the dynamic viscosity. This gave an analytical approximation of the achieved wall shear stress (τ) within the cultured vascular graft. To validate this approximation and the assumption of a laminar regime within the vascular graft, in silico simulations were used to assess the local fluid dynamics within the vascular graft and graft frame interior. Briefly, the CAD model of the graft frame was meshed and exported to a computational fluid dynamics (CFD) solver (ANSYS Fluent). Dynamic culture with a wide range of flow rates was simulated under steady-state flow and Newtonian rheological conditions, after which the calculated wall shear stress on the interior graft wall was analyzed and compared to the aforementioned analytical solution (Figures S4 and S5).

2.9. Tissue Culture of Vascular Endothelial Cells Under Dynamic Conditions

Tubular electrospun scaffolds were cut to 7.5 cm length and biofunctionalized with 20 µg/mL FN as described previously. After inserting the graft frame into the culture chamber, 2×10^6 vECs/mL were seeded into the tubular scaffold. In order to achieve homogeneous cell adhesion across the entire tube, the culture chamber was placed horizontally and rotated every 15 minutes over 45 ° for 3 h at 37 °C and 5% CO_2. The culture chamber was consecutively connected to the rest of the bioreactor setup and filled with 70 mL culture medium, supplemented with 1% penicillin-streptomycin and 1% PrimocinTM (ant-pm-1, InvivoGen, San Diego, CA, USA). The seeded cells were allowed to proliferate under static conditions during the first three days, after which the flow rate was slowly increased over the course of two days, as shown in Figure 2e. Subsequently, the tubular construct was cultured under constant flow for seven days.

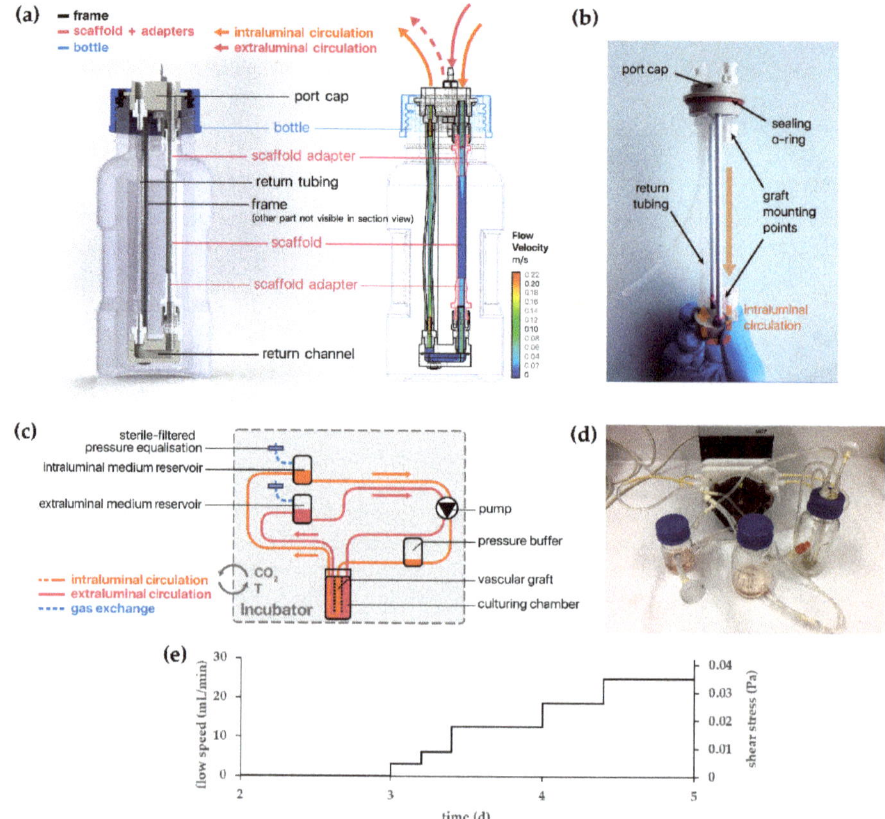

Figure 2. (**a**) A cross-sectional schematic representation of the culturing chamber and its parts. The wireframe model on the right is overlaid by the results of an in silico simulation and shows the flow velocity when the system is perfused with a flow rate of $Q = 20$ mL/min. (**b**) This photograph shows the graft frame (without scaffold), once it is taken out of the culturing chamber. (**c**) A schematic representation of the entire bioreactor setup, showing the circulation and connections to the medium reservoirs and pressure buffer/bubble trap. (**d**) A photograph showing the assembled bioreactor setup with all the components for the intraluminal circulation. (**e**) Applied perfusion flow speed as function of time with the corresponding wall shear stress.

2.10. Immunofluorescence Staining

In order to examine the protein coating, the biofunctionalized scaffolds were stained using DCN mouse monoclonal IgG_1 (1:200; sc-73896, Santa Cruz Biotechnology, Dallas, TX, USA) and FN polyclonal rabbit IgG (1:500; F3648, Sigma-Aldrich) antibodies. For fluorescence labeling, AlexaFluor 488 anti-mouse IgG (1:250; A-11001, Thermo Fisher Scientific) and AlexaFluor 546 anti-rabbit IgG (1:250; A-11035, Thermo Fisher Scientific) were used as secondary antibodies.

Cells cultured on the scaffolds were stained as follows: after washing once with PBS, the cell-seeded scaffolds were fixed with 4% paraformaldehyde (P6148, Sigma-Aldrich). In order to reduce nonspecific binding, the samples were incubated with 2% goat serum-containing block solution for 30 min. Afterwards, the cells were incubated over night at 4 °C with the following antibodies: Vascular endothelial cadherin (VE-cadherin) monoclonal mouse IgG_{2B} (1:500, MAB9381, R&D systems, Minneapolis, MN, USA), VEGFR2 polyclonal rabbit IgG (1:75, ab2349, Abcam, Cambridge, UK), PECAM-1 monoclonal mouse IgG_1 (1:100, sc-71872, Santa Cruz), von Willebrand factor (vWF) polyclonal rabbit IgG (1:200, A0082, Dako, Glostrup, Denmark), and vinculin monoclonal mouse IgG$_1$ (1:500, MAP3574, Millipore, Burlington, MA, USA). F-actin was stained for 45 min in the dark with Alexa Fluor 647 Phalloidin (1:500, A22287, Thermo Fisher Scientific). Subsequently, samples were incubated with the appropriate secondary antibodies (AlexaFluor 488 anti-mouse IgG, AlexaFluor 546 anti-rabbit IgG, and AlexaFluor 488 anti-mouse IgG2b (all 1:250; Thermo Fisher Scientific)).

Finally, nuclei were stained with DAPI (1:50) for 15 min in the dark. Images were obtained by using a fluorescence microscope (Cell Observer, Carl Zeiss AG, Oberkochen, Germany).

2.11. Examination of the Cell Coverage on the Tubular Scaffolds

The cell coverage of the inner wall of the tubular constructs was investigated using MTT (3-(4,5-dimethylthiazol-2-yl)-2,5-diphenyltetrazolium bromide) (M2128-1G, Sigma-Aldrich). After culturing with vECs, the constructs were incubated for 20 min with 1 mg/mL MTT at 37 °C and 5% CO_2. The insoluble purple formazan produced by the cellular reduction of MTT was then examined macroscopically.

2.12. Image Analysis

FN and DCN coating were quantified by measuring the relative pixel intensity (RPI) of the immunofluorescence images. To assess protein expression in the experiments, the area within a defined fluorescence intensity threshold was measured and normalized to the cell number. The cell count in the static experiments was quantified by counting the DAPI-stained cell nuclei per area. The quantification of the adherent ECFCs in the dynamic experiment was performed by measuring the DAPI-stained area normalized to the total area. All images were analyzed using ImageJ [58].

2.13. Scanning Electron Microscopy of Cells

Prior to SEM imaging of the scaffolds with cells, a critical point drying step was performed. First, cells were fixed for 60 min with 4% paraformaldehyde (PFA)/ 25% glutaraldehyde in PBS. Subsequently, a series of ethanol solutions in ascending concentration up to 100% was carried out to remove water. Critical point drying was done with a CPD 030 (Bal-Tec AG, Balzers, Liechtenstein) according to the manufacturer's protocol. Prior to imaging, the specimens were platina-coated (SCD050, Bal-Tec AG) for one minute at 0.05 mbar and rinsed with Argon after the coating process. SEM imaging was performed with a SU8030 (Hitachi, Tokyo, Japan) and an Auriga® 40 (Zeiss, Oberkochen, Germany).

For SEM imaging of the monocytes and macrophages, the cells were cultured for two days on uncoated (w/o), DCN- or FN-coated scaffolds, followed by preparation (as described in Reference [62]) and imaging with a JCM 6000 Benchtop (JEOL, Freising, Germany).

2.14. Statistical Analysis

Except stated otherwise, data are presented as mean ± standard deviation. For the immune data, GraphPad Prism (GraphPad Software, San Diego, CA, USA) was used to determine statistical significance between two groups using a one-way ANOVA/Kruskal–Wallis test. For the other data, a one-way ANOVA/Fisher's Least Significant Difference test was performed. A Welch's t-test was performed to compare between two data groups using OriginPro (OriginLab, Northampton, MA, USA). Probability values of 95%, 99%, 99.9%, and 99.99% were used to determine significance.

3. Results

3.1. Biofunctionalization Does Not Impact the Mechanical Properties of Electrospun Tubular Constructs

Electrospinning was used to fabricate 110-mm long tubular scaffolds with an inner diameter of 5 mm and a thickness of 0.40 ± 0.06 mm (Figure 3a). In order to modulate the cell–material interaction, the surface was biofunctionalized with FN, DCN, or FN + DCN. The impact of the biofunctionalization on the morphological and mechanical properties of the material was investigated (Figure 3). Fiber and pore size analysis of the SEM images revealed no significant alteration due to protein adsorption (Figure 3e). Higher magnifications of the SEM images showed distribution of the proteins on the fibers. While DCN formed randomly distributed aggregates on the TPCU scaffolds, FN coating showed a network-like deposition in the nanometer range, which was also seen in the FN + DCN-coated samples, in which clearly recognizable aggregates were deposited on the protein network (Figure 3b, white arrows). Biofunctionalization utilizing both proteins individually and in combination was confirmed by IF staining. DCN IF staining revealed a more heterogeneous distribution of DCN in combination with FN than alone (Figure 3c, white arrows). The contact angle of the scaffolds was not significantly changed by the adsorption of either FN or DCN in comparison with the uncoated scaffolds. A significantly higher swelling ratio was observed of scaffolds that had been coated with FN + DCN (Figure 3e; control: 93.7% ± 7.7% versus FN + DCN: 117.1% ± 8.7%, $p < 0.05$). Overall, biofunctionalization had no significant influence on the mechanical properties (Figure 3e). The ultimate tensile strength ranged from 21.1 ± 3.5 MPa (DCN) to 22.1 ± 3.7 MPa (FN). Burst pressures were in the range between 3124 ± 466 mmHg (FN + DCN) to 3326 ± 78 mmHg (controls). Interestingly, the elastic modulus of the samples coated with FN + DCN showed a lower value compared to the controls, although this was not statistically significant (3.7 ± 0.5 MPa FN + DCN versus 4.8 ± 0.6 MPa controls, $p = 0.125$).

We compared the mechanical properties (elastic modulus and burst pressure) of our electrospun scaffolds with autologous grafts, which are today's gold standard for vascular bypass surgeries, using data obtained from literature (Table 2) [65]. The elastic modulus of our constructs (4.8 ± 0.6 MPa) was slightly higher than that of saphenous veins (2.25–4.2 MPa) [66,67] and of iliofemoral arteries (1.54 MPa) and veins (3.11 MPa) [68]. However, compared with an internal mammary artery (8 MPa) and a femoral artery (FA, 10.5 MPa)—used for popliteal bypass surgery—our engineered scaffolds showed a lower elastic modulus [66,69,70]. Regarding the burst pressure, engineered scaffolds (3326 ± 78 mmHg) lied within the range of a saphenous vein (1250–3900 mmHg) [66,67,71,72] and an internal mammary artery (2000–3196 mmHg) [66,71]. Konig et al. recommends for a TEGV a minimum burst pressure of 1700 mmHg [71]. We can therefore argue that our constructs have suitable mechanical properties to serve as a vascular graft or TEGV.

Table 2. Mechanical properties of the electrospun constructs and native blood vessels.

Graft Type	Elastic Modulus (MPa)	Burst Pressure (mmHg)	Ref.
Electrospun vascular graft	4.8 ± 0.6	3326 ± 78	-
Saphenous vein	4.2	1680–3900	[66]
Saphenous vein	2.25	1250	[67]
Saphenous vein	NA	1680	[73]
Saphenous vein	NA	2200	[72]
Saphenous vein	NA	1599	[71]
Internal mammary artery	NA	3196	[71]
Internal mammary artery	8	2000	[66]
Femoral artery	9–12	NA	[69]
Iliofemoral artery	1.54	NA	[68]
Iliofemoral vein	3.11	NA	[68]

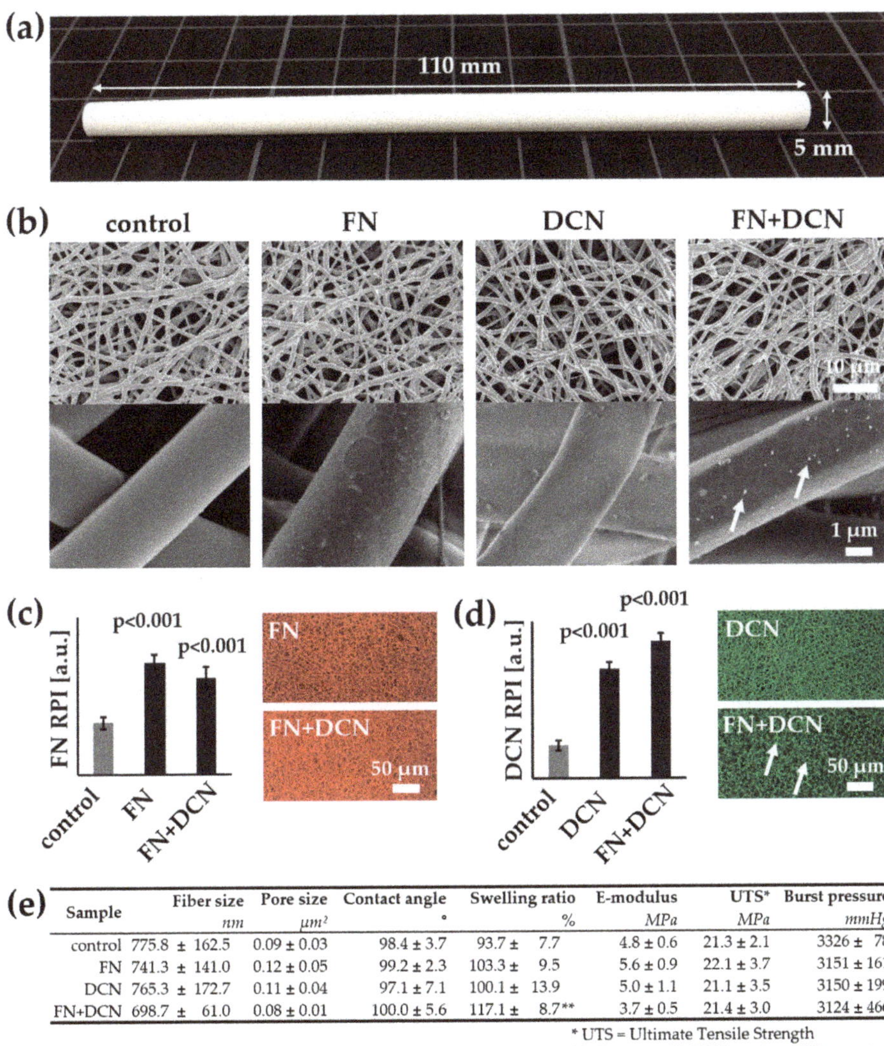

Sample	Fiber size	Pore size	Contact angle	Swelling ratio	E-modulus	UTS*	Burst pressure
	nm	μm²	°	%	MPa	MPa	mmHg
control	775.8 ± 162.5	0.09 ± 0.03	98.4 ± 3.7	93.7 ± 7.7	4.8 ± 0.6	21.3 ± 2.1	3326 ± 78
FN	741.3 ± 141.0	0.12 ± 0.05	99.2 ± 2.3	103.3 ± 9.5	5.6 ± 0.9	22.1 ± 3.7	3151 ± 161
DCN	765.3 ± 172.7	0.11 ± 0.04	97.1 ± 7.1	100.1 ± 13.9	5.0 ± 1.1	21.1 ± 3.5	3150 ± 199
FN+DCN	698.7 ± 61.0	0.08 ± 0.01	100.0 ± 5.6	117.1 ± 8.7**	3.7 ± 0.5	21.4 ± 3.0	3124 ± 466

* UTS = Ultimate Tensile Strength
** One-way ANOVA, n=4, p<0.05 vs control

Figure 3. Morphological and mechanical characterization of the tubular biofunctionalized scaffolds: (**a**) Electrospun tubular scaffolds were fabricated with a length of 110 mm, an inner diameter of 5 mm, and a thickness of 0.40 ± 0.06 mm. (**b**) SEM images of control and biofunctionalized scaffolds: Scaffolds coated with FN show a network-like structure on the fibers. Aggregates deposited on the FN + DCN-coated samples are indicated by white arrows. (**c**,**d**) The coating of FN, DCN, or FN + DCN in combination was confirmed with IF staining: FN (red) and DCN (green). The white arrows indicate aggregates deposited on the FN + DCN-coated samples. Two-tailed *t*-test vs. control, n = 3, RPI = relative pixel intensity. (**e**) Fiber and pore size analysis shows no significant difference between the biofunctionalized scaffolds and the controls. Mechanical properties are not influenced by the protein coating. One-way ANOVA, n = 4, *p* < 0.05 vs. control.

3.2. Decorin and Fibronectin Coating of the Scaffolds Does Not Induce a Disadvantageous Immune Response

The effect of DCN- or FN-coated TPCU scaffolds on immune cells was investigated in order to estimate their suitability as vascular graft material. The immune response of a combination coating was not required as the immune system would not react differently to the presence of both proteins in one coating. The performed immunological evaluation followed the normal sequence of immune activation [9], starting with PMNs that are followed by monocytes, which differentiate into macrophages at the site of injury, and finally T cells that become activated (Figure 4a).

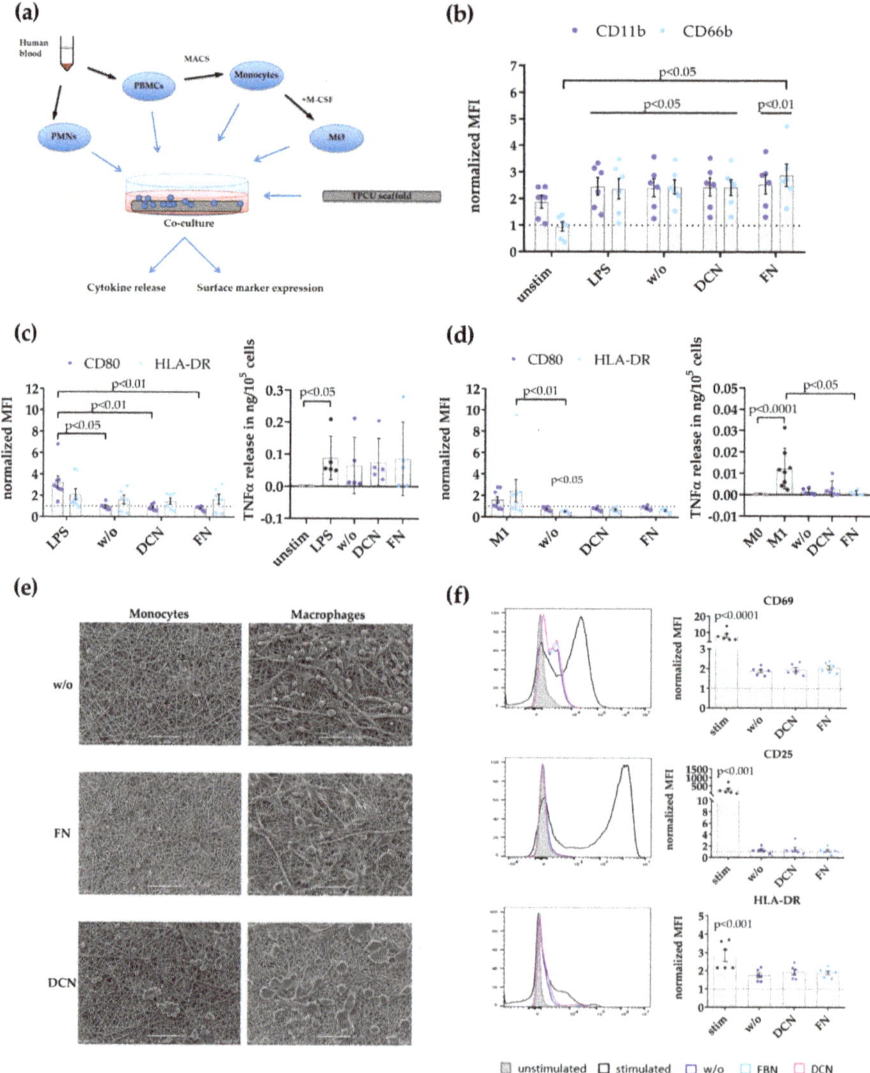

Figure 4. Immune response profile of FN- and DCN-coated planar scaffolds: (**a**) Schematic overview of the analysis steps and used immune cell assays. Polymorph nuclear cells (PMNs) and peripheral blood mononuclear cells (PBMCs) were isolated from human blood. Monocytes were acquired from PBMCs by magnetic separation via CD14 beads. Monocytes were differentiated into M0 macrophages (MØ) by stimulation with 50 ng/mL of macrophage colony-stimulating factor (M-CSF) for 7 days. (**b**) Surface expression of activation markers CD11b and CD66b by PMNs after 4 h: Displayed are the mean fluorescence intensities (MFI) normalized to unstimulated PMNs after isolation as mean ± SEM

(standard error of the mean) for unstimulated (unstim) and lipopolysaccharide (LPS)-stimulated cells, as well as PMNs cultured on the uncoated (w/o), DCN-coated (DCN), and FN-coated (FN) scaffolds determined with flow cytometry. Kruskal–Wallis test, n = 6. (**c**) Surface expression of activation markers CD80 and human leukocyte antigen DR isotype (HLA-DR), and tumor necrosis factor alpha (TNFα) release by monocytes. Shown are the MFI normalized to unstimulated monocytes as mean ± SEM for LPS-stimulated cells as well as monocytes cultured on uncoated (w/o), DCN-coated (DCN), and FN-coated (FN) scaffolds. Kruskal–Wallis test, n = 6–8. The TNF release is depicted in ng/10^5 cells as mean ± SEM for unstimulated (unstim) and LPS-stimulated cells as well as monocytes cultured on the uncoated (w/o), DCN-coated (DCN), and FN-coated (FN) scaffolds. Kruskal–Wallis test, n = 5. (**d**) Surface expression of activation markers CD80 and HLA-DR, and TNFα release by macrophage: Displayed is the MFI normalized to unstimulated M0 macrophages as mean ± SEM for macrophages differentiated to M1 and as well as cells cultured on uncoated (w/o), DCN-coated (DCN), and FN-coated (FN) scaffolds. Kruskal–Wallis test, n = 6–8. The TNFα release is shown in ng/10^5 cells as mean ± SEM for unstimulated M0 macrophages; macrophages differentiated to M1; and as well as cells cultured on the uncoated (w/o), DCN-coated (DCN), and FN-coated (FN) scaffolds. Kruskal–Wallis test, n = 6–9. (**e**) Representative SEM images of monocytes (left) and macrophages (right) on uncoated (w/o) and with biofunctionalized scaffolds (DCN and FN). Scale bars represent 50 μm. (**f**) Expression of activation markers CD69, CD25, and HLA-DR on CD3+ T cells in whole PBMC co-cultures: Shown are representative histograms (left) and the surface expression levels as MFI normalized to unstimulated T cells as mean ± SEM (right) for αCD3/αCD28-stimulated T cells (stim) as well as T cells cultured on uncoated (w/o), DCN-coated, and FN-coated scaffold. Kruskal–Wallis test, n = 6.

Initially, the expression of known PMN activation markers, the integrin CD11b, and the adhesion molecule CD66b was analyzed (Figure 4b). The normalized mean fluorescence intensity (MFI) for CD11b (stim 2.461 ± 0.3323, p = 0.0179; w/o 2.406 ± 0.3393, p = 0.0378; DCN 2.442 ± 0.3361, p = 0.0217; FN 2.549 ± 0.3644, p < 0.0090; all versus unstim 0 hours 1 ± 0) and CD66b (stim 2.372 ± 0.3875, p = 0.0453; w/o 2.448 ± 0.2728, p = 0.0414; DCN 2.431 ± 0.3041, p = 0.0453; FN: 2.893 ± 0.4239, p = 0.0073; all versus unstim 0 h 1 ± 0) was significantly increased on PMNs after LPS stimulation (positive control) and, after culture on the uncoated/coated scaffolds, compared to the level of PMNs directly after isolation (dotted line, set to 1). Additionally, PMNs on FN-coated TPCU scaffolds displayed a significantly higher CD66b expression compared with the unstimulated controls (FN 2.893 ± 0.4239 versus unstim 4 h 0.9438 ± 0.1723, p < 0.0345).

In a next step, monocyte responses were studied by flow cytometry analysis of the activation markers CD80 and HLA-DR (Figure 4c). The expression level for the co-stimulatory molecule CD80 was significantly upregulated only on LPS-stimulated monocytes compared with all other experimental groups (LPS 3.254 ± 0.5533 versus w/o 0.9592 ± 0.1342, p = 0.0143; versus DCN 0.8888 ± 0.1209, p = 0.0046; versus FN 0.8325 ± 0.08414, p = 0.0018). No significant differences in HLA-DR expression were detectable between the tested conditions. Additionally, no enhanced TNFα release of monocytes cultured on the uncoated/coated scaffolds was measured in contrast to a significantly elevated secretion in the LPS-stimulated controls compared to the unstimulated controls (LPS 0.08859 ± 0.03039 versus unstim 0.0005580 ± 0.0002111, p = 0.0228).

Then, macrophages (M0 type) generated in vitro by M-CSF were screened for signs of activation or polarization (Figure 4d). M0 (unstimulated) and M1 macrophages (IFNγ/LPS-stimulated) were used as control groups. Enhanced CD80 and HLA-DR expression and increase of TNFα secretion are hallmarks of pro-inflammatory M1 macrophages. There was no difference in the CD80 expression level between M0 macrophages (dotted line, set to 1) and all other experimental groups. The expression of HLA-DR by macrophages on uncoated scaffolds was significantly decreased compared with the M0 and M1 control settings (w/o 0.5220 ± 0.05753 versus M0 1 ± 0, p = 0.0106; versus M1 2.453 ± 1.040, p = 0.0049). Whereas M1 macrophages significantly elevated their TNFα release compared with M0 macrophages (M1 0.01229 ± 0.003333 versus M0 0.0002707 ± 0.00004142, p < 0.0001), no enhancement in pro-inflammatory cytokine release was measurable in all other experimental groups. Macrophages on the FN-coated scaffolds actually decreased their TNFα release compared with the M1 controls

(FN 0.0009826 ± 0.0004063 versus M1 0.01229 ± 0.003333, p = 0.0432). Complementary to the analysis of changes in surface marker and pro-inflammatory cytokine release by monocytes and macrophages, scanning electron microscopy was applied to assess the effects of co-culture on their morphology (Figure 4e). Scanning electron microscopy images were taken after the cells were cultured for two days on the different scaffold groups. Monocytes and macrophages on the DCN-coated scaffolds formed clusters of preferentially rounded cells. Macrophages cultured on uncoated or FN-coated scaffolds displayed more diverse shapes in contrast with cells grown on the DCN-coated TPCU scaffolds.

The potential activation of T cells was determined by flow cytometry analysis of known activation markers CD69, CD25, and HLA-DR [74] after culturing complete human PBMCs on either uncoated or coated scaffolds (Figure 4f). However, only anti-CD3/anti-CD28 stimulated T cells (stim; positive control) significantly elevated the expression level for CD69 (stim 7.956 ± 1.319 versus unstim 1 ± 0, p < 0.0001), CD25 (stim 265.6 ± 101.5 versus unstim 1 ± 0, p = 0.0008), and HLA-DR (stim 2.824 ± 0.3099 versus unstim 1 ± 0, p = 0.0001) compared with the level of the unstimulated controls (dotted line, set to 1). No significant increase in T cell activation marker expression was observed in any other experimental group.

3.3. Simulation of Endothelial Progenitor Cell Homing Using Endothelial Colony Forming Cells

3.3.1. ECFCs Show Altered VEGFR2 and PECAM-1 Expression Patterns on FN + DCN-Coated TPCU Scaffolds Under Static Culture Conditions

ECFCs were seeded on the biofunctionalized planar scaffolds and cultured under static conditions for 24 and 48 h. The amount of adherent ECFCs was significantly higher on samples coated with FN (24 h: 257 ± 57 cells/mm^2 versus control with 137 ± 46 cells/mm^2, p < 0.01; 48 h: 301 v 64 cells/mm^2 versus control with 52 ± 32 cells/mm^2, p < 0.001) and FN + DCN (24 h: 243 ± 63 cells/mm^2 versus control with 137 ± 46 cells/mm^2, p < 0.01; 48 h: 292 ± 54 cells/mm^2 versus control with 52 ± 32 cells/mm^2, p < 0.001) when compared with the uncoated samples (controls) throughout the entire culture period (Figure 5a). No significant difference of adherent cells was observed between FN coating and FN + DCN coating (24 h: p = 0.656; 48 h: p = 756). DCN coating did not show any significant difference in cell density in comparison with the uncoated controls (24 h: 105 ± 40 cells/mm^2 versus control with 137 ± 46 cells/mm^2, p = 0.340; 48 h: 30 ± 11 cells/mm^2 versus control with 52 ± 32 cells/mm^2, p = 0.460).

SEM analyses revealed that the ECFCs on the control and DCN-coated TPCU scaffolds had attained a spherical shape after 24 h whereas those on TPCU scaffolds that were coated with FN and FN + DCN showed a stretched morphology (Figure 5b). Immunofluorescence staining of samples 24 h after seeding (Figure 5c,d) identified a significantly lower PECAM-1 expression in ECFCs on FN + DCN-coated samples in comparison with FN coating (0.64 ± 0.30 versus 0.90 ± 0.25, p < 0.05). After 48 h, this effect tended to reverse, although the difference was not significant (0.70 ± 0.15 versus 0.54 ± 0.23, p = 0.073). A similar and statistically not significant tendency was detected for the fluorescence intensity of vWF. No significant changes were observed in VE-cadherin or vinculin expression. VEGFR2 expression was significantly decreased in cells cultured on FN-coated scaffolds when compared with cells grown on FN + DCN-coated scaffolds after 24 h (0.64 ± 0.11 versus 0.29 ± 0.16, p < 0.01). After 48 h, this effect vanished (0.28 ± 0.17 versus 0.28 ± 0.15, p = 0.942).

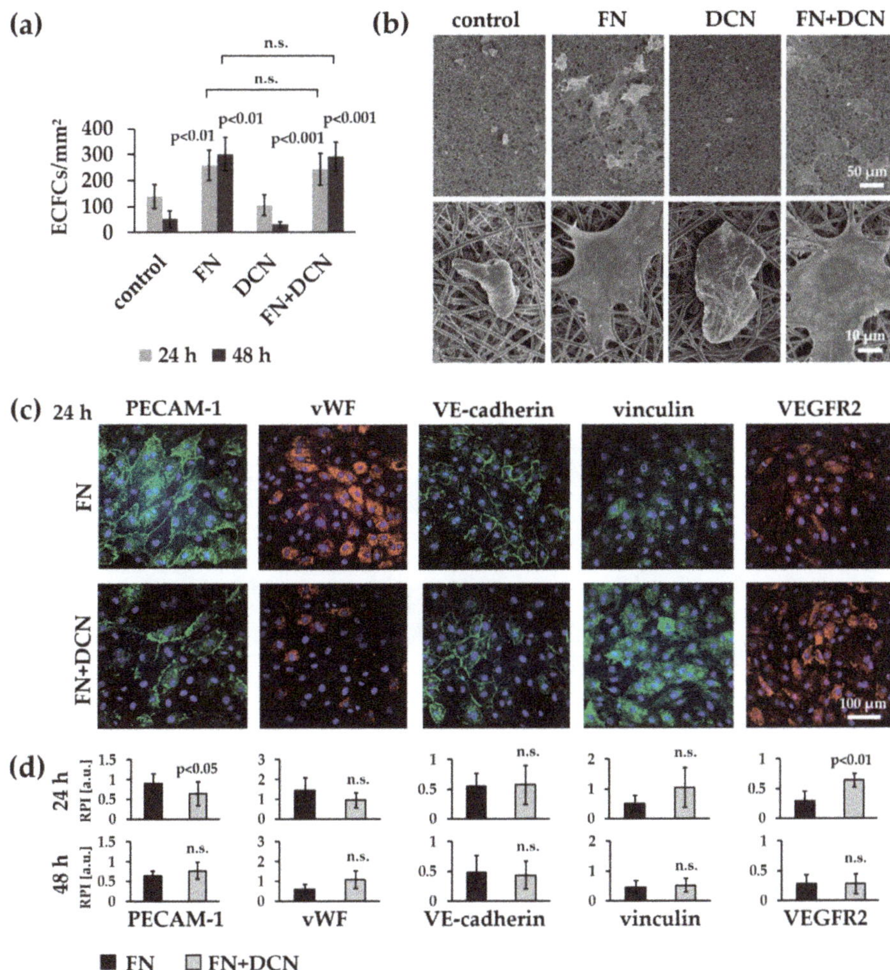

Figure 5. Static experiments of human ECFCs on FN-, DCN-, or FN + DCN-coated scaffolds: (**a**) Attachment and proliferation of the human ECFCs after 24 h and 48 h. Cells on FN and FN + DCN coating show a significantly higher proliferation when compared with cells gown on DCN and controls. Two-tailed *t*-test, compared to controls, n = 5, n.s. = not significant. (**b**) SEM images and (**c**) Immunofluorescence staining of ECFCs 24 h after seeding on ECM protein-coated scaffolds: Cells on FN and FN + DCN show a spread morphology in contrast to DCN coating and controls. (**d**) Semiquantitative fluorescence intensity analysis (relative pixel intensity (arbitrary units)) of cells on FN and FN + DCN shows no significant difference for the endothelial cell type marker von Willebrand factor (vWF) as well as vinculin and vascular endothelial cadherin (VE-cadherin). Platelet endothelial cell adhesion molecule (PECAM-1) expression is significantly decreased and VEGFR2 expression is significantly increased on FN + DCN-coated scaffolds after 24 h. Two-tailed *t*-test, n = 6, n.s. = not significant.

3.3.2. FN + DCN-Coating Attracts ECFCs Under Dynamic Culture Conditions

After ECFC seeding under static conditions, the cell-seeded scaffolds were dynamically cultured on a roller mixer for 24 h (Figure 6a). This approach was performed to reflect more closely the in vivo conditions. The analysis of the adherent cells showed a significantly increased cell number on the

FN + DCN-coated samples when compared with the controls and DCN-coated samples (5.7% ± 4.4% versus DCN coating with 1.0% ± 0.8%, $p < 0.05$ and versus control with 0.6% ± 0.7%, $p < 0.05$). The FN coating led to a nonsignificant decrease of adherent cells compared to FN + DCN coating (Figure 6b; 3.4% ± 1.5% versus 5.7% ± 4.4%, $p = 0.226$). Cells on all samples showed comparable PECAM-1 and vWF expression levels (Figure 6c). Distinct differences were observed in the cell morphology. F-actin staining helped visualizing the spread cells on the FN- and FN + DCN-coated scaffolds and cells with a more rounded morphology on the control samples and DCN-coated scaffolds (Figure 6c).

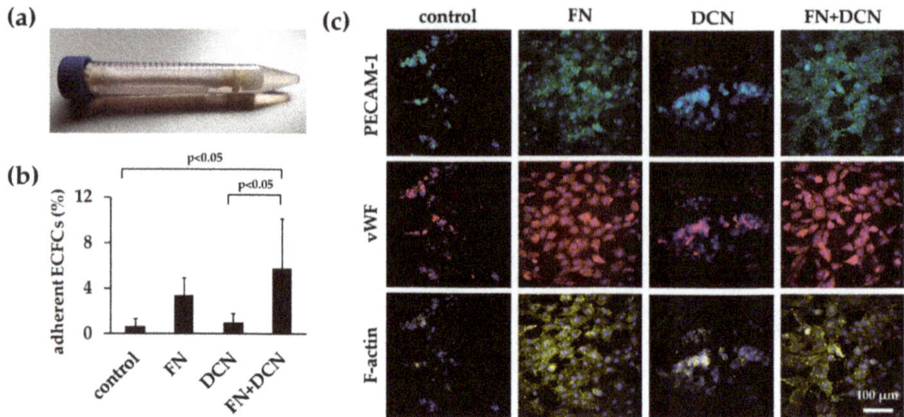

Figure 6. In vitro simulation of in vivo processes: ECFC attraction under dynamic conditions. (**a**) ECFCs were seeded into tubular constructs and cultured for 24 h on a roller mixer. (**b**) Adherent cells after 24 h on control scaffolds and on DCN-, FN-, and FN + DCN-coated scaffolds. FN + DCN coating shows a significantly higher cell number when compared with DCN coating and controls. One-way ANOVA, n = 4. (**c**) PECAM-1 (green), vWF (red), and F-actin (yellow) expression in ECFCs. Cells on FN and FN + DCN show a more spread morphology in contrast to the DCN and control samples.

3.4. In Vitro Tissue Engineering Approach Using Vascular Endothelial Cells

3.4.1. vECs Form an Endothelial Layer on FN- and FN + DCN-Coated Scaffolds Under Static Culture Conditions

vECs were seeded on the biofunctionalized planar constructs and cultured for 1, 4, and 7 days in order to investigate endothelialization (Figure 7a). One day after seeding, the cell number for all conditions was not significantly different. On day 4, vECs significantly increased proliferation on FN coating (78 ± 26 cells/mm^2 versus control with 8 ± 7 cells/mm^2, $p < 0.01$) and FN + DCN coating (55 ± 27 cells/mm^2 versus control with 8 ± 7 cells/mm^2, $p < 0.05$), while the VEC count on the DCN-coated samples had slightly decreased (7 ± 5 cells/mm^2 versus control with 8 ± 7 cells/mm^2, $p < 0.931$). This trend continued until day 7, on which a significantly increased cell count was detected for FN coating (186 ± 47 cells/mm^2 versus control with 16 ± 16 cells/mm^2, $p < 0.001$) and FN + DCN coating (135 ± 50 cells/mm^2 versus control with 16 ± 16 cells/mm^2, $p < 0.01$) in comparison with the uncoated controls. DCN coating of the TPCU scaffolds showed no improvement when compared with the control samples. Over the entire period of the experiment, the cell count was not significantly different between FN and FN + DCN coating.

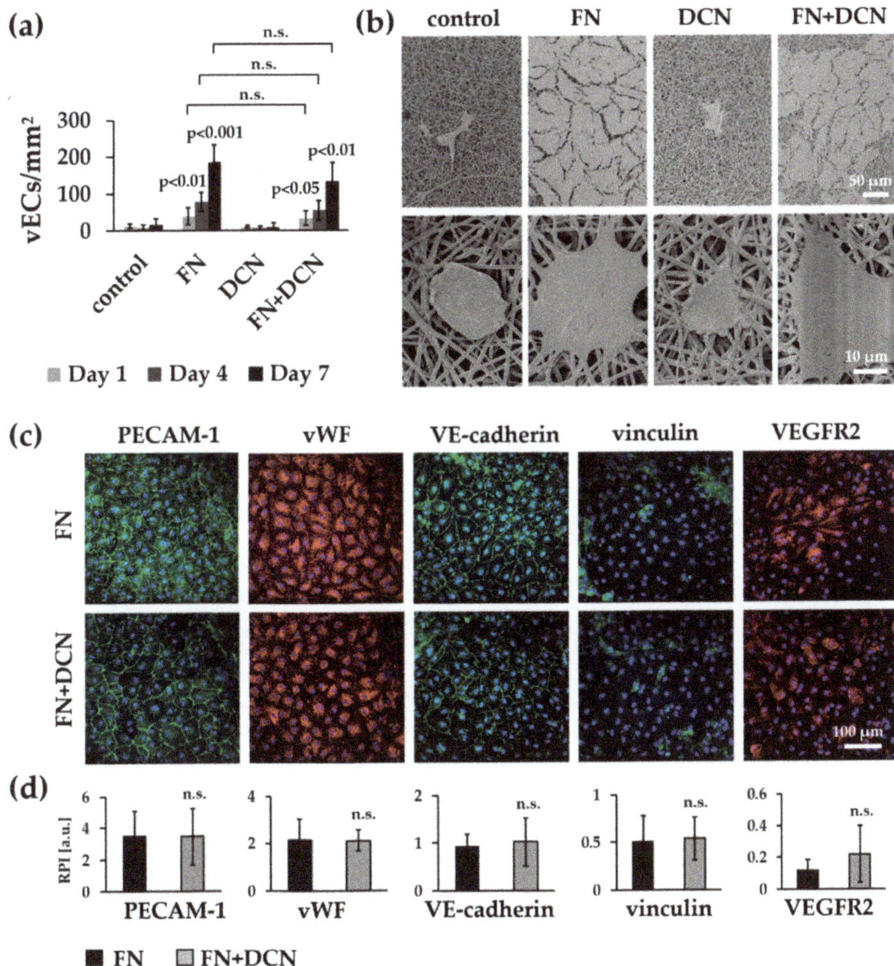

Figure 7. Static cell culture experiments of vECs on FN- and DCN-coated scaffolds: (**a**) Attachment and proliferation of vECs after 1, 4, and 7 days. vECs on FN and FN + DCN coating show a significantly higher proliferation rate compared with cells gown on DCN coating or control scaffolds. Two-tailed *t*-test, compared with control samples, n = 3, n.s. = not significant. (**b**) SEM images and (**c**) IF staining of vECs 7 days after seeding on ECM-coated scaffolds. Cells on FN and FN + DCN coating show a spread morphology in contrast with cells on DCN coating and control samples. (**d**) Semiquantitative fluorescence intensity analysis (relative pixel intensity (a.u.)) of cells on FN and FN + DCN coating shows no significant difference for PECAM-1, vWF, vinculin, or VE-cadherin expression. Two-tailed *t*-test, n = 5, n.s. = not significant.

While vECs on the control and DCN-coated scaffolds showed a spherical shape after 7 days as assessed using SEM, on FN and FN + DCN-coated scaffolds, vECs were stretched out and formed an almost confluent endothelial cell layer (Figure 7b). IF staining confirmed the expression of the endothelial cell type-specific markers PECAM-1, vWF, and VE-cadherin in the vECs on both FN and FN + DCN coating (Figure 7c). Semiquantitative analysis of fluorescence intensities revealed no significant differences of marker expression between FN and FN + DCN coating (Figure 7d). Vinculin expression was comparable in vECs on both coatings. With regard to VEGFR2, an increased fluorescence intensity in cells grown on the FN + DCN-coated samples was observed. However, due to a high variation in expression levels of individual experiments, no statistical significance between cells grown on FN or FN + DCN coating could be determined.

In summary, our data showed that DCN coating of the TPCU scaffolds did not have a substantial advantage when aiming for an increased VEC proliferation or an improved cell–cell or cell–material interaction. For this reason, only FN biofunctionalized TPCU scaffolds were used for the following in vitro tissue engineering experiments.

3.4.2. vECs Cultured in a Custom-Made Bioreactor Under Flow Form a Confluent and Aligned Cell Layer on FN-Biofunctionalized TPCU

After successful implementation of the developed bioreactor system, we aimed to test whether the FN-biofunctionalized TPCU scaffolds can be endothelialized under dynamic conditions. vECs were seeded into the tubular TPCU scaffolds, and after an initial culture for three days under static conditions to allow cell attachment, a flow was employed that was stepwise increased to 25 mL/min within 1.5 days (Figure 2e). Under this flow, which causes a shear stress of about 0.03 Pa, the vEC-seeded FN-biofunctionalized scaffolds were cultured for seven days. Metabolic activity assessment using an MTT assay showed that a large part of the inner wall of our construct was covered with living cells, as indicated by the purple formazan stain (Figure 8a). IF staining and SEM further revealed a layer of confluent vECs that were aligned in the direction of flow (Figure 8b,c).

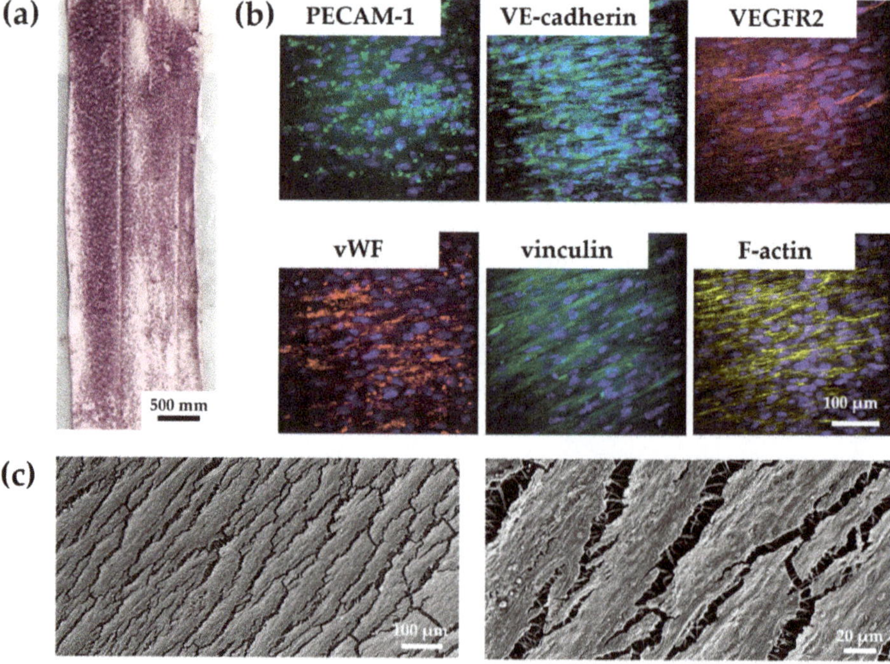

Figure 8. Tissue-engineering approach with vascular endothelial cells cultured for 7 days on FN-biofunctionalized electrospun tubular TPCU scaffolds under dynamic conditions: (**a**) Inner wall of the tubular construct shows living vECs indicated by the purple formazan stain. (**b**) PECAM-1, vWF, VE-cadherin, vinculin, VEGFR2, and F-actin expression were detected. vECs show an aligned morphology. (**c**) SEM confirms vECs that had aligned with the flow to which they were exposed to during the dynamic culture in the bioreactor.

We confirmed the expression of the endothelial cell markers PECAM-1, vWF, and VE-cadherin. However, PECAM-1 and VE-cadherin did not appear to be located on the cell membrane as usual. Vinculin and VEGFR2 were also detected in the cells. Nevertheless, the staining of VEGFR2 showed only a weak signal.

4. Discussion

Due to a proven biocompatibility and biostability at body temperature [55,57], we selected for this study a novel thermoplastic polycarbonate urethane for the fabrication of a TEVG. At first, scaffolds were produced by electrospinning of the TPCU and were disinfected with 70% ethanol. Microbiological studies showed that ethanol treatment did not achieved 100% sterility (Figure S3; 2 out of 9 plates showed germ growth). We are aware that disinfection with ethanol does not necessarily inactivate all forms of microorganisms [75]; therefore, for the clinical translation, a more efficient sterilization method should be considered.

After disinfection, scaffolds were then biofunctionalized by adsorption of FN and DCN, either alone or in combination. The adsorbed proteins did not impact elastic modulus or burst pressure of the tubular constructs (Figure 3). We demonstrated that the biomechanical properties of our constructs were comparable to native vascular tissue (Table 2).

The ability to mimic the nanofibrous topography of the ECM makes electrospinning a powerful method for cardiovascular tissue-engineering applications. Several studies have already described the influence of fiber and pore size on cell adhesion, cell migration, proliferation, and differentiation, as well as cell–cell interaction [76–78]. In native blood vessels, the ECs are located on the basal lamina, a mixture of defined ECM proteins that form a network and bind cells [79]. The literature describes a wide range of pore and fiber diameters (1–1000 nm) from different vessels, depending on the position and physical properties of the vessel [80]. The main collagen component of the basal lamina is collagen type IV. It forms fibers that range from 20 to 52 nm [80–82]. In our study, the fiber diameters were between 699 ± 61 nm and 776 ± 163 nm, which is much higher compared to the collagen type IV fibers in native vessels. However, other studies developing electrospun vascular grafts reported comparable [83] or even larger fiber sizes [84,85] on which a functional endothelium was formed [84]. The pore size strongly depends on the vessel type and ranges between 5 nm and 8 µm [80,82,86–89]. Our constructs showed pore sizes between 0.08 ± 0.01 µm^2 and 0.12 ± 0.05 µm^2, which lies in the range of a native vessel.

Several studies have already described that FN improves the endothelialization of vascular grafts [19,48,49,51]. In our study, we observed a fibrous-like structure of the coated FN (Figure 3b). This phenomenon can be interpreted as material-driven fibrillogenesis, first described by Salmeron-Sanchez et al. [3]. In the human body, FN matrix assembly is a cell-mediated process [90] that influences cell growth, cell differentiation, and cell–cell interaction [76–78,90,91]. It has been shown that the adhesion of FN on poly (ethyl acrylate) (PEA) can lead to a spontaneous organization of FN into protein networks. It has also been shown that cell-free material-induced FN fibrillogenesis influences the maintenance and differentiation of stem cells [3,92]. Furthermore, it was described that the FN network has an increased ability to store growth factors [93]. To the best of our knowledge, our study is the first to show that material-driven fibrillogenesis can be observed on electrospun TPCU fibers. We presume that the surface properties, such as hydrophobicity and polarity, are comparable to those of PEA. Whether the FN network has a significant advantage in terms of cell behavior or growth factor binding compared to dispersed, coated FN molecules would need further investigation.

In addition to FN coating, in this study, we also used DCN coating. We observed that, after coating on the TPCU, DCN was randomly distributed in aggregates on the fibers (Figure 3b). Since DCN does not form fibrils, this coating behavior was expected. Even larger, globular DCN aggregates were observed on the FN + DCN samples (Figure 3b,d). Interestingly, these aggregates were predominantly seen on the FN fibrils and not on the TPCU itself. It is known that DCN interacts with FN [94,95]. Furthermore, the interaction of proteins with materials is determined by the geometrical, chemical, and electrical properties of the substrate [96]. In this respect, it can be hypothesized that the DCN prefers the FN surface more than the hydrophobic polyurethane surface. Interestingly, we observed a significantly increased swelling ratio for FN + DCN (Figure 3e). This was not the case with individually FN- or DCN-coated TPCU. Depending on the surface properties of the material and the interaction with other proteins, the conformation, orientation, and bioactivity of a protein can also be influenced [96–98]. With this in mind, one can assume that both DCN and FN in combination can have a different bioactivity [99].

In contrast to our previous findings using poly (ethylene glycol) dimethacrylate-poly (L-lactide) (PEGdma-PLA) or a blend of poly-ε-caprolacton and gelatin [25,100], we identified a cell-repellent effect of the DCN-coated TCPU electrospun scaffolds for both human ECFCs and human vECs. As already discussed, cells prefer to adhere to hydrophilic surfaces [101]. Since the TCPU itself is highly hydrophobic (control: 98.4 ± 3.7 °), it can cause a cell-repellent effect. DCN alone was not able to diminish this effect (Figure 5a,b). Cell adhesion is influenced by cell-adhesive peptides such as the RGD sequence. Since DCN does not contain these sequences, as it is the case with FN, we assume that at least this integrin-based cell–material interaction cannot be mediated by DCN. It has been described that DCN can even partially inhibit cell adhesion; however, this has only been observed with fibroblasts and not with endothelial cells [28,102]. Hinderer et al. observed an attraction of ECFCs to DCN-coated PEGdma-PLA [25]. A direct comparison with this study is therefore difficult, since this polymer has different surface properties, which influence the amount and orientation of the adsorbed DCN and thus may have an altered impact on cell behavior [96]. FN coating reversed the cell-repellent effect of the TCPU, both with and without DCN (Figure 5). We can therefore conclude that the cell attraction and proliferation is supported by FN but not affected by DCN [99,103].

Scaffolds should in general exhibit a low immunogenicity and at the same time support tissue regenerative processes. The evaluation of the immune response profiles of the analyzed control and ECM-coated scaffolds excluded any major adverse effects, with only minor innate activation characteristics. Co-culturing PMNs, as the first cells of an innate immune response, induced an activated cell phenotype regarding the expression of CD11b and CD66b. Monocytes were incompletely activated after co-culturing with the scaffold as indicated by only a weak tendency to upregulate the HLA-DR expression and to increase their TNFα release. From the literature, it is well known that the upregulation of CD80 and HLA-DR would be a hallmark of M1 macrophages [62,104] and that the fiber and pore size of electrospun scaffolds could impact the macrophage polarization state [105]. When analyzing the potential impact of the TPCU scaffolds on macrophage polarization, no clear trend to drive the process into a specific macrophage subtype could be determined. Also, the coating by either DCN or FN did not trigger a specific type of macrophage polarization. In contrast, co-culture studies with soluble recombinant DCN demonstrated that macrophages responded with an upregulated CD80 expression as well an increased secretion of TNFα and IL-10 [25]. The absent responses in the present study may result from the far lower amount of protein present on the coated scaffolds in comparison with the high protein amounts available within solutions or even by conformational changes. Not surprisingly, adaptive T cell responses were also not detected. T cells on scaffolds simply showed a trend to upregulate CD69 and HLA-DR without significant changes.

A functional endothelium is mainly characterized by cell–cell junctions [106]. As PECAM-1 is the most abundant component of the EC junction, which contributes to the maintenance of the EC permeability barrier, its expression is essential for a functional EC layer [107]. In our study, the ECFCs on FN coating revealed a significantly increased PECAM-1 expression after 24 h compared with ECFCs cultured on FN + DCN-coated scaffolds. In contrast, the VEGFR2 expression was significantly decreased in the ECFCs on FN coating after 24 h compared with FN + DCN coating. It has been reported that VEGFR2 is highly expressed in early endothelial precursor cells but not in all mature ECs [108,109]. For example, PECAM-1 is less expressed in endothelial progenitor cells, as it is typically associated with a more mature EC phenotype [110]. Interestingly, DCN has been reported to stimulate the maintenance of undifferentiated progenitor cells [111], and FN promotes endothelial cell differentiation [112]. Therefore, we hypothesize that the FN + DCN coating in our experiments kept the ECFCs in a precursor cell state compared with the culture on only FN. It may also be possible that a direct interaction of DCN with VEGFR2 leads to its upregulation. A positive feedback loop between VEGF and VEGFR2 has been described [113]. Whether DCN has the same effect remains to be confirmed.

Since DCN exerts many other functions, an indirect regulation of VEGFR2 is also conceivable [34,114]. Mazor et al. showed that the matrix metalloproteinase-1 (MMP-1) promotes the expression of VEGFR2 [115]. The core protein of DCN in turn is able to stimulate the expression of MMP-1 [116,117]. Furthermore, Murakami et al. reported that increased concentrations of the fibroblast growth factor (FGF) led to an increase in VEGFR2 levels [118]. DCN, in turn, can bind to FGF and can increase its activity [119]. It was also described that VEGFR2 expression is regulated by the disruption of the c-MET receptor tyrosine kinase [120]. As an antagonistic ligand of c-MET, DCN is able to inhibit its activity and thus might indirectly promote VEGFR2 expression [38]. We have already discussed the hypothesis that DCN in interaction with FN may exhibit an altered bioactivity. This would explain why DCN, which was adsorbed on the TPCU scaffold surface, impacted ECs in combination with FN but did not without [96–98]. The reason for VEGFR2 upregulation can also be due to FN. It might be possible that, in combination with DCN, its conformation and function is also changed [96–98]. It has been shown that conformational remodeling of the FN matrix selectively regulates VEGF signaling [121]. VEGF in turn regulates VEGFR2 expression [113]. By binding to VEGF, FN can promote full phosphorylation and activation of VEGFR2 [122]. Interestingly, after 48 h, the difference between FN and FN + DCN coating for both the PECAM-1 and VEGFR2 expression had vanished (Figure 5d). With regard to VEGFR2, a short half-life of the receptor is described, which enables ECs to adapt quickly to changes in the extracellular environment [118,123]. This leads to the question of how long the biofunctionalized DCN coating was fully biologically active in our study. Due to its natural presence in the body, it can be easily degraded [124]. We showed that DCN acts on ECFCs for at least 24 h under static conditions. The culture of vECs over 7 days under static conditions revealed the same expression of PECAM-1 and VEGFR2 on FN and FN + DCN coating (Figure 7). This observation supports the assumption that the DCN was only active for a short period of time and that its effect had disappeared after 7 days. In addition, it is possible that the vECs are not as sensitive to DCN, as we have observed with the ECFCs. Several studies have described an increase in VEGFR2 expression during differentiation and expansion of endothelial progenitor cells [109,125]. At the same time, VEGFR2 expression was relatively low during the proliferation phase [126]. Since the vECs are mature cells, it can be assumed that the externally changed conditions do not affect the VEFGR2 expression significantly. Nevertheless, in this study, we successfully showed that vECs formed an endothelium on biofunctionalized FN-coated constructs after 7 days of culture whereas DCN-coated TPCU scaffolds did not show a significant effect on cell proliferation.

In our TEVG experiments using a custom-made bioreactor, we observed a unidirectional cell orientation in the direction of the flow. The response of ECs to shear stress is well studied [127–129]. It has been shown that, under flow, the morphology of vECs changes from a cobblestone (static) to an elongated form and that vECs align in the direction of the flow in only 24 h [127]. The hemodynamic forces can modulate not only the phenotype but also the gene expression of the cells. In this context, the correct flow is of great importance for a properly functioning endothelium [130]. In our study, IF staining revealed the expression of vWF, PECAM-1, and VE-cadherin. However, PECAM-1 and VE-cadherin were not located on the cell membrane as usually seen. VEGFR2 expression was quite weak, and the F-actin staining revealed a rather fibroblast-like cell morphology. We hypothesize that the vECs underwent endothelial-mesenchymal transition (EndMT). ECs, which undergo EndMT, lose the expression of the characteristic surface endothelial markers PECAM-1, VE-cadherin, and VEGFR2 [39,131,132]. Mahmoud et al. showed that the EndMT can be induced under low shear stress (0.4 Pa) [133]. In our approach, the cells experienced a wall shear stress of about 0.03 Pa, which is slightly lower than a venous wall shear stress (0.06 Pa) [134]. In silico simulations of our dynamic bioreactor culture confirmed laminar flow conditions along a large part of the vascular wall using the applied parameters. Another reason for the fibroblast-like phenotype could be that ECs are highly plastic [135,136]. Therefore, culturing ECs in vitro in an artificial environment can lead to cell dedifferentiation [136,137]. This highlights the importance of fine-tuning the culture conditions to create a functional TEGV.

5. Conclusions

In the present study, we successfully engineered a TPCU electrospun vascular graft which combines appropriate mechanical properties with a highly bioactive surface for the attraction of ECs. The FN biofunctionalization was characterized by a material-driven fibrillogenesis, which might have a positive impact on FN functionality [3]. To imitate the physiological conditions of a blood vessel, a bioreactor for in vitro tissue culture was designed and manufactured. vECs seeded on the FN-functionalized constructs formed a confluent and functional endothelium under static and dynamic conditions. In contrast, DCN-biofunctionalized TPCU scaffolds had a cell-repellent effect on vECs and ECFCs, most likely due to the high hydrophobic properties of the TPCU. However, since DCN has been shown to inhibit the adhesion of fibroblasts, it remains a promising protein for the functionalization of vascular grafts [29].

The challenge for the future will be to combine the advantages of different proteins and to thus increase the selectivity, functionality, and stability of a biofunctionalized vascular graft while keeping the complexity of the coating as low as possible.

Supplementary Materials: The following are available online at http://www.mdpi.com/2073-4409/9/3/778/s1, Figure S1: Mechanical characterization of the electrospun TPCU scaffolds and biocompatibility of the materials, Figure S2: Cytotoxicity tests of the materials, Figure S3: Microbiological studies of the ethanol disinfected electrospun TPCU scaffolds, Figure S4: The part of the culture chamber that was considered for CFD simulations, Figure S5: The Poiseuille values (developed wall shear stress value) within the scaffold for different flow rates.

Author Contributions: Conceptualization, R.D., D.V., C.W., S.H., U.A.S., M.S. (Martina Seifert), and K.S.-L.; methodology, R.D., D.V., C.W., and M.S. (Maria Schneider); resources, K.S.-L., M.S. (Martina Seifert), L.K., G.L., and M.W.; writing—original draft preparation, R.D.; writing—review and editing, D.V., C.W., M.S. (Martina Seifert), and K.S.-L.; visualization, R.D., D.V., and C.W.; supervision, K.S.-L., M.S. (Martina Seifert), S.H., M.S. (Maria Schneider), and L.K.; project administration, K.S.-L., U.A.S., and M.S. (Martina Seifert); funding acquisition, K.S.-L., U.A.S., and M.S. (Martina Seifert). All authors have read and agreed to the published version of the manuscript.

Funding: This research was funded by the Deutsche Forschungsgemeinschaft (SCHE701/14-1 to K.S.-L., STO359/13-1 to U.A.S., and SE657/12-1 to M.S.). R.D. was funded by the doctoral program of the University Tübingen "Intelligente Prozess- und Materialentwicklung in der Biometrics (IPMB)" that is supported by the MWK Baden-Württemberg.

Acknowledgments: The authors are thankful to Rebecca Haupt for her support in electrospinning, Elke Nadler and Kathrin Stadelmann for the SEM imaging and scientific advice, Elsa Arefaine for the microbiological studies, and Germano Piccirillo for his scientific advice.

Conflicts of Interest: The authors declare no conflict of interest.

References

1. Catto, V.; Farè, S.; Freddi, G.; Tanzi, M.C. Vascular Tissue Engineering: Recent Advances in Small Diameter Blood Vessel Regeneration. *ISRN Vasc. Med.* **2014**, *2014*, 923030. [CrossRef]

2. Causes of Death. Available online: https://www.who.int/data/gho/data/themes/topics/causes-of-death/GHO/causes-of-death (accessed on 29 December 2019).

3. Salmerón-Sánchez, M.; Rico, P.; Moratal, D.; Lee, T.T.; Schwarzbauer, J.E.; García, A.J. Role of Material-Driven Fibronectin Fibrillogenesis in Cell Differentiation. *Biomaterials* **2011**, *32*, 2099–2105. [CrossRef] [PubMed]

4. Sánchez, P.F.; Brey, E.M.; Carlos Briceño, J.C. Endothelialization Mechanisms in Vascular Grafts. *J. Tissue Eng. Regen. Med.* **2018**, *12*, 2164–2178. [CrossRef] [PubMed]

5. L'Heureux, N.; Dusserre, N.; Marini, A.; Garrido, S.; De la Fuente, L.; McAllister, T. Technology Insight: The Evolution of Tissue-Engineered Vascular Grafts - From Research to Clinical Practice. *Nat. Clin. Pract. Cardiovasc. Med.* **2007**, *4*, 389–395. [CrossRef] [PubMed]

6. Ercolani, E.; Del Gaudio, C.; Bianco, A. Vascular Tissue Engineering of Small-Diameter Blood Vessels: Reviewing the Electrospinning Approach. *J. Tissue Eng. Regen. Med.* **2015**, *9*, 861–888. [CrossRef] [PubMed]

7. Seifu, D.G.; Purnama, A.; Mequanint, K.; Mantovani, D. Small-Diameter Vascular Tissue Engineering. *Nat. Rev. Cardiol.* **2013**, *10*, 410–421. [CrossRef]

8. Ravi, S.; Qu, Z.; Chaikof, E.L. Polymeric Materials for Tissue Engineering of Arterial Substitutes. *Vascular* **2009**, *17*, S45–S54. [CrossRef]

9. Julier, Z.; Park, A.J.; Briquez, P.S.; Martino, M.M. Promoting Tissue Regeneration by Modulating the Immune System. *Acta Biomater.* **2017**, *53*, 13–28. [CrossRef]

10. Hinderer, S.; Brauchle, E.; Schenke-Layland, K. Generation and Assessment of Functional Biomaterial Scaffolds for Applications in Cardiovascular Tissue Engineering and Regenerative Medicine. *Adv. Healthc. Mater.* **2015**, *4*, 2326–2341. [CrossRef]

11. Ndreu, A.; Nikkola, L.; Ylikauppilar, H.; Ashammakhi, N.; Hasirci, V. Electrospun Biodegradable Nanofibrous Mats for Tissue Engineering. *Nanomedicine* **2008**, *3*, 45–60. [CrossRef]

12. Li, M.; Mondrinos, M.J.; Gandhi, M.R.; Ko, F.K.; Weiss, A.S.; Lelkes, P.I. Electrospun Protein Fibers as Matrices for Tissue Engineering. *Biomaterials* **2005**, *26*, 5999–6008. [CrossRef]

13. Boland, E.D.; Matthews, J.A.; Pawlowski, K.J.; Simpson, D.G.; Wnek, G.E.; Bowlin, G.L. Electrospinning Collagen and Elastin: Preliminary Vascular Tissue Engineering. *Front. Biosci.* **2004**, *9*, 1422–1432. [CrossRef] [PubMed]

14. Zhang, M.; Wang, Z.; Wang, Z.; Feng, S.; Xu, H.; Zhao, Q.; Wang, S.; Fang, J.; Qiao, M.; Kong, D. Immobilization of Anti-CD31 Antibody on Electrospun Poly(E{open}-Caprolactone) Scaffolds through Hydrophobins for Specific Adhesion of Endothelial Cells. *Colloids Surf. B Biointerfaces* **2011**, *85*, 32–39. [CrossRef]

15. Markway, B.D.; McCarty, O.J.T.; Marzec, U.M.; Courtman, D.W.; Hanson, S.R.; Hinds, M.T. Capture of Flowing Endothelial Cells Using Surface-Immobilized Anti-Kinase Insert Domain Receptor Antibody. *Tissue Eng.-Part C Methods* **2008**, *14*, 97–105. [CrossRef]

16. Kanie, K.; Narita, Y.; Zhao, Y.; Kuwabara, F.; Satake, M.; Honda, S.; Kaneko, H.; Yoshioka, T.; Okochi, M.; Honda, H.; et al. Collagen Type IV-Specific Tripeptides for Selective Adhesion of Endothelial and Smooth Muscle Cells. *Biotechnol. Bioeng.* **2012**, *109*, 1808–1816. [CrossRef]

17. Li, J.; Ding, M.; Fu, Q.; Tan, H.; Xie, X.; Zhong, Y. A Novel Strategy to Graft RGD Peptide on Biomaterials Surfaces for Endothelization of Small-Diamater Vascular Grafts and Tissue Engineering Blood Vessel. *J. Mater. Sci. Mater. Med.* **2008**, *19*, 2595–2603. [CrossRef]

18. Edlund, U.; Sauter, T.; Albertsson, A.-C. Covalent VEGF Protein Immobilization on Resorbable Polymeric Surfaces. *Polym. Adv. Technol.* **2011**, *22*, 166–171. [CrossRef]

19. De Visscher, G.; Mesure, L.; Meuris, B.; Ivanova, A.; Flameng, W. Improved Endothelialization and Reduced Thrombosis by Coating a Synthetic Vascular Graft with Fibronectin and Stem Cell Homing Factor SDF-1α. *Acta Biomater.* **2012**, *8*, 1330–1338. [CrossRef] [PubMed]

20. Schleicher, M.; Hansmann, J.; Elkin, B.; Kluger, P.J.; Liebscher, S.; Huber, A.J.T.; Fritze, O.; Schille, C.; Müller, M.; Schenke-Layland, K.; et al. Oligonucleotide and Parylene Surface Coating of Polystyrene and EPTFE for Improved Endothelial Cell Attachment and Hemocompatibility. *Int. J. Biomater.* **2012**, *2012*, 397813. [CrossRef] [PubMed]

21. Strahm, Y.; Flueckiger, A.; Billinger, M.; Meier, P.; Mettler, D.; Weisser, S.; Schaffner, T.; Hess, O. Endothelial-Cell-Binding Aptamer for Coating of Intracoronary Stents. *J. Invasive Cardiol.* **2010**, *22*, 481–487.

22. Suuronen, E.J.; Zhang, P.; Kuraitis, D.; Cao, X.; Melhuish, A.; McKee, D.; Li, F.; Mesana, T.G.; Veinot, J.P.; Ruel, M. An Acellular Matrix-Bound Ligand Enhances the Mobilization, Recruitment and Therapeutic Effects of Circulating Progenitor Cells in a Hindlimb Ischemia Model. *FASEB J.* **2009**, *23*, 1447–1458. [CrossRef] [PubMed]

23. Tardif, K.; Cloutier, I.; Miao, Z.; Lemieux, C.; St-Denis, C.; Winnik, F.M.; Tanguay, J.F. A Phosphorylcholine-Modified Chitosan Polymer as an Endothelial Progenitor Cell Supporting Matrix. *Biomaterials* **2011**, *32*, 5046–5055. [CrossRef] [PubMed]

24. Melchiorri, A.J.; Hibino, N.; Fisher, J.P. Strategies and Techniques to Enhance the in Situ Endothelialization of Small-Diameter Biodegradable Polymeric Vascular Grafts. *Tissue Eng. Part B. Rev.* **2013**, *19*, 292–307. [CrossRef] [PubMed]

25. Hinderer, S.; Sudrow, K.; Schneider, M.; Holeiter, M.; Layland, S.L.; Seifert, M.; Schenke-Layland, K. Surface Functionalization of Electrospun Scaffolds Using Recombinant Human Decorin Attracts Circulating Endothelial Progenitor Cells. *Sci. Rep.* **2018**, *8*, 110. [CrossRef]

26. Zhang, W.; Ge, Y.; Cheng, Q.; Zhang, Q.; Fang, L.; Zheng, J. Decorin Is a Pivotal Effector in the Extracellular Matrix and Tumour Microenvironment. *Oncotarget* **2018**, *9*, 5480–5491. [CrossRef]

27. Chen, S.; Young, M.F.; Chakravarti, S.; Birk, D.E. Interclass Small Leucine-Rich Repeat Proteoglycan Interactions Regulate Collagen Fibrillogenesis and Corneal Stromal Assembly. *Matrix Biol.* **2014**, *35*, 103–111. [CrossRef]

28. Fiedler, L.R.; Schönherr, E.; Waddington, R.; Niland, S.; Seidler, D.G.; Aeschlimann, D.; Eble, J.A. Decorin Regulates Endothelial Cell Motility on Collagen I through Activation of Insulin-like Growth Factor I Receptor and Modulation of A2β1 Integrin Activity. *J. Biol. Chem.* **2008**, *283*, 17406–17415. [CrossRef]

29. Fiedler, L.R.; Eble, J.A. Decorin Regulates Endothelial Cell-Matrix Interactions during Angiogenesis. *Cell Adh. Migr.* **2009**, *3*, 3–6. [CrossRef]

30. Zafiropoulos, A.; Nikitovic, D.; Katonis, P.; Tsatsakis, A.; Karamanos, N.K.; Tzanakakis, G.N. Decorin-Induced Growth Inhibition Is Overcome through Protracted Expression and Activation of Epidermal Growth Factor Receptors in Osteosarcoma Cells. *Mol. Cancer Res.* **2008**, *6*, 785–794. [CrossRef]

31. Nili, N.; Cheema, A.N.; Giordano, F.J.; Barolet, A.W.; Babaei, S.; Hickey, R.; Eskandarian, M.R.; Smeets, M.; Butany, J.; Pasterkamp, G.; et al. Decorin Inhibition of PDGF-Stimulated Vascular Smooth Muscle Cell Function: Potential Mechanism for Inhibition of Intimal Hyperplasia after Balloon Angioplasty. *Am. J. Pathol.* **2003**, *163*, 869–878. [CrossRef]

32. D'Antoni, M.L.; Risse, P.A.; Ferraro, P.; Martin, J.G.; Ludwig, M.S. Effects of Decorin and Biglycan on Human Airway Smooth Muscle Cell Adhesion. *Matrix Biol.* **2012**, *31*, 101–112. [CrossRef]

33. De Lange Davies, C.; Melder, R.J.; Munn, L.L.; Mouta-Carreira, C.; Jain, R.K.; Boucher, Y. Decorin Inhibits Endothelial Migration and Tube-like Structure Formation: Role of Thrombospondin-1. *Microvasc. Res.* **2001**, *62*, 26–42. [CrossRef]

34. Järveläinen, H.; Sainio, A.; Wight, T.N. Pivotal Role for Decorin in Angiogenesis. *Matrix Biol.* **2015**, *43*, 15–26. [CrossRef]

35. Riessen, R.; Isner, J.M.; Blessing, E.; Loushin, C.; Nikol, S.; Wight, T.N. Regional Differences in the Distribution of the Proteoglycans Biglycan and Decorin in the Extracellular Matrix of Atherosclerotic and Restenotic Human Coronary Arteries. *Am. J. Pathol.* **1994**, *144*, 962–974.

36. Salisbury, B.G.; Wagner, W.D. Isolation and Preliminary Characterization of Proteoglycans Dissociatively Extracted from Human Aorta. *J. Biol. Chem.* **1981**, *256*, 8050–8057.

37. Khan, G.A.; Girish, G.V.; Lala, N.; di Guglielmo, G.M.; Lala, P.K. Decorin Is a Novel VEGFR-2-Binding Antagonist for the Human Extravillous Trophoblast. *Mol. Endocrinol.* **2011**, *25*, 1431–1443. [CrossRef]

38. Goldoni, S.; Humphries, A.; Nyström, A.; Sattar, S.; Owens, R.T.; McQuillan, D.J.; Ireton, K.; Iozzo, R.V. Decorin Is a Novel Antagonistic Ligand of the Met Receptor. *J. Cell Biol.* **2009**, *185*, 743–754. [CrossRef]

39. Van Meeteren, L.A.; Ten Dijke, P. Regulation of Endothelial Cell Plasticity by TGF-β. *Cell Tissue Res.* **2012**, *347*, 177–186. [CrossRef]

40. Järvinen, T.A.H.; Ruoslahti, E. Target-Seeking Antifibrotic Compound Enhances Wound Healing and Suppresses Scar Formation in Mice. *Proc. Natl. Acad. Sci. USA* **2010**, *107*, 21671–21676. [CrossRef]

41. Liverani, L.; Killian, M.S.; Boccaccini, A.R. Fibronectin Functionalized Electrospun Fibers by Using Benign Solvents: Best Way to Achieve Effective Functionalization. *Front. Bioeng. Biotechnol.* **2019**, *7*, 68. [CrossRef]

42. Campos, D.M.; Gritsch, K.; Salles, V.; Attik, G.N.; Grosgogeat, B. Surface Entrapment of Fibronectin on Electrospun PLGA Scaffolds for Periodontal Tissue Engineering. *Biores. Open Access* **2014**, *3*, 117–126. [CrossRef]

43. Regis, S.; Youssefian, S.; Jassal, M.; Phaneuf, M.; Rahbar, N.; Bhowmick, S. Integrin A5β1-Mediated Attachment of NIH/3T3 Fibroblasts to Fibronectin Adsorbed onto Electrospun Polymer Scaffolds. *Polym. Eng. Sci.* **2014**, *54*, 2587–2594. [CrossRef]

44. Monchaux, E.; Vermette, P. Effects of Surface Properties and Bioactivation of Biomaterials on Endothelial Cells. *Front. Biosci.-Sch.* **2010**, *2*, 239–255.

45. Lenselink, E.A. Role of Fibronectin in Normal Wound Healing. *Int. Wound J.* **2015**, *12*, 313–316. [CrossRef]

46. Grinnell, F. Fibronectin and Wound Healing. *J. Cell. Biochem.* **1984**, *26*, 107–116. [CrossRef]

47. Sgarioto, M.; Vigneron, P.; Patterson, J.; Malherbe, F.; Nagel, M.D.; Egles, C. Collagen Type I Together with Fibronectin Provide a Better Support for Endothelialization. *Comptes Rendus-Biol.* **2012**, *335*, 520–528. [CrossRef]

48. Ardila, D.; Liou, J.-J.; Maestas, D.; Slepian, M.; Badowski, M.; Wagner, W.; Harris, D.; Vande Geest, J. Surface Modification of Electrospun Scaffolds for Endothelialization of Tissue-Engineered Vascular Grafts Using Human Cord Blood-Derived Endothelial Cells. *J. Clin. Med.* **2019**, *8*, 185. [CrossRef]

49. Tersteeg, C.; Roest, M.; Mak-Nienhuis, E.M.; Ligtenberg, E.; Hoefer, I.E.; de Groot, P.G.; Pasterkamp, G. A Fibronectin-Fibrinogen-Tropoelastin Coating Reduces Smooth Muscle Cell Growth but Improves Endothelial Cell Function. *J. Cell. Mol. Med.* **2012**, *16*, 2117–2126. [CrossRef]

50. Ota, T.; Sawa, Y.; Iwai, S.; Kitajima, T.; Ueda, Y.; Coppin, C.; Matsuda, H.; Okita, Y. Fibronectin-Hepatocyte Growth Factor Enhances Reendothelialization in Tissue-Engineered Heart Valve. *Ann. Thorac. Surg.* **2005**, *80*, 1794–1801. [CrossRef]

51. Wang, X.; Liu, T.; Chen, Y.; Zhang, K.; Maitz, M.F.; Pan, C.; Chen, J.; Huang, N. Extracellular Matrix Inspired Surface Functionalization with Heparin, Fibronectin and VEGF Provides an Anticoagulant and Endothelialization Supporting Microenvironment. *Appl. Surf. Sci.* **2014**, *320*, 871–882. [CrossRef]

52. Hoenig, M.R.; Campbell, G.R.; Campbell, J.H. Vascular Grafts and the Endothelium. *Endothel. J. Endothel. Cell Res.* **2006**, *13*, 385–401. [CrossRef]

53. Matsuzaki, Y.; John, K.; Shoji, T.; Shinoka, T. The Evolution of Tissue Engineered Vascular Graft Technologies: From Preclinical Trials to Advancing Patient Care. *Appl. Sci.* **2019**, *9*, 1274. [CrossRef]

54. Popov, G.; Vavilov, V.; Yukina, G.; Popryadukhin, P.; Dobrovolskaya, I.; Ivan'kova, E.; Yudin, V. Long-Term Functioning Aneurysmal Free Tissue-Engineered Vascular Graft Based on Composite Bi-Layered (PLLA/FPL) Scaffold. *Eur. J. Vasc. Endovasc. Surg.* **2019**, *58*, e817–e818. [CrossRef]

55. Kutuzova, L.; Athanasopulu, K.; Schneider, M.; Kandelbauer, A.; Kemkemer, R.; Lorenz, G. In Vitro Bio-Stability Screening of Novel Implantable Polyurethane Elastomers: Morphological Design and Mechanical Aspects. *Curr. Dir. Biomed. Eng.* **2018**, *4*, 535–538. [CrossRef]

56. Broadwater, S.J.; Roth, S.L.; Price, K.E.; Kobašlija, M.; McQuade, D.T. One-Pot Multi-Step Synthesis: A Challenge Spawning Innovation. *Org. Biomol. Chem.* **2005**, *3*, 2899–2906. [CrossRef]

57. Athanasopulu, K.; Kutuzova, L.; Thiel, J.; Lorenz, G.; Kemkemer, R. Enhancing the Biocompatibility of Siliconepolycarbonate Urethane Based Implant Materials. *Curr. Dir. Biomed. Eng.* **2019**, *5*, 453–455. [CrossRef]

58. Schindelin, J.; Arganda-Carreras, I.; Frise, E.; Kaynig, V.; Longair, M.; Pietzsch, T.; Preibisch, S.; Rueden, C.; Saalfeld, S.; Schmid, B.; et al. Fiji: An Open-Source Platform for Biological-Image Analysis. *Nat. Methods* **2012**, *9*, 676–682. [CrossRef]

59. Laterreur, V.; Ruel, J.; Auger, F.A.; Vallières, K.; Tremblay, C.; Lacroix, D.; Tondreau, M.; Bourget, J.M.; Germain, L. Comparison of the Direct Burst Pressure and the Ring Tensile Test Methods for Mechanical Characterization of Tissue-Engineered Vascular Substitutes. *J. Mech. Behav. Biomed. Mater.* **2014**, *34*, 253–263. [CrossRef]

60. Hinderer, S.; Seifert, J.; Votteler, M.; Shen, N.; Rheinlaender, J.; Schäffer, T.E.; Schenke-Layland, K. Engineering of a Bio-Functionalized Hybrid off-the-Shelf Heart Valve. *Biomaterials* **2014**, *35*, 2130–2139. [CrossRef]

61. Becker, M.; Schneider, M.; Stamm, C.; Seifert, M. A Polymorphonuclear Leukocyte Assay to Assess Implant Immunocompatibility. *Tissue Eng. Part C Methods* **2019**, *25*, 500–511. [CrossRef]

62. Schneider, M.; Stamm, C.; Brockbank, K.G.M.; Stock, U.A.; Seifert, M. The Choice of Cryopreservation Method Affects Immune Compatibility of Human Cardiovascular Matrices. *Sci. Rep.* **2017**, *7*, 1–14. [CrossRef]

63. Pusch, J.; Votteler, M.; Göhler, S.; Engl, J.; Hampel, M.; Walles, H.; Schenke-Layland, K. The Physiological Performance of a Three-Dimensional Model That Mimics the Microenvironment of the Small Intestine. *Biomaterials* **2011**, *32*, 7469–7478. [CrossRef] [PubMed]

64. Piccirillo, G.; Carvajal Berrio, D.A.; Laurita, A.; Pepe, A.; Bochicchio, B.; Schenke-Layland, K.; Hinderer, S. Controlled and Tuneable Drug Release from Electrospun Fibers and a Non-Invasive Approach for Cytotoxicity Testing. *Sci. Rep.* **2019**, *9*, 1–10. [CrossRef] [PubMed]

65. Al-Sabti, H.A.; Al Kindi, A.; Al-Rasadi, K.; Banerjee, Y.; Al-Hashmi, K.; Al-Hinai, A. Saphenous Vein Graft vs. Radial Artery Graft Searching for the Best Second Coronary Artery Bypass Graft. *J. Saudi Heart Assoc.* **2013**, *25*, 247–254. [CrossRef]

66. Stekelenburg, M.; Rutten, M.C.M.; Snoeckx, L.H.E.H.; Baaijens, F.P.T. Dynamic Straining Combined with Fibrin Gel Cell Seeding Improves Strength of Tissue-Engineered Small-Diameter Vascular Grafts. *Tissue Eng. Part A* **2009**, *15*, 1081–1089. [CrossRef]

67. Soletti, L.; Hong, Y.; Guan, J.; Stankus, J.J.; El-Kurdi, M.S.; Wagner, W.R.; Vorp, D.A. A Bilayered Elastomeric Scaffold for Tissue Engineering of Small Diameter Vascular Grafts. *Acta Biomater.* **2010**, *6*, 110–122. [CrossRef]

68. Pukacki, F.; Jankowski, T.; Gabriel, M.; Oszkinis, G.; Krasinski, Z.; Zapalski, S. The Mechanical Properties of Fresh and Cryopreserved Arterial Homografts. *Eur. J. Vasc. Endovasc. Surg.* **2000**, *20*, 21–24. [CrossRef]

69. Porter, T.R.; Taylor, D.O.; Fields, J.; Cycan, A.; Akosah, K.; Mohanty, P.K.; Pandian, N.G. Direct in Vivo Evaluation of Pulmonary Arterial Pathology in Chronic Congestive Heart Failure with Catheter-Based Intravascular Ultrasound Imaging. *Am. J. Cardiol.* **1993**, *71*, 754–757. [CrossRef]

70. Abbott, W.M. Prosthetic Above-Knee Femoral-Popliteal Bypass: Indications and Choice of Graft. *Semin. Vasc. Surg.* **1997**, *10*, 3–7.

71. Konig, G.; McAllister, T.N.; Dusserre, N.; Garrido, S.A.; Iyican, C.; Marini, A.; Fiorillo, A.; Avila, H.; Wystrychowski, W.; Zagalski, K.; et al. Mechanical Properties of Completely Autologous Human Tissue Engineered Blood Vessels Compared to Human Saphenous Vein and Mammary Artery. *Biomaterials* **2009**, *30*, 1542–1550. [CrossRef]

72. Lamm, P.; Juchem, G.; Milz, S.; Schuffenhauer, M.; Reichart, B. Autologous Endothelialized Vein Allograft: A Solution in the Search for Small-Caliber Grafts in Coronary Artery Bypass Graft Operations. *Circulation* **2001**, *104*, I-108. [CrossRef]

73. L'Heureux, N.; Pâquet, S.; Labbé, R.; Germain, L.; Auger, F.A. A Completely Biological Tissue-Engineered Human Blood Vessel. *FASEB J.* **1998**, *12*, 47–56.

74. Caruso, A.; Licenziati, S.; Corulli, M.; Canaris, A.D.; De Francesco, M.A.; Fiorentini, S.; Peroni, L.; Fallacara, F.; Dima, F.; Balsari, A.; et al. Flow Cytometric Analysis of Activation Markers on Stimulated T Cells and Their Correlation with Cell Proliferation. *Cytometry* **1997**, *27*, 71–76. [CrossRef]

75. Lerouge, S. Introduction to Sterilization: Definitions and Challenges. In *Sterilisation of Biomaterials and Medical Devices*; Elsevier: Amsterdam, The Netherlands, 2012; pp. 1–19.

76. Ameer, J.M.; Anil Kumar, P.R.; Kasoju, N. Strategies to Tune Electrospun Scaffold Porosity for Effective Cell Response in Tissue Engineering. *J. Funct. Biomater.* **2019**, *10*, 30. [CrossRef]

77. Bružauskaitė, I.; Bironaitė, D.; Bagdonas, E.; Bernotienė, E. Scaffolds and Cells for Tissue Regeneration: Different Scaffold Pore Sizes—Different Cell Effects. *Cytotechnology* **2016**, *68*, 355–369. [CrossRef]

78. Murphy, C.M.; O'Brien, F.J. Understanding the Effect of Mean Pore Size on Cell Activity in Collagen-Glycosaminoglycan Scaffolds. *Cell Adhes. Migr.* **2010**, *4*, 377–381. [CrossRef]

79. Arends, F.; Lieleg, O. Biophysical Properties of the Basal Lamina: A Highly Selective Extracellular Matrix. In *Composition and Function of the Extracellular Matrix in the Human Body*; InTech: London, UK, 2016.

80. Liliensiek, S.J.; Nealey, P.; Murphy, C.J. Characterization of Endothelial Basement Membrane Nanotopography in Rhesus Macaque as a Guide for Vessel Tissue Engineering. *Tissue Eng. Part A* **2009**, *15*, 2643–2651. [CrossRef]

81. Sage, H. Collagens of Basement Membranes. *J. Investig. Dermatol.* **1982**, *79*, 51s–59s. [CrossRef]

82. Abrams, G.A.; Murphy, C.J.; Wang, Z.Y.; Nealey, P.F.; Bjorling, D.E. Ultrastructural Basement Membrane Topography of the Bladder Epithelium. *Urol. Res.* **2003**, *31*, 341–346.

83. Inoguchi, H.; Kwon, I.K.; Inoue, E.; Takamizawa, K.; Maehara, Y.; Matsuda, T. Mechanical Responses of a Compliant Electrospun Poly(L-Lactide-Co-ε- Caprolactone) Small-Diameter Vascular Graft. *Biomaterials* **2006**, *27*, 1470–1478. [CrossRef]

84. Nottelet, B.; Pektok, E.; Mandracchia, D.; Tille, J.-C.; Walpoth, B.; Gurny, R.; Möller, M. Factorial Design Optimization and *in Vivo* Feasibility of Poly(ε-Caprolactone)-Micro- and Nanofiber-Based Small Diameter Vascular Grafts. *J. Biomed. Mater. Res. Part A* **2009**, *89A*, 865–875. [CrossRef] [PubMed]

85. Vaz, C.M.; van Tuijl, S.; Bouten, C.V.C.; Baaijens, F.P.T. Design of Scaffolds for Blood Vessel Tissue Engineering Using a Multi-Layering Electrospinning Technique. *Acta Biomater.* **2005**, *1*, 575–582. [CrossRef] [PubMed]

86. Hironaka, K.; Makino, H.; Yamasaki, Y.; Ota, Z. Renal Basement Membranes by Ultrahigh Resolution Scanning Electron Microscopy. *Kidney Int.* **1993**, *43*, 334–345. [CrossRef] [PubMed]

87. Takeuchi, T.; Gonda, T. Distribution of the Pores of Epithelial Basement Membrane in the Rat Small Intestine. *J. Vet. Med. Sci.* **2004**, *66*, 695–700. [CrossRef] [PubMed]

88. Yurchenco, P.D.; Ruben, G.C. Basement Membrane Structure in Situ: Evidence for Lateral Associations in the Type IV Collagen Network. *J. Cell Biol.* **1987**, *105*, 2559–2568. [CrossRef]

89. Howat, W.J.; Holmes, J.A.; Holgate, S.T.; Lackie, P.M. Basement Membrane Pores in Human Bronchial Epithelium: A Conduit for Infiltrating Cells? *Am. J. Pathol.* **2001**, *158*, 673–680. [CrossRef]

90. Wierzbicka-Patynowski, I.; Schwarzbauer, J.E. Cell-Surface Transglutaminase Promotes Fibronectin Assembly via Interaction with the Gelatin-Binding Domain of Fibronectin: A Role in TGFbeta-Dependent Matrix Deposition. *J. Cell Sci.* **2003**, *116*, 3269–3276. [CrossRef]

91. Sevilla, C.A.; Dalecki, D.; Hocking, D.C. Regional Fibronectin and Collagen Fibril Co-Assembly Directs Cell Proliferation and Microtissue Morphology. *PLoS ONE* **2013**, *8*, e77316. [CrossRef]

92. Rico, P.; Mnatsakanyan, H.; Dalby, M.J.; Salmerón-Sánchez, M. Material-Driven Fibronectin Assembly Promotes Maintenance of Mesenchymal Stem Cell Phenotypes. *Adv. Funct. Mater.* **2016**, *26*, 6563–6573. [CrossRef]

93. Llopis-Hernández, V.; Cantini, M.; González-García, C.; Cheng, Z.A.; Yang, J.; Tsimbouri, P.M.; García, A.J.; Dalby, M.J.; Salmerón-Sánchez, M. Material-Driven Fibronectin Assembly for High-Efficiency Presentation of Growth Factors. *Sci. Adv.* **2016**, *2*, e1600188. [CrossRef]

94. Schmidt, G.; Hausser, H.; Kresse, H. Interaction of the Small Proteoglycan Decorin with Fibronectin. Involvement of the Sequence NKISK of the Core Protein. *Biochem. J.* **1991**, *280*, 411–414. [CrossRef] [PubMed]

95. Winnemöller, M.; Schmidt, G.; Kresse, H. Influence of Decorin on Fibroblast Adhesion to Fibronectin. *Eur. J. Cell Biol.* **1991**, *54*, 10–17. [PubMed]

96. Dee, K.C.; Puleo, D.A.; Bizios, R. Protein-Surface Interactions. In *An Introduction to Tissue-Biomaterial Interaction*; Wiley: Hoboken, NJ, USA, 2002; pp. 37–51.

97. Thyparambil, A.A.; Wei, Y.; Latour, R.A. Experimental Characterization of Adsorbed Protein Orientation, Conformation, and Bioactivity. *Biointerphases* **2015**, *10*, 019002. [CrossRef] [PubMed]

98. Latour, R.A. Molecular Simulation of Protein-Surface Interactions. In *Biological Interactions on Material Surfaces*; Puleo, D.A., Bizios, R., Eds.; Springer US: Berlin/Heidelberg, Germany, 2009; pp. 73–74.

99. Lebaron, R.G.; Athanasiou, K.A. Extracellular Matrix Cell Adhesion Peptides: Functional Applications in Orthopedic Materials. *Tissue Eng.* **2000**, *6*, 85–103. [CrossRef]

100. Hinderer, S.; Schesny, M.; Bayrak, A.; Ibold, B.; Hampel, M.; Walles, T.; Stock, U.A.; Seifert, M.; Schenke-Layland, K. Engineering of Fibrillar Decorin Matrices for a Tissue-Engineered Trachea. *Biomaterials* **2012**, *33*, 5259–5266. [CrossRef]

101. Ishizaki, T.; Saito, N.; Takai, O. Correlation of Cell Adhesive Behaviors on Superhydrophobic, Superhydrophilic, and Micropatterned Superhydrophobic/Superhydrophilic Surfaces to Their Surface Chemistry. *Langmuir* **2010**, *26*, 8147–8154. [CrossRef]

102. Schmidt, G.; Robenek, H.; Harrach, B.; Glössl, J.; Nolte, V.; Hörmann, H.; Richter, H.; Kresse, H. Interaction of Small Dermatan Sulfate Proteoglycan from Fibroblasts with Fibronectin. *J. Cell Biol.* **1987**, *104*, 1683–1691. [CrossRef]

103. Zhao, J.; Mitrofan, C.G.; Appleby, S.L.; Morrell, N.W.; Lever, A.M.L. Disrupted Endothelial Cell Layer and Exposed Extracellular Matrix Proteins Promote Capture of Late Outgrowth Endothelial Progenitor Cells. *Stem Cells Int.* **2016**, *2016*, 1406304. [CrossRef]

104. Brown, B.N.; Badylak, S.F. Expanded Applications, Shifting Paradigms and an Improved Understanding of Host-Biomaterial Interactions. *Acta Biomater.* **2013**, *9*, 4948–4955. [CrossRef]

105. Wang, Z.; Cui, Y.; Wang, J.; Yang, X.; Wu, Y.; Wang, K.; Gao, X.; Li, D.; Li, Y.; Zheng, X.L.; et al. The Effect of Thick Fibers and Large Pores of Electrospun Poly(ε-Caprolactone) Vascular Grafts on Macrophage Polarization and Arterial Regeneration. *Biomaterials* **2014**, *35*, 5700–5710. [CrossRef]

106. Dejana, E.; Orsenigo, F.; Molendini, C.; Baluk, P.; McDonald, D.M. Organization and Signaling of Endothelial Cell-to-Cell Junctions in Various Regions of the Blood and Lymphatic Vascular Trees. *Cell Tissue Res.* **2009**, *335*, 17–25. [CrossRef] [PubMed]

107. Lertkiatmongkol, P.; Liao, D.; Mei, H.; Hu, Y.; Newman, P.J. Endothelial Functions of Platelet/Endothelial Cell Adhesion Molecule-1 (CD31). *Curr. Opin. Hematol.* **2016**, *23*, 253–259. [CrossRef] [PubMed]

108. Yamaguchi, T.P.; Dumont, D.J.; Conlon, R.A.; Breitman, M.L.; Rossant, J. Flk-1, an Fit-Related Receptor Tyrosine Kinase Is an Early Marker for Endothelial Cell Precursors. *Development* **1993**, *118*, 489–498. [PubMed]

109. Smadja, D.M.; Bièche, I.; Helley, D.; Laurendeau, I.; Simonin, G.; Muller, L.; Aiach, M.; Gaussem, P. Increased VEGFR2 Expression during Human Late Endothelial Progenitor Cells Expansion Enhances in Vitro Angiogenesis with Up-Regulation of Integrin A6. *J. Cell. Mol. Med.* **2007**, *11*, 1149–1161. [CrossRef]

110. Kusuma, S.; Zhao, S.; Gerecht, S. The Extracellular Matrix Is a Novel Attribute of Endothelial Progenitors and of Hypoxic Mature Endothelial Cells. *FASEB J.* **2012**, *26*, 4925–4936. [CrossRef]

111. Ichii, M.; Frank, M.B.; Iozzo, R.V.; Kincade, P.W. The Canonical Wnt Pathway Shapes Niches Supportive of Hematopoietic Stem/Progenitor Cells. *Blood* **2012**, *119*, 1683–1692. [CrossRef]

112. Wijelath, E.S.; Rahman, S.; Murray, J.; Patel, Y.; Savidge, G.; Sobel, M. Fibronectin Promotes VEGF-Induced CD34+ Cell Differentiation into Endothelial Cells. *J. Vasc. Surg.* **2004**, *39*, 655–660. [CrossRef]

113. Shen, B.Q.; Lee, D.Y.; Gerber, H.P.; Keyt, B.A.; Ferrara, N.; Zioncheck, T.F. Homologous Up-Regulation of KDR/Flk-1 Receptor Expression by Vascular Endothelial Growth Factor in Vitro. *J. Biol. Chem.* **1998**, *273*, 29979–29985. [CrossRef]

114. Neill, T.; Schaefer, L.; Iozzo, R.V. Decorin: A Guardian from the Matrix. *Am. J. Pathol.* **2012**, *181*, 380–387. [CrossRef]

115. Mazor, R.; Alsaigh, T.; Shaked, H.; Altshuler, A.E.; Pocock, E.S.; Kistler, E.B.; Karin, M.; Schmid-Schönbein, G.W. Matrix Metalloproteinase-1-Mediated up-Regulation of Vascular Endothelial Growth Factor-2 in Endothelial Cells. *J. Biol. Chem.* **2013**, *288*, 598–607. [CrossRef]

116. Huttenlocher, A.; Werb, Z.; Tremble, P.; Huhtala, P.; Rosenberg, L.; Damsky, C.H. Decorin Regulates Collagenase Gene Expression in Fibroblasts Adhering to Vitronectin. *Matrix Biol.* **1996**, *15*, 239–250. [CrossRef]

117. Schönherr, E.; Schaefer, L.; O'Connell, B.C.; Kresse, H. Matrix Metalloproteinase Expression by Endothelial Cells in Collagen Lattices Changes during Co-Culture with Fibroblasts and upon Induction of Decorin Expression. *J. Cell. Physiol.* **2001**, *187*, 37–47. [CrossRef]

118. Murakami, M.; Nguyen, L.T.; Hatanaka, K.; Schachterle, W.; Chen, P.Y.; Zhuang, Z.W.; Black, B.L.; Simons, M. FGF-Dependent Regulation of VEGF Receptor 2 Expression in Mice. *J. Clin. Investig.* **2011**, *121*, 2668–2678. [CrossRef] [PubMed]

119. Penc, S.F.; Pomahac, B.; Winkler, T.; Dorschner, R.A.; Eriksson, E.; Herndon, M.; Gallo, R.L. Dermatan Sulfate Released after Injury Is a Potent Promoter of Fibroblast Growth Factor-2 Function. *J. Biol. Chem.* **1998**, *273*, 28116–28121. [CrossRef] [PubMed]

120. Chen, T.T.; Filvaroff, E.; Peng, J.; Marsters, S.; Jubb, A.; Koeppen, H.; Merchant, M.; Ashkenazi, A. MET Suppresses Epithelial VEGFR2 via Intracrine VEGF-Induced Endoplasmic Reticulum-Associated Degradation. *EBioMedicine* **2015**, *2*, 406–420. [CrossRef] [PubMed]

121. Chen, S.; Chakrabarti, R.; Keats, E.C.; Chen, M.; Chakrabarti, S.; Khan, Z.A. Regulation of Vascular Endothelial Growth Factor Expression by Extra Domain B Segment of Fibronectin in Endothelial Cells. *Investig. Ophthalmol. Vis. Sci.* **2012**, *53*, 8333–8343. [CrossRef]

122. Wijelath, E.S.; Rahman, S.; Namekata, M.; Murray, J.; Nishimura, T.; Mostafavi-Pour, Z.; Patel, Y.; Suda, Y.; Humphries, M.J.; Sobel, M. Heparin-II Domain of Fibronectin Is a Vascular Endothelial Growth Factor-Binding Domain. *Circ. Res.* **2006**, *99*, 853–860. [CrossRef]

123. Kou, R.; SenBanerjee, S.; Jain, M.K.; Michel, T. Differential Regulation of Vascular Endothelial Growth Factor Receptors (VEGFR) Revealed by RNA Interference: Interactions of VEGFR-1 and VEGFR-2 in Endothelial Cell Signaling. *Biochemistry* **2005**, *44*, 15064–15073. [CrossRef]

124. Scott, R.A.; Paderi, J.E.; Sturek, M.; Panitch, A. Decorin Mimic Inhibits Vascular Smooth Muscle Proliferation and Migration. *PLoS ONE* **2013**, *8*, e82456. [CrossRef]

125. Harding, A.; Cortez-Toledo, E.; Magner, N.L.; Beegle, J.R.; Coleal-Bergum, D.P.; Hao, D.; Wang, A.; Nolta, J.A.; Zhou, P. Highly Efficient Differentiation of Endothelial Cells from Pluripotent Stem Cells Requires the MAPK and the PI3K Pathways. *Stem Cells* **2017**, *35*, 909–919. [CrossRef]

126. Bryan, B.A.; Walshe, T.E.; Mitchell, D.C.; Havumaki, J.S.; Saint-Geniez, M.; Maharaj, A.S.; Maldonado, A.E.; D'Amore, P.A. Coordinated Vascular Endothelial Growth Factor Expression and Signaling during Skeletal Myogenic Differentiation. *Mol. Biol. Cell* **2008**, *19*, 994–1006. [CrossRef] [PubMed]

127. Zeng, Y.; Zhang, X.F.; Fu, B.M.; Tarbell, J.M. The Role of Endothelial Surface Glycocalyx in Mechanosensing and Transduction. In *Advances in Experimental Medicine and Biology*; Springer New York LLC: New York, NY, USA, 2018; Volume 1097, pp. 1–27.

128. Ballermann, B.J.; Dardik, A.; Eng, E.; Liu, A. Shear Stress and the Endothelium. *Kidney Int.* **1998**, *54*, S100–S108. [CrossRef] [PubMed]

129. Sato, M.; Kataoka, N.; Ohshima, N. Response of Vascular Endothelial Cells to Flow Shear Stress: Phenomenological Aspect. In *Biomechanics*; Springer Japan: Tokyo, Japan, 1996; pp. 3–27.

130. Zhou, J.; Li, Y.S.; Chien, S. Shear Stress-Initiated Signaling and Its Regulation of Endothelial Function. *Arterioscler. Thromb. Vasc. Biol.* **2014**, *34*, 2191–2198. [CrossRef]
131. Sánchez-Duffhues, G.; García de Vinuesa, A.; ten Dijke, P. Endothelial-to-Mesenchymal Transition in Cardiovascular Diseases: Developmental Signaling Pathways Gone Awry. *Dev. Dyn.* **2018**, *247*, 492–508. [CrossRef] [PubMed]
132. Moonen, J.R.A.J.; Lee, E.S.; Schmidt, M.; Maleszewska, M.; Koerts, J.A.; Brouwer, L.A.; Van Kooten, T.G.; Van Luyn, M.J.A.; Zeebregts, C.J.; Krenning, G.; et al. Endothelial-to-Mesenchymal Transition Contributes to Fibro-Proliferative Vascular Disease and Is Modulated by Fluid Shear Stress. *Cardiovasc. Res.* **2015**, *108*, 377–386. [CrossRef]
133. Mahmoud, M.M.; Serbanovic-Canic, J.; Feng, S.; Souilhol, C.; Xing, R.; Hsiao, S.; Mammoto, A.; Chen, J.; Ariaans, M.; Francis, S.E.; et al. Shear Stress Induces Endothelial-To-Mesenchymal Transition via the Transcription Factor Snail. *Sci. Rep.* **2017**, *7*, 1–12. [CrossRef]
134. Melchiorri, A.J.; Bracaglia, L.G.; Kimerer, L.K.; Hibino, N.; Fisher, J.P. In Vitro Endothelialization of Biodegradable Vascular Grafts Via Endothelial Progenitor Cell Seeding and Maturation in a Tubular Perfusion System Bioreactor. *Tissue Eng.-Part C Methods* **2016**, *22*, 663–670. [CrossRef]
135. Dejana, E.; Hirschi, K.K.; Simons, M. The Molecular Basis of Endothelial Cell Plasticity. *Nat. Commun.* **2017**, *8*, 1–11. [CrossRef]
136. Lacorre, D.A.; Baekkevold, E.S.; Garrido, I.; Brandtzaeg, P.; Haraldsen, G.; Amalric, F.; Girard, J.P. Plasticity of Endothelial Cells: Rapid Dedifferentiation of Freshly Isolated High Endothelial Venule Endothelial Cells Outside the Lymphoid Tissue Microenvironment. *Blood* **2004**, *103*, 4164–4172. [CrossRef]
137. Nguyen, M.T.X.; Okina, E.; Chai, X.; Tan, K.H.; Hovatta, O.; Ghosh, S.; Tryggvason, K. Differentiation of Human Embryonic Stem Cells to Endothelial Progenitor Cells on Laminins in Defined and Xeno-Free Systems. *Stem Cell Rep.* **2016**, *7*, 802–816. [CrossRef]

 *cells*

Review

Cardiac Progenitor Cells from Stem Cells: Learning from Genetics and Biomaterials

Sara Barreto [1], Leonie Hamel [2], Teresa Schiatti [1], Ying Yang [1] and Vinoj George [1,*]

[1] Guy Hilton Research Centre, School of Pharmacy & Bioengineering, Keele University, Staffordshire ST4 7QB, UK; s.barreto-francisco@keele.ac.uk (S.B.); teresa.schiatti@gmail.com (T.S.); y.yang@keele.ac.uk (Y.Y.)
[2] RCSI Bahrain, P.O. Box 15503, Adliya, Bahrain; 18211372@rcsi-mub.com
* Correspondence: v.george@keele.ac.uk

Received: 18 October 2019; Accepted: 21 November 2019; Published: 28 November 2019

Abstract: Cardiac Progenitor Cells (CPCs) show great potential as a cell resource for restoring cardiac function in patients affected by heart disease or heart failure. CPCs are proliferative and committed to cardiac fate, capable of generating cells of all the cardiac lineages. These cells offer a significant shift in paradigm over the use of human induced pluripotent stem cell (iPSC)-derived cardiomyocytes owing to the latter's inability to recapitulate mature features of a native myocardium, limiting their translational applications. The iPSCs and direct reprogramming of somatic cells have been attempted to produce CPCs and, in this process, a variety of chemical and/or genetic factors have been evaluated for their ability to generate, expand, and maintain CPCs in vitro. However, the precise stoichiometry and spatiotemporal activity of these factors and the genetic interplay during embryonic CPC development remain challenging to reproduce in culture, in terms of efficiency, numbers, and translational potential. Recent advances in biomaterials to mimic the native cardiac microenvironment have shown promise to influence CPC regenerative functions, while being capable of integrating with host tissue. This review highlights recent developments and limitations in the generation and use of CPCs from stem cells, and the trends that influence the direction of research to promote better application of CPCs.

Keywords: cardiac progenitor cells; induced pluripotent stem cells; transdifferentiation; direct reprogramming; genetic engineering; cardiac tissue engineering; biomaterials

---

1. Cardiac Regeneration—A Problem to Solve or A Solution with Promise?

With morbidity rates associated with cardiovascular diseases in the decline in the developed world from improved treatments and pharmacological intervention, scientists and clinicians have been approaching therapies recently for these diseases with vigor. However, there is still no reliable therapy for acute cardiac conditions like myocardial infarction (MI), which account for nearly half of all cardiovascular deaths in the industrialized world [1,2]. Regenerative medicine-based strategies for infarcted myocardium have shown promise in preclinical animal models as well as early clinical trials [3]. Whilst these have demonstrated some physiological improvements in ventricular function, they were associated with very low cell retention after some weeks, suggesting a paracrine effect of transplanted cells rather than functional integration within the damaged tissue [4].

The heart was long viewed as a post-mitotic or terminally differentiated organ with no ability to regenerate or repair, a dogma that has been challenged abundantly in recent years [5,6]. Cardiac regeneration, following injury, is still an unresolved debate over whether it is attributed to dedifferentiation and proliferation of resident cardiomyocytes or from an inherent trigger in differentiation of cardiac stem or progenitor cells in putative cell niches within the heart [7–11]. The turnover of cardiomyocytes in the adult heart is around 1% per year which is insufficient to counter the loss caused by MI that can lead to loss of around 1 billion cardiomyocytes [12]. Therefore, the only long-term solution relies on heart transplantation, but this does not come without its own issues such as insufficient number of donors coupled with the requirement for a life-long immunosuppressive therapy. This catapulted research towards cell-based therapies for cardiac regeneration [13]. Cardiomyocytes are the main cardiac cell type that is lost in cardiovascular disorders, like heart failure, myocardial infarction, and ischemia, and therefore, replacing these cells could potentially restore heart function. However, transplanting cardiomyocytes to repair diseased hearts has shown to yield only transient responses as most cells are eventually lost in the host environment [14,15]. This is because cardiomyocytes have very limited proliferative ability and as a result, they are unable to repopulate and replenish the damaged tissue efficiently [16,17]. Furthermore, other cell types like smooth muscle cells, and endothelial cells can suffer from collateral damage and their functional renewal is vital for effective heart regeneration [18]. This puts emphasis on the role of a precursor cell type capable of extensive expansion and differentiation into the key cell players of cardiac regeneration.

Even though some level of cell turnover has been observed in the adult heart, cells with self-renewal or potency capabilities are generally considered lacking in this tissue [19]. Nevertheless, several studies report the evidence of a progenitor population from resident cardiac stem cells (CSCs) in the heart, called Cardiac Progenitor Cells (CPCs) [20–23]. In contrast to terminally differentiated cardiomyocytes, CPCs are highly proliferative and can theoretically differentiate into all the necessary cardiac cell types for effective reconstitution of damaged cardiac tissue and promoting its neovascularization [14,18,20, 21,24–31]. Therefore, CPCs present a more effective cell source than cardiomyocytes for cell-based regenerative strategies. However, the application of CPCs has not been straight-forward particularly in chronic infarcts, where CPCs are associated with senescence, decreased telomerase activity and increased apoptosis [7]. Cell therapy using CPCs generally involve transplantation of in vitro-expanded CPC populations which in turn yield mild improvements in cardiac function [32]. However, long term prognosis with such therapies are poor owing to reduced cell viability and inefficient engraftment into the host tissue. This is compounded by the somewhat hostile microenvironment created by MI, from scar formation and associated inflammatory or tissue alterations, which compromises the effectiveness of such therapies [33–35]. There are also reports that the administration of CPCs predisposed the risk of cardiac arrhythmias and teratoma formation [36]. Therefore, better understanding of the CPC cell behavior in dynamic pathophysiological microenvironments could aid in developing strategies to optimize their contribution to cardiac repair.

Various approaches have been developed to generate CPCs ex vivo, in the hope of obtaining reliable source of cells that can trigger mechanisms of cardiac regeneration. For example, CPCs from the heart tissue (also known as putative CPCs) can be isolated and expanded in vitro [27,37–40]. However, such cells are hard to access and are present in low numbers in the tissue, making them extremely challenging to harvest and realize their potential [41]. Pluripotent stem cells, such as embryonic stem cells (ESCs) and induced pluripotent stem cells (iPSCs), are thought to be a superior alternative cell source since they could potentially provide an unlimited supply of cardiac progenitor cells. However, ESC-based therapy faces several challenges like immunogenicity, high risk of tumor formation and the characteristic ethical concerns, which have prevented their clinical application [42,43]. On the other hand, iPSCs avoid the ethical issues associated with ESCs and allows for the development of patient-specific derived CPCs, which represents an advantage over other cell sources in the creation of immune-compatible cardiac therapies [44,45]. However, with issues surrounding the safety of iPSC-based therapies, in terms of the potential risk of tumor formation associated with such therapies or immune rejection of iPS-derived cells from a common donor, scientists are looking at reprogramming from a different perspective [46–48]. Reprogramming patient somatic cells into other cell types, bypassing the step of stem cell generation, can potentially overcome issues with translating iPSC technology. This process is known as direct cellular reprogramming or transdifferentiation, and might represent a more robust approach to rapidly generate sufficient numbers of CPCs from somatic cells for therapy [49].

This review focuses on the ongoing progress and limitations in generating CPCs from iPSCs and through direct reprogramming. It will start by providing a concise introduction about the various cardiac progenitor cells identified in embryonic and adult heart tissues. The review will then move towards discussing reprogramming approaches that were successful in generating CPCs and the functionality of these CPC-derived cells. Strategies to improve efficiencies of current protocols and tissue engineering advances to mimic CPC microenvironment and in vivo applications of CPCs will also be evaluated. Finally, the review will finish with a summary of existing challenges and limitations and future directions for CPC research, hopefully convincing readers it is a promising strategy for cardiac regeneration (Figure 1).

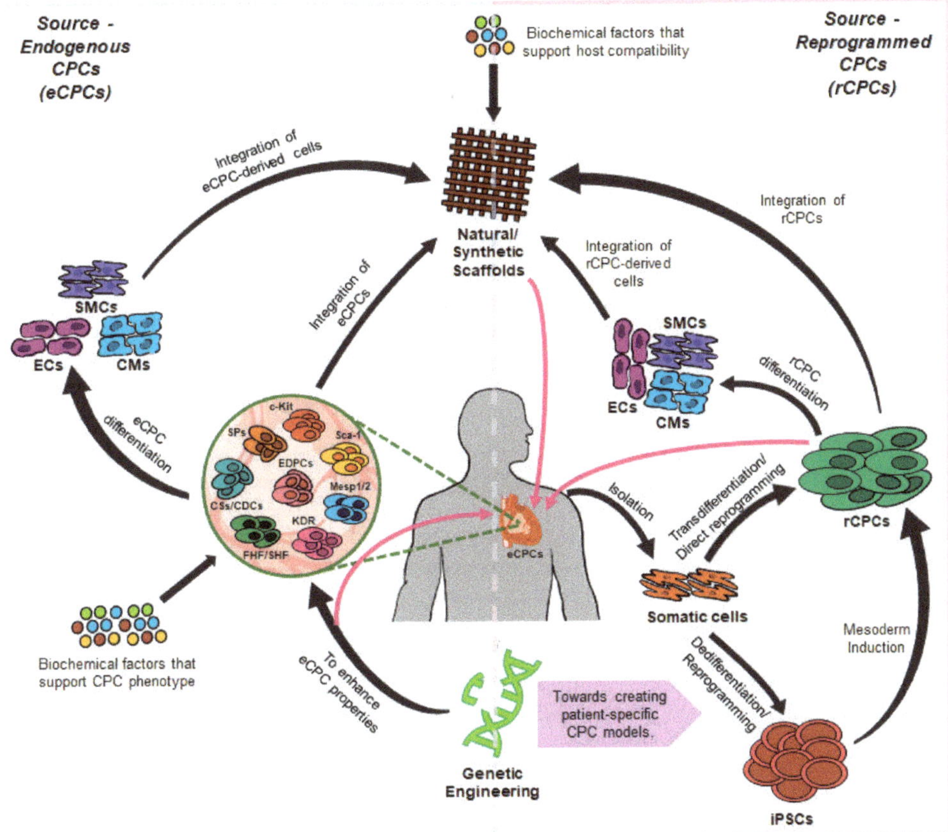

Figure 1. The interplay between genetics and biomaterials for understanding Cardiac Progenitor Cells (CPCs) biology, function, and its regenerative applications. eCPCs (endogenous CPCs), rCPCs (reprogrammed CPCs), iPSCs (induced Pluripotent Stem Cells), SPs (Side Population-derived CPCs), CSs/CDCs (Cardiospheres/Cardiosphere-Derived Cells), EDPCs (Epicardium-derived CPCs), FHF/SHF (First Heart Field-/Second Heart Field-derived CPCs) CMs (Cardiomyocytes), SMCs (Smooth Muscle Cells), ECs (Endothelial Cells).

2. Cardiac Progenitor Cells (CPCs) In Vivo

Progenitor cells are distinct from embryonic stem cells as they have a predetermined differentiation fate and therefore, their ability to self-renew and differentiate into other cell types is restricted [19]. CPCs generate cells of the three cardiac lineages: cardiomyocytes, smooth muscle cells and endothelial cells. These cells are also responsible for the maintenance of cardiac homeostasis under physiological and pathological conditions [50]. Several studies have identified and isolated multiple CPC populations from distinct stages of cardiac development and heart locations. These cells are collectively characterized according to their cell surface and genetic marker expression profiles. The various CPCs reported to date are described below and their specific features are summarized in Table 1.

2.1. c-KIT-Expressing CPCs

The first-ever detected CPCs were isolated from female rats and were characterized by the expression of the stem cell surface marker c-KIT [28]. These CPCs are present throughout the ventricular and atrial myocardium, particularly in the atria and the ventricular apex [28]. These progenitor cells also express the cardiac transcription factors NKX2.5, GATA4, and MEF2C, and are negative for hematopoietic lineage markers, such as CD45, CD34, CD3, CD14, CD16, CD19, CD20 and CD56 [50–52]. They are self-renewing, clonogenic and are able to differentiate into the three cardiac cell types in vitro and in vivo [28,53]. The c-KIT receptor binds to the Stem Cell Factor (SCF) which leads to the activation of the signaling pathways Phosphoinositide 3-kinase/Protein Kinase B (PI3K/AKT) and p38 Mitogen-Activated Protein Kinase (MAPK) [54,55]. These pathways regulate a variety of CPC functions like self-renewal, proliferation, survival, and migration [54–57]. Even though c-KIT CPCs contribute to the generation of cardiomyocytes at earlier stages of embryonic development and right after birth, this ability is mostly lost in the adult heart and very low percentages of new cardiomyocytes seem to originate from these CPCs [58–60]. Therefore, the improvement of cardiac function by c-KIT CPCs might be a result of paracrine factors rather than the production of de novo cardiomyocytes [58,61]. Furthermore, c-KIT expression is considered necessary but not sufficient to define CPCs [62].

2.2. SCA1-Expressing CPCs

Another CPC population present in adult hearts expresses the Stem Cell Antigen 1 (SCA1). The cells were first identified in adult mouse hearts [11] and are predominantly located in the atrium, the intra-atrial septum, the atrium-ventricular boundary and scattered within the epicardial layer [37]. SCA1 is a cell surface protein of the Ly6 gene family and it was initially used to isolate hematopoietic stem cells [63]. Additionally, SCA1 is widely expressed by stem and progenitor cells from a variety of tissues, including the heart, and it has roles in cell survival, proliferation and differentiation [63]. Several studies have shown that SCA1 CPCs are negative for hematopoietic lineage markers and are able to differentiate into the three cardiac lineages [11,64]. These CPCs also have the ability of homing in response to injury and contribute to neovascularization in vivo [11,65,66]. Although this CPC population seems promising for cardiac regeneration, their translational relevance is not without caveats. First, all the SCA1 CPC populations identified to date display different gene expression profiles and distinct differentiation potential [37,66–71]. In addition, several studies have shown that the benefits resulted from the transplantation of these CPCs might be predominantly due to paracrine mechanisms as these cells differentiate into cardiomyocytes with very low efficiency [66,68,70,71]. Finally, SCA1 is only present in murine cells and a human ortholog of SCA1 has yet to be identified [63]. Therefore, the nature of the epitope target of SCA1 in humans and the nature of regeneration of the associated CPC population have yet to be determined.

2.3. MESP1/2-Expressing CPCs

During the development of mesoderm, embryonic cells express the transcription factor Mesoderm Posterior Protein 1/2 (MESP1/2), which is essential for proper cell migration [15,72–74]. MESP1/2 expression marks the first step in the commitment of the nascent mesoderm into the myocardial lineages, and it describes the first population of multipotent cardiac progenitor cells that produce the various cardiac cell types of the heart [72,75]. Although MESP1/2 CPCs show increased cardiac potential, in comparison to other CPC types, they are not irreversibly committed to the cardiac fate [76]. Consequently, there is a possibility that these cells will differentiate into derivates of the paraxial mesoderm and skeletal muscle [77,78]. Furthermore, MESP1/2 is only transiently expressed during embryonic development, which increases the difficulty of tracking the expansion and differentiation of the CPCs [79,80].

2.4. KDR/FLK1-Expressing CPCs

During cell movement from the primitive streak to the anterior regions of the embryo, the precardiac mesodermal cells start to express a receptor for Vascular Endothelial Growth Factor (VEGF) called KDR/FLK1 [20,81]. These FLK1-expressing progenitor cells have the ability to generate cells from both hematopoietic and cardiac lineages [20,81–83]. Selection between these two lineages is determined by the levels of FLK1 activity [81]. For example, high expression of FLK1 promotes differentiation towards hematopoietic lineages, whereas low or absent FLK1 expression stimulates cells to follow the cardiac fate [81,82]. These negative FLK1-expressing cells further generate a second $FLK1^+$ cell population that represents the first multipotent cardiac progenitor cells that are permanently committed to the cardiogenic fate [20,29]. Because KDR/FLK1 displays a broad expression, it is frequently used in combination with other cardiac markers, such as Platelet-Derived Growth Factor-alpha (PDGFRα), C-X-C chemokine Receptor type 4 (CXCR4) and sometimes MESP1/2, to enrich for CPCs [79,80].

2.5. CPCs from the First and Second Heart Fields

The cardiac mesoderm contains two unique progenitor cell pools that give rise to the primary and secondary heart fields [20,22]. The two fields develop sequentially and display distinctive molecular profiles that lead to the formation of different heart components. CPCs of the first heart field (FHF) express the transcription factor NKX2.5, whereas CPCs from the second heart field (SHF) express the transcription factor ISL1 [15,22,40,75,84]. FHF-derived CPCs are more difficult to isolate owing to a lack of unique markers except for NKX2.5 [84]. The hyperpolarisation-activated nucleotide-gated cation channel HCN4 has been suggested as an additional marker for FHF, however, this marker might isolate a more restricted CPC that preferentially generates cells of the conduction system [85–87]. Regardless of the markers, FHF-CPCs predominantly differentiate into cardiomyocytes and have some tendency towards smooth muscle lineages [21]. On the other hand, ISL1 CPCs can generate cells of all the three cardiac lineages and they are responsible for producing most of the cardiomyocytes (around 40%) during heart development [22,30,40]. In addition, these CPCs have also been identified in the adult heart, specifically in the outflow tract, atria and right ventricle [30,40].

2.6. Epicardium-Derived CPCs

Several studies have demonstrated that a specific CPC population present in the postnatal and adult heart is derived from the epicardium. They express the transcription factor Wilms tumor 1 (WT1) and are originally derived from CPCs of the second heart field [88]. Additionally, these CPCs emerge from epicardial cells that migrate into the myocardium and undergo epithelial-to-mesenchymal transition (EMT) [88,89]. The epicardial-derived CPCs can differentiate into several different cell types such as coronary smooth muscle cells, cardiomyocytes, endothelial cells, perivascular and cardiac interstitial fibroblasts, albeit with varying efficiencies [51,88–93]. Even though WT1 CPCs could potentially be an additional cell source for cardiac regeneration, these cells seem to share some characteristics with c-KIT CPCs: they participate in cardiomyocyte formation during cardiac development but are present at extremely low levels in the adult heart [9,88,90,92]. Stimulatory factors like peptide thymosin beta 4 (Tβ4) can potentially reactivate the developmental program of adult epicardial cells, however, the reactivated cells still exhibit distinct phenotype from their embryonic counterparts, raising doubts about their cardiogenic potential [92].

2.7. Side Population-Derived CPCs

Side populations (SPs) have been detected in various tissues, including the heart, and are enriched for stem and progenitor cell activity [38,94–98]. These cells are generally identified in vitro by their ability to export the DNA Hoechst dye from their nuclei when stained [94,95]. This dye efflux is performed by an ATP (Adenosine Triphosphate)-binding cassette transporter (also known as ABC transporter protein) that is present in their cellular membranes [94,95]. The phenomenon causes the side population cells to contain much lower concentrations of the dye in their nuclei compared to other cell types, allowing for their isolation using cell sorting techniques [95]. The main ABC transporter protein used to identify cardiac SPs is the ABCG2, which was demonstrated to have a role in stem cell proliferation and differentiation and is expressed in SP cells during early development and in the postnatal heart [38,94,95,99]. These cardiac SPs can be found in the perivascular and interstitial areas of the heart, and display self-renewal, homing and multipotency [94,97,100–102]. Noseda et al. (2015) demonstrated that cardiac SPs, co-expressing SCA1 and PDGFRα, displayed high clonogenicity and multilineage potential [103]. They particurlaly demonstrated that clones derived from cardiac SPs subjected to long-term propagation (more than 10 months and 300 doublings) still resembled freshly isolated SP cells, showing maintainance of phenotype, self-renewal and tri-lineage capacity and absence of replicative senescence. However, cardiac SPs exhibit a few disadvantages that could potentially prevent their clinical application. For instance, the differentiation potential of human SPs has not been thoroughly investigated [38]. In addition, the multipotency of SPs might be attributed to their heterogeneous nature as they are composed of several subpopulations with distinct differentiation potential (cardiac, hematopoietic and mesenchymal) [38,104]. Therefore, it is still inconclusive on which markers can predict the SP subpopulation with the best cardiac potential.

2.8. Cardiosphere-Derived CPCs

Cardiospheres contain a mixture of stromal, mesenchymal and progenitor cells that are isolated from human heart biopsy cultures [39,52]. They represent a niche-like environment containing a mixture of cells, with cardiac-committed cells in the center and supporting cells, such as mesenchymal and endothelial progenitor cells, in the periphery of the spherical cluster [105,106]. Many cells can be harvested from these cell clusters and they are called cardiosphere-derived cells (CDCs) [52,105]. However, like in the case of c-KIT and epicardial CPCs, the regenerative potential of CDCs is questionable as it has been shown that cardiac repair by these cells mainly results from paracrine mechanisms rather than cell generation [105].

Table 1. Types of CPCs identified in the heart tissue.

CPC Type	Marker Expression	Differential Potential	Functionality of the Differentiated Cells	Applied to Disease In Vivo	Concerns	Ref.
c-KIT	$Ki67^+$ $NKX2.5^+$ $GATA4/5^+$ $MEF2C^+$ $TBX5^+$ $CD45^-$ $CD34^-$ $CD31^{+/-}$	-Differentiation trend towards CMs *, ** -Few fibroblasts * -ECs *	In vitro: -Atrial and ventricular CMs and cells of the conduction system * -CMs show a disorganized structure, no sarcomeres, and smaller size than their adult counterparts *, ** In vivo: -CMs couple with host cells and display spontaneous beating and striated structures *, **	-Formation of structural and functional CMs and contribution to coronary vessels in MI rats ** -Reconstitution of a myocardial wall that encompassed up to 70% of LV in MI rats ***	-CPC population is heterogeneous with cells at distinct stage of differentiation and with different commitment to the cardiac lineages *, ** -Differentiated cells show an immature phenotype *, ** -No consensus regarding the regenerative capability of c-KIT CPCs and their lineage marker expression *, ** -Distinct differential potential between neonatal and adult $c-KIT^+$ CPCs and between species *, ** -Benefits are mainly a result of paracrine factors *, **	[21,27,28,31,53,58,59,107,108]
SCA1	$ISL1^+$ $c-KIT^{+/-}$ $PDGFR\alpha^+$ $CD105^+$ $CD90^+$ $CD44^+$ $GATA4^+$ $MEF2C^+$ $NKX2.5^{+/-}$ $TEF-1^+$ $CD31^{+/-}$ $CD34^-$ $ABCG2^+$	-CMs, SMCs, and ECs *, ** -Foetal $SCA1^+$ CPCs tend to differentiate into ECs, whereas adult CPCs have more efficiency towards CMs **	In vitro: -CMs display spontaneous beating, myofilaments and expressed connexin 43 *, ** -Immature CMs and SMCs *, ** -ECs form tube-like structures *, ** -Foetal $SCA1^+$ CPCs exhibit more spontaneous beating than adult $SCA1^+$ CPCs ** In vivo: -ECs contribute to capillaries and CMs display defined striated structures *	-Knockdown of SCA1 led to larger LV volume, increased infarct rate and limited angiogenesis in MI mice * -$SCA1^+/CD31^-$ cell population numbers increased in the LV following MI * -Transplantation of $SCA1^+/CD31^-$ in MI mice attenuates adverse LV remodeling *	-No human homolog of SCA1 identified ** -SCA1 does not discriminate between proliferating and differentiating cells *, ** -$SCA1^+$ CPCs represent a heterogeneous population with subpopulations displaying different lineage potential *, ** -Distinct potency between neonatal and adult $SCA1^+$ CPCs ** -Differentiation into CMs requires co-culture with adult/neonatal CMs ** -Benefits are mainly a result of paracrine factors *, **	[11,21,37,64–67,69,70,109]
KDR/FLK1$^{low/-}$	T^+ $MESP1^+$ $c-KIT^-$ $GATA4^+$ $TBX5^{+/}$ $NKX2.5^{+/-}$ $CD31^{+/-}$ $SL1^{+/-}$ SMA^+ $PDGFR\alpha^+$	-Highest efficiency for SMCs, followed by CMs and then ECs *, ** -$KDR^+/CXCR4^+$ has better efficiency towards CMs *	In vitro: -CMs display spontaneous Beating *, ** -Predominantly atrial and ventricular CMs ** -Few pacemaker and conduction system cells * -Electrical coupling is observed ** -ECs display LDL-uptake capacity ** -ECs and SMCs form tube-like structures ** In vivo: -Human ESC-derived KDR^+ CPCs differentiate into CMs and ECs **	-Human ESC-derived KDR^+ progenitors increased ejection fraction in infarcted hearts of NOD/SCID mice **	-Hematopoietic tendency *, ** -FLK1/KDR marks two populations with distinct cardiac potential that develop at different temporal stages of mesoderm differentiation *	[20,29,79,82,110]

Table 1. *Cont.*

CPC Type	Marker Expression	Differential Potential	Functionality of the Differentiated Cells	Applied to Disease In Vivo	Concerns	Ref.
MESP1/2	SSEA1$^+$ OCT4$^+$ T$^+$ KDR$^+$ ISL1$^+$ TBX5/6/18/20$^+$ GATA4/6$^+$ NKX2.5$^+$ MEF2C$^+$ MYOCD$^+$ PDGFRα/β^+ CXCR4$^+$ WNT8A$^+$ FGF8$^+$ HAND2$^+$	-More efficiency towards SMCs and ECs *, ** -Some CMs *, **	In vitro: -Formation of ventricular CMs * -CMs express sarcomeric structures when co-cultured with human cardiac fibroblasts and CMs ** In vivo: -CMs display organized myofibrillar striations and express CX43, and SMCs and ECs form tube-like structures and contribute to neovasculogenesis *	-Murine ESC-derived MESP1 CPCs decreased LV-EDV, scar size, and improved LV ejection fraction, stroke volume and cardiac function in MI mice hearts *	-Not fully committed to the cardiac lineages *, ** -Not thoroughly investigated as CPCs *, ** -MESP1 marks a mixed population of CPCs with different multilineage differentiation potential *, ** -MESP1 CPC might be a subset of KDR$^+$/PDGFRα^+ cells *, ** -MESP1 is transiently expressed, making it difficult to track the expansion and differentiation of the CPCs *, **	[72,76,79,80,111]
From First Heart Field (FHF)	NKX2.5$^+$ HAND1$^+$ TBX5$^+$ HCN4$^+$	-More efficiency towards CMs *, ** -Some SMCs *, **	In vitro: -Atrial, left ventricle and conduction myocytes *, ** -Presence of both mature and immature CMs * -Some spontaneous beating *, ** -Most CMs display a ventricular-like action potential * -Some atrial-like and nodal-like action potentials are formed * In vivo: -ESC-derived CPCs differentiate into SMCs and CMs, which display beating and form myofibrils *	-Not yet applied in vivo in a disease context	-Difficult to identify and characterized due to lack of markers *, ** -FHF have limited potency *, ** -Not thoroughly investigated as CPCs *, **	[21,84–86]
From Second Heart Field (SHF)	ISL1$^+$ c-KIT$^{-/+}$ NKX2.5$^{+/-}$ TBX1$^+$ GATA4$^+$ KDR$^{+/-}$ FGF8/10$^+$ FOXH1$^+$ MEF2C$^+$ WT1$^+$	-Majority to CMs, including pacemaker *, ** -Some cardiac fibroblasts, SMCs and ECs *, ** -ISL1$^+$/KDR$^+$ into ECs and SMCs * -NKX2.5$^+$/ISL1$^+$ into CMs *, ** -NKX2.5$^+$/KDR$^+$ into SMCs *	In vitro: -Remarkable contribution to the sino-atrial node * -Only a few towards atrial-ventricular node * -CMs exhibit synchronized calcium transients * In vivo: -Contribution to the coronary arterial system * -SMCs are in the most proximal outflow tract * -ESC-derived ISL1$^+$ CPCs differentiate into pacemaker and ventricular CMs, SMCs and ECs * -Knockdown of ISL1 led to a reduction in cardiac tissue formation and affects CPC proliferation, survival and migration *	-Not yet applied in vivo in a disease context	-Majority of contribution to the conduction system is restricted to the sino-atrial node * -EC and SMC contribution is limited to the proximal area of the great vessels * -Embryo-derived SHF show a significant reduction in differentiation into CMs and tripotency was rare *	[22,30,40,83,84,112,113]

Table 1. *Cont.*

CPC Type	Marker Expression	Differential Potential	Functionality of the Differentiated Cells	Applied to Disease In Vivo	Concerns	Ref.
Epicardial-derived	WT1[+] TBX18[+] SLUG RALDH2 SCA1[+] PDGFRα[+]	-Vascular SMCs *, ** -CMs under certain in vitro conditions *, ** -Some cardiac fibroblasts (perivascular and interstitial) *, **	In vitro: -SMCs and fibroblasts *,** -Atrial and ventricular CMs, with striated cytoarchitecture, spontaneous contraction, native calcium oscillations and electrical coupling * In vivo: -SMCs contribute to the coronary arteries * -Differentiation into fibroblasts, SMCs and coronary endothelial cells; CMs can be formed when subjected to the stimulation of exogenous factors *	-Epicardial-derived CPCs increased vessel formation and stimulate angiogenesis in murine MI models * -Epicardial-derived CPC conditioned medium reduced infarcted size and improved heart function in MI mice models *,** -Priming of the epicardium with Tβ4 prior to injury led to enhanced migration of epicardial-derived CPCs and generation of CMs in MI mice *	-Epicardial-derived CPCs descend from NKX2.5-and ISL1-expressing cells *, ** -No EC differentiation *, ** -Epicardial-derived CPCs are difficult to culture in Vitro *, ** -No consensus about the level of contribution of the epicardium in cardiac repair *,**	[88–91,114–117]
Side Population (SP)	ABCG2[+] SCA1[+] CD34[+/−] CD31[+/−] c-KIT[−] NKX2.5[+/−] GATA4[+/−] MEF2C[+] CD45[−] VE-cadherin[−]	-Fibroblasts & SMCs *, ** -SCA1[+]/CD31[−] SPs into CMs * -SCA1[+]/CD31[+] SPs + VEGF into ECs * -CD45[−] SPs into ECs *	In vitro: -CMs show spontaneous beating and striations on staining * -Electrical coupling is observed when SPs are co-cultured with adult CMs * In vivo: -Differentiation into CMs, forming striated sarcomere structures, SMCs, ECs, and fibroblasts *, ***	-Cardiac SP numbers are significantly increased, particularly in the left ventricle, following acute ischemia ** -Myocardial injury facilitated migration and homing of cardiac SPs *, ***	-Hematopoietic differentiation tendency * -Low percentage of CMs reach advanced maturity *, ** -Contradictory results between different studies on the maturity of the SP-derived CMs *, ** -SPs represent an extremely heterogeneous population * -Complete differentiation requires both cell-intrinsic and -extrinsic factors *	[38,94,96,97,100–102,104,118]
Cardiosphere (CS)-derived cells (CDCs)	KDR[+] c-KIT[+] SCA1[+] CD34[+/−] CD45[−] CD133[−] NKX2.5[+] GATA4[+] ISL1[+] CD105[+]/CD31[+]/CD90[+]/c-KIT[−] supporting cells	CMs, SMCs & ECs *, **	In vitro: -CMs display spontaneous beating, but lack sarcomeric structure * -Differentiation into ECs and SMCs with VEGF treatment *, ** In vivo: -Differentiation into SMCs and ECs, some potential towards CMs lineages *, ** -Formation of tubular-like structures *	-Transplantation of CDCs/CSs improved cell survival, engraftment and LV ejection fraction, stimulated angiogenesis, inhibited adverse LV remodeling and reduced infarct size in the infarcted mice **	-Human CSs/CDCs require co-culture with adult CMs to stimulate contraction and advance maturity ** -Stemness decreases in monolayer cultures ** -CSs/CDCs represent a mixed cell population *, ** -Benefits result from paracrine factors *, ** -Low CDC engraftment and differentiation efficiency ** -Different markers used, which isolate cells with distinct differentiation potential *, **	[39,106,119,120]

CMs: Cardiomyocytes; SMCs: Smooth Muscle Cells; ECs: Endothelial Cells; MI: Myocardial Infarction; LV: Left Ventricle; EDV: End-Diastolic Volume; LDL: Low Density Lipoprotein; ESC: Embryonic Stem Cell; NOD/SCID: Non-Obese Diabetic/Severe Combined Immunodeficient; VEGF: Vascular Endothelial Growth Factor; *, Mouse; **, Human; ***, Rat.

3. Generation of CPCs from Human iPSCs

Native CPCs are present in very low numbers in the heart tissue, and therefore, a larger source of cells is required for efficient cardiac regeneration [41]. The reprogramming of human adult somatic cells into embryonic stem cell-like cells (known as iPSCs) using defined factors opened new possibilities for the generation of patient-specific pluripotent cells. In turn, human iPSCs could potentially offer an unlimited source of differentiated cells and in the process, offer the chance to recreate the development process of CPCs in vitro [121]. This section will provide a detailed description and assessment of current methods used to induce, expand and maintain CPCs derived from iPSCs.

Several techniques have been developed to modulate cardiac differentiation in iPSCs (Table 2). However, the efficiencies for cardiac differentiation can vary considerably between iPSC lines [121–124]. Regardless of the type of culture employed, the first step in all protocols involves the dedifferentiation of a chosen cell type into a pluripotent state using conventional reprogramming factors, such as OCT4, SOX2, KFL4 and c-MYC [44,45]. Once pluripotency has been achieved, the following step is to induce cardiac differentiation of the iPSCs. Different methods have been employed to accomplish differentiation of iPSCs into cardiomyocytes: embryoid body (EB); monolayer-based cultures supplemented with growth factors, serum or small molecules, matrices, and co-culture with visceral endodermal stromal (END2) layers [15,122,125]. Recent protocols utilise a monolayer culture with a serum-free medium, such as mTeSR1 or E8 medium, which maintains iPSC pluripotency and self-renewal in a feeder-free culture [126,127]. Unfortunately, these studies predominantly focused on the generation of iPSC-cardiomyocytes and not necessarily the homogeneity of CPC population entering the cardiac lineages.

In addition to the nature of the pluripotent culture employed, the type and timing of growth factors and/or small molecules added throughout the protocol affects cardiomyocyte differentiation efficiency. Early differentiation protocols only employed growth factors that modulate key signaling pathways involved in cardiomyogenesis, like Bone Morphogenic Protein (BMP), Activin/Nodal and Fibroblast Growth Factor (FGF) signaling pathways [15,128]. Such factors included Activin A, BMP2/4 and FGF2 which induce cardiac mesoderm formation in iPSCs [15,29,122,128,129]. Lian et al. (2012) demonstrated that iPSC differentiation towards cardiac lineages could be accomplished by exclusively using small molecule modulators of the Wingless (WNT) signaling pathway [130]. Minami et al. (2012) also showed that combining analogous WNT modulators during the early and middle stages of the cardiac differentiation process can further enhance the protocol's efficiency [131]. Many protocols rely on adding a Glycogen Synthase Kinase (GSK3) inhibitor, such as CHIR99021 (CHIR), to the medium for 24 h to activate the canonical WNT signaling [126,130,132]. Induction of the canonical WNT signaling stimulates the expression of the mesoderm marker Brachyury (T) in undifferenced iPSCs, initiating mesoderm induction [126,132]. Once T^+ cells have been established, the WNT signaling is then suppressed to direct the mesodermal cells towards the cardiac fate [126]. Several inhibitory molecules can be used, like XAV939, inhibitor of WNT production (IWP), inhibitor of WNT response (IWR) or an exogenous β-catenin small hairpin RNA (shRNA). After 3/4 days of WNT signaling suppression, iPSC-derived T^+ mesodermal cells begin to express cardiac transcription factors, like NKX2.5, ISL1, FLK1, and PDGFRα, which transitions into the CPC population.

More recent studies have been successful in generating CPCs from iPSCs using a single small molecule, potentially reducing costs, time and labor. For instance, the immunosuppressant cyclosporin-A (CSA) was shown to stimulate differentiation of FLK1-positive mesodermal cells into $FLK1^+/CXCR4^+/VE$-cadherin$^-$ CPCs and cardiomyocytes [133,134]. When CSA was added to the medium, the CPC and cardiomyocyte yield was 10 to 20 times higher compared to untreated cells. Additionally, the generated cardiomyocytes exhibited molecular, structural and functional properties of adult cardiomyocytes. However, additional factors and/or other protocols may be required to produce cells from the other cardiac lineages as $FLK1^+/CXCR4^+/VE$-cadherin$^-$ CPCs have an exceptionally low endothelial potential and cannot differentiate into smooth muscle cells [110,133]. Furthermore, the study used co-culture with END2 cells to induce cardiac differentiation in iPSCs, which prevents reproducibility of the protocol due to the presence of END2-derived growth factors at unknown concentrations [135]. Another study also demonstrated that treating human iPSCs with the cardiogenic small molecule isoxazole (ISX-9) for 7 days stimulated the expression of CPC markers [136]. These CPCs expressed NKX2.5, GATA4, ISL1, and MEF2C and were able to generate cardiomyocytes, smooth muscle cells and endothelial cells under basal differentiation conditions. Furthermore, ISX-9 seems to modulate key signaling pathways involved in cardiomyogenesis, like VEGF, Activin A and canonical and non-canonical WNT signaling. The study also demonstrated that the small molecule might participate in CPC generation by upregulating activators involved in both canonical and non-canonical WNT pathways in a temporal and sequential manner (WNT3A at day 3 and WNT5A and WNT11 at day 7, respectively).

Therefore, the application of iPSC technology in CPC research has great prospects for improving current cardiac regeneration approaches through the development of novel cell therapies, disease models and drug screens. However, most studies using iPSCs in cardiac regeneration predominantly focus on producing cardiomyocytes and improving their maturation [15,126,130,137–142]. Current knowledge about associating this with the generation of iPSC-CPCs, however, remain limited.

Table 2. Protocols producing CPCs as target cells or as intermediate cells from iPSCs.

Pluripotent Culture	Protocol		CPC-Associated Markers Identified	CPCs as Target or Intermediate	Differentiation and Functionality Potential	Limitations	Ref.
	Mesoderm Differentiation	Cardiac Specification					
Mouse iPSCs on feeder-layers and human iPSCs in hESC culture medium without bFGF	Differentiation medium with 20% FBS + gelatin-coated plates + AA between day 2 and 6		NKX2.5$^+$ TBX5$^+$ & FLK1$^+$ CXCR4$^+$	Intermediate	-Synchronous beating and better-organized striated myofilaments in CMs	-AA is not able to promote mesodermal differentiation and CM proliferation -No reports on CPC potential into SMCs and ECs	[143]
Human iPSCs in monolayer culture (mTeSR1 + Matrigel-coated plates)	ROCK inhibitor (Y27632) for 1 day and DMEM/F12/B27-vitamin A + BMP4 + AA + CHIR for 3 days		SSEA1$^+$ MESP1/2$^+$ ISL1$^+$	Target	-Differentiation into the three cardiac lineages under specific differentiation media -80% efficiency towards CMs, and 90% into SMCs and ECs -Synchronized beating and presence of organized sarcomeric structures	-Both early and late CPC-related markers were co-expressed in the generated CPCs -Repeated passaging leads to a decrease in CPC proliferation rate -Only one iPSC line was tested	[144]
Human iPSCs on inactivated MEFs followed by feeder depletion culture in Matrigel	BMP4 for 3 days and +/− Activin A + bFGF from day 1 until day 3	DKK1 + VEGF + SB +/− Dorsomorphin/Noggin at day 3	KDR$^+$ PDGFRα^+	Intermediate	-Low yield of CMs (11%)	-iPSC line variability affects protocol's efficiency and optimal growth factor concentrations -Presence of the CPC population does not always predict efficient differentiation to CMs	[128]
Mouse iPSCs in DMEM with 15% FCS on feeder layers	Differentiation medium with 10% FCS + type IV collagen-coated dishes/OP9 cell sheets for 96–108 h	FLK1$^+$ mesodermal cells co-cultured on OP9 cells + differentiation medium + cyclosporin-A	FLK1$^+$ CXCR4$^+$ VE-cadherin$^-$	Target	-Synchronous beating -Pacemaker and ventricular action potentials -Myofilaments formation with transverse Z-bands -Presence of ion channels (Cav3.2, HCN4 and kir2.1) and intercalated disks	-CPCs were only isolated from mouse iPSCs -Differentiation efficiency was different for various iPSC lines -Incomplete human CM maturation	[133]
Human iPSCs on SNL feeder cells and Matrigel-coated plates	Co-culture on END-2 cells + cyclosporin-A at day 8			Target			
Human iPSCs on inactivated MEFs with KO-DMEM medium	Serum-free medium (RPMI/B27) + BMP2 + SU5402 for 6 days		OCT4$^+$ SSEA1$^+$ MESP1$^+$ TBX5$^+$ TBX6$^+$ TBX18$^+$ GATA4$^+$ MEF2C$^+$ NKX2.5$^+$ ISL1$^+$ TBX20$^+$	Target	-Differentiation towards CMs, SMCs and ECs under specific conditions -Arranged sarcomeric organization and gap junctions when CPCs were co-cultured with either fibroblasts + FCS, cardiac fibroblasts + CMs or conditioned medium -Trend towards ventricular CMs	-Only one iPSC line was tested -SSEA1$^+$ CPCs can differentiate into multiple cardiac lineages, like FHF, SHF, epicardium and cardiac neural crest in the presence of FGF signals	[145]

Table 2. *Cont.*

Pluripotent Culture	Protocol		CPC-Associated Markers Identified	CPCs as Target or Intermediate	Differentiation and Functionality Potential	Limitations	Ref.
	Mesoderm Differentiation	Cardiac Specification					
Murine iPSCs on inactivated MEFs	Feeder-free culture on gelatin-coated plates + BIO	IMDM with 15% FCS	FLK1[+] MESP1[+] NKX2.5[+]	Target	-Presence of CM, EC and SMC markers	-Incomplete CM maturation -Functionality of the differentiated cells in in vitro conditions needs further assessment	[146]
Human iPSCs on Matrigel-coated plates	E8 medium + ROCK inhibitor for 24 h and RPMI/B27-insulin + CHIR for 48 h/ 4 days		TBX5[+] NKX2.5[+] CORIN[+] HCN4[+] GATA4[+]	Target	-FHF: mainly differentiates into left ventricular (90%) and some atrial CMs (10%) -Presence of ion channels (Kir2.1) and higher contraction velocity	-4 different CPC populations identified with distinct differentiation potential -Isolation of the CPC populations was performed via a double transgene reporter -Expression of TBX5 and NKX2.5 dynamically changed during differentiation culture, except for the double negative (TBX5[−]/NKX2.5[−]) cell population	[147]
			TBX5[+] NKX2.5[−] HCN4[+] GATA4[+] WT1[+] TBX18[+] KDR[+] PECAM1[+]	Target	-Epicardial progenitors: contribute to nodal (80%) and some atrial CMs -Formation of tight junctions and expression of the ion channel KCNJ3 -Some potential towards fibroblasts, SMCs and ECs		
			TBX5[−] NKX2.5[+] GATA4[+] MEF2C[+] ISL1[+]	Target	-SHF: differentiation predominantly into atrial (90%) and some nodal and ventricular CMs -Atrial CMs displayed slower beating rates -Some potential towards SMCs and ECs		
			TBX5[−] NKX2.5[−] KDR[+] PECAM1[+]	Target	-Endothelial potential -Formation of tube-like structures under VEGF		
Human iPSCs on inactivated MEFs followed by EB suspension culture	BMP4 for 4 days	IWR1/IWP1 for 2 days	NKX2.5[+] ISL1[+] GATA4[+] MEF2C[+]	Intermediate	-Low percentage of CMs -Organized sarcomeric structures -Normal calcium transient rhythm	-The CPCs were only identified when using human ESCs -Embryonic action potentials -CPC was an intermediate state during differentiation into CMs	[148]
Human iPSCs on MEFs	DMEM/F12 with 20% FBS + AA + EB plating on gelatin-coated dishes at day 7	MEFs for 24 h and BMP2 + SU5402 for 4 days in RPMI/B27-vitamin A	ISL1[+] NKX2.5[+] KDR[+] MESP1[+] TXB20[+] GATA4[+]	Target	-Differentiation towards myocytes and vascular lineages under specific conditions	-Differentiation trend and CM maturation in vitro were not fully assessed	[149]
Human iPSCs on Synthemax-coated plates in E8 medium then mTeSR1/E8 + ROCK inhibitor for 24 h	Albumin-free RPMI + CHIR for 24 h	RPMI + IWP2 for 2 days at day 3 + basal medium at day 5	ISL1[+] NKX2.5[+] KDR[+]	Intermediate	-Spontaneous contraction and well-organized sarcomere filaments -Development of ventricular action potentials -Spontaneous calcium transients and connexin 43 expression in CMs	-No information about differentiation potential towards ECs and SMCs	[150]

Table 2. *Cont.*

Pluripotent Culture	Protocol		CPC-Associated Markers Identified	CPCs as Target or Intermediate	Differentiation and Functionality Potential	Limitations	Ref.
	Mesoderm Differentiation	Cardiac Specification					
Human iPSCs on Matrigel in MEF-CM supplemented with bFGF	RPMI/B27-insulin + Activin A for 24 h + BMP4 and bFGF for 4 days	RPMI/B27-insulin + DKK1 for 2 days	MESP1$^+$ KDR$^+$ ISL1$^+$ NKX2.5$^+$	Intermediate	-Sarcomere formation -Ventricular and pacemaker action potentials -CM yield varied between 4 and 34%	-Protocol efficiency and CM differentiation and maturation is affected by cell line variability -Incomplete CM maturation -CPC was an intermediate state during differentiation into CMs	[140]
Human iPSCs in Geltrex with E8 medium using spheroid culture	RPMI/B27-insulin + CHIR + BMP4 for 48 h	XAV939 for 48 h at day 4	ISL1$^+$ TBX1$^+$ FGF10$^+$ FGF8$^+$ CXCR4$^+$ (SHF)	Target	-38% efficiency towards CMs -More potential to generate SMCs, ECs and fibroblasts	-No information about the functionality of the differentiated cells -Only one hiPSC line was tested	[151]
			ISL1$^+$ HCN4$^+$ TBX5$^+$ GATA4$^+$ CXCR4$^-$ (FHF)	Target	-62% efficiency towards CMs -Low levels of EC and fibroblast markers		
Human PSCs on Matrigel/Synthemax-coated plates in mTeSR1/E8 medium with ROCK inhibitor	CHIR in RPMI basal medium for 24 h	IWP2/IWP4 in RPMI basal medium from day 3 to day 5 + LaSR basal or RPMI/Vc/Ins with ROCK inhibitor at day 6 + CHIR for 48 h from day 7	WT1$^+$ TBX18$^+$ TCF21$^+$ ALDH1A2$^-$ KDR$^+$	Target		-Epicardial progenitor cells are derived from a more multipotent CPC population (PDGFRα^+/ ISL1$^+$/NKX2.5$^+$/GATA4$^+$/TBX5$^+$) -Format size of the culture (i.e., 96-well or 6-well plate) affects maturity of the epicardial cells -Different protocols lead to the formation of mesodermal cells expressing distinct markers (PDGFRα^+/KDR$^+$ and ISL1$^+$/NKX2.5$^+$) -Epicardial progenitor cells exhibit multiple origins	[152]
	Albumin-free RPMI + CHIR for 24 h	RPMI + IWP2 for 2 days at day 3 + RPMI/Vc/Ins with ROCK inhibitor for 24 h at day 6 + CHIR in RPMI//Vc/Ins for 48 h at day 7		Target	-Differentiation towards fibroblasts and SMCs -Fibroblasts and SMCs display fibroid spindle-like shape and a fusiform appearance, respectively -Formation of mature epithelial-like sheets with tight junctions (cobblestone morphology and expression of ZO1 along cell borders) -SMCs display calcium transients and contractibility		[153]
Human iPSCs on inactivated MEFs	StemPro-34 medium + BMP4 for 24 h + BMP4, Activin A and bFGF from day 1 until day 3	StemPro-34 medium + Matrigel-coated plates + BMP4 + CHIR + SB + VEGF for 2 days		Target			[154]
Human iPSCs in CDM + BSA + Activin A + FGF2 on gelatin-coated plates	CDM + PVA + FGF2 + LY294002 + BMP4 for 36 h and CDM + PVA + FGF2 + BMP4 for 3.5 days	CDM + PVA + BMP4 + WNT3A + RA for 10 days		Target			[155]
Human iPSCs in E8 medium and monolayer culture on vitronectin-coated plates	S12-insulin medium + CHIR for 24 h	S12-insulin medium + IWR1 for 48 h at day 3 and RA + CHIR between day 5 and 8		Target			[156]

Table 2. *Cont.*

Protocol			CPC-Associated Markers Identified	CPCs as Target or Intermediate	Differentiation and Functionality Potential	Limitations	Ref.
Pluripotent Culture	**Mesoderm Differentiation**	**Cardiac Specification**					
Murine iPSCs in inactivated MEFs in SCM	SCM-LIF + AA at day 2	Puromycin at day 6 for 3 days	NKX2.5$^+$ c-KIT$^+$ FLK1$^+$ SCA1$^+$	Target	-Differentiation potential towards ventricular CMs, SMCs and ECs -Sarcomeric organization and intracellular coupling observed	-Presence of CPCs expressing different sets of markers -Application of a plasmid system for CPC enrichment	[157]
Human iPSCs on MEFs followed by suspension culture in ESC culture medium	Gelatin-or human laminin211-coated plates + IMDM-serum and CHIR + BIO for 3 days	KY02111 +/− XAV939 or IWP2 from day 3 until day 9	NKX2.5$^+$ GATA4$^+$	Intermediate	-Predominantly ventricular CMs and 16% pacemaker cells -Spontaneous beating, sarcomere myofilaments, Z-bands, ion channels (HERG and KCNQ1) intercalated disks observed	-Mechanism of canonical WNT inhibition by KY02111 not fully understood -Protocol efficiency is affected by the presence of serum and cytokines -No differentiation into SMCs and ECs	[131]
Human iPSCs in E8 medium on Synthemax/Matrigel-coated plates	CDM3 medium (RPMI basal medium + AA + rice-derived RHA) + CHIR for 2 days	CDM3 medium + WNT-C59 for 48 h at day 2	MESP1$^+$ KDR$^+$ ISL1$^+$ GATA4$^+$ NKX2.5$^+$ TBX5$^+$ MEF2C$^+$	Intermediate	-Formation of atrial, ventricular and nodal CMs	-Presence of unspecified CMs, without a defined subtype -Incomplete CM maturation -No differentiation into SMCs and ECs -CPC was an intermediate state during differentiation into CMs	[137]
Human iPSCs in mTeSR1 + ROCK inhibitor on Matrigel/Synthemax	Pre-treatment with CHIR/BIO for 3 days	RPMI/B27-insulin + Activin A for 24 h + BMP4 for 4 days	ISL1$^+$ NKX2.5$^+$	Intermediate	-High yield of CMs -Normal sarcomere organization with transverse Z-bands -Presence of intercalated disks -Maturation trend towards ventricular CMs (80–90%) Some atrial-like action potential (10%) and absence of nodal-like potentials -Some formation of SMCs	-Optimal BMP4 concentration varies with different cell lines -Heterogenous activation of the canonical WNT signaling upon CHIR treatment in transgenic iPSC lines -Requirement of long periods of time (>60 days) to reach advanced CM maturity -Greater efficiency observed in studies using transgenic models	[126,130]
Transgenic iPSC lines carrying lentiviral integrated β-catenin shRNA	CHIR in RPMI/B27-insulin for 24 h	Doxycycline at 36 h post-CHIR addition	ISL1$^+$ NKX2.5$^+$ TBX5$^+$ WT1$^+$	Intermediate			
Non-transgenic hiPSC lines		IWP4 or IWP2 at day 3	Not reported	-			[130]
		IWP2 at day 3	ISL1$^+$ NKX2.5$^+$	Intermediate			[126]
Human iPSCs on vitronectin-coated plates in mTeSR1 + ROCK inhibitor for 24 h	RPMI/B27-insulin + ISX-9 for 7 days		NKX2.5$^+$ GATA4$^+$ ISL1$^+$ MEF2C$^+$	Target	-Differentiation potential towards CMs, ECs, and SMCs in vitro and in vivo -CMs displayed myofilaments, mitochondria and glycogen particles -Formation of tube-like structures and LDL-uptake in ECs -ECs, and SMCs formed vascular structures in vivo	-The exact mechanisms by which ISX-9 induces the expression of cardiac transcription factors is unclear -No reports about electric coupling between generated CMs and endogenous CMs in vivo -No information about the electrophysiology of CMs	[136]

Table 2. *Cont.*

Pluripotent Culture	Protocol		CPC-Associated Markers Identified	CPCs as Target or Intermediate	Differentiation and Functionality Potential	Limitations	Ref.
	Mesoderm Differentiation	Cardiac Specification					
Human iPSCs on Matrigel in mTeSR1 + ROCK inhibitor	CHIR in RPMI/B27-insulin for 24 h + bFGF	IWP2 from day 3 to day 5	MESP1+ T+ GATA4+ ISL1+ NKX2.5+ TBX1+ HAND2+ at day 2–3 & KDR+ PDGFRα+ at day 4–5	Intermediate	-Formation of SHF-derived CPCs -Differentiation trend into fibroblasts, which exhibited characteristics of fetal ventricular fibroblasts	-Stage-specific progenitors were generated with this protocol -Differentiation potential was limited to fibroblasts -The fibroblasts generated might represent just one of the populations of cardiac fibroblasts present in the native heart -Only one hiPSC line was tested (line variability effects need further assessment)	[158]
Human iPSCs in feeder-free (Geltrex) monolayer culture	RPMI + PVA + BMP4 + FGF2 for 2 days	RPMI-insulin + 20% FBS/human serum for 2 days	MESP1+ ISL1+ NKX2.5+	Intermediate	-Robust contraction -Striated sarcomeres and gap junction formation -High yield of CMs (64–89%) -Presence of physiological calcium transients and functional electrical coupling -Differentiation trend into ventricular CMs	-FBS is undefined -Incomplete CM maturation -CPC was an intermediate state during differentiation into CMs	[122]
		RPMI-insulin + 20% HSA + AA for 2 days		Intermediate			
		RPMI-insulin + 20% HSA + AA for 2 days		Intermediate			

hiPSCs: human iPSCs; (h)ESC(s): (human) Embryonic Stem Cell(s); b(FGF): (basic) Fibroblast Growth Factor; FBS: Foetal Bovine Serum; AA: Ascorbic Acid; CM(s): Cardiomyocyte(s); SMC(s): Smooth Muscle Cell(s); EC(s): Endothelial Cell(s); DMEM/F12/B27: Dulbecco's Modified Eagle Medium/Ham's F12 Nutrient Mixture/B27 serum supplement; BMP: Bone Morphogenic Protein; CHIR: CHIR99021; MEF(s): Murine Embryonic Fibroblast(s); DKK1: Dickkopf WNT signaling Pathway Inhibitor 1; VEGF: Vascular Endothelial Growth Factor; SB: SB-431542; FCS: Foetal Calf Serum; OP9: Mouse bone marrow-derived stromal cells; SNL: Mouse Fibroblast STO cell line-derived feeder cells; END-2: Visceral Endodermal Stromal cells; KO-DMEM: KnockOut DMEM; RPMI/B27: Roswell Park Memorial Institute/B27; FHF: First Heart Field; SHF: Second Heart Field; BIO: 6-bromoindirubin-3'-oxime; IMDM: Iscove's Modified Dulbecco's Medium; EB: Embryoid Body; IWR: Inhibitor of WNT Response; IWP: Inhibitor of WNT Production; MEF-CM: MEF-Conditioned Medium; LaSR: advanced DMEM/F12 with ascorbic acid; RPMI/Vc/Ins: RPMI with Ascorbic Acid (Vc) and Insulin (Ins); ZO1: Zonula Occludens-1/Tight junction protein-1; CDM: Chemically Defined Medium; BSA: Bovine Serum Albumin; PVA: Polyvinyl Alcohol; RA: Retinoic Acid; S12: Chemically Defined S12 Differentiation Medium; SCM: Stem Cell Medium; LIF: Leukaemia Inhibitor Factor; RHA: Recombinant Human Albumin; shRNA: small hairpin RNA; ISX-9: isoxazole; HSA: Human Serum Albumin.

4. Direct Reprogramming into CPCs

The discovery of iPSC reprogramming prompted studies to evaluate if it would be possible to reprogram somatic cells directly into other cell types without an iPSC intermediate stage, a process known as transdifferentiation or direct reprogramming. Transdifferentiation has shown to be a much quicker process than dedifferentiation into iPSCs, with the former taking only a few days to achieve, whereas the latter can last up to 3 weeks plus differentiation time to produce the desired cell lineages. With the added advantage of avoiding potential cumulative mutation or epigenetic changes, generally associated during complex iPSC reprogramming processes, direct reprogramming of somatic cells can potentially offer a simpler, faster and safer alternative to generate cells compared to iPSC dedifferentiation [41]. Most transdifferentiation studies in the cardiac field involve the generation of fully differentiated cardiac cells, particularly cardiomyocytes, rather than cardiac progenitor cells [159–172]. Potentially, using transdifferentiation protocols to generate CPCs might be a superior approach for regenerative medicine applications. This section focuses on the current approaches that are associated with producing CPCs from direct reprogramming.

4.1. Partial Somatic Cell Reprogramming into CPCs

Some studies have developed transdifferentiation protocols that involve a transient stage of pluripotency of somatic cells before they continue into CPC fates. The use of reprogramming factors (OCT4, SOX2, KLF4 and C-MYC) seems to be enough to initiate resetting of epigenetic memory of somatic cells towards a stem cell path (partial reprogramming), but the factors alone are insufficient to directly activate cardiac lineage-specific genes for directed differentiation [159]. In order to achieve lineage commitment, signaling molecules involved in cardiogenesis, like BMPs, WNT modulators and FGFs, need to be activated in the cultures [14,159], similar to differentiation protocols for cardiomyocytes from iPSCs. One study demonstrated that secondary mouse embryonic fibroblasts can be converted into CPCs using a technique developed by Wang et al. (2014) called Cell Activation and Signaling-Directed (CASD) lineage conversion [165], which combines reprogramming and cardiac-specific factors to induce cell activation and direct cell fate towards cardiogenesis, respectively [14]. Zhang et al. (2016) transiently exposed the mouse fibroblasts to reprogramming medium containing doxycycline and JAK inhibitor 1 (JI1) for 6 days, and then to transdifferentiation medium with CHIR99021 and JI1 for 2 days to induce cardiac differentiation. Following this, the cells are treated with a mixture of CHIR99021, BMP4, Activin A, and SU5402 (inhibitor of FGF, VEGF, and PDGF signaling) for 3 days. The obtained CPCs from this protocol expressed the proliferative marker Ki-67, the typical cardiac transcription factors GATA4, MEF2C, TBX5 and NKX2.5, and the cell surface molecules FLK1 and PDGFRα and were capable of producing cells from the three cardiac lineages. Efe et al. (2011) also demonstrated that transient expression of pluripotent markers (OCT4, SOX2, KLF4 and C-MYC) followed by exposure to chemically defined media containing BMP4 and the JAK inhibitor JI1 induced cardiac conversion of mouse embryonic and tail-tip fibroblasts [159]. JI1 was added to the reprogramming media for 9 days and from day 9, BMP4 was added and the media was subsequently changed to RPMI supplemented with N2 and B27 lacking vitamin A for 5 additional days. This protocol upregulated the expression of several CPC markers such as NKX2.5, GATA4, and FLK1 by day 9/10.

Wang et al. (2014) were able to significantly reduce the number of reprogramming factors to successfully stimulate cardiac transformation in mouse fibroblasts [165]. This protocol involved a single transcription factor (OCT4) and a cocktail of small molecules: an activin A/TGF-β receptor (ALK4/5/7) inhibitor (SB431542), GSK inhibitor (CHIR), Lysine (K)-Specific Demethylase 1 (LSD1/KDM1) inhibitor (parnate/tranylcypromine) and an adenylyl cyclase activator (forskolin). Mouse fibroblasts were first exposed to the reprogramming media, containing the small molecules, for 15 days. This was followed by media change to RPMI supplemented with N2 and B27 lacking vitamin A and addition of BMP4 during the first 5 days. CPC markers, like FLK1, MESP1, ISL1, GATA4, and Ki-67, can be detected around days 15–20. These cells went on to differentiate into cardiomyocytes, endothelial cells and smooth muscle cells under specific conditions. Another study developed an entirely chemical reprogramming protocol that utilised a larger combination of small molecules compared to Wang et al. (2014): CHIR, the ALK5 inhibitor RepSox, forskolin, the histone deacetylase (HDAC) 1 inhibitor valproic acid (VPA), parnate and the retinoid pathway activator TTNPB [173]. Mouse fibroblasts were exposed to the reprogramming cocktail for 16 days and CPC markers could be detected around day 8-20. The markers identified included SCA1, ABCG2, WT1, FLK1, and MESP1, demonstrating that the protocol can generate CPC populations. Most of the studies described protocols predominantly focused on their ability to generate cardiomyocytes from somatic cells using some iPS factors, and whilst CPCs were observed in some of these studies, their characteristics were not necessarily a focus of their attention and would warrant some investigation in their potency independently.

4.2. Direct Somatic Reprogramming into CPCs

Direct reprogramming of somatic cells involves the transdifferentiation into other cell types without an iPSC intermediate stage. One study showed that CPCs can be directly generated from adult mouse fibroblasts from different tissues (cardiac, lung and tail tip) using a 11- (MESP1, MESP2, GATA4, GATA6, BAF60C, SRF, ISL1, NKX2.5, IRX4, TBX5 and TBX20) or a 5- Factor (MEF2C, TBX5, GATA4 NKX2.5, BAF60C) reprogramming protocol [24]. Both protocols led to the formation of CPCs expressing the genes *NKX2.5*, *MEF2C*, *MESP1*, *TBX20*, *IRX4*, and the cell surface protein CXCR4, independently of factor combination and tissue origin of the fibroblasts. The CPCs also showed downregulation of fibroblasts-specific genes, such as *FSP1*, and could differentiate into the three cardiac lineages. Furthermore, adding a canonical WNT activator, and a JAK/STAT activator during the reprogramming process can increase the protocol efficiency, leading to the production of more CPCs. Even though the 11-factor and 5-factor protocols generated CPCs with comparable characteristics, they differ in the amount of CPC colonies generated, with the former producing more, and in the expression of smooth muscle cell and endothelial cell markers in CPC-differentiated cells, with the 5-factor protocol-based CPCs generating more of these markers than the 11-factor system. Another study showed that human dermal fibroblasts can be directly reprogrammed into CPCs by overexpressing the genes *MESP1* and *ETS2* [174]. In this specific reprogramming protocol, human dermal fibroblasts are converted into CPCs through a 4-day co-expression of ETS2 and MESP1 using lentiviral vectors, which is then followed by Activin A and BMP2 treatment for another 2 days. Human ETS2 is a transcription factor involved in development, apoptosis and oncogenic transformation and when co-expressed with MESP1, induces the expression of BMP2, initiates the Activin A/Nodal signaling and stimulates the emergence of CD31/PECAM-1 (endothelial cells) and KDR cells (CPCs). ETS2 could potentially be substituted by other ETS transcripts, such as ETS1, FLI1, ETV1, ETV5, ERG and ETV that are also highly abundant in the developing heart, and might function similarly to ETS2 in reprogramming human somatic cells into CPCs.

All these protocols described required the use of viral vectors, usually lentiviruses, to deliver the reprogramming factors into cells. This implied host cell genome changes which could potentially affect its suitability for translational applications. One method that addresses this concern is through the delivery of reprogramming proteins, related to transcription factors, directly into cells. These proteins can modulate the gene expression of cells to convert them into other cell types. For example, using a nonviral-based protein delivery system with the cardiac transcription factors GATA4, HAND2, MEF2C, and TBX5 induces reprogramming of human dermal fibroblasts into CPCs [41]. Additionally, adding growth factors such as BMP4, Activin A and basic Fibroblast Growth Factor (bFGF) can further stimulate and sustain potency towards a CPC state. This combination increased the cellular expression of CPC markers (FLK1 and ISL1) and decreased the expression of fibroblast-specific markers (COL1A2 and FSP1). Furthermore, the protocol demonstrated high efficiency in direct transdifferentiation, converting more than 80% of the human dermal fibroblasts into CPCs.

4.3. Somatic Reprogramming into Cardiospheres

Recent studies have shown that adult skin fibroblasts from mouse and human can be converted into cardiospheres that, in turn, have the potential to generate CPCs [175,176]. For this, the skin cells were first reprogrammed with the Yamanaka factors SOX2, KLF4 and OCT4 overnight, followed by media change to Knockout Serum Replacement-based media for 18 days and finally treatment with the GSK3 inhibitor BIO and Oncostatin for 2 days [175,176]. The resulted cardiospheres resembled endogenous cardiospheres formed from the cellular outgrowth of cardiac explants in vitro [39], but produced a higher number of MESP1, ISL1-, and NKX2.5- expressing cells [175,176]. On passaging, the cardiospheres became enriched with CPCs expressing c-KIT, FLK1 and CXCR4, which were able to differentiate into cardiomyocytes [175]. However, human cardiospheres do not display spontaneous beating and fail to propagate in vitro compared to mouse cardiopsheres, suggesting different signaling pathways being utilized for somatic reprogramming into cardiospheres in both mice and humans [175,176].

4.4. In Vivo Direct Reprogramming

One exciting potential of direct reprogramming is its application *in vivo*, in which endogenous cardiac cells would be directly converted into CPCs to repair the damaged myocardium. This approach could represent an improvement in promoting cardiac regeneration as it bypasses the several issues associated with cellular transplantation [166,177]. In addition, it avoids the need for cell harvesting, expansion, maintenance, and/or effective delivery systems, which are current challenges faced by cellular in vitro methods. In vivo direct reprogramming takes advantage of the heart native environment that might contain extracellular matrix proteins and growth factors that could make cells more permissive for functional reprogramming and lead to the formation of more mature cardiac cells [160,177–180]. In a study using an in vivo zebrafish model [181], cardiac ventricular injury induced the expression of Notch and RALDH2 in atrial cardiomyocytes, which caused the cells to lose their sarcomeric organization and re-express CPC transcription factors, such as GATA4, HAND2, NKX2.5, TBX5, TBX20 and MEF2. Once these dedifferentiated atrial cardiomyocytes reached the ventricle, they further expressed ventricle-specific markers, like Iroquois Homeobox Protein Ziro 1 (IRX1A) and ventricular Myosin Heavy Chain (vMHC), and differentiated into ventricular cardiomyocytes. Another study demonstrated that adult murine atrial and ventricular cardiomyocytes can acquire properties of CPCs through spontaneous dedifferentiation in vitro [182]. The dedifferentiated cardiomyocytes gave rise to CPCs that expressed the cardiac markers c-KIT, GATA4, and NKX2.5, self-organised into cardiospheres and were able to differentiate into functional cardiomyocytes and endothelial cells [182]. These results were further investigated by Zhang et al. (2015) in vivo using a MI mouse model [183]. They specifically analysed DNA methylome changes during cardiomyocyte dedifferentiation and observed that cardiomyocyte-specific genes, like Myosin Light Chain Kinase 3 (*MYLK3*) and Myosin Heavy Chain 6 and 7 (*MYH6* and *MYH7*), became hypermethylated (repressed), whereas cell cycle and proliferation genes, such as Epiregulin (*EREG*) and SRY-box 4 (*SOX4*), were hypomethylated (upregulated) in the generated CPCs. This concept could potentially be applied in in vivo CPC reprogramming. However, the molecular mechanisms involved in somatic cell dedifferentiation are not fully elucidated and more information is needed to identify the factors responsible.

Although in vivo reprogramming shows great potential, it has only been employed to derive fully differentiated cardiac cells, specifically cardiomyocytes, and not CPCs as such [160,177–180,184,185]. Therefore, even though direct reprogramming seems to be a suitable approach to generate CPCs, there are still some issues that influence its application in regenerative therapeutics. These include sub-optimal efficiencies in transdifferentiation protocols for CPC generation and lack of in-depth characteristics of CPC potency, differentiation potential and functionality of their derivatives.

5. In Vitro Culture of CPCs Derived Through Reprogramming Protocols

Establishing reprogramming protocols to generate CPCs from iPSCs and somatic cells is essential to advance CPC research for cardiac regeneration. However, the field also faces issues regarding the isolation, propagation, and expansion of CPCs in vitro. This section focuses on the current methods that have been successful in isolating, expanding and maintaining CPCs in vitro.

5.1. Isolation of CPCs

Isolation of CPCs is usually performed based on their characteristic gene expression patterns and surface markers (see Table 1). For example, ISL1 and NKX2.5 genes are frequently used to identify CPCs [186]. However, these genes are transiently expressed in cells which can lead to the isolation of a heterogeneous cell population containing various CPCs with distinct self-renewal and differentiation potential [186]. When using only cell surface markers, a combination of at least two markers is frequently used as a single surface marker seems insufficient to discriminate a CPC signature. For instance, Nsair et al. (2012) demonstrated that the co-expression of two cell surface markers, FLT1 (VEGFR1) and FLT4 (VEGFR3) specifically identifies ISL1/NKX2.5-expressing CPCs [187]. This combination was also shown to be more effective in identifying homogenous CPC populations (approximately 89% pure) compared to other combinations, such as FLK1 alone or FLK1 with PDGFRα. Furthermore, the isolated CPCs were able to differentiate into the three cardiac lineages and engraft into the host tissue. One study by Nelson et al. (2008) used the cell surface markers CXCR4 and FLK1 to isolate a more restricted CPC from a heterogeneous FLK1 positive population [188]. Zhou et al. (2017) also demonstrated that the marker SIX2 is able to target temporally distinct cell subpopulations from second heart field-associated CPCs [189]. One very recent study (Torán et al., 2019) used proteomic and genomic approaches to comprehensively characterize the proteome of human adult c-KIT CPCs [190]. It was demonstrated that these CPCs highly express 4 surface markers: GPR4 (G protein-coupled receptor 4), CACNG7 (calcium voltage-gated channel auxiliary subunit gamma 7), CDH5 (VE-cadherin) and F11R (F11 receptor) in comparison to mesenchymal stem cells, human dermal fibroblasts and cardiac fibroblasts. More research, however, will be required to further clarify the role of these proteins in CPC functions.

Thus, new markers are continuously being discovered to isolate specific CPC populations. However, they are frequently identified in CPCs derived from neonatal/adult tissue but fewer in ESC-CPCs and iPSC-CPCs [107,133,134,190–192]. Further validation of such markers is vital to assign a common signature that accurately identifies these cells.

5.2. Expansion and Maintenance of iPSC-CPCs

The maintenance of β-catenin concentration seems to be an efficient method for CPC expansion in vitro [187,193]. Applying GSK3 inhibitors, like WNT3A, CHIR, or 6-bromoindirubin-3′-oxime/BIO, can promote CPC expansion and suppress myocytic differentiation, leading to the formation of a relatively homogenous CPC colony [193]. Furthermore, the combination of a WNT/β-catenin inhibitor (IQ-1) and a ROCK inhibitor (Thiazovivn) is also able to expand CPCs in a feeder-free medium for a minimum of 4 weeks, while maintaining their multipotent state (more than 90% remained multipotent) [187]. IQ-1 is a selective β-catenin inhibitor that targets the signaling mediated by the protein's interaction with p300. This suppresses p300 pro-differentiation function and stimulates a pluripotency state. Furthermore, WNT signaling seems to interact with other signaling pathways, such as Notch and FGF signaling, to stimulate the expansion of CPCs [194,195]. For example, activation of the Notch signaling by Notch1 represses expansion, self-renewal and β-catenin activity in CPCs [195]. Activation of both WNT and FGF signaling pathways enhances ISL1 CPCs in a cooperative manner [194]. Therefore, using biomolecules that inhibit and activate the Notch and FGF signaling, respectively, together with WNT activators might facilitate CPC expansion. Notably, inhibition of FGF signaling has also been demonstrated to enhance CPC expansion, but this inhibition is suggested to affect only a subset of CPCs (expressing SCA1) [196] and therefore, warrants further investigation.

Several studies have shown that persistent inhibition of the BMP signaling enhances expansion of CPCs and prevents their differentiation [186,197,198]. For example, the BMP inhibitor Gremlin 2 (GREM2), whose expression initiates in NKX2.5$^+$ CPCs after cardiac mesoderm specification and follows cardiac lineage differentiation, promotes proliferation of CPCs from iPSCs by suppressing the BMP4 receptor activity [197]. This effect was demonstrated to be consistent across distinct iPSCs lines and independent of the differentiation method used. However, GREM2 is also able to induce differentiation of CPCs into the cardiac cell subtypes. Therefore, timing and potency of this BMP antagonist may need careful evaluation to CPCs and avoid spontaneous differentiation. Notably, GREM2 appears to only increase the number of KDR^{low} and NKX2.5$^+$ CPCs in vitro, and its function seems to be lost in the adult heart. Ao et al. (2012) used a second-generation BMP inhibitor called Dorsomorphin homologue 1 (DMH1) that was able to enrich CPCs, expressing Branchyury, MESP1 and ISL1 markers, from pluripotent cells [198]. Additionally, DMH1 was shown to be a more selective inhibitor of BMP type 1 receptors compared with other BMP inhibitors. This selective inhibition is, therefore, best applied during early stages of cardiac differentiation (pre-mesoderm and cardiac mesoderm stages) in order to increase the number of CPCs.

Another molecule that enhances CPC expansion in vitro is Ascorbic acid (AA) [143]. AA was shown to enhance the expansion of isolated iPSC-derived FLK1$^+$/CXCR4$^+$ CPCs through the MEK-ERK1/2 pathway by promoting collagen synthesis. However, the effects of AA on other CPC types need to be evaluated before AA can be used as a universal factor for efficient CPC expansion. Birket et al. (2015) used a cocktail of molecules modulators of the FGF, VEGF, PDGF, BMP, Nodal, AKT and hedgehog signaling pathways (SU5402, DMH1, SB431542, Insulin-Like Growth Factor 1 (IGF1) and Smoothened Agonist (SAG)) that was capable of expanding CPCs for more than 40 population doublings [186]. However, this study used MYC-transduced iPSC lines and consequently, the method needs further assessment using CPCs derived from non-transgenic iPSCs. Bao and colleagues (2017) developed two protocols, with and without serum, to maintain self-renewal and stimulate expansion of human iPSC-derived epicardial CPCs for long periods of time [152,153]. Both methods involve the addition of a TGF-β inhibitor, such as SB431542 or A83-01, to the medium. The epicardial CPCs can either be cultured in LaSR basal medium, which contains albumin, or in RPMI with ascorbic acid and insulin (RPMI/Vc/Ins), a xeno-free/chemically defined medium. Cells kept in LaSR basal medium can be maintained for up to 2 months, whereas CPCs in RPMI/Vc/Ins can be cultured for approximately 24 days before they start undergoing epithelial-to-mesenchymal transition (EMT) and lose their morphology. The use of a gentler dissociating buffer (Versene) also seemed to improve expansion efficiency of the CPCs from human pluripotent stem cells (iPSCs and ESCs) after 8 passages [152]. One study developed a Good Manufacturing Practice (GMP)-compatible system for the expansion of CPCs, using stirred tank bioreactors and microcarrier technology [199]. Human CPCs from three different donors were inoculated with microcarriers (Cytodex 1 coated with CELLstart^{TM}CTS^{TM}) for up to 7 days. The microcarrier-based stirred cultures lead to a cell suspension increase of 3-fold and greater cell viabilities compared with standard static T-flask monolayers. Furthermore, the CPCs in the culture system expressed the markers CD44, CD105, CD166, KDR, GATA4, and TBX5. This method provides tight control of environmental cues to mimic physiological conditions, which could potentially improve the production of high-quality CPCs for therapeutic applications.

5.3. Expansion and Maintenance of Transdifferentiated CPCs

CPCs derived from direct reprogramming of somatic cells seem to have similar requirements as iPSCs-CPCs for expansion and maintenance. For instance, adding a canonical WNT activator and a JAK/STAT activator to the cultures was shown to maintain the proliferative and multipotent state of the CPCs for several passages (over 20 and 30 passages for a 5- and 11- Factor reprogramming protocol, respectively) without continuous expression of the reprogramming factors [24]. However, CPC maintenance and expansion potential can be negatively affected when utilising somatic cells from tissues other than cardiac tissue, like lung and skin tissues. Furthermore, fibroblast-derived CPCs can be alternatively expanded and maintained using a combination of signaling molecules (BMP4, Activin A, CHIR, and SU5402) that synergistically repress cardiac differentiation and sustain CPC self-renewal [14]. In this case, the CPCs' undifferentiated morphology, gene expression pattern and cell surface molecule expression remain the same for more than 18 passages regardless of the tissue origin of the donor cells.

Overall, the requirements for in vitro culture of CPCs involved the precise temporal activation and suppression of several signaling pathways. It remains challenging to expand CPCs while maintaining their self-renewal and multipotent differentiation potential as the process is extremely complex, preventing the development of standard conditions yet. This can be more complicated when considering CPCs derived from iPSCs and direct reprogramming and their associated characteristics [186,200–202]. Therefore, more comparative studies of current protocols will be imperative to establish standard in vitro culture conditions that are optimal for the isolation, expansion and maintenance of specific CPCs.

6. Strategies to Improve CPC Reprogramming

Strategies for producing CPCs are still developing with time. Whilst the concept of CPC generation through reprogramming or transdifferentiation has taken precedence to produce desired cardiac lineages, the protocols suffer from poor efficiency or lack of mechanistic insight to achieve the target population and desired functional improvement. Strategies to accelerate proliferation and extend replicative lifespan of CPCs are being essentially employed to understand and potentially overcome the inherent limitations of patient CPC populations derived from compromised, aged, or damaged myocardium. With developments in genetic engineering approaches and factors, such as CRISPR gene editing, epigenetic modulators and/or microRNAs, and its significance in cardiac development, there seems scope for applying this in the field of CPC regeneration and address some of the current limitations. This section will describe examples of such strategies in the context of CPCs.

6.1. Genetic Engineering with PIM1

Genetic engineering with PIM1, has been applied in CPCs to enhance their properties, like proliferation, survival and differentiation [203]. Pro-viral insertion site for the moloney murine leukemia virus (*PIM1*), a proto-oncogene serine/threonine-protein kinase, is highly expressed in bone marrow, tumor cells and fetal heart and is associated with many signaling pathways, mostly related to anti-cell apoptosis and cell cycle regulation [204]. Mohsin and colleagues (2013) genetically modified patient-derived human CPCs (hCPCs) with PIM1 kinase (termed hCPCeP) to increase proliferation, telomere length, survival and decrease expression of cellular senescence markers, rejuvenating the phenotypic and functional properties of hCPCs, in an effort to ameliorate the cumulative effects of age and disease [205]. The PIM1-engineered cells also showed increased commitment to the three cardiac lineages [203]. Interestingly, the effect of PIM1 in hCPCeP normalizes after several rounds of passaging, consistent with the notion that PIM-1 can transiently increase mitosis coupled with telomere stability (increased TERT activity) and without resultant oncogenic transformation through a c-MYC synergy. These properties of hCPCeP can be modulated by targeted localization of PIM1 in mitochondrial or nuclear components, conferring an optimal stem cell trait irrespective of patient-associated cell heterogeneity [206]. Furthermore, intramyocardial injection of hCPCeP into cardiomyopathic challenged-SCID mice demonstrate increased cellular engraftment and differentiation with improved vasculature and reduced infarct size [203]. Similar results were also observed when using murine CPCs [207] but these earlier studies relied largely on viral delivery methods to induce PIM1 overexpression. In an alternative strategy, a non-viral modified plasmid-minicircle (MC) was used as a vehicle to deliver PIM1 into mouse CPCs (mCPCs) in vitro and the myocardium in vivo [208]. Mice with PIM1-MC injection showed increased protection compared to control groups measured by ejection fraction at 3- and 7-days post injury, supporting the potential of a non-cell based therapeutic approach for treatment of ischemic heart disease and MI.

6.2. CRISPR in Context with CPCs

In an effort to identify previously unknown regulators of cardiomyocyte differentiation from human ESCs (hESCs) through quantitative proteomics, Murry lab [209] demonstrated that DAB2 (Disabled 2) plays a functional role in cardiac lineage specification towards cardiomyocytes by being preferentially upregulated in CPCs. CRISPR/Cas9 deletion of Dab2 in zebrafish embryos was used to show increase in WNT/β-catenin signaling and consequent decrease in cardiomyocyte number, suggesting that inhibiting WNT/β-catenin signaling by DAB2 (or analogous inhibitors like the Dickkopf WNT signaling Pathway Inhibitor 1 (DKK1)) can be crucial in maintaining cardiomyocyte numbers from CPCs in the developing heart. Supporting this mechanism, the same lab, using antisense knockdown and CRISPR/Cas9 mutagenesis in hESCs and zebrafish, went on to demonstrate that Alpha Protein Kinase 2 (ALPK2) is temporally expressed during specification of CPCs (but not in endocardial-like endothelial cells), and cardiac commitment through negative regulation of WNT/β-catenin signaling [210]. In a more recent study [211], CRISPR-mediated ablation of *Furin* gene in mouse CPCs, whose product is a natural target of *Nkx2.5* repression during heart development, produces abnormalities in embryo characterized by reduced proliferation of CPCs and their premature differentiation, suggesting *Furin* mediates some aspects of *Nkx2.5* function in heart and is necessary for CPC differentiation. This role of *Furin* in the maturation of CPCs is, in part, mediated by the modulation of the BMP pathway by *Nkx2.5*. Therefore, genetic engineering using CRISPR has been pivotal in recent years to identify mechanisms associated with CPCs and continue to show promise with a perpetual trend in CRISPR advances.

6.3. Epigenetic Modulators

Distinct cell types display different epigenetic profiles that leads to differential gene expression. Cellular reprogramming is associated with changes in the epigenetic signature of cells. During these epigenetic transitions, proteins called epigenetic modulators bind to specific regions of the chromatin and regulate the transcription of genes. Therefore, inhibition and/or overexpression of these modulators might affect cardiac reprogramming efficiency [41,212]. For example, knockdown of the polycomb ring finger pro-oncogene *Bmi1* in several fibroblast types (murine embryonic, neonatal and adult tip tail and adult cardiac fibroblasts) results in the activation of core cardiac transcription factors, such as GATA4, ISL1 and TBX20, which converts the cells into cardiomyocytes [212]. Additionally, Zhou et al. (2016) demonstrated that silencing of *Bmi1* allowed for efficient cardiomyocyte reprogramming using just two factors (MEF2C and TBX5). The induced cardiomyocytes displayed features of advanced maturity, such as contractile activity, sarcomere structures and periodic calcium oscillation. Therefore, it would be useful to investigate the role of *Bmi1* in the context of CPC reprogramming, considering the significance of ISL1 upregulation under *Bmi1* depletion. Another epigenetic modulator that could potentially be employed in CPC reprogramming is the BAF chromatin remodeling protein BAF60A. BAF60A is thought to have a role in the maintenance of CPC self-renewal thought interaction with TBX1 [213,214]. TBX1 seems to recruit BAF60A onto the promoter region of *WNT5A* gene, upregulating its expression in CPCs [214]. WNT5A is a non-canonical WNT pathway ligand that is highly expressed in CPCs derived from the SHF, and it cooperates with another non-canonical WNT ligand, called WNT11, to induce development of CPCs from the two heart fields [215]. Accurate identification of the cellular epigenetic barriers could potentially reduce the number of reprogramming factors employed to generate CPCs and ultimately, lead to faster and safer protocols.

6.4. MicroRNAs

MicroRNAs are short non-coding RNA molecules that bind to messenger RNA and repress gene expression. MicroRNAs show a promising alternative to traditional reprogramming protocols as they are easily delivered and display low toxicity in animal models [184]. In addition, several microRNA transcripts can be packed into a single delivery vector, which could potentially increase reprogramming efficiency. However, most studies have mainly examined the use of microRNAs in converting somatic cells directly into cardiomyocytes, not CPCs as such [180,184,216]. Nevertheless, microRNAs have been shown to modulate CPC functions [217–219] (see Table 3). Sirish et al. (2012) investigated the miRNA expression changes in CPC development [219]. They identified 8 differentially expressed microRNAs (miR-103, -130a, -17, -130b, -208b, -185, -200b and -486) in mouse neonatal and adult LIN$^-$/c-KIT$^+$ CPCs. The target proteins of microRNAs were predicted to be predominantly involved in cell proliferation, with a few proteins having roles in cell organisation, development, metabolic process, adhesion, homeostasis, activation, communication, and motility. The group also demonstrated that overexpression of the miR-17-92 cluster, which targets cell cycle proteins, in adult CPCs increased their proliferative capacity by 2-fold in vivo. Two studies showed that the microRNAs miR-1, -499 and -204 repress proliferation and stimulate differentiation in committed SCA1$^+$ CPCs [217,218]. Additionally, Xiao et al. (2012) revealed that inhibition of miR-204 suppressed CPC differentiation and promoted proliferation without affecting cell viability [218]. A study in 2016 identified several microRNAs that regulate cardiac fate, like let-7, miR-18, miR-302 and the miR-17-92 cluster, in MESP1$^+$ CPCs [220]. It was also shown that the CPCs were particularly enriched for the miR-322/-503 cluster which targets the CUG-binding protein Elav-like family member 1 (CELF1). Ectopic CELF1 expression promoted neural lineage-specification at the expense of cardiomyocyte differentiation in the CPCs. Therefore, miR-322/-503 may be a key regulator in promoting the cardiac program in early mesodermal cells by cross-suppressing other lineages. Garate et al. (2018) investigated the expression of microRNAs during the differentiation of human pluripotent stem cells (hPSCs) towards mesoderm and cardiac cells [221]. They found several microRNA families (miR-302, C19MC, miR-17/92 and miR-26) that were highly expressed in EpCAM/CD326-negative and NCAM/CD56-positive mesoendodermal progenitor cells (MPCs) [222]. The microRNA families identified were speculated to be associated with the epithelial to mesenchymal transition occurring during the development of mesoderm. However, the specific roles of the microRNAs in CPCs will need to be determined as MPCs are able to generate all the mesoendodermal lineages, including cardiovascular, hematoendothelial and mesenchymal. One very recent study by Cheng et al. (2019) showed that the ischemic heart secretes microRNAs (miR-1a, miR-133a, miR-208a and miR-499) that mobilised LIN$^-$/c-KIT$^+$ bone marrow progenitor cells (BM PCs) into the site of injury, where they proliferated and promoted vascularisation [223]. These results demonstrated the principle of employing microRNAs to target endogenous progenitor cells to enhance ischemic cardiovascular repair. Therefore, as molecular mechanisms regulated by microRNAs during CPC development get explored more, they offer a suitable choice of target for improving CPC generation from iPSCs or for transdifferentiation.

Table 3. Role of microRNAs in CPC biology.

CPC Property			MiRNA Involved	Target Protein/Pathway	Mechanism	Ref.
Proliferation			miR-21	PTEN	Inhibit negative regulators of cell proliferation	[224]
			miR-218	SFRP2		
			miR-548c	MEIS1		
			miR-509			
			miR-23b			
			miR-204	ATF2	Repress proliferation-related transcription factors and induces differentiation	[225]
			miR-1	HDAC4		
				HAND2		
			miR-200b	GATA4		
			miR-17-92 cluster	Not reported	Increases proliferation rate	[219]
Differentiation	CMs		miR-133	NELFA	Suppresses cardiogenesis	[226]
			miR-218	SFRP2	Inhibits a negative regulator of cell proliferation	[227]
			miR-142	MEF2C	Suppresses CM formation	[228]
			miR-1	DLL1	Increases NKX2.5 and Myogenin expression	[229]
			miR-499	ROD1	Suppresses inhibitory factors of cardiac differentiation	[224,230]
				SOX6		
			miR-708	N-RAS		[231]
			miR-322-503 cluster	CELF1		[220]
	SMCs		miR-22	EVI1	Inhibits negative regulators of SMC marker gene expression and of SMC transcription factors	[232]
			miR-29a	YY1		[233]
		miR-669a		MYOD	Increases CPC differentiation potential by preventing skeletal myogenesis	[234]
		miR-669q				
Migration			miR-206	TIMP3	Suppresses a metalloproteinase inhibitor	[235]
			miR-21	PTEN	Promotes migration of SCA1$^+$ CPCs (not fully clear)	[236]
Apoptosis			miR-21	BIM	Inhibit apoptotic activators	[237]
				PDCD4		
			miR-24	BIM		
			miR-221			
Necrotic Cell Death			miR-155	RIP1	Inhibits necrosis activators	[238]
Vascular Remodeling			miR-221	c-KIT	Inhibit endothelial cell migration and proliferation	[239]
			miR-222	eNOS		
Cell Repolarization			miR-1	KCNE1	Reduce potassium current in hyperglycemia conditions	[240]
			miR-133	KCNQ1		

CM(s): Cardiomyocyte(s); SMC(s): Smooth Muscle Cell(s); PTEN: Phosphatase and Tensin Homolog; SFRP2: secreted Frizzled-Related Protein 2; MEIS1: Meis Homeobox 1; ATF2: Activating Transcription Factor 2; HDAC4: Histone Deacetylase 4; NELFA: Negative Elongation Factor-A; DLL1: Delta-Like protein 1; ROD1: Regulator of Differentiation 1; N-RAS: Neuroblastoma RAS Viral Oncogene Homolog; CELF1: CUG-binding Protein Elav-like Family Member 1; EVI1: Ecotropic Virus Integration Site 1 Protein Homolog; YY1: Transcription Factor Yin Yang 1; MYOD: Myoblast Determination Protein 1; TIMP3: Tissue Inhibitor of Metalloproteinase 3; BIM: BCL2-like Protein 11; PDCD4: Programmed Cell Death 4; RIP1: Receptor-Interacting Protein Kinase 1; eNOS: endothelial Nitric Oxide Synthase; KCN-E1/-Q1: Potassium Voltage-Gated Channel Subfamily E Member 1/Subfamily Q Member 1.

7. Tissue Engineering with CPCs and CPC-Derived Cardiomyocytes

Several studies have demonstrated that the cells generated from CPCs, particularly cardiomyocytes, display an immature phenotype similar to that of embryonic cardiac cells [3]. However, when the CPCs are transplanted into a host environment, the differentiated cells reach a more advanced maturity, such as greater organisation of sarcomeres and formation of gap junctions (in the case of cardiomyocytes) and development of tubular-like structures (for smooth muscle and endothelial cells) [11,24,27,28]. Furthermore, CPCs seem to have distinct differentiation potential in vitro and in vivo [96,111]. This could mean that the microenvironment of the heart might have a key role in CPC functions. Stem and progenitor cells reside in specific tissue microenvironments, called niches, which provide protection and support to the cells [241]. A way to potentially enhance CPC regenerative potential could be to mimic their microenvironment. Cardiac tissue engineering aims to achieve this goal by combining multiple microenvironment components, such as cells, extracellular matrix (ECM) and biochemical factors like BMP2, VEGF, bFGF, DKK1, and IGF1, to create cardiac tissue constructs. Therefore, determining the ideal matrix for supporting CPCs and their derivatives is paramount. In principle, the scaffold matrix should be biodegradable, immune-privileged, provide electrical and/or mechanical properties for cell coupling and assembly, and support vascularisation [242,243]. Two types of materials are typically employed in the production of scaffolds for tissue engineering: natural matrices and synthetic matrices. This section will describe different types of scaffolds that have been used in combination with CPCs and CPC-derived cardiomyocytes (Table 4).

7.1. Natural Scaffolds

Natural matrices have the advantage of being composed of native ECM cues that modulate cell behavior [243,244]. These scaffolds can comprise pure ECM elements, like hydrogels made from natural materials such as fibrin, alginate, gelatin, and collagen, or acellular tissue which displays the biochemical and biomechanical properties (tensile strength and composition) of the native ECM tissue [245,246]. Three independent studies used a fibrin patch seeded with CPCs (murine and human) to develop a tissue construct, which was then tested in vivo [247–249]. Vallée et al. (2012) specifically utilized BMP2-primed murine ESCs seeded onto fibrin matrices as single cells, small cluster and embryoid bodies [249]. These constructs were then engrafted onto myocardial infarcted rat hearts, which led to a reduction in remodeling and deterioration of cardiac functions. Seeded cells were identified by the expression of the cardiac genes MESP1, NKX2.5, MEF2C, TBX6 and GATA4, speculating a CPC-related population. The transplanted cells were also able to colonize the outer connective tissue where they differentiated into cardiomyocytes and promoted neovascularization. The results from Vallée et al. (2012) encouraged two other studies to apply their tissue engineering approach with human CPCs [247,248]. Bellamy et al. (2015) and Menasché et al. (2015) seeded human CPCs, expressing the markers SSEA1 and ISL1, in a fibrinogen patch [247,248]. The two studies differed in the number of CPCs used, Bellamy et al. (2015) used 700,000 cells whereas Menasché et al. (2015) used 4 million cells; and in the in vivo model chosen, myocardial infarction rats and a 68-year-old patient suffering from severe heart failure, respectively. Improvement of contractility and attenuation of ventricular remodeling was observed in both studies. It was also shown that these benefits were likely a result of paracrine factors secreted by the transplanted CPCs rather than de novo generation of tissue. Gaetani and colleagues (2012 and 2015) used 3D printing with SCA1$^+$/CD105$^+$ fetal CPCs, which are referred to as human fetal cardiomyocyte progenitor cells (hCMPCs), and three types of natural scaffolds (pure, RGD-modified alginate and a hyaluronic acid/gelatin-based matrix) [250,251]. The hCMPCs were able to migrate from the scaffolds, colonize the surrounding areas and form tubular-like structures [250,251]. Another study by Christoforou et al. (2013) used murine iPSC-derived CPCs mixed within a fibrin/Matrigel hydrogel that were applied in polydimethylsiloxane (PDMS) molds and cultured for 14 days in vitro [157]. These CPCs expressed NKX2.5, GATA4, c-KIT and either FLK1 or SCA1 and differentiated into mature cardiomyocytes that aligned into unidirectional myofilament and displayed abundant electromechanical connections. This study also concluded that accessibility to oxygen and nutrients within tissue constructs greatly affects integration of the implanted cells.

Native ECM generally comprise of various components such as glycosaminoglycans (GAGs), collagen, fibrinogen, hyaluronic acid and hydroxyapatite (HA) [246]. To mimic this, recent studies have applied natural scaffolds generated from the decellularisation of tissues. This technique removes any cells present in the tissue while preserving its original 3D architecture and ECM. Two studies have combined decellularised scaffolds with iPSC-derived CPCs [252,253]. Lu et al. (2013) used human iPSC-CPCs that were positive (low) for the marker KDR and negative for c-KIT to repopulate a whole decellularised mouse heart [252]. The CPCs differentiated into cardiomyocytes, endothelial cells and smooth muscle cells, and efficiency to a specific lineage could be changed with the addition of growth factors. The recellularised scaffolds displayed vessel-like structures, spontaneous contraction, uniform wave propagation in some regions, and the ECM seemed to stimulate proliferation of CPCs and formation of wider myofilaments of cardiomyocytes. However, drawbacks of this study included the uneven recellularisation of the heart constructs which led to weaker mechanical forces and incomplete synchronization, and inability to generate cells of the conduction system and cardiac fibroblasts. Although natural scaffolds retain the ultrastructure and biological information of the native tissue, there is a risk of immunological reaction, disease transmission (in case of animal-derived materials) and are generally variable in their physical properties [243,245].

7.2. Synthetic Scaffolds

The ideal synthetic scaffold should be biocompatible, degradable, display a surface that allows for cell attachment, migration and differentiation, and a macrostructure that supports cell growth and nutrient and waste exchange [245]. Structure and properties of synthetic scaffolds, like the associated mechanics, chemistry and degradation rate, can be easily customised for the type of cells being used [243,245,246]. Two studies employed self-assembling peptide nanofibres with CPCs and tested the constructs in vivo [254,255]. Both studies used two distinct experimental designs: Padin-Iruegas et al. (2009) seeded adult rat Lin$^-$ c-KIT$^+$ CPCs onto nanofibres tethered with IGF1, whereas Tokunaga et al. (2010) used adult mouse SCA1$^+$ CPCs mixed with Puramatrix® (3D Matrix, Ltd.) (no tethered factors). The CPCs in Tokunaga et al. (2010) nanofibres minimally contributed to de novo cardiomyocyte generation and had no differentiation potential towards endothelial lineages [255]. The benefits observed were associated to effects from paracrine signaling. On the other hand, Padin-Iruegas et al. (2009) showed that continued IGF1 release from nanofibres enhanced CPC survival and proliferation, and stimulated differentiation into cardiomyocytes, smooth muscle cells and endothelial cells [254]. Additionally, the regenerated cardiomyocytes were able to couple with resident cardiomyocytes, and the smooth muscle cells and endothelial cells formed vascular structures. These studies demonstrated that functionalising self-assembling peptide nanofibres can potentially support long-term CPC survival, proliferation and differentiation, and lead to a more robust maturity of the CPC-derived cells, especially if applied in the human CPC context. Li et al. (2011) used a solution made of mouse cardiosphere-derived cells and degradable poly(N-isopropylacrylamide) hydrogel and performed in vitro testing of the effects of scaffold stiffness and presence/absence of collagen on the cells' functions [256]. The hydrogels with medium stiffness and collagen were optimal for cardiosphere-derived cells differentiation into cardiomyocytes, which displayed the highest expression of maturation genes (MYH6 and cTNT). Unfortunately, there were no reports on the effects of the hydrogels on cardiosphere-derived cells differentiation potential towards smooth muscle cells and endothelial cells. Liu et al. (2015) also employed nanofibres with CPCs, but they used poly(l-lactic acid) and mouse ESC-derived CPCs [257]. These CPCs were positive for ISL1 and GATA4 and differentiated into the three cardiac lineages in both in vitro and in vivo conditions. Additionally, differentiation potential towards endothelial lineages was improved in vivo compared to that of in vitro. The scaffolds supported CPC survival, engraftment, proliferation and integration with the host tissue, and stimulated the expression of intercellular coupling proteins (connexin 43) and maturation of cardiomyocytes.

One study used a novel concept called "scaffold-in-scaffold" to promote human CPC growth and differentiation in vitro [258]. The aim of this approach was to create a structure with different physical characteristics to better mimic the ECM microarchitecture. The multitexture 3D scaffold was composed of a polyethylene glycol diacrylate (PEGDa) woodpile and a softer PEGDa hydrogel. Human LIN⁻ SCA1⁺ CPCs seeded on these scaffolds highly differentiated into cardiomyocytes, which aligned in an orderly manner. However, robust cardiomyocyte maturation, such as sarcomeric organisation and formation of gap junctions, was not achieved. In addition, there were no reports on the differentiation potential towards other cardiac lineages.

Synthetic biomaterials are a great promise to constructing 3D microenvironments with adjustable features. However, they still come with a few limitations, such as poor biocompatibility, incomplete polymer degradation, and some toxicity, that will need to be addressed systematically to achieve better cellular responses.

A significant trend that has been popular with human Pluripotent Stem Cell-Cardiomyocytes (hPSC-CMs), has been the implementation of electrically-compatible scaffolds or biomaterials (in 2D or 3D) compatible with standard electrophysiology measurements to stimulate hPSC-CM electrical behavior and consequently its mature electrophysiological phenotypes (see Table 5). This would be a strategy for exploration with CPCs as we improve our understanding of the CPC niche. Furthermore, while most of the studies described above employed ESC-derived or putative CPCs on scaffolds, studies using patient-specific CPCs from iPSCs or from transdifferentiation in engineered scaffolds to model phenotypes are very rare. Therefore, with potential improvements in cardiac tissue engineering and mechanistic understanding of responses in situ, the CPC niche can be exploited to assess normal and disease-associated cardiac cell behavior to produce better regenerative outcomes (Figure 2).

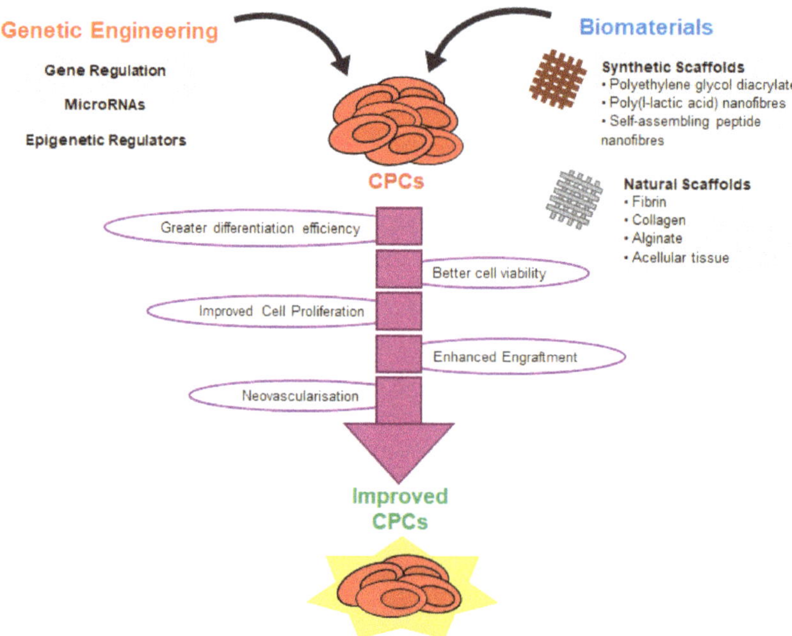

Figure 2. Promising strategies to improve CPC characteristics and functionality. Strategies for producing CPCs to date through reprogramming or transdifferentiation has been associated with poor efficiency or lack of mechanistic insight to achieve the target population and desired functional improvement. With a range of tools for genetic engineering or gene modulation, and with advances in tissue engineering approaches, new strategies have been applied in this field to accelerate proliferation, enhance differentiation, extend replicative lifespan or improve functionality or engraftment of CPCs (detailed in Sections 6 and 7).

Table 4. Cardiac tissue engineering strategies with biomaterials using CPCs.

Scaffold Biomaterial	Experimental Design	Outcome	Limitations	Ref.
Fibrin patch	SSEA1[+] and ISL1[+] hESCs-CPCs mixed in fibrinogen, and scaffolds were then transplanted into myocardial infarction rats	-Improved contractility and decrease in adverse ventricular remodeling -Increased angiogenesis and attenuation of fibrosis	-Poor long-term cell engraftment -Functional improvements resulted from paracrine signaling	[247]
	Same process as above, except the scaffolds were delivered surgically on the infarct area of a 68-year-old patient suffering from severe heart failure	-No observation of ventricular arrhythmias -Decreased in adverse ventricular remodeling	-Presence of T-cell response 3 months post-implantation -Absence of neovascularization in patch-treated area	[248]
	mESCs were primed with BMP2 for 36 h and seeded into fibrin matrices The constructs were then implanted onto normal or infarcted rat left ventricles	-Efficient cell engraftment -Attenuation of left ventricle dilation -Promotion of neovascularization	-Rapid inflammation-driven degradation of scaffolds -Unclear whether neovascularization was due to in situ cell differentiation or endogenous EC recruitment	[249]
Polyethylene glycol diacrylate woodpile (PEGDa-Wp) and PEGDa hydrogel.	Human adult LIN[−]/SCA1[+] CPCs were seeded in a PEGDa hydrogel and the mixture was then cultured onto a PEGDa-Wp	-Benefits on cell assembly and alignment -Induction of cell spatial-ordered multilayer organization and differentiation towards a CM phenotype	-Incomplete maturation of CMs -No differentiation into SMCs and ECs -No in vivo testing of the scaffolds	[258]
Poly(l-lactic acid) Nanofibres	mESC-derived ISL1[+]/GATA4[+] CPCs were seeded onto nanofibres After 7 days of in vitro differentiation, the scaffolds were implanted subcutaneously in the dorsal area of athymic nude mice	-Enhancement of cell attachment, extension and differentiation in vitro -Improvement of cell survival, integration and commitment to the three cardiac lineages in vivo -Induction of angiogenesis in vivo	-Poor in vitro differentiation into ECs -Unclear whether neovascularization was due to paracrine factors or CPC-derived SMCs and ECs	[257]
Tissue Printing using Sodium Alginate	Human SCA1[+] CPCs were mixed with alginate matrixes, including an RGD-modified alginate, which were then used to print porous and non-porous scaffolds	-Porosity preserved viability and proliferation and increased cardiac commitment of CPCs -CPCs migrated from the construct and formed tubular-like structures	-Incomplete maturation of the differentiated cells -No in vivo testing of the scaffolds	[250]
Porcine- and human-derived myocardial matrices	Human SCA1[+] CPCs were seeded onto porcine and human ECM Scaffolds were injected into the left ventricular free wall of healthy hearts of Sprague Dawley rats	-Porcine-derived ECM was more efficient at promoting CPC differentiation, whereas human-derived ECM promoted CPC proliferation	-Variation in ECM properties due to distinct decellularised methods used, patient-to-patient variability and tissue age	[259]
3D-printed hyaluronic acid/gelatin-based matrix	Human SCA1[+] CPCs were printed together with the matrix The cell-loaded patches were transplanted in myocardial infarction mice	-Reduction of adverse remodeling and fibrosis -Long-term CPC survival and engraftment -Formation of vessel-like structures within the scaffold in vivo	-Absence of neovascularization in the infarcted region -Incomplete maturation of CMs in vivo	[251]
Collagen/Matrigel hydrogels	Human SCA1[+] CPCs were encapsulated in collagen/Matrigel hydrogels which were cultured in either stress-free or unidirectional constrained conditions	-Enhanced cardiac differentiation and matrix remodeling -Constrained hydrogels stabilized CPC viability, attachment and proliferation -Static strain stimulated actin fiber formation and cell alignment	-Differentiation trend towards CMs -Incomplete maturation of CMs -No CPC differentiation into SMCs and ECs -No in vivo testing	[260]

Table 4. *Cont.*

Scaffold Biomaterial	Experimental Design	Outcome	Limitations	Ref.
Decellularised porcine ventricular ECM	Human Foetal and adult SCA1[+] CPCs were resuspended in porcine myocardial matrix and collagen type I solutions The cell/matrix mixtures were injected into the left ventricular wall of Sprague Dawley rats	-The myocardial matrix improved CPCs adhesion, survival, proliferation and cardiac commitment both in vitro and in vivo -Foetal CPCs survived better than adult CPCs in vivo	-Rats were euthanized 30 min post-implantation, preventing assessment of long-term effects on cell survival, migration and cardiac function	[261]
	Same procedure as above, exceptions: use of adult rat c-KIT[+] CPCs and no in vivo implantation	-The cardiac ECM improved cardiac commitment, cell survival, proliferation and adhesion	-Differentiation trend towards CMs. -Low differentiation efficiency towards ECs and SMCs	[262]
Whole decellularised mouse heart	hiPSC- and hESC-derived KDR[+]/c-KIT[−] CPCs were seeded into a whole decellularised mouse heart The repopulated hearts were perfused with VEGF and DKK1 or VEGF and bFGF	-Efficient control of in situ iPSC-CPC differentiation -Advanced CM maturation -Development of vessel-like structures and spontaneous contraction for both iPSC-and ESC-CPC constructs	-Scattered regions of uncoupled cells -Insufficient mechanical force generation and incomplete electrical synchronization of the constructs	[252]
	FLT1 (VEGFR1)[+]/PDGFRα[+] hESC-CPCs were seeded onto decellularised mice hearts, which were implanted subcutaneously into SCID mice	-In situ generation of CMs, SMCs and ECs -Formation of a vascular network and higher expression of CM markers in vivo	-In vivo differentiated ECs were not ubiquitously distributed in the decellularised scaffold -Absence of beating populations	[263]
Whole decellularised rat heart	hESC-derived KDR[+]/PDGFRα[+] CPCs were expanded in a stirred-suspension bioreactor and seeded onto perfusion-decellularised *Wistar* rat hearts containing immobilized bFGF	-Improved CPC retention, proliferation and cardiac differentiation potential -Spontaneous and synchronous contractions -Advanced CM maturation	-Growth factor immobilization prevents spatiotemporal control -No in vivo testing	[264]
Whole decellularised human heart	Human adult c-KIT[+] CPCs from human cardiac biopsies were cultured onto perfused-decellularised heart ventricles	-Increased CPC growth and stimulated differentiation towards cardiac lineages in vitro	-Poor CPC infiltration into the matrix -No electrical signal propagation. -No in vivo testing	[265]
Rat and pig collagen matrix and decellularised left ventricle ECM	iPSC-CPCs were cultured on rat or pig collagen matrices and decellularised ECM CPCs were also co-cultured with ECs and CMs	-Enhanced expression of contractile protein gene expression -Cell communication was observed in co-cultures	-No results reported on CPC proliferation and differentiation -No information about the CPC markers	[253]
3D-bioprinted patch containing decellularised porcine ventricular ECM	Bioinks composed of decellularised ECM, human neonatal c-KIT[+] CPCs and gelatin methacrylate were used to print patches, which were implanted onto the epicardial surface of the right ventricle of Sprague Dawley rat hearts	-Good CPC retention and viability in the scaffolds -Enhanced cardiogenic differentiation and angiogenic potential -Presence of vascularization in the patches in vivo	-Main purpose of the patch was to improve the paracrine release from the CPCs -No influence in SMC differentiation	[266]
Foetal and adult rat decellularised ventricle ECM	Immortalized adult mouse LIN[−]/SCA1[+] CPCs were seeded onto embryonic, neonatal and adult rat ECM	-Good CPC retention, motility and viability -Remodeling of the supporting ECM -Enhanced production of cardiac repair factors	-No evidence of CPC differentiation -No in vivo testing	[267]
Decellularised murine embryonic heart	Day 5 and 9 mESC-CPCs were then seeded onto the decellularised scaffolds	-Day 5 progenitors formed spontaneously beating constructs in the scaffolds	-Mixed cell population isolated -Not all cell populations led to functional maturation	[268]

Table 4. *Cont.*

Scaffold Biomaterial	Experimental Design	Outcome	Limitations	Ref.
Decellularised human pericardium-derived microporous scaffold	Human SCA1$^+$ CPCs were seeded onto 3D microporous pericardium scaffolds, which were then implanted subcutaneously into Wistar rats	-Improved CPC migration, survival, proliferation and differentiation -Reduction of immunological response and enhanced angiogenesis	-No influence in CPC differentiation towards SMCs	[269]
Self-assembling peptide nanofibers	Adult LIN$^-$/c-KIT$^+$ rat CPCs were seeded onto IGF1-tethered nanofibres CPCs and scaffolds were injected into myocardial infarction rats	-Enhanced CPC survival, proliferation and differentiation into CMs -Improved angiogenesis, recruitment of resident CPCs and attenuation of ventricle dilation	-Growth factor immobilization prevents spatiotemporal control -Newly formed CMs were derived from resident CPCs -CPCs were not cultured on the scaffolds prior to implantation	[254]
	Adult mouse SCA1$^+$ CPCs were mixed with Puramatrix® complex and injected into the border area of the myocardium in myocardial infarction mice	-Reduction of the infarct area and attenuation of ventricular dilation. -Enhanced neovascularization	-No CPC differentiation towards ECs -Functional improvements resulted from paracrine signaling -Poor CPC engraftment	[255]
RDG-modified collagen and porous gelatin solid foam	Human adult CS-CPC were grown as secondary CSs, which were seeded onto the scaffolds	-Enhanced cell migration and ECM production -Increased CPC cardiogenic potential, cell retention and adherence	-Cardiac commitment trend towards CMs -Distinct scaffold morphologies promoted different biological processes	[270]
Degradable Poly(*N*-isopropylacrylamide) hydrogel	Mouse CDCs were added into hydrogel solutions, with or without collagen and containing different stiffness	-Preservation of CDC proliferation -Stimulation of differentiation into mature cardiac cells in hydrogels with medium stiffness and collagen	-No differentiation into ECs and SMCs -No in vivo testing	[256]
Biodegradable gelatin	Human CDCs were seeded onto bFGF immobilized gelatin hydrogels, which were implanted in the epicardium of immunosuppressed myocardial infarction pigs	-Enhanced angiogenesis, cell engraftment -Reduction of the infarct area and attenuation of adverse ventricular remodeling	-Growth factor immobilization prevents spatiotemporal control -No differentiation into ECs and SMCs	[271]
Fibrinogen/Matrigel mixture and PDMS molds	NKX2.5$^+$/c-KIT$^+$/either FLK1$^+$ or SCA1$^+$ iPSC-CPCs were mixed in a fibrinogen/Matrigel hydrogel and applied into PDMS molds	-Spontaneous and synchronous contraction -Highly organized sarcomere structures and robust electromechanical connections	-Improper nutrient access within the construct -No differentiation potential towards SMCs and ECs -No in vivo testing	[157]
Collagen sponge	CPCs were seeded onto collagen sponges and then transplanted into rat hearts with atrioventricular conduction block	-Enhanced vascularization -Gap junction formation -Differentiation into CMs, conduction cells and ECs	-No information about the functionality of the CPC-derived cells	[272]

(h/m)ESC(s): (human/murine) Embryonic Stem Cell(s); BMP: Bone Morphogenic Protein; EC(s): Endothelial Cell(s); CM(s): Cardiomyocyte(s); SMC(s): Smooth Muscle Cell(s); ECM: Extracellular Matrix; VEGF: Vascular Endothelial Growth Factor; DKK1: Dickkopf WNT Signaling Pathway Inhibitor 1; bFGF: basic Fibroblast Growth Factor; hiPSC(s): human induced Pluripotent Stem Cell(s); SCID: Severe Combined Immunodeficiency; IGF1: Insulin-like Growth Factor; CS(s): Cardiosphere(s); CDC(s): Cardiosphere-Derived Cell(s); PDMS: polydimethylsiloxane.

Table 5. In vitro cardiac tissue engineering techniques with biomaterials to stimulate and record hPSC-CM electrical activity.

Cells	Biomaterial/Scaffold	Platform	Stimulation	Electrophysiology	Ref.
hiPSC-CMs	Graphene substrate	2D	FET (current pulse with f = 1 Hz) For calcium: voltage ramp from −80 to +60 mV at 20 mV/s	-Enhanced electrophysiological properties: $RP = -40.54 \pm 1.72$ mV $AP = 75.24 \pm 3.91$ mV $CV = 5.34 \pm 1.60$ cm/s I_{Ca2+} density $= -9.31 \pm 2.35$ pA/pF $I_{Ca2+,L}$ density $= -2.47 \pm 0.6$ pA/pF I_k density $= 46.24 \pm 8.45$ pA/pF I_{kr} density $= 36.57 \pm 5.84$ pA/pF Ca^{2+} transients: Amplitude intensity $= 1.69 \pm 0.20$ u Upstroke velocity 3.09 ± 0.99 u/s Decay velocity (50%) $= 0.84 \pm 0.29$ s	[273]
iCell® CMs & hESC-CMs	Reduced graphene oxide (rGO)	2D	Light: intensity >1 mW/mm², duration 40-ms-2-Hz light pulses and 3-s step of light	-Optical stimulation on rGO substrates improves CMs electrophysiology -rGO increases AP peaks frequency -On rGO CMs contraction frequency increases with light intensity	[274]
Neonatal Sprague Dawley rat vCMs	Electrospun gelatine + PCL nanofibres	3D	FET (1–3 V, 50-ms-long pulses at 1–2 Hz)	-Electrical stimulation results in regularly spaced spikes (f = 1–2 Hz) with shape and width consistent with CM extracellular signals -NE increases electrical activity and frequency of calcium transients	[275]
hiPSC-CMs	PLGA electrospun aligned nanofibres	3D	Not applied	-Enhanced CM maturity and electrical activity -CM drug (E4031) response showed higher electrophysiological homogeneity -L-ANFs increased FP amplitude, number of electrically active cells, synchronization and anisotropic propagation of the electrical signal	[276]
hESC-CMs & hiPSC-CMs	Type I collagen gel template suture (Biowires)	3D	Electrical field with daily and progressively frequency increase: low frequency ramp-up regimen (from 1 to 3 Hz) or high frequency ramp-up regimen (from 1 to 6 Hz)	-Electrical stimulation enhanced electrical activity frequency -High frequency increased electrophysiological properties, contractile activity, synchronization and CV -High frequency decreased excitation threshold and variability in AP duration -High frequency improved CM response to caffeine and Ca^{2+} handling properties: $I_{ERG} = 0.81 \pm 0.09$ pA/pF $I_{K1} = 1.53 \pm 0.25$ pA/pF	[277]

Table 5. *Cont.*

Cells	Biomaterial/Scaffold	Platform	Stimulation	Electrophysiology	Ref.
hESC-CMs	MEA coated with collagen type I + agarose layer	2D	Anti-arrhythmic and pro-arrhythmic drugs	-Pharmacological stimulation influences CMs electrophysiology -FPD and CT are dependent on the dose of arrhythmogenic drugs: E-4031 & Astemizole increased FPD Flecainide & Terfenadine decreased FPD Flecainide, Astemizole & Terfenadine increased CT and of safe drugs: Verapamil & Lidocine decreased FPD Lidocine slightly increased CT	[278]
hiPSC-CMs	MEA coated with hydrogel containing fluorescence microbeads	2D	Electrical: periodic voltage pulses (biphasic square waves with pulse width = 4 ms, f = 0.2 Hz, peak-to-peak amplitude = 4 V) Pharmacological: drug exposure (NE and Blebbistatin)	-Good electrical coupling of CMs (FP = 9–35 μV and CV = 16 cm/s) -Electrical pacing promoted synchronized contraction (f = 11 bpm) -Recorded impedance increased with cell attachment and at each contraction -Blebbistan inhibited beating activity and has no effect on FP -NE increased CV and contraction spikes rate	[279]

hiPSC(s): human induced Pluripotent Stem Cell(s); hESC(s): human Embryonic Stem Cell(s); (v)CM(s): (ventricular) Cardiomyocyte(s); FET: Field Effect Transistor; f: frequency; RP: Resting Potential; AP: Action Potential; CV: Conduction Velocity; PCL: Polycaprolactone; NE: Norepinephrine; PLGA: Poly(lactic-co-glycolic) acid; L-ANFs: Low-density nanofibres; FP: Field Potential; FPD: Field Potential Duration; CT: Condition Time; MEA: Micro-Electrode Array; IDE: Interdigitated electrode.

8. In Vivo Applications of Human CPCs

The end-goal of in vitro and animal in vivo studies in CPC research is to provide enough evidence regarding the efficacy and safety of cell therapies for further application in human trials. This is not without the caveat that, despite promising results from in vitro and animal models, the translation to clinical trials still suffer from serious inefficiencies in desirable outcomes over long term, costing billions of dollars in the process [280]. Even though there is not yet an agreement on the CPC population that displays the best regenerative capacity, a variety of CPCs have been used or are being used in clinical trials, which are summarized in Table 6.

The first-ever clinical trial using CPCs, called SCIPIO (Stem Cell Infusion in Patients with Ischemic cardiOmyopathy) used human LIN⁻ c-KIT⁺ CPCs to improve postinfarction left ventricular dysfunction. However, this study has now been retracted due to concerns about the randomisation and lack of integrity of certain data [281,282]. In 2012, the randomised phase I trial CADUCEUS (CArdiosphere-Derived aUtologous stem CELLs to reverse ventricUlar dySfunction) employed cardiosphere-derived cells to reduce scarring after myocardial infarction [283]. These cells were obtained from endomyocardial biopsy specimens and were transplanted into patients 1.5–3 months post-myocardial infarction using intracoronary infusion. The results showed that the cells led to an improvement in viable heart tissue and a reduction of scarring. Differentiation potential of cardiosphere-derived cells towards cardiac lineages remained to be elucidated and thus, it is likely that the benefits observed in the CADUCEUS study were a result of paracrine factors. In the same year, another phase I trial called ALCADIA (AutoLogous human CArdiac-Derived stem cell to treat Ischemic cArdiomyopathy) used autologous human CPCs in combination with a controlled released of bFGF in patients suffering from ischemic cardiomyopathy and heart failure [284,285]. These CPCs expressed the mesenchymal surface markers CD105 and CD90 and were also derived from endomyocardial biopsy specimens. The cells were injected intramyocardially and a biodegradable gelatin hydrogel sheet containing bFGF was then implanted on the epicardium, which covered the injection sites areas. However, as in the case of the CADUCEUS study, the benefits observed, such as attenuation of adverse ventricular remodelling and neovascularisation, were probably due to paracrine mechanisms as there was no compelling evidence that the employed CPCs can differentiate into cardiomyocytes in vivo [284,286]. A more recent trial published in 2018, named ESCORT (Transplantation of Human Embryonic Stem Cell-derived Progenitors in Severe Heart Failure), used hESC-derived CPCs, expressing the markers SSEA1/CD15 and ISL1, embedded in a fibrin gel [287]. The scaffold was then delivered onto the epicardium of the infarct area. The aim of the study was to confirm the safety and feasibility of the therapy rather than evaluating its regenerative effects in the patients. Further investigation will be needed to thoroughly assess the benefits of the fibrin gel patch in severe heart failure.

There are also reports on phase I and II clinical trials assessing the use of autologous cardiosphere-derived cells in paediatric patients suffering from hypoplastic left heart syndrome [288,289]. The phase I TICAP (Transcoronary Infusion of CArdiac Progenitor cells in patients with single ventricle physiology) demonstrated that the approach was safe and feasible for improving cardiac function after 18 months [288]. The safety of the therapy was also analysed at 36 months post-transplantation [290]. There was no tumour formation and the initial observed benefits were enhanced, with attenuation of ventricular stiffness and improvement of ventriculoarterial coupling. The results obtained from TICAP were further confirmed by the phase II PERSEUS (Cardiac Progenitor Cell Infusion to Treat Univentricular Heart Disease) [289]. Furthermore, the therapy is currently being tested in a phase III trial (APOLLON) [291] and applied in paediatric patients diagnosed with dilated cardiomyopathy (phase I trial TICAP-DCM: Transcoronary Infusion of CArdiac Progenitor cells in paediatric Dilated CardioMyopathy) [292], for which results are still waiting.

Most trials involving CPCs come with limitations in employing small sample sizes or lack of blinded assessment, which ultimately leads to inconclusive results regarding the therapies' efficiency in recovering from cardiac disorders. In addition, it is still inconclusive whether the positive results are attributed to intracoronary infusion of CPCs themselves or from paracrine factors as speculated by some trials. It will, therefore, be imperative to perform future clinical trials with a broader assessment of study subjects and an established human reproducible model to better explore the CPCs' regenerative capacity in human hearts.

Table 6. Past and ongoing clinical trials using CPCs.

Clinical Trial Name	Phase	Start/End Date	CPC Type	Delivery of Cells	Biomaterial Added	Results	Ref.
CADUCEUS prospective, randomized trial	I	2009–2012	CDCs	Direct injection via catheter	none	LVEF unchanged at 12 months Scar size decreased 12.3% at 12 months Regional contractility and systolic wall thickening increased	[283,293]
ALCADIA Open-label, non-randomized trial	I	2010–2013	CDCs	Direct injection via catheter	Biodegradable gelatin hydrogel sheet containing 200 μg of bFGF planted onto epicardium covering the injection site	LVEF increase 12% at 6 months Scar size decrease 3.3% at 6 months	[285]
ALLSTAR Open-label cohort (PI), double-blinded, randomized, placebo-controlled study (PII)	I/II	2012–2019	CDCs	Direct injection via catheter	none	Terminated (follow-up activities were ceased)	[294]
ESCORT Open-label trial	I	2013–2018	ESC-derived ISL1$^+$/CD15$^+$	Epicardial patch via coronary artery bypass procedure	Fibrin gel patch containing progenitor cells	LVEF increase of 12.5% No arrhythmias, or tumor formation	[287]
CAREMI Double blinded, randomized, placebo-controlled trial	I/II	2014–2016	CDCs	Direct injection via catheter	none	Infarct size decreased to 15.6% at 12 months LVEF increase of 7.7% at 12 months	[295]
DYNAMIC Open-label trial, randomized, double-blinded, placebo-controlled trial	I	2014–ongoing	CDCs	Direct injection via catheter to multi-vessel areas of heart	none	Ongoing	[296]
CONCERT-HF Randomized, double-blinded, placebo-controlled trial	II	2015–ongoing	c-KIT$^+$	Direct injection via catheter	none	Ongoing (paused on 29.10.18, re-approved 06.02.2019)	[297]
TICAP Open-label trial, non-randomized	I	2011–2013	CDCs	Direct injection via catheter	none	RVEF increase of around 8.0% at 18 and 36 months No tumor formation	[288,290]
PERSEUS Open-label trial, randomized	II	2013–2016	CDCs	Direct injection via catheter	none	LVEF increase of 6.4% at 3 months Reduction in scar size	[289]
APOLLON Randomized, single-blinded	III	2016 & Unknown	CDCs	Direct injection via catheter	none	Unknown status (last update was September 2017)	[291]
TICAP-DCM Randomized	I	2017–ongoing	CDCs	Direct injection via catheter	none	Recruiting	[292]
REGRESS-HFpEF Randomized, double-blinded, placebo-controlled trial	II	2017–ongoing	CDCs	Direct injection via catheter	none	Ongoing	[298]

CDCs: Cardiosphere-Derived Cells; ESC: Embryonic Stem Cell; bFGF: basic Fibroblast Growth Factor; LVEF: Left Ventricular Ejection Fraction; RVEF: Right Ventricular Ejection Fraction.

9. Current Challenges and Limitations

There is still a lot of debate on the effect that CPCs play a role in cardiac regeneration and repair in the context of diseases like MI, demonstrating increased left ventricular ejection fraction, decreased infarct size, and an increase in hemodynamic function following infusion of autologous CPCs. Even though there is a growing emphasis on the application of CPCs for cardiac regeneration, its impact is still obscure, particularly owing to its heterogeneous nature and mechanistic silencing from deep-rooted complexities associated with the nature of the cardiomyopathic disease. For example, there is still no consensus regarding which CPC population is the ideal cell type for cell-based regenerative therapies and which combination of markers accurately characterise CPCs. Additionally, the characteristic epigenetic, gene, protein and secretome profiles of most CPCs remain unclear [19,41]. This could elucidate how phenotypes and genotypes of CPCs alter throughout their development and their effects on self-renewal and potency potentials. Furthermore, not many studies have investigated and compared the therapeutic efficacy of different CPCs. The ideal CPC type should be able to tolerate autologous transplantation, expand extensively in vitro, differentiate into mature cardiac cell subtypes and integrate with the host cells [299].

Viral transduction remains the main approach applied in most reprogramming processes (both in vitro and in vivo) as it shows the greatest efficiency. However, this is associated with a risk of genome integration and activation of oncogenic genes. In addition, the currently developed protocols require the use of both reprogramming and growth factors which substantially increases their complexity and final cost. It is, therefore, imperative to develop a more effective and simpler gene transfer methods that ensure cell therapies are safe and display a good cost-benefit ratio.

Furthermore, the populations of CPC-derived cells are heterogeneous and frequently represent immature cells, which could potentially lead to arrhythmias, lower long-term stability and poor integration when transplanted [3,300]. The mechanisms involved in cardiac lineage subtype specification will need to be fully investigated and optimised to produce purer and more mature populations of the desired cell types from the CPCs. With the growing pace of CRISPR strategies and its potential to address limitations associated with genetic control and regulation, it will not be surprising that this will be applied to CPCs for this purpose in the very near future.

Epigenetic profiles seem to strongly affect reprogramming efficiencies for both iPSCs and transdifferentiation technologies. For example, using cells from non-cardiac tissue organs or aged tissue negatively affects the cardiogenesis capability of iPSCs [301]. The success of reprogramming a cell fate relies on the ability to overcome the several epigenetic barriers present in somatic cells. The more distinct the donor somatic cells are from the cardiac tissue, the higher the number of epigenetic barriers that need to be overcome and consequently, the harder it is to reprogram the cells. Therefore, understanding the epigenetic regulatory mechanisms involved CPC formation might be vital to improving reprogramming efficiency.

Another limitation in CPC research is that many studies have been performed in rodent models, which display distinct cardiac anatomy and physiology from the human heart. Additionally, current techniques developed using animal cells will need to be further validated for human context. For example, the direct reprogramming protocol involving the three core cardiac genes GATA4, MEF2C and TBX5 (also known as GMT) was demonstrated to induce mouse fibroblasts into cardiac cells, but it was insufficient to convert human fibroblasts [164].

For future preclinical trials, the relationship between the number of CPCs and their effects on cardiac regeneration and the appropriate frequency of administration of each cell therapy needs to be further investigated [299]. In addition, molecules and/or cells are very often directly injected into the heart during open-surgery. This is an invasive approach that could cause additional injury and pain to the patients. Other less invasive methods, such as intracoronary and intravenous injection, have been employed to deliver cells to the heart. However, these techniques rely on correct homing of cells into the damaged tissue, and very often the delivered cells become trapped in other organs [302,303].

Consequently, other delivery systems that are less aggressive and show the best efficacy and safety need to be developed before CPCs can be applied in regenerative medicine strategies.

There are sufficient reports that support the existence of CPCs within specialized niche structures in the myocardium [241]. For therapeutic applications, these CPCs can be isolated and cultured in vitro, prior to transplantation into the affected heart or, the local microenvironment can be modulated to recruit CPCs to the infarct area. Current biomaterial strategies (discussed in Tables 4 and 5) have exploited both these methods for functional improvements but do not report complete recovery under physiological conditions or pathological insults. This is evident in the lack of clinical trials with CPCs using biomaterials (Table 6). This offers an opportunity to integrate engineering with mechanistic modulation (perhaps through genetic engineering) to contextualize CPC behavior with disease factors.

The difficulties described above rely, to some extent, on the incomplete understanding of the heart development and cardiac regeneration processes. Increasing this knowledge will clarify the precise stoichiometry of the cardiac factors and optimal culture conditions to accurately mimic the development of CPCs in vitro.

10. Final Thoughts—Controversies Surrounding CPCs

It does seem that the debate surrounding CPCs and adult heart repair is taking a full circle—it is there, it is not there, it is there, etc? With the first evidence in rodents supporting the notion of c-KIT$^+$ cells from bone marrow or adult heart to replace damaged myocardial tissue, from Piero Anversa's lab, and subsequent retractions of 31 papers from his group owing to unreliable data, it has encouraged the field to challenge the theory by more robust techniques in mouse models [58,61,304,305]. Results from such studies showed that cardiomyocyte generation from a c-KIT$^+$ cells was an extremely rare event. Notably, more recently, the data from Li et al. (2018) showed compelling evidence to support endogenous stem cell to myocyte conversion in embryonic but not in adult heart [306].

Ironically, a more recent work in 2019, by Narino et al., has demonstrated that c-KIT expression labels a heterogeneous cardiac cell population, with cells low in c-KIT expression enriched for CSCs while c-KIT high expressers having endothelial/mast cell differentiation potential [307]. This study went on to show that adult c-KIT-labeled CSCs in mouse "can be myogenic" and help to regenerate after injury and to counteract effects of aging on cardiac structure and function, thus boldly suggesting that CSCs as the bonafide endogenous source of cardiomyocytes in healthy/pathological heart. Consequently, they identified c-KIT haploinsufficiency, generated usually in lineage-tracing studies, prevents efficient labeling of true CSCs on one hand while affecting the regenerative potential of these cells on the other, which perhaps could have been the oversight in the rival camp. Nevertheless, irrespective of the c-KIT controversy, there is no denying that animal studies and clinical trials have appreciated the benefit of a range of cell types for CPCs from many different sources through cellular transplantation approaches ([308–310] and Tables 1, 4 and 6). Furthermore, there is an emerging theory that injected/infused CPCs can induce a reconditioning of the injured heart through paracrine signalling or that these cells stimulate an acute inflammatory response when these cells die and are cleared, resulting in a secondary acute healing response [311].

Therefore, as implied in Table 1, there is still no consensus on an endogenous CPC type that is critical for myocardial repair and regeneration but there is growing consensus that regeneration associated with these CPCs are not robust enough to repair severe myocardial damage such as in MI (commented in [307]). While this review does not offer to bias the reader for one or the other theory, in light of these recent studies, it offers the field impetus to interrogate other strategies and CPC sources (like from stem cells or transdifferentiation of somatic cells) to provide mechanistic insights into how CPCs can be more functionally significant in the context of cardiac regenerative medicine.

11. Future Directions

Heart failure patients are typically elderly, and suffer from chronic cardiomyopathies and associated complications like diabetes, hypertension, etc. Notably, they possess CPCs with compromised regenerative potential, insufficient to recover lost cardiac function [312,313]. The propensity of CPCs to affect cardiac repair is influenced by several factors, including genetics [314], epigenetic dysregulation [315], environmental stress [315], disease progression and pathogenesis [316,317], heart load [318], medication, and aging [319,320]. Nevertheless, discovery of CPC characteristics has revolutionized the conceptual view of treatment for heart disease, supported by the capacity of CPCs to form functionally integrated cardiomyocytes and vasculature [321]. Therefore, it is rational to enhance potential of CPCs from the adult or reprogrammed cell sources prior to adoptive transfer into a damaged myocardium. Hence, CPC research is gaining momentum to improve its feasibility for cardiac regenerative therapeutics. Advances in this field are progressing towards combining optimised reprogramming approaches from iPSCs and somatic cells with tissue engineering strategies. This will undoubtedly bring advances in genomics, epigenomics, and proteomics of CPCs and their differentiated counterparts, to realise their full potential. Future regenerative approaches might bring together genetic engineering (a very tested strategy in iPSCs), the addition of multiple stimuli (mechanical, electrical and biochemical factors) and tissue engineering approaches to develop a meticulously controlled system that maximises CPC regenerative capacity, and that could potentially be applied in cell therapy, disease modelling, and drug screening (Figure 1).

Funding: This research was funded by the Engineering and Physical Sciences Research Council (EPSRC), UK, grant number EP/L015072/1 and the APC was funded by the same.

Conflicts of Interest: The authors declare no conflict of interest. The funders had no role in the design of the study; in the collection, analyses, or interpretation of data; in the writing of the manuscript, or in the decision to publish the results.

References

1. Go, A.S.; Mozaffarian, D.; Roger, V.L.; Benjamin, E.J.; Berry, J.D.; Blaha, M.J.; Dai, S.; Ford, E.S.; Fox, C.S.; Franco, S.; et al. Executive summary: Heart disease and stroke statistics—2014 Update: A report from the American Heart Association. *Circulation* **2014**, *129*, 399–410. [CrossRef] [PubMed]

2. Fuster, V. Global burden of cardiovascular disease. *J. Am. Coll. Cardiol.* **2014**, *64*, 520–522. [CrossRef] [PubMed]

3. Witman, N.; Sahara, M. Cardiac progenitor cells in basic biology and regenerative medicine. *Stem Cells Int.* **2018**, *2018*, 1–9. [CrossRef] [PubMed]

4. Madonna, R.; Van Laake, L.W.; Davidson, S.M.; Engel, F.B.; Hausenloy, D.J.; Lecour, S.; Leor, J.; Perrino, C.; Schulz, R.; Ytrehus, K.; et al. Position paper of the European society of cardiology working group cellular biology of the heart: Cell-based therapies for myocardial repair and regeneration in ischemic heart disease and heart failure. *Eur. Heart J.* **2016**, *37*, 1789–1798. [CrossRef]

5. Bergmann, O.; Zdunek, S.; Felker, A.; Salehpour, M.; Alkass, K.; Bernard, S.; Sjostrom, S.L.; Szewczykowska, M.; Jackowska, T.; dos Remedios, C.; et al. Dynamics of cell generation and turnover in the human heart. *Cell* **2015**, *161*, 1566–1575. [CrossRef]

6. Graham, E.; Bergmann, O. Dating the heart: Exploring cardiomyocyte renewal in humans. *Physiology* **2017**, *32*, 33–41. [CrossRef]

7. Urbanek, K.; Torella, D.; Sheikh, F.; De Angelis, A.; Nurzynska, D.; Silvestri, F.; Beltrami, C.A.; Bussani, R.; Beltrami, A.P.; Quaini, F.; et al. Myocardial regeneration by activation of multipotent cardiac stem cells in ischemic heart failure. *Proc. Natl. Acad. Sci. USA* **2005**, *102*, 8692–8697. [CrossRef]

8. Bloomekatz, J.; Galvez-Santisteban, M.; Chi, N.C. Myocardial plasticity: Cardiac development, regeneration and disease. *Curr. Opin. Genet. Dev.* **2016**, *40*, 120–130. [CrossRef]

9. van Berlo, J.H.; Molkentin, J.D. An emerging consensus on cardiac regeneration. *Nat. Med.* **2014**, *20*, 1386–1393. [CrossRef]

10. Liang, S.X.; Phillips, W.D. Migration of resident cardiac stem cells in myocardial infarction: Migration of cardiac stem cells. *Anat. Rec.* **2013**, *296*, 184–191. [CrossRef]

11. Oh, H.; Bradfute, S.B.; Gallardo, T.D.; Nakamura, T.; Gaussin, V.; Mishina, Y.; Pocius, J.; Michael, L.H.; Behringer, R.R.; Garry, D.J.; et al. Cardiac progenitor cells from adult myocardium: Homing, differentiation, and fusion after infarction. *Proc. Natl. Acad. Sci. USA* **2003**, *100*, 12313–12318. [CrossRef] [PubMed]

12. Laflamme, M.A.; Murry, C.E. Heart regeneration. *Nature* **2011**, *473*, 326–335. [CrossRef] [PubMed]

13. Müller, P.; Lemcke, H.; David, R. Stem cell therapy in heart diseases—Cell types, mechanisms and improvement strategies. *Cell. Physiol. Biochem.* **2018**, *48*, 2607–2655. [CrossRef] [PubMed]

14. Zhang, Y.; Cao, N.; Huang, Y.; Spencer, C.I.; Fu, J.; Yu, C.; Liu, K.; Nie, B.; Xu, T.; Li, K.; et al. Expandable cardiovascular progenitor cells reprogrammed from fibroblasts. *Cell Stem Cell* **2016**, *18*, 368–381. [CrossRef] [PubMed]

15. Burridge, P.W.; Keller, G.; Gold, J.D.; Wu, J.C. Production of de novo cardiomyocytes: Human pluripotent stem cell differentiation and direct reprogramming. *Cell Stem Cell* **2012**, *10*, 16–28. [CrossRef]

16. Lam, J.T.; Moretti, A.; Laugwitz, K.-L. Multipotent progenitor cells in regenerative cardiovascular medicine. *Pediatric Cardiol.* **2009**, *30*, 690–698. [CrossRef]

17. Bergmann, O.; Bhardwaj, R.D.; Bernard, S.; Zdunek, S.; Barnabé-Heider, F.; Walsh, S.; Zupicich, J.; Alkass, K.; Buchholz, B.A.; Druid, H.; et al. Evidence for cardiomyocyte renewal in humans. *Science* **2009**, *324*, 98–102. [CrossRef]

18. Ebrahimi, B. Cardiac progenitor reprogramming for heart regeneration. *Cell Regen.* **2018**, *7*, 1–6. [CrossRef]

19. Birket, M.J.; Mummery, C.L. Pluripotent stem cell derived cardiovascular progenitors—A developmental perspective. *Dev. Biol.* **2015**, *400*, 169–179. [CrossRef]

20. Kattman, S.J.; Huber, T.L.; Keller, G.M. Multipotent Flk-1+ cardiovascular progenitor cells give rise to the cardiomyocyte, endothelial, and vascular smooth muscle lineages. *Dev. Cell* **2006**, *11*, 723–732. [CrossRef]

21. Wu, S.M.; Fujiwara, Y.; Cibulsky, S.M.; Clapham, D.E.; Lien, C.; Schultheiss, T.M.; Orkin, S.H. Developmental origin of a bipotential myocardial and smooth muscle cell precursor in the mammalian heart. *Cell* **2006**, *127*, 1137–1150. [CrossRef] [PubMed]

22. Cai, C.-L.; Liang, X.; Shi, Y.; Chu, P.-H.; Pfaff, S.L.; Chen, J.; Evans, S. Isl1 identifies a cardiac progenitor population that proliferates prior to differentiation and contributes a majority of cells to the heart. *Dev. Cell* **2003**, *5*, 877–889. [CrossRef]

23. Henning, R.J. Stem cells in cardiac repair. *Future Cardiol.* **2011**, *7*, 99–117. [CrossRef] [PubMed]

24. Lalit, P.A.; Salick, M.R.; Nelson, D.O.; Squirrell, J.M.; Shafer, C.M.; Patel, N.G.; Saeed, I.; Schmuck, E.G.; Markandeya, Y.S.; Wong, R.; et al. Lineage reprogramming of fibroblasts into proliferative induced cardiac progenitor cells by defined factors. *Cell Stem Cell* **2016**, *18*, 354–367. [CrossRef] [PubMed]

25. Christoforou, N.; Miller, R.A.; Hill, C.M.; Jie, C.C.; McCallion, A.S.; Gearhart, J.D. Mouse ES cell–derived cardiac precursor cells are multipotent and facilitate identification of novel cardiac genes. *J. Clin. Investig.* **2008**, *118*, 894–903. [CrossRef]

26. Miyamoto, S.; Kawaguchi, N.; Ellison, G.M.; Matsuoka, R.; Shin'oka, T.; Kurosawa, H. Characterization of long-term cultured c-kit $^+$ cardiac stem cells derived from adult rat hearts. *Stem Cells Dev.* **2010**, *19*, 105–116. [CrossRef]

27. Bearzi, C.; Rota, M.; Hosoda, T.; Tillmanns, J.; Nascimbene, A.; De Angelis, A.; Yasuzawa-Amano, S.; Trofimova, I.; Siggins, R.W.; LeCapitaine, N.; et al. Human cardiac stem cells. *Proc. Natl. Acad. Sci. USA* **2007**, *104*, 14068–14073. [CrossRef]

28. Beltrami, A.P.; Barlucchi, L.; Torella, D.; Baker, M.; Limana, F.; Chimenti, S.; Kasahara, H.; Rota, M.; Musso, E.; Urbanek, K.; et al. Adult cardiac stem cells are multipotent and support myocardial regeneration. *Cell* **2003**, *114*, 763–776. [CrossRef]

29. Yang, L.; Soonpaa, M.H.; Adler, E.D.; Roepke, T.K.; Kattman, S.J.; Kennedy, M.; Henckaerts, E.; Bonham, K.; Abbott, G.W.; Linden, R.M.; et al. Human cardiovascular progenitor cells develop from a KDR+ embryonic-stem-cell-derived population. *Nature* **2008**, *453*, 524–528. [CrossRef]

30. Moretti, A.; Caron, L.; Nakano, A.; Lam, J.T.; Bernshausen, A.; Chen, Y.; Qyang, Y.; Bu, L.; Sasaki, M.; Martin-Puig, S.; et al. Multipotent embryonic Isl1+ progenitor cells lead to cardiac, smooth muscle, and endothelial cell diversification. *Cell* **2006**, *127*, 1151–1165. [CrossRef]

31. Tallini, Y.N.; Greene, K.S.; Craven, M.; Spealman, A.; Breitbach, M.; Smith, J.; Fisher, P.J.; Steffey, M.; Hesse, M.; Doran, R.M.; et al. c-Kit expression identifies cardiovascular precursors in the neonatal heart. *Proc. Natl. Acad. Sci. USA* **2009**, *106*, 1808–1813. [CrossRef] [PubMed]

32. Boyle, A.J.; Schulman, S.P.; Hare, J.M. Stem cell therapy for cardiac repair: Ready for the next step. *Circulation* **2006**, *114*, 339–352. [CrossRef] [PubMed]

33. Vanhoutte, D.; Schellings, M.; Pinto, Y.; Heymans, S. Relevance of matrix metalloproteinases and their inhibitors after myocardial infarction: A temporal and spatial window. *Cardiovasc. Res.* **2006**, *69*, 604–613. [CrossRef] [PubMed]

34. Deddens, J.C.; Sadeghi, A.H.; Hjortnaes, J.; van Laake, L.W.; Buijsrogge, M.; Doevendans, P.A.; Khademhosseini, A.; Sluijter, J.P.G. Modeling the human scarred heart in vitro: Toward new tissue engineered models. *Adv. Healthc. Mater.* **2017**, *6*, 1600571. [CrossRef]
35. Dobaczewski, M.; Gonzalez-Quesada, C.; Frangogiannis, N.G. The extracellular matrix as a modulator of the inflammatory and reparative response following myocardial infarction. *J. Mol. Cell. Cardiol.* **2010**, *48*, 504–511. [CrossRef]
36. Le, T.Y.L.; Thavapalachandran, S.; Kizana, E.; Chong, J.J. New developments in cardiac regeneration. *Heart Lung Circ.* **2017**, *26*, 316–322. [CrossRef]
37. Van Vliet, P.; Roccio, M.; Smits, A.M.; van Oorschot, A. a. M.; Metz, C.H.G.; van Veen, T. a. B.; Sluijter, J.P.G.; Doevendans, P.A.; Goumans, M.-J. Progenitor cells isolated from the human heart: A potential cell source for regenerative therapy. *Neth Heart J.* **2008**, *16*, 163–169. [CrossRef]
38. Sandstedt, J.; Jonsson, M.; Kajic, K.; Sandstedt, M.; Lindahl, A.; Dellgren, G.; Jeppsson, A.; Asp, J. Left atrium of the human adult heart contains a population of side population cells. *Basic Res. Cardiol.* **2012**, *107*, 255. [CrossRef]
39. Messina, E.; De Angelis, L.; Frati, G.; Morrone, S.; Chimenti, S.; Fiordaliso, F.; Salio, M.; Battaglia, M.; Latronico, M.V.G.; Coletta, M.; et al. Isolation and expansion of adult cardiac stem cells from human and murine heart. *Circ. Res.* **2004**, *95*, 911–921. [CrossRef]
40. Laugwitz, K.-L.; Moretti, A.; Lam, J.; Gruber, P.; Chen, Y.; Woodard, S.; Lin, L.-Z.; Cai, C.-L.; Lu, M.M.; Reth, M.; et al. Postnatal isl1+ cardioblasts enter fully differentiated cardiomyocyte lineages. *Nature* **2005**, *433*, 647–653. [CrossRef]
41. Li, X.-H.; Li, Q.; Jiang, L.; Deng, C.; Liu, Z.; Fu, Y.; Zhang, M.; Tan, H.; Feng, Y.; Shan, Z.; et al. Generation of functional human cardiac progenitor cells by high-efficiency protein transduction: Protein-generated cardiac progenitor cells. *Stem Cells Transl. Med.* **2015**, *4*, 1415–1424. [CrossRef] [PubMed]
42. Pessina, A.; Gribaldo, L. The key role of adult stem cells: Therapeutic perspectives. *Curr. Med. Res. Opin.* **2006**, *22*, 2287–2300. [CrossRef] [PubMed]
43. Cedar, S. The function of stem cells and their future roles in healthcare. *Br. J. Nurs.* **2006**, *15*, 104–107. [CrossRef] [PubMed]
44. Takahashi, K.; Tanabe, K.; Ohnuki, M.; Narita, M.; Ichisaka, T.; Tomoda, K.; Yamanaka, S. Induction of pluripotent stem cells from adult human fibroblasts by defined factors. *Cell* **2007**, *131*, 861–872. [CrossRef] [PubMed]
45. Takahashi, K.; Yamanaka, S. Induction of pluripotent stem cells from mouse embryonic and adult fibroblast cultures by defined factors. *Cell* **2006**, *126*, 663–676. [CrossRef]
46. Cyranoski, D. 'Reprogrammed' stem cells approved to mend human hearts for the first time. *Nature* **2018**, *557*, 619–620. [CrossRef]
47. Mandai, M.; Watanabe, A.; Kurimoto, Y.; Hirami, Y.; Morinaga, C.; Daimon, T.; Fujihara, M.; Akimaru, H.; Sakai, N.; Shibata, Y.; et al. Autologous induced stem-cell–derived retinal cells for macular degeneration. *N. Engl. J. Med.* **2017**, *376*, 1038–1046. [CrossRef]
48. Ronen, D.; Benvenisty, N. Genomic stability in reprogramming. *Curr. Opin. Genet. Dev.* **2012**, *22*, 444–449. [CrossRef]
49. Margariti, A.; Kelaini, S.; Cochrane, A. Direct reprogramming of adult cells: Avoiding the pluripotent state. *Stem Cells Cloning: Adv. Appl.* **2014**, *7*, 19. [CrossRef]
50. Xu, J.; Lian, W.; Li, L.; Huang, Z. Generation of induced cardiac progenitor cells via somatic reprogramming. *Oncotarget* **2017**, *8*, 29442. [CrossRef]
51. Sassoli, C. Cardiac progenitor cells as target of cell and growth factor-based therapies for myocardial regeneration. *J. Stem Cell Res. Ther.* **2013**, *9*, 004. [CrossRef]
52. Le, T.; Chong, J. Cardiac progenitor cells for heart repair. *Cell Death Discov.* **2016**, *2*, 16052. [CrossRef] [PubMed]
53. Ellison, G.M.; Vicinanza, C.; Smith, A.J.; Aquila, I.; Leone, A.; Waring, C.D.; Henning, B.J.; Stirparo, G.G.; Papait, R.; Scarfò, M.; et al. Adult c-kitpos cardiac stem cells are necessary and sufficient for functional cardiac regeneration and repair. *Cell* **2013**, *154*, 827–842. [CrossRef] [PubMed]
54. Edling, C.E.; Hallberg, B. c-Kit—A hematopoietic cell essential receptor tyrosine kinase. *Int. J. Biochem. Cell Biol.* **2007**, *39*, 1995–1998. [CrossRef] [PubMed]

55. Vajravelu, B.N.; Hong, K.U.; Al-Maqtari, T.; Cao, P.; Keith, M.C.L.; Wysoczynski, M.; Zhao, J.; Moore IV, J.B.; Bolli, R. c-Kit promotes growth and migration of human cardiac progenitor cells via the PI3K-AKT and MEK-ERK pathways. *PLoS ONE* **2015**, *10*, e0140798. [CrossRef]

56. Kuang, D.; Zhao, X.; Xiao, G.; Ni, J.; Feng, Y.; Wu, R.; Wang, G. Stem cell factor/c-kit signaling mediated cardiac stem cell migration via activation of p38 MAPK. *Basic Res. Cardiol.* **2008**, *103*, 265–273. [CrossRef]

57. Ayach, B.B.; Yoshimitsu, M.; Dawood, F.; Sun, M.; Arab, S.; Chen, M.; Higuchi, K.; Siatskas, C.; Lee, P.; Lim, H.; et al. Stem cell factor receptor induces progenitor and natural killer cell-mediated cardiac survival and repair after myocardial infarction. *Proc. Natl. Acad. Sci. USA* **2006**, *103*, 2304–2309. [CrossRef]

58. van Berlo, J.H.; Kanisicak, O.; Maillet, M.; Vagnozzi, R.J.; Karch, J.; Lin, S.-C.J.; Middleton, R.C.; Marbán, E.; Molkentin, J.D. c-Kit+ cells minimally contribute cardiomyocytes to the heart. *Nature* **2014**, *509*, 337–341. [CrossRef]

59. Jesty, S.A.; Steffey, M.A.; Lee, F.K.; Breitbach, M.; Hesse, M.; Reining, S.; Lee, J.C.; Doran, R.M.; Nikitin, A.Y.; Fleischmann, B.K.; et al. c-Kit+ precursors support postinfarction myogenesis in the neonatal, but not adult, heart. *Proc. Natl. Acad. Sci. USA* **2012**, *109*, 13380–13385. [CrossRef]

60. Zaruba, M.-M.; Soonpaa, M.; Reuter, S.; Field, L.J. Cardiomyogenic potential of c-Kit $^+$–Expressing cells derived from neonatal and adult mouse hearts. *Circulation* **2010**, *121*, 1992–2000. [CrossRef]

61. Sultana, N.; Zhang, L.; Yan, J.; Chen, J.; Cai, W.; Razzaque, S.; Jeong, D.; Sheng, W.; Bu, L.; Xu, M.; et al. Resident c-kit+ cells in the heart are not cardiac stem cells. *Nat. Commun.* **2015**, *6*, 8701. [CrossRef] [PubMed]

62. Vicinanza, C.; Aquila, I.; Scalise, M.; Cristiano, F.; Marino, F.; Cianflone, E.; Mancuso, T.; Marotta, P.; Sacco, W.; Lewis, F.C.; et al. Adult cardiac stem cells are multipotent and robustly myogenic: C-Kit expression is necessary but not sufficient for their identification. *Cell Death Differ.* **2017**, *24*, 2101–2116. [CrossRef] [PubMed]

63. Holmes, C.; Stanford, W.L. Concise review: Stem cell antigen-1: Expression, function, and enigma. *Stem Cells* **2007**, *25*, 1339–1347. [CrossRef] [PubMed]

64. Matsuura, K.; Nagai, T.; Nishigaki, N.; Oyama, T.; Nishi, J.; Wada, H.; Sano, M.; Toko, H.; Akazawa, H.; Sato, T.; et al. Adult cardiac sca-1-positive cells differentiate into beating cardiomyocytes. *J. Biol. Chem.* **2004**, *279*, 11384–11391. [CrossRef]

65. Tateishi, K.; Ashihara, E.; Takehara, N.; Nomura, T.; Honsho, S.; Nakagami, T.; Morikawa, S.; Takahashi, T.; Ueyama, T.; Matsubara, H.; et al. Clonally amplified cardiac stem cells are regulated by Sca-1 signaling for efficient cardiovascular regeneration. *J. Cell Sci.* **2007**, *120*, 1791–1800. [CrossRef]

66. Wang, X.; Hu, Q.; Nakamura, Y.; Lee, J.; Zhang, G.; From, A.H.L.; Zhang, J. The role of the Sca-1 $^+$/CD31 $^-$ cardiac progenitor cell population in postinfarction left ventricular remodeling. *Stem Cells* **2006**, *24*, 1779–1788. [CrossRef]

67. Van Vliet, P.; Smits, A.M.; De Boer, T.P.; Korfage, T.H.; Metz, C.H.G.; Roccio, M.; Van Der Heyden, M.A.G.; Van Veen, T.A.B.; Sluijter, J.P.G.; Doevendans, P.A.; et al. Foetal and adult cardiomyocyte progenitor cells have different developmental potential. *J. Cell. Mol. Med.* **2010**, *14*, 861–870. [CrossRef]

68. Huang, C.; Gu, H.; Yu, Q.; Manukyan, M.C.; Poynter, J.A.; Wang, M. Sca-1+ cardiac stem cells mediate acute cardioprotection via paracrine factor SDF-1 following myocardial ischemia/reperfusion. *PLoS ONE* **2011**, *6*, e29246. [CrossRef]

69. Takamiya, M.; Haider, K.H.; Ashraf, M. Identification and characterization of a novel multipotent sub-population of Sca-1+ cardiac progenitor cells for myocardial regeneration. *PLoS ONE* **2011**, *6*, e25265. [CrossRef]

70. Uchida, S.; De Gaspari, P.; Kostin, S.; Jenniches, K.; Kilic, A.; Izumiya, Y.; Shiojima, I.; grosse Kreymborg, K.; Renz, H.; Walsh, K.; et al. Sca1-derived cells are a source of myocardial renewal in the murine adult heart. *Stem Cell Rep.* **2013**, *1*, 397–410. [CrossRef]

71. Ge, Z.; Lal, S.; Le, T.Y.L.; dos Remedios, C.; Chong, J.J.H. Cardiac stem cells: Translation to human studies. *Biophys. Rev.* **2015**, *7*, 127–139. [CrossRef] [PubMed]

72. David, R.; Jarsch, V.B.; Schwarz, F.; Nathan, P.; Gegg, M.; Lickert, H.; Franz, W.-M. Induction of MesP1 by Brachyury(T) generates the common multipotent cardiovascular stem cell. *Cardiovasc. Res.* **2011**, *92*, 115–122. [CrossRef] [PubMed]

73. Liu, Y.; Schwartz, R.J. Transient Mesp1 expression: A driver of cardiac cell fate determination. *Transcription* **2013**, *4*, 92–96. [CrossRef] [PubMed]

74. Kitajima, S.; Takagi, A.; Inoue, T.; Saga, Y. MesP1 and MesP2 are essential for the development of cardiac mesoderm. *Development* **2000**, *127*, 3215–3226.

75. Habib, M.; Caspi, O.; Gepstein, L. Human embryonic stem cells for cardiomyogenesis. *J. Mol. Cell. Cardiol.* **2008**, *45*, 462–474. [CrossRef]

76. Wu, S.M.; Chien, K.R.; Mummery, C. Origins and fates of cardiovascular progenitor cells. *Cell* **2008**, *132*, 537–543. [CrossRef]

77. Saga, Y.; Miyagawa-Tomita, S.; Takagi, A.; Kitajima, S.; Miyazaki, J.; Inoue, T. MesP1 is expressed in the heart precursor cells and required for the formation of a single heart tube. *Development* **1999**, *126*, 3437–3447.

78. Chan, S.S.-K.; Shi, X.; Toyama, A.; Arpke, R.W.; Dandapat, A.; Iacovino, M.; Kang, J.; Le, G.; Hagen, H.R.; Garry, D.J.; et al. Mesp1 patterns mesoderm into cardiac, hematopoietic, or skeletal myogenic progenitors in a context-dependent manner. *Cell Stem Cell* **2013**, *12*, 587–601. [CrossRef]

79. Bondue, A.; Tännler, S.; Chiapparo, G.; Chabab, S.; Ramialison, M.; Paulissen, C.; Beck, B.; Harvey, R.; Blanpain, C. Defining the earliest step of cardiovascular progenitor specification during embryonic stem cell differentiation. *J. Cell Biol.* **2011**, *192*, 751–765. [CrossRef]

80. Liu, Y.; Chen, L.; Diaz, A.D.; Benham, A.; Xu, X.; Wijaya, C.S.; Fa'ak, F.; Luo, W.; Soibam, B.; Azares, A.; et al. Mesp1 marked cardiac progenitor cells repair infarcted mouse hearts. *Sci. Rep.* **2016**, *6*, 31457. [CrossRef]

81. Ema, M. Deletion of the selection cassette, but not cis-acting elements, in targeted Flk1-lacZ allele reveals Flk1 expression in multipotent mesodermal progenitors. *Blood* **2006**, *107*, 111–117. [CrossRef] [PubMed]

82. Kouskoff, V.; Lacaud, G.; Schwantz, S.; Fehling, H.J.; Keller, G. Sequential development of hematopoietic and cardiac mesoderm during embryonic stem cell differentiation. *Proc. Natl. Acad. Sci. USA* **2005**, *102*, 13170–13175. [CrossRef] [PubMed]

83. Bu, L.; Jiang, X.; Martin-Puig, S.; Caron, L.; Zhu, S.; Shao, Y.; Roberts, D.J.; Huang, P.L.; Domian, I.J.; Chien, K.R. Human ISL1 heart progenitors generate diverse multipotent cardiovascular cell lineages. *Nature* **2009**, *460*, 113–117. [CrossRef] [PubMed]

84. Buckingham, M.; Meilhac, S.; Zaffran, S. Building the mammalian heart from two sources of myocardial cells. *Nat. Rev. Genet.* **2005**, *6*, 826–835. [CrossRef] [PubMed]

85. Liang, X.; Wang, G.; Lin, L.; Lowe, J.; Zhang, Q.; Bu, L.; Chen, Y.; Chen, J.; Sun, Y.; Evans, S.M. HCN4 dynamically marks the first heart field and conduction system precursors. *Circ. Res.* **2013**, *113*, 399–407. [CrossRef] [PubMed]

86. Später, D.; Abramczuk, M.K.; Buac, K.; Zangi, L.; Stachel, M.W.; Clarke, J.; Sahara, M.; Ludwig, A.; Chien, K.R. A HCN4+ cardiomyogenic progenitor derived from the first heart field and human pluripotent stem cells. *Nat. Cell Biol.* **2013**, *15*, 1098–1106. [CrossRef]

87. Garcia-Frigola, C.; Shi, Y.; Evans, S.M. Expression of the hyperpolarization-activated cyclic nucleotide-gated cation channel HCN4 during mouse heart development. *Gene Expr. Patterns* **2003**, *3*, 777–783. [CrossRef]

88. Zhou, B.; Ma, Q.; Rajagopal, S.; Wu, S.M.; Domian, I.; Rivera-Feliciano, J.; Jiang, D.; von Gise, A.; Ikeda, S.; Chien, K.R.; et al. Epicardial progenitors contribute to the cardiomyocyte lineage in the developing heart. *Nature* **2008**, *454*, 109–113. [CrossRef]

89. van Tuyn, J.; Atsma, D.E.; Winter, E.M.; van der Velde-van Dijke, I.; Pijnappels, D.A.; Bax, N.A.M.; Knaän-Shanzer, S.; Gittenberger-de Groot, A.C.; Poelmann, R.E.; van der Laarse, A.; et al. Epicardial cells of human adults can undergo an epithelial-to-mesenchymal transition and obtain characteristics of smooth muscle cells in vitro. *Stem Cells* **2007**, *25*, 271–278. [CrossRef]

90. Cai, C.-L.; Martin, J.C.; Sun, Y.; Cui, L.; Wang, L.; Ouyang, K.; Yang, L.; Bu, L.; Liang, X.; Zhang, X.; et al. A myocardial lineage derives from Tbx18 epicardial cells. *Nature* **2008**, *454*, 104–108. [CrossRef]

91. Smart, N.; Bollini, S.; Dubé, K.N.; Vieira, J.M.; Zhou, B.; Davidson, S.; Yellon, D.; Riegler, J.; Price, A.N.; Lythgoe, M.F.; et al. De novo cardiomyocytes from within the activated adult heart after injury. *Nature* **2011**, *474*, 640–644. [CrossRef] [PubMed]

92. Bollini, S.; Vieira, J.M.N.; Howard, S.; Dubè, K.N.; Balmer, G.M.; Smart, N.; Riley, P.R. Re-activated adult epicardial progenitor cells are a heterogeneous population molecularly distinct from their embryonic counterparts. *Stem Cells Dev.* **2014**, *23*, 1719–1730. [CrossRef] [PubMed]

93. Smart, N.; Dubé, K.N.; Riley, P.R. Epicardial progenitor cells in cardiac regeneration and neovascularisation. *Vasc. Pharmacol.* **2013**, *58*, 164–173. [CrossRef] [PubMed]

94. Martin, C.M.; Meeson, A.P.; Robertson, S.M.; Hawke, T.J.; Richardson, J.A.; Bates, S.; Goetsch, S.C.; Gallardo, T.D.; Garry, D.J. Persistent expression of the ATP-binding cassette transporter, Abcg2, identifies cardiac SP cells in the developing and adult heart. *Dev. Biol.* **2004**, *265*, 262–275. [CrossRef]

95. Pfister, O.; Oikonomopoulos, A.; Sereti, K.-I.; Sohn, R.L.; Cullen, D.; Fine, G.C.; Mouquet, F.; Westerman, K.; Liao, R. Role of the ATP-binding cassette transporter *Abcg2* in the phenotype and function of cardiac side population cells. *Circ. Res.* **2008**, *103*, 825–835. [CrossRef]

96. Liang, S.X.; Tan, T.Y.L.; Gaudry, L.; Chong, B. Differentiation and migration of Sca1+/CD31− cardiac side population cells in a murine myocardial ischemic model. *Int. J. Cardiol.* **2010**, *138*, 40–49. [CrossRef]

97. Oyama, T.; Nagai, T.; Wada, H.; Naito, A.T.; Matsuura, K.; Iwanaga, K.; Takahashi, T.; Goto, M.; Mikami, Y.; Yasuda, N.; et al. Cardiac side population cells have a potential to migrate and differentiate into cardiomyocytes in vitro and in vivo. *J. Cell Biol.* **2007**, *176*, 329–341. [CrossRef]

98. Zhou, S.; Schuetz, J.D.; Bunting, K.D.; Colapietro, A.-M.; Sampath, J.; Morris, J.J.; Lagutina, I.; Grosveld, G.C.; Osawa, M.; Nakauchi, H.; et al. The ABC transporter Bcrp1/ABCG2 is expressed in a wide variety of stem cells and is a molecular determinant of the side-population phenotype. *Nat. Med.* **2001**, *7*, 1028–1034. [CrossRef]

99. Alfakir, M.; Dawe, N.; Eyre, R.; Tyson-Capper, A.; Britton, K.; Robson, S.C.; Meeson, A.P. The temporal and spatial expression patterns of ABCG2 in the developing human heart. *Int. J. Cardiol.* **2012**, *156*, 133–138. [CrossRef]

100. Hierlihy, A.M.; Seale, P.; Lobe, C.G.; Rudnicki, M.A.; Megeney, L.A. The post-natal heart contains a myocardial stem cell population. *FEBS Lett.* **2002**, *530*, 239–243. [CrossRef]

101. Pfister, O.; Mouquet, F.; Jain, M.; Summer, R.; Helmes, M.; Fine, A.; Colucci, W.S.; Liao, R. CD31− but not CD31+ cardiac side population cells exhibit functional cardiomyogenic differentiation. *Circ. Res.* **2005**, *97*, 52–61. [CrossRef] [PubMed]

102. Yoon, J.; Choi, S.-C.; Park, C.-Y.; Shim, W.-J.; Lim, D.-S. Cardiac side population cells exhibit endothelial differentiation potential. *Exp. Mol. Med.* **2007**, *39*, 653–662. [CrossRef] [PubMed]

103. Noseda, M.; Harada, M.; McSweeney, S.; Leja, T.; Belian, E.; Stuckey, D.J.; Abreu Paiva, M.S.; Habib, J.; Macaulay, I.; de Smith, A.J.; et al. PDGFRα demarcates the cardiogenic clonogenic Sca1+ stem/progenitor cell in adult murine myocardium. *Nat. Commun* **2015**, *6*, 6930. [CrossRef] [PubMed]

104. Yamahara, K.; Fukushima, S.; Coppen, S.R.; Felkin, L.E.; Varela-Carver, A.; Barton, P.J.R.; Yacoub, M.H.; Suzuki, K. Heterogeneic nature of adult cardiac side population cells. *Biochem. Biophys. Res. Commun.* **2008**, *371*, 615–620. [CrossRef]

105. Chimenti, I.; Smith, R.R.; Li, T.-S.; Gerstenblith, G.; Messina, E.; Giacomello, A.; Marbán, E. Relative roles of direct regeneration versus paracrine effects of human cardiosphere-derived cells transplanted into infarcted mice. *Circ. Res.* **2010**, *106*, 971–980. [CrossRef]

106. Li, T.-S.; Cheng, K.; Lee, S.-T.; Matsushita, S.; Davis, D.; Malliaras, K.; Zhang, Y.; Matsushita, N.; Smith, R.R.; Marbán, E. Cardiospheres recapitulate a niche-like microenvironment rich in stemness and cell-matrix interactions, rationalizing their enhanced functional potency for myocardial repair. *Stem Cells* **2010**, *28*, 2088–2098. [CrossRef]

107. He, J.-Q.; Vu, D.M.; Hunt, G.; Chugh, A.; Bhatnagar, A.; Bolli, R. Human cardiac stem cells isolated from atrial appendages stably express c-kit. *PLoS ONE* **2011**, *6*, e27719. [CrossRef]

108. Hesse, M.; Fleischmann, B.K.; Kotlikoff, M.I. Concise Review: The role of c-kit expressing cells in heart repair at the neonatal and adult stage: C-kit + cells in heart repair. *Stem Cells* **2014**, *32*, 1701–1712. [CrossRef]

109. Freire, A.G.; Nascimento, D.S.; Forte, G.; Valente, M.; Resende, T.P.; Pagliari, S.; Abreu, C.; Carvalho, I.; Nardo, P.D.; Pinto-do-Ó, P. Stable phenotype and function of immortalized Lin− Sca-1+ cardiac progenitor cells in long-term culture: A step closer to standardization. *Stem Cells Dev.* **2014**, *23*, 1012–1026. [CrossRef]

110. Yamashita, J.K.; Takano, M.; Hiraoka-Kanie, M.; Shimazu, C.; Peishi, Y.; Yanagi, K.; Nakano, A.; Inoue, E.; Kita, F.; Nishikawa, S.-I. Prospective identification of cardiac progenitors by a novel single cell-based cardiomyocyte induction. *FASEB J.* **2005**, *19*, 1534–1536. [CrossRef]

111. Lescroart, F.; Chabab, S.; Lin, X.; Rulands, S.; Paulissen, C.; Rodolosse, A.; Auer, H.; Achouri, Y.; Dubois, C.; Bondue, A.; et al. Early lineage restriction in temporally distinct populations of Mesp1 progenitors during mammalian heart development. *Nat. Cell Biol.* **2014**, *16*, 829–840. [CrossRef] [PubMed]

112. Fuentes, T.I.; Appleby, N.; Tsay, E.; Martinez, J.J.; Bailey, L.; Hasaniya, N.; Kearns-Jonker, M. Human neonatal cardiovascular progenitors: Unlocking the secret to regenerative ability. *PLoS ONE* **2013**, *8*, e77464. [CrossRef] [PubMed]

113. Sun, Y.; Liang, X.; Najafi, N.; Cass, M.; Lin, L.; Cai, C.-L.; Chen, J.; Evans, S.M. Islet 1 is expressed in distinct cardiovascular lineages, including pacemaker and coronary vascular cells. *Dev. Biol.* **2007**, *304*, 286–296. [CrossRef] [PubMed]

114. Chong, J.J.H.; Reinecke, H.; Iwata, M.; Torok-Storb, B.; Stempien-Otero, A.; Murry, C.E. Progenitor cells identified by PDGFR-alpha expression in the developing and diseased human heart. *Stem Cells Dev.* **2013**, *22*, 1932–1943. [CrossRef]

115. Chong, J.J.H.; Chandrakanthan, V.; Xaymardan, M.; Asli, N.S.; Li, J.; Ahmed, I.; Heffernan, C.; Menon, M.K.; Scarlett, C.J.; Rashidianfar, A.; et al. Adult cardiac-resident MSC-like stem cells with a proepicardial origin. *Cell Stem Cell* **2011**, *9*, 527–540. [CrossRef]

116. Wessels, A.; Pérez-Pomares, J.M. The epicardium and epicardially derived cells (EPDCs) as cardiac stem cells: Epicardially derived cells as cardiac stem cells. *Anat. Rec. Part. A: Discov. Mol. Cell. Evol. Biol.* **2004**, *276A*, 43–57. [CrossRef]

117. Smits, A.; Riley, P. Epicardium-derived heart repair. *J. Dev. Biol.* **2014**, *2*, 84–100. [CrossRef]

118. Emmert, M.Y.; Emmert, L.S.; Martens, A.; Ismail, I.; Schmidt-Richter, I.; Gawol, A.; Seifert, B.; Haverich, A.; Martin, U.; Gruh, I. Higher frequencies of BCRP+ cardiac resident cells in ischaemic human myocardium. *Eur. Heart J.* **2013**, *34*, 2830–2838. [CrossRef]

119. Smith, R.R.; Barile, L.; Cho, H.C.; Leppo, M.K.; Hare, J.M.; Messina, E.; Giacomello, A.; Abraham, M.R.; Marbán, E. Regenerative potential of cardiosphere-derived cells expanded from percutaneous endomyocardial biopsy specimens. *Circulation* **2007**, *115*, 896–908. [CrossRef]

120. Ye, J.; Boyle, A.; Shih, H.; Sievers, R.E.; Zhang, Y.; Prasad, M.; Su, H.; Zhou, Y.; Grossman, W.; Bernstein, H.S.; et al. Sca-1+ cardiosphere-derived cells are enriched for Isl1-expressing cardiac precursors and improve cardiac function after myocardial injury. *PLoS ONE* **2012**, *7*, e30329. [CrossRef]

121. Klein, D. iPSCs-based generation of vascular cells: Reprogramming approaches and applications. *Cell. Mol. Life Sci.* **2018**, *75*, 1411–1433. [CrossRef] [PubMed]

122. Burridge, P.W.; Thompson, S.; Millrod, M.A.; Weinberg, S.; Yuan, X.; Peters, A.; Mahairaki, V.; Koliatsos, V.E.; Tung, L.; Zambidis, E.T. A universal system for highly efficient cardiac differentiation of human induced pluripotent stem cells that eliminates interline variability. *PLoS ONE* **2011**, *6*, e18293. [CrossRef] [PubMed]

123. Kim, K.; Doi, A.; Wen, B.; Ng, K.; Zhao, R.; Cahan, P.; Kim, J.; Aryee, M.J.; Ji, H.; Ehrlich, L.I.R.; et al. Epigenetic memory in induced pluripotent stem cells. *Nature* **2010**, *467*, 285–290. [CrossRef] [PubMed]

124. Chin, M.H.; Mason, M.J.; Xie, W.; Volinia, S.; Singer, M.; Peterson, C.; Ambartsumyan, G.; Aimiuwu, O.; Richter, L.; Zhang, J.; et al. Induced pluripotent stem cells and embryonic stem cells are distinguished by gene expression signatures. *Cell Stem Cell* **2009**, *5*, 111–123. [CrossRef] [PubMed]

125. Mummery, C.L.; Zhang, J.; Ng, E.S.; Elliott, D.A.; Elefanty, A.G.; Kamp, T.J. Differentiation of human embryonic stem cells and induced pluripotent stem cells to cardiomyocytes: A methods overview. *Circ. Res.* **2012**, *111*, 344–358. [CrossRef] [PubMed]

126. Lian, X.; Zhang, J.; Azarin, S.M.; Zhu, K.; Hazeltine, L.B.; Bao, X.; Hsiao, C.; Kamp, T.J.; Palecek, S.P. Directed cardiomyocyte differentiation from human pluripotent stem cells by modulating Wnt/β-catenin signaling under fully defined conditions. *Nat. Protoc.* **2013**, *8*, 162–175. [CrossRef] [PubMed]

127. Olmer, R.; Haase, A.; Merkert, S.; Cui, W.; Paleček, J.; Ran, C.; Kirschning, A.; Scheper, T.; Glage, S.; Miller, K.; et al. Long term expansion of undifferentiated human iPS and ES cells in suspension culture using a defined medium. *Stem Cell Res.* **2010**, *5*, 51–64. [CrossRef]

128. Kattman, S.J.; Witty, A.D.; Gagliardi, M.; Dubois, N.C.; Niapour, M.; Hotta, A.; Ellis, J.; Keller, G. Stage-specific optimization of activin/nodal and BMP signaling promotes cardiac differentiation of mouse and human pluripotent stem cell lines. *Cell Stem Cell* **2011**, *8*, 228–240. [CrossRef]

129. Drowley, L.; Koonce, C.; Peel, S.; Jonebring, A.; Plowright, A.T.; Kattman, S.J.; Andersson, H.; Anson, B.; Swanson, B.J.; Wang, Q.-D.; et al. Human induced pluripotent stem cell-derived cardiac progenitor cells in phenotypic screening: A transforming growth factor-β type 1 receptor kinase inhibitor induces efficient cardiac differentiation: iPSC-derived cardiac progenitors for phenotypic screening. *Stem Cells Transl. Med.* **2016**, *5*, 164–174.

130. Lian, X.; Hsiao, C.; Wilson, G.; Zhu, K.; Hazeltine, L.B.; Azarin, S.M.; Raval, K.K.; Zhang, J.; Kamp, T.J.; Palecek, S.P. Robust cardiomyocyte differentiation from human pluripotent stem cells via temporal modulation of canonical Wnt signaling. *Proc. Natl. Acad. Sci. USA* **2012**, *109*, E1848–E1857. [CrossRef]

131. Minami, I.; Yamada, K.; Otsuji, T.G.; Yamamoto, T.; Shen, Y.; Otsuka, S.; Kadota, S.; Morone, N.; Barve, M.; Asai, Y.; et al. A Small molecule that promotes cardiac differentiation of human pluripotent stem cells under defined, cytokine- and xeno-free conditions. *Cell Rep.* **2012**, *2*, 1448–1460. [CrossRef] [PubMed]

132. Kempf, H.; Olmer, R.; Kropp, C.; Rückert, M.; Jara-Avaca, M.; Robles-Diaz, D.; Franke, A.; Elliott, D.A.; Wojciechowski, D.; Fischer, M.; et al. Controlling expansion and cardiomyogenic differentiation of human pluripotent stem cells in scalable suspension culture. *Stem Cell Rep.* **2014**, *3*, 1132–1146. [CrossRef] [PubMed]

133. Fujiwara, M.; Yan, P.; Otsuji, T.G.; Narazaki, G.; Uosaki, H.; Fukushima, H.; Kuwahara, K.; Harada, M.; Matsuda, H.; Matsuoka, S.; et al. Induction and enhancement of cardiac cell differentiation from mouse and human induced pluripotent stem cells with cyclosporin-A. *PLoS ONE* **2011**, *6*, e16734. [CrossRef] [PubMed]

134. Yan, P.; Nagasawa, A.; Uosaki, H.; Sugimoto, A.; Yamamizu, K.; Teranishi, M.; Matsuda, H.; Matsuoka, S.; Ikeda, T.; Komeda, M.; et al. Cyclosporin-A potently induces highly cardiogenic progenitors from embryonic stem cells. *Biochem. Biophys. Res. Commun.* **2009**, *379*, 115–120. [CrossRef] [PubMed]

135. Uosaki, H.; Andersen, P.; Shenje, L.T.; Fernandez, L.; Christiansen, S.L.; Kwon, C. Direct contact with endoderm-like cells efficiently induces cardiac progenitors from mouse and human pluripotent stem cells. *PLoS ONE* **2012**, *7*, e46413. [CrossRef] [PubMed]

136. Xuan, W.; Wang, Y.; Tang, Y.; Ali, A.; Hu, H.; Maienschein-Cline, M.; Ashraf, M. Cardiac progenitors induced from human induced pluripotent stem cells with cardiogenic small molecule effectively regenerate infarcted hearts and attenuate fibrosis. *Shock* **2018**, *50*, 627–639. [CrossRef]

137. Burridge, P.W.; Matsa, E.; Shukla, P.; Lin, Z.C.; Churko, J.M.; Ebert, A.D.; Lan, F.; Diecke, S.; Huber, B.; Mordwinkin, N.M.; et al. Chemically defined generation of human cardiomyocytes. *Nat. Methods* **2014**, *11*, 855–860. [CrossRef]

138. Zhang, J.; Wilson, G.F.; Soerens, A.G.; Koonce, C.H.; Yu, J.; Palecek, S.P.; Thomson, J.A.; Kamp, T.J. Functional cardiomyocytes derived from human induced pluripotent stem cells. *Circ. Res.* **2009**, *104*, e30–e41. [CrossRef]

139. Gai, H. Generation and characterization of functional cardiomyocytes using induced pluripotent stem cells derived from human fibroblasts. *Cell Biol. Int.* **2009**, *33*, 1184–1193. [CrossRef]

140. Uosaki, H.; Fukushima, H.; Takeuchi, A.; Matsuoka, S.; Nakatsuji, N.; Yamanaka, S.; Yamashita, J.K. Efficient and scalable purification of cardiomyocytes from human embryonic and induced pluripotent stem cells by VCAM1 surface expression. *PLoS ONE* **2011**, *6*, e23657. [CrossRef]

141. Zwi, L.; Caspi, O.; Arbel, G.; Huber, I.; Gepstein, A.; Park, I.-H.; Gepstein, L. Cardiomyocyte differentiation of human induced pluripotent stem cells. *Circulation* **2009**, *120*, 1513–1523. [CrossRef] [PubMed]

142. Denning, C.; Borgdorff, V.; Crutchley, J.; Firth, K.S.A.; George, V.; Kalra, S.; Kondrashov, A.; Hoang, M.D.; Mosqueira, D.; Patel, A.; et al. Cardiomyocytes from human pluripotent stem cells: From laboratory curiosity to industrial biomedical platform. *Biochim. Et. Biophys. Acta Mol. Cell Res.* **2016**, *1863*, 1728–1748. [CrossRef] [PubMed]

143. Cao, N.; Liu, Z.; Chen, Z.; Wang, J.; Chen, T.; Zhao, X.; Ma, Y.; Qin, L.; Kang, J.; Wei, B.; et al. Ascorbic acid enhances the cardiac differentiation of induced pluripotent stem cells through promoting the proliferation of cardiac progenitor cells. *Cell Res.* **2012**, *22*, 219–236. [CrossRef] [PubMed]

144. Cao, N.; Liang, H.; Huang, J.; Wang, J.; Chen, Y.; Chen, Z.; Yang, H.-T. Highly efficient induction and long-term maintenance of multipotent cardiovascular progenitors from human pluripotent stem cells under defined conditions. *Cell Res.* **2013**, *23*, 1119–1132. [CrossRef] [PubMed]

145. Blin, G.; Nury, D.; Stefanovic, S.; Neri, T.; Guillevic, O.; Brinon, B.; Bellamy, V.; Rücker-Martin, C.; Barbry, P.; Bel, A.; et al. A purified population of multipotent cardiovascular progenitors derived from primate pluripotent stem cells engrafts in postmyocardial infarcted nonhuman primates. *J. Clin. Investig.* **2010**, *120*, 1125–1139. [CrossRef] [PubMed]

146. Mauritz, C.; Martens, A.; Rojas, S.V.; Schnick, T.; Rathert, C.; Schecker, N.; Menke, S.; Glage, S.; Zweigerdt, R.; Haverich, A.; et al. Induced pluripotent stem cell (iPSC)-derived Flk-1 progenitor cells engraft, differentiate, and improve heart function in a mouse model of acute myocardial infarction. *Eur. Heart J.* **2011**, *32*, 2634–2641. [CrossRef] [PubMed]

147. Zhang, J.Z.; Termglinchan, V.; Shao, N.-Y.; Itzhaki, I.; Liu, C.; Ma, N.; Tian, L.; Wang, V.Y.; Chang, A.C.Y.; Guo, H.; et al. A human iPSC double-reporter system enables purification of cardiac lineage subpopulations with distinct function and drug response profiles. *Cell Stem Cell* **2019**, *24*, 802–811. [CrossRef]

148. Ren, Y.; Lee, M.Y.; Schliffke, S.; Paavola, J.; Amos, P.J.; Ge, X.; Ye, M.; Zhu, S.; Senyei, G.; Lum, L.; et al. Small molecule Wnt inhibitors enhance the efficiency of BMP-4-directed cardiac differentiation of human pluripotent stem cells. *J. Mol. Cell. Cardiol.* **2011**, *51*, 280–287. [CrossRef]

149. Moretti, A.; Bellin, M.; Jung, C.B.; Thies, T.-M.; Takashima, Y.; Bernshausen, A.; Schiemann, M.; Fischer, S.; Moosmang, S.; Smith, A.G.; et al. Mouse and human induced pluripotent stem cells as a source for multipotent Isl1 [+] cardiovascular progenitors. *FASEB J.* **2010**, *24*, 700–711. [CrossRef]

150. Lian, X.; Bao, X.; Zilberter, M.; Westman, M.; Fisahn, A.; Hsiao, C.; Hazeltine, L.B.; Dunn, K.K.; Kamp, T.J.; Palecek, S.P. Chemically defined, albumin-free human cardiomyocyte generation. *Nat. Methods* **2015**, *12*, 595–596. [CrossRef]

151. Andersen, P.; Tampakakis, E.; Jimenez, D.V.; Kannan, S.; Miyamoto, M.; Shin, H.K.; Saberi, A.; Murphy, S.; Sulistio, E.; Chelko, S.P.; et al. Precardiac organoids form two heart fields via Bmp/Wnt signaling. *Nat. Commun.* **2018**, *9*, 3140. [CrossRef] [PubMed]

152. Bao, X.; Lian, X.; Qian, T.; Bhute, V.J.; Han, T.; Palecek, S.P. Directed differentiation and long-term maintenance of epicardial cells derived from human pluripotent stem cells under fully defined conditions. *Nat. Protoc.* **2017**, *12*, 1890–1900. [CrossRef] [PubMed]

153. Bao, X.; Lian, X.; Hacker, T.A.; Schmuck, E.G.; Qian, T.; Bhute, V.J.; Han, T.; Shi, M.; Drowley, L.; Plowright, A.T.; et al. Long-term self-renewing human epicardial cells generated from pluripotent stem cells under defined xeno-free conditions. *Nat. Biomed. Eng.* **2017**, *1*, 1–12. [CrossRef] [PubMed]

154. Witty, A.D.; Mihic, A.; Tam, R.Y.; Fisher, S.A.; Mikryukov, A.; Shoichet, M.S.; Li, R.-K.; Kattman, S.J.; Keller, G. Generation of the epicardial lineage from human pluripotent stem cells. *Nat. Biotechnol.* **2014**, *32*, 1026–1035. [CrossRef]

155. Iyer, D.; Gambardella, L.; Bernard, W.G.; Serrano, F.; Mascetti, V.L.; Pedersen, R.A.; Talasila, A.; Sinha, S. Robust derivation of epicardium and its differentiated smooth muscle cell progeny from human pluripotent stem cells. *Development* **2015**, *142*, 1528–1541. [CrossRef]

156. Zhao, J.; Cao, H.; Tian, L.; Huo, W.; Zhai, K.; Wang, P.; Ji, G.; Ma, Y. Efficient differentiation of TBX18[+]/WT1[+] epicardial-like cells from human pluripotent stem cells using small molecular compounds. *Stem Cells Dev.* **2017**, *26*, 528–540. [CrossRef]

157. Christoforou, N.; Liau, B.; Chakraborty, S.; Chellapan, M.; Bursac, N.; Leong, K.W. Induced pluripotent stem cell-derived cardiac progenitors differentiate to cardiomyocytes and form biosynthetic tissues. *PLoS ONE* **2013**, *8*, e65963. [CrossRef]

158. Zhang, J.; Tao, R.; Campbell, K.F.; Carvalho, J.L.; Ruiz, E.C.; Kim, G.C.; Schmuck, E.G.; Raval, A.N.; da Rocha, A.M.; Herron, T.J.; et al. Functional cardiac fibroblasts derived from human pluripotent stem cells via second heart field progenitors. *Nat. Commun.* **2019**, *10*, 2238. [CrossRef]

159. Efe, J.A.; Hilcove, S.; Kim, J.; Zhou, H.; Ouyang, K.; Wang, G.; Chen, J.; Ding, S. Conversion of mouse fibroblasts into cardiomyocytes using a direct reprogramming strategy. *Nat. Cell Biol.* **2011**, *13*, 215–222. [CrossRef]

160. Qian, L.; Huang, Y.; Spencer, C.I.; Foley, A.; Vedantham, V.; Liu, L.; Conway, S.J.; Fu, J.; Srivastava, D. In vivo reprogramming of murine cardiac fibroblasts into induced cardiomyocytes. *Nature* **2012**, *485*, 593–598. [CrossRef]

161. Ieda, M.; Fu, J.-D.; Delgado-Olguin, P.; Vedantham, V.; Hayashi, Y.; Bruneau, B.G.; Srivastava, D. Direct reprogramming of fibroblasts into functional cardiomyocytes by defined factors. *Cell* **2010**, *142*, 375–386. [CrossRef] [PubMed]

162. Fu, J.-D.; Stone, N.R.; Liu, L.; Spencer, C.I.; Qian, L.; Hayashi, Y.; Delgado-Olguin, P.; Ding, S.; Bruneau, B.G.; Srivastava, D. Direct reprogramming of human fibroblasts toward a cardiomyocyte-like state. *Stem Cell Rep.* **2013**, *1*, 235–247. [CrossRef] [PubMed]

163. Qian, L.; Berry, E.C.; Fu, J.; Ieda, M.; Srivastava, D. Reprogramming of mouse fibroblasts into cardiomyocyte-like cells in vitro. *Nat. Protoc.* **2013**, *8*, 1204–1215. [CrossRef] [PubMed]

164. Wada, R.; Muraoka, N.; Inagawa, K.; Yamakawa, H.; Miyamoto, K.; Sadahiro, T.; Umei, T.; Kaneda, R.; Suzuki, T.; Kamiya, K.; et al. Induction of human cardiomyocyte-like cells from fibroblasts by defined factors. *Proc. Natl. Acad. Sci. USA* **2013**, *110*, 12667–12672. [CrossRef] [PubMed]

165. Wang, H.; Cao, N.; Spencer, C.I.; Nie, B.; Ma, T.; Xu, T.; Zhang, Y.; Wang, X.; Srivastava, D.; Ding, S. Small molecules enable cardiac reprogramming of mouse fibroblasts with a single factor, Oct4. *Cell Rep.* **2014**, *6*, 951–960. [CrossRef]

166. Mathison, M.; Gersch, R.P.; Nasser, A.; Lilo, S.; Korman, M.; Fourman, M.; Hackett, N.; Shroyer, K.; Yang, J.; Ma, Y.; et al. In vivo cardiac cellular reprogramming efficacy is enhanced by angiogenic preconditioning of the infarcted myocardium with vascular endothelial growth factor. *J Am Heart Assoc* **2012**, *1*, e005652. [CrossRef]

167. Protze, S.; Khattak, S.; Poulet, C.; Lindemann, D.; Tanaka, E.M.; Ravens, U. A new approach to transcription factor screening for reprogramming of fibroblasts to cardiomyocyte-like cells. *J. Mol. Cell. Cardiol.* **2012**, *53*, 323–332. [CrossRef]

168. Addis, R.C.; Ifkovits, J.L.; Pinto, F.; Kellam, L.D.; Esteso, P.; Rentschler, S.; Christoforou, N.; Epstein, J.A.; Gearhart, J.D. Optimization of direct fibroblast reprogramming to cardiomyocytes using calcium activity as a functional measure of success. *J. Mol. Cell. Cardiol.* **2013**, *60*, 97–106. [CrossRef]

169. Christoforou, N.; Chellappan, M.; Adler, A.F.; Kirkton, R.D.; Wu, T.; Addis, R.C.; Bursac, N.; Leong, K.W. Transcription factors MYOCD, SRF, Mesp1 and SMARCD3 enhance the cardio-inducing effect of GATA4, TBX5, and MEF2C during direct cellular reprogramming. *PLoS ONE* **2013**, *8*, e63577. [CrossRef]

170. Hirai, H.; Katoku-Kikyo, N.; Keirstead, S.A.; Kikyo, N. Accelerated direct reprogramming of fibroblasts into cardiomyocyte-like cells with the MyoD transactivation domain. *Cardiovasc. Res.* **2013**, *100*, 105–113. [CrossRef]

171. Ifkovits, J.L.; Addis, R.C.; Epstein, J.A.; Gearhart, J.D. Inhibition of TGFβ signaling increases direct conversion of fibroblasts to induced cardiomyocytes. *PLoS ONE* **2014**, *9*, e89678. [CrossRef] [PubMed]

172. Wang, L.; Liu, Z.; Yin, C.; Asfour, H.; Chen, O.; Li, Y.; Bursac, N.; Liu, J.; Qian, L. Stoichiometry of Gata4, Mef2c, and Tbx5 influences the efficiency and quality of induced cardiac myocyte reprogramming. *Circ. Res.* **2015**, *116*, 237–244. [CrossRef] [PubMed]

173. Fu, Y.; Huang, C.; Xu, X.; Gu, H.; Ye, Y.; Jiang, C.; Qiu, Z.; Xie, X. Direct reprogramming of mouse fibroblasts into cardiomyocytes with chemical cocktails. *Cell Res.* **2015**, *25*, 1013–1024. [CrossRef] [PubMed]

174. Islas, J.F.; Liu, Y.; Weng, K.-C.; Robertson, M.J.; Zhang, S.; Prejusa, A.; Harger, J.; Tikhomirova, D.; Chopra, M.; Iyer, D.; et al. Transcription factors ETS2 and MESP1 transdifferentiate human dermal fibroblasts into cardiac progenitors. *Proc. Natl. Acad. Sci. USA* **2012**, *109*, 13016–13021. [CrossRef]

175. Xu, J.-Y.; Lee, Y.-K.; Ran, X.; Liao, S.-Y.; Yang, J.; Au, K.-W.; Lai, W.-H.; Esteban, M.A.; Tse, H.-F. Generation of induced cardiospheres via reprogramming of skin fibroblasts for myocardial regeneration: Induced cardiospheres for myocardial regeneration. *Stem Cells* **2016**, *34*, 2693–2706. [CrossRef]

176. Lian, W.; Jia, Y.; Li, L.; Huang, Z.; Xu, J. Generation of induced cardiospheres via reprogramming of mouse skin fibroblasts. *Curr. Protoc. Stem Cell Biol.* **2018**, *46*, e59. [CrossRef]

177. Song, K.; Nam, Y.-J.; Luo, X.; Qi, X.; Tan, W.; Huang, G.N.; Acharya, A.; Smith, C.L.; Tallquist, M.D.; Neilson, E.G.; et al. Heart repair by reprogramming non-myocytes with cardiac transcription factors. *Nature* **2012**, *485*, 599–604. [CrossRef]

178. Sadahiro, T.; Yamanaka, S.; Ieda, M. Direct cardiac reprogramming: Progress and challenges in basic biology and clinical applications. *Circ. Res.* **2015**, *116*, 1378–1391. [CrossRef]

179. Srivastava, D.; DeWitt, N. In vivo cellular reprogramming: The next generation. *Cell* **2016**, *166*, 1386–1396. [CrossRef]

180. Nam, Y.-J.; Song, K.; Luo, X.; Daniel, E.; Lambeth, K.; West, K.; Hill, J.A.; DiMaio, J.M.; Baker, L.A.; Bassel-Duby, R.; et al. Reprogramming of human fibroblasts toward a cardiac fate. *Proc. Natl. Acad. Sci. USA* **2013**, *110*, 5588–5593. [CrossRef]

181. Zhang, R.; Han, P.; Yang, H.; Ouyang, K.; Lee, D.; Lin, Y.-F.; Ocorr, K.; Kang, G.; Chen, J.; Stainier, D.Y.R.; et al. In vivo cardiac reprogramming contributes to zebrafish heart regeneration. *Nature* **2013**, *498*, 497–501. [CrossRef] [PubMed]

182. Zhang, Y.; Li, T.-S.; Lee, S.-T.; Wawrowsky, K.A.; Cheng, K.; Galang, G.; Malliaras, K.; Abraham, M.R.; Wang, C.; Marbán, E. Dedifferentiation and proliferation of mammalian cardiomyocytes. *PLoS ONE* **2010**, *5*, e12559. [CrossRef] [PubMed]

183. Zhang, Y.; Zhong, J.F.; Qiu, H.; Robb MacLellan, W.; Marbán, E.; Wang, C. Epigenomic reprogramming of adult cardiomyocyte-derived cardiac progenitor cells. *Sci. Rep.* **2015**, *5*, 17686. [CrossRef]

184. Jayawardena, T.M.; Egemnazarov, B.; Finch, E.A.; Zhang, L.; Payne, J.A.; Pandya, K.; Zhang, Z.; Rosenberg, P.; Mirotsou, M.; Dzau, V.J. MicroRNA-Mediated In Vitro and In Vivo Direct Reprogramming of Cardiac Fibroblasts to Cardiomyocytes. *Circ. Res.* **2012**, *110*, 1465–1473. [CrossRef] [PubMed]

185. Ma, H.; Wang, L.; Yin, C.; Liu, J.; Qian, L. In vivo cardiac reprogramming using an optimal single polycistronic construct: Figure 1. *Cardiovasc. Res.* **2015**, *108*, 217–219. [CrossRef] [PubMed]

186. Birket, M.J.; Ribeiro, M.C.; Verkerk, A.O.; Ward, D.; Leitoguinho, A.R.; den Hartogh, S.C.; Orlova, V.V.; Devalla, H.D.; Schwach, V.; Bellin, M.; et al. Expansion and patterning of cardiovascular progenitors derived from human pluripotent stem cells. *Nat. Biotechnol.* **2015**, *33*, 970–979. [CrossRef]

187. Nsair, A.; Schenke-Layland, K.; Van Handel, B.; Evseenko, D.; Kahn, M.; Zhao, P.; Mendelis, J.; Heydarkhan, S.; Awaji, O.; Vottler, M.; et al. Characterization and therapeutic potential of induced pluripotent stem cell-derived cardiovascular progenitor cells. *PLoS ONE* **2012**, *7*, e45603. [CrossRef]

188. Nelson, T.J.; Faustino, R.S.; Chiriac, A.; Crespo-Diaz, R.; Behfar, A.; Terzic, A. CXCR4$^+$/FLK-1$^+$ biomarkers select a cardiopoietic lineage from embryonic stem cells. *Stem Cells* **2008**, *26*, 1464–1473. [CrossRef]

189. Zhou, Z.; Wang, J.; Guo, C.; Chang, W.; Zhuang, J.; Zhu, P.; Li, X. Temporally distinct Six2 -positive second heart field progenitors regulate mammalian heart development and disease. *Cell Rep.* **2017**, *18*, 1019–1032. [CrossRef]

190. Torán, J.L.; López, J.A.; Gomes-Alves, P.; Aguilar, S.; Torroja, C.; Trevisan-Herraz, M.; Moscoso, I.; Sebastião, M.J.; Serra, M.; Brito, C.; et al. Definition of a cell surface signature for human cardiac progenitor cells after comprehensive comparative transcriptomic and proteomic characterization. *Sci. Rep.* **2019**, *9*, 4647. [CrossRef]

191. Ardehali, R.; Ali, S.R.; Inlay, M.A.; Abilez, O.J.; Chen, M.Q.; Blauwkamp, T.A.; Yazawa, M.; Gong, Y.; Nusse, R.; Drukker, M.; et al. Prospective isolation of human embryonic stem cell-derived cardiovascular progenitors that integrate into human fetal heart tissue. *Proc. Natl. Acad. Sci. USA* **2013**, *110*, 3405–3410. [CrossRef] [PubMed]

192. Skelton, R.J.P.; Costa, M.; Anderson, D.J.; Bruveris, F.; Finnin, B.W.; Koutsis, K.; Arasaratnam, D.; White, A.J.; Rafii, A.; Ng, E.S.; et al. SIRPA, VCAM1 and CD34 identify discrete lineages during early human cardiovascular development. *Stem Cell Res.* **2014**, *13*, 172–179. [CrossRef] [PubMed]

193. Qyang, Y.; Martin-Puig, S.; Chiravuri, M.; Chen, S.; Xu, H.; Bu, L.; Jiang, X.; Lin, L.; Granger, A.; Moretti, A.; et al. The renewal and differentiation of Isl1+ cardiovascular progenitors are controlled by a Wnt/β-catenin pathway. *Cell Stem Cell* **2007**, *1*, 165–179. [CrossRef] [PubMed]

194. Cohen, E.D.; Wang, Z.; Lepore, J.J.; Lu, M.M.; Taketo, M.M.; Epstein, D.J.; Morrisey, E.E. Wnt/β-catenin signaling promotes expansion of Isl-1–positive cardiac progenitor cells through regulation of FGF signaling. *J. Clin. Investig.* **2007**, *117*, 1794–1804. [CrossRef]

195. Kwon, C.; Qian, L.; Cheng, P.; Nigam, V.; Arnold, J.; Srivastava, D. A regulatory pathway involving Notch1/β-catenin/Isl1 determines cardiac progenitor cell fate. *Nat. Cell Biol.* **2009**, *11*, 951–957. [CrossRef]

196. Rosenblatt-Velin, N.; Lepore, M.G.; Cartoni, C.; Beermann, F.; Pedrazzini, T. FGF-2 controls the differentiation of resident cardiac precursors into functional cardiomyocytes. *J. Clin. Investig.* **2005**, *115*, 1724–1733. [CrossRef]

197. Bylund, J.B.; Trinh, L.T.; Awgulewitsch, C.P.; Paik, D.T.; Jetter, C.; Jha, R.; Zhang, J.; Nolan, K.; Xu, C.; Thompson, T.B.; et al. Coordinated proliferation and differentiation of human-induced pluripotent stem cell-derived cardiac progenitor cells depend on bone morphogenetic protein signaling regulation by GREMLIN 2. *Stem Cells Dev.* **2017**, *26*, 678–693. [CrossRef]

198. Ao, A.; Hao, J.; Hopkins, C.R.; Hong, C.C. DMH1, a novel BMP small molecule inhibitor, increases cardiomyocyte progenitors and promotes cardiac differentiation in mouse embryonic stem cells. *PLoS ONE* **2012**, *7*, e41627. [CrossRef]

199. Gomes-Alves, P.; Serra, M.; Brito, C.; Ricardo, C.P.; Cunha, R.; Sousa, M.F.; Sanchez, B.; Bernad, A.; Carrondo, M.J.T.; Rodriguez-Borlado, L.; et al. In vitro expansion of human cardiac progenitor cells: Exploring 'omics tools for characterization of cell-based allogeneic products. *Transl. Res.* **2016**, *171*, 96–110.e3. [CrossRef]

200. Dyer, L.A.; Makadia, F.A.; Scott, A.; Pegram, K.; Hutson, M.R.; Kirby, M.L. BMP signaling modulates hedgehog-induced secondary heart field proliferation. *Dev. Biol.* **2010**, *348*, 167–176. [CrossRef]

201. Gude, N.; Muraski, J.; Rubio, M.; Kajstura, J.; Schaefer, E.; Anversa, P.; Sussman, M.A. Akt promotes increased cardiomyocyte cycling and expansion of the cardiac progenitor cell population. *Circ. Res.* **2006**, *99*, 381–388. [CrossRef] [PubMed]

202. Li, T.-S.; Cheng, K.; Malliaras, K.; Matsushita, N.; Sun, B.; Marbán, L.; Zhang, Y.; Marbán, E. Expansion of human cardiac stem cells in physiological oxygen improves cell production efficiency and potency for myocardial repair. *Cardiovasc. Res.* **2011**, *89*, 157–165. [CrossRef] [PubMed]

203. Mohsin, S.; Khan, M.; Toko, H.; Bailey, B.; Cottage, C.T.; Wallach, K.; Nag, D.; Lee, A.; Siddiqi, S.; Lan, F.; et al. Human cardiac progenitor cells engineered with Pim-I kinase enhance myocardial repair. *J. Am. Coll. Cardiol.* **2012**, *60*, 1278–1287. [CrossRef] [PubMed]

204. Eichmann, A.; Yuan, L.; Bréant, C.; Alitalo, K.; Koskinen, P.J. Developmental expression of Pim kinases suggests functions also outside of the hematopoietic system. *Oncogene* **2000**, *19*, 1215–1224. [CrossRef]

205. Mohsin, S.; Khan, M.; Nguyen, J.; Alkatib, M.; Siddiqi, S.; Hariharan, N.; Wallach, K.; Monsanto, M.; Gude, N.; Dembitsky, W.; et al. Rejuvenation of human cardiac progenitor cells with Pim-1 kinase. *Circ. Res.* **2013**, *113*, 1169–1179. [CrossRef]

206. Samse, K.; Emathinger, J.; Hariharan, N.; Quijada, P.; Ilves, K.; Völkers, M.; Ormachea, L.; De La Torre, A.; Orogo, A.M.; Alvarez, R.; et al. Functional effect of Pim1 depends upon intracellular localization in human cardiac progenitor cells. *J. Biol. Chem.* **2015**, *290*, 13935–13947. [CrossRef]

207. Fischer, K.M.; Cottage, C.T.; Wu, W.; Din, S.; Gude, N.A.; Avitabile, D.; Quijada, P.; Collins, B.L.; Fransioli, J.; Sussman, M.A. Enhancement of myocardial regeneration through genetic engineering of cardiac progenitor cells expressing Pim-1 kinase. *Circulation* **2009**, *120*, 2077–2087. [CrossRef]

208. Liu, N.; Wang, B.J.; Broughton, K.M.; Alvarez, R.; Siddiqi, S.; Loaiza, R.; Nguyen, N.; Quijada, P.; Gude, N.; Sussman, M.A. PIM1-minicircle as a therapeutic treatment for myocardial infarction. *PLoS ONE* **2017**, *12*, e0173963. [CrossRef]

209. Hofsteen, P.; Robitaille, A.M.; Chapman, D.P.; Moon, R.T.; Murry, C.E. Quantitative proteomics identify DAB2 as a cardiac developmental regulator that inhibits WNT/β-catenin signaling. *Proc. Natl. Acad. Sci. USA* **2016**, *113*, 1002–1007. [CrossRef]

210. Hofsteen, P.; Robitaille, A.M.; Strash, N.; Palpant, N.; Moon, R.T.; Pabon, L.; Murry, C.E. ALPK2 promotes cardiogenesis in zebrafish and human pluripotent stem cells. *iScience* **2018**, *2*, 88–100. [CrossRef]

211. Dupays, L.; Towers, N.; Wood, S.; David, A.; Stuckey, D.J.; Mohun, T. Furin, a transcriptional target of NKX2-5, has an essential role in heart development and function. *PLoS ONE* **2019**, *14*, e0212992. [CrossRef] [PubMed]

212. Zhou, Y.; Wang, L.; Vaseghi, H.R.; Liu, Z.; Lu, R.; Alimohamadi, S.; Yin, C.; Fu, J.-D.; Wang, G.G.; Liu, J.; et al. Bmi1 is a key epigenetic barrier to direct cardiac reprogramming. *Cell Stem Cell* **2016**, *18*, 382–395. [CrossRef] [PubMed]

213. Lei, I.; Liu, L.; Sham, M.H.; Wang, Z. SWI/SNF in cardiac progenitor cell differentiation: SWI/SNF in Cardiac Progenitors. *J. Cell. Biochem.* **2013**, *114*, 2437–2445. [CrossRef] [PubMed]

214. Chen, L.; Fulcoli, F.G.; Ferrentino, R.; Martucciello, S.; Illingworth, E.A.; Baldini, A. Transcriptional control in cardiac progenitors: Tbx1 interacts with the BAF chromatin remodeling complex and regulates Wnt5a. *PLoS Genet.* **2012**, *8*, e1002571. [CrossRef] [PubMed]

215. Cohen, E.D.; Miller, M.F.; Wang, Z.; Moon, R.T.; Morrisey, E.E. Wnt5a and Wnt11 are essential for second heart field progenitor development. *Development* **2012**, *139*, 1931–1940. [CrossRef] [PubMed]

216. Muraoka, N.; Yamakawa, H.; Miyamoto, K.; Sadahiro, T.; Umei, T.; Isomi, M.; Nakashima, H.; Akiyama, M.; Wada, R.; Inagawa, K.; et al. MiR-133 promotes cardiac reprogramming by directly repressing Snai1 and silencing fibroblast signatures. *EMBO J.* **2014**, *33*, 1565–1581. [CrossRef]

217. Sluijter, J.P.G.; van Mil, A.; van Vliet, P.; Metz, C.H.G.; Liu, J.; Doevendans, P.A.; Goumans, M.-J. MicroRNA-1 and -499 regulate differentiation and proliferation in human-derived cardiomyocyte progenitor cells. *Arterioscler. Thromb. Vasc. Biol.* **2010**, *30*, 859–868. [CrossRef]

218. Xiao, J.; Liang, D.; Zhang, H.; Liu, Y.; Zhang, D.; Liu, Y.; Pan, L.; Chen, X.; Doevendans, P.A.; Sun, Y.; et al. MicroRNA-204 is required for differentiation of human-derived cardiomyocyte progenitor cells. *J. Mol. Cell. Cardiol.* **2012**, *53*, 751–759. [CrossRef]

219. Sirish, P.; López, J.E.; Li, N.; Wong, A.; Timofeyev, V.; Young, J.N.; Majdi, M.; Li, R.A.; Chen, H.V.; Chiamvimonvat, N. MicroRNA profiling predicts a variance in the proliferative potential of cardiac progenitor cells derived from neonatal and adult murine hearts. *J. Mol. Cell. Cardiol.* **2012**, *52*, 264–272. [CrossRef]

220. Shen, X.; Soibam, B.; Benham, A.; Xu, X.; Chopra, M.; Peng, X.; Yu, W.; Bao, W.; Liang, R.; Azares, A.; et al. miR-322/-503 cluster is expressed in the earliest cardiac progenitor cells and drives cardiomyocyte specification. *Proc. Natl. Acad. Sci. USA* **2016**, *113*, 9551–9556. [CrossRef]

221. Garate, X.; La Greca, A.; Neiman, G.; Blüguermann, C.; Santín Velazque, N.L.; Moro, L.N.; Luzzani, C.; Scassa, M.E.; Sevlever, G.E.; Romorini, L.; et al. Identification of the miRNAome of early mesoderm progenitor cells and cardiomyocytes derived from human pluripotent stem cells. *Sci. Rep.* **2018**, *8*, 8072. [CrossRef] [PubMed]

222. Evseenko, D.; Zhu, Y.; Schenke-Layland, K.; Kuo, J.; Latour, B.; Ge, S.; Scholes, J.; Dravid, G.; Li, X.; MacLellan, W.R.; et al. Mapping the first stages of mesoderm commitment during differentiation of human embryonic stem cells. *Proc. Natl. Acad. Sci. USA* **2010**, *107*, 13742–13747. [CrossRef] [PubMed]

223. Cheng, M.; Yang, J.; Zhao, X.; Zhang, E.; Zeng, Q.; Yu, Y.; Yang, L.; Wu, B.; Yi, G.; Mao, X.; et al. Circulating myocardial microRNAs from infarcted hearts are carried in exosomes and mobilise bone marrow progenitor cells. *Nat. Commun.* **2019**, *10*, 959. [CrossRef] [PubMed]

224. Li, B.; Meng, X.; Zhang, L. microRNAs and cardiac stem cells in heart development and disease. *Drug Discov. Today* **2019**, *24*, 233–240. [CrossRef]

225. Castellan, R.F.P.; Meloni, M. Mechanisms and therapeutic targets of cardiac regeneration: Closing the age gap. *Front. Cardiovasc. Med.* **2018**, *5*, 7. [CrossRef]

226. Carè, A.; Catalucci, D.; Felicetti, F.; Bonci, D.; Addario, A.; Gallo, P.; Bang, M.-L.; Segnalini, P.; Gu, Y.; Dalton, N.D.; et al. MicroRNA-133 controls cardiac hypertrophy. *Nat. Med.* **2007**, *13*, 613–618. [CrossRef]

227. Wang, Y.; Liu, J.; Cui, J.; Sun, M.; Du, W.; Chen, T.; Ming, X.; Zhang, L.; Tian, J.; Li, J.; et al. MiR218 modulates wnt signaling in mouse cardiac stem cells by promoting proliferation and inhibiting differentiation through a positive feedback loop. *Sci. Rep.* **2016**, *6*, 20968. [CrossRef]

228. Chen, Z.-Y.; Chen, F.; Cao, N.; Zhou, Z.-W.; Yang, H.-T. miR-142-3p contributes to early cardiac fate decision of embryonic stem cells. *Stem Cells Int.* **2017**, *2017*, 1–10. [CrossRef]

229. Ivey, K.N.; Muth, A.; Arnold, J.; King, F.W.; Yeh, R.-F.; Fish, J.E.; Hsiao, E.C.; Schwartz, R.J.; Conklin, B.R.; Bernstein, H.S.; et al. MicroRNA regulation of cell lineages in mouse and human embryonic stem cells. *Cell Stem Cell* **2008**, *2*, 219–229. [CrossRef]

230. Purvis, N.; Bahn, A.; Katare, R. The role of microRNAs in cardiac stem cells. *Stem Cells Int.* **2015**, *2015*, 1–10. [CrossRef]

231. Deng, S.; Zhao, Q.; Zhou, X.; Zhang, L.; Bao, L.; Zhen, L.; Zhang, Y.; Fan, H.; Liu, Z.; Yu, Z. Neonatal heart-enriched miR-708 promotes differentiation of cardiac progenitor cells in rats. *Int. J. Mol. Sci.* **2016**, *17*, 875. [CrossRef] [PubMed]

232. Yang, F.; Chen, Q.; He, S.; Yang, M.; Maguire, E.M.; An, W.; Afzal, T.A.; Luong, L.A.; Zhang, L.; Xiao, Q. miR-22 is a novel mediator of vascular smooth muscle cell phenotypic modulation and neointima formation. *Circulation* **2018**, *137*, 1824–1841. [CrossRef] [PubMed]

233. Jin, M.; Wu, Y.; Wang, Y.; Yu, D.; Yang, M.; Yang, F.; Feng, C.; Chen, T. MicroRNA-29a promotes smooth muscle cell differentiation from stem cells by targeting YY1. *Stem Cell Res.* **2016**, *17*, 277–284. [CrossRef] [PubMed]

234. Crippa, S.; Cassano, M.; Messina, G.; Galli, D.; Galvez, B.G.; Curk, T.; Altomare, C.; Ronzoni, F.; Toelen, J.; Gijsbers, R.; et al. miR669a and miR669q prevent skeletal muscle differentiation in postnatal cardiac progenitors. *J. Cell Biol.* **2011**, *193*, 1197–1212. [CrossRef] [PubMed]

235. Limana, F.; Esposito, G.; D'Arcangelo, D.; Di Carlo, A.; Romani, S.; Melillo, G.; Mangoni, A.; Bertolami, C.; Pompilio, G.; Germani, A.; et al. HMGB1 attenuates cardiac remodelling in the failing heart via enhanced cardiac regeneration and miR-206-mediated inhibition of TIMP-3. *PLoS ONE* **2011**, *6*, e19845. [CrossRef] [PubMed]

236. Zhou, Q.; Sun, Q.; Zhang, Y.; Teng, F.; Sun, J. Up-regulation of miRNA-21 expression promotes migration and proliferation of Sca-1+ cardiac stem cells in mice. *Med. Sci. Monit.* **2016**, *22*, 1724–1732. [CrossRef] [PubMed]

237. Hu, S.; Huang, M.; Nguyen, P.K.; Gong, Y.; Li, Z.; Jia, F.; Lan, F.; Liu, J.; Nag, D.; Robbins, R.C.; et al. Novel microRNA prosurvival cocktail for improving engraftment and function of cardiac progenitor cell transplantation. *Circulation* **2011**, *124*, S27–S34. [CrossRef]

238. Liu, J.; van Mil, A.; Vrijsen, K.; Zhao, J.; Gao, L.; Metz, C.H.G.; Goumans, M.-J.; Doevendans, P.A.; Sluijter, J.P.G. MicroRNA-155 prevents necrotic cell death in human cardiomyocyte progenitor cells via targeting RIP1. *J. Cell. Mol. Med.* **2011**, *15*, 1474–1482. [CrossRef]

239. Urbich, C.; Kuehbacher, A.; Dimmeler, S. Role of microRNAs in vascular diseases, inflammation, and angiogenesis. *Cardiovasc. Res.* **2008**, *79*, 581–588. [CrossRef]

240. Li, Y.; Yang, C.-M.; Xi, Y.; Wu, G.; Shelat, H.; Gao, S.; Cheng, J.; Geng, Y.-J. MicroRNA-1/133 targeted dysfunction of potassium channels KCNE1 and KCNQ1 in human cardiac progenitor cells with simulated hyperglycemia. *Int. J. Cardiol.* **2013**, *167*, 1076–1078. [CrossRef]

241. Mauretti, A.; Spaans, S.; Bax, N.A.M.; Sahlgren, C.; Bouten, C.V.C. Cardiac progenitor cells and the interplay with their microenvironment. *Stem Cells Int.* **2017**, *2017*, 1–20. [CrossRef] [PubMed]

242. Gaetani, R.; Rizzitelli, G.; Chimenti, I.; Barile, L.; Forte, E.; Ionta, V.; Angelini, F.; Sluijter, J.P.G.; Barbetta, A.; Messina, E.; et al. Cardiospheres and tissue engineering for myocardial regeneration: Potential for clinical application. *J. Cell. Mol. Med.* **2010**, *14*, 1071–1077. [CrossRef] [PubMed]

243. Vunjak-Novakovic, G.; Tandon, N.; Godier, A.; Maidhof, R.; Marsano, A.; Martens, T.P.; Radisic, M. Challenges in cardiac tissue engineering. *Tissue Eng. Part. B: Rev.* **2010**, *16*, 169–187. [CrossRef] [PubMed]

244. Hwang, N.S.; Varghese, S.; Elisseeff, J. Controlled differentiation of stem cells. *Adv. Drug Deliv. Rev.* **2008**, *60*, 199–214. [CrossRef]

245. Mendelson, K.; Schoen, F.J. Heart valve tissue engineering: Concepts, approaches, progress, and challenges. *Ann. Biomed. Eng.* **2006**, *34*, 1799–1819. [CrossRef]

246. Dawson, E.; Mapili, G.; Erickson, K.; Taqvi, S.; Roy, K. Biomaterials for stem cell differentiation. *Adv. Drug Deliv. Rev.* **2008**, *60*, 215–228. [CrossRef]

247. Bellamy, V.; Vanneaux, V.; Bel, A.; Nemetalla, H.; Emmanuelle Boitard, S.; Farouz, Y.; Joanne, P.; Perier, M.-C.; Robidel, E.; Mandet, C.; et al. Long-term functional benefits of human embryonic stem cell-derived cardiac progenitors embedded into a fibrin scaffold. *J. Heart Lung Transplant.* **2015**, *34*, 1198–1207. [CrossRef]

248. Menasché, P.; Vanneaux, V.; Hagège, A.; Bel, A.; Cholley, B.; Cacciapuoti, I.; Parouchev, A.; Benhamouda, N.; Tachdjian, G.; Tosca, L.; et al. Human embryonic stem cell-derived cardiac progenitors for severe heart failure treatment: First clinical case report: Figure 1. *Eur. Heart J.* **2015**, *36*, 2011–2017. [CrossRef]

249. Vallée, J.-P.; Hauwel, M.; Lepetit-Coiffé, M.; Bei, W.; Montet-Abou, K.; Meda, P.; Gardier, S.; Zammaretti, P.; Kraehenbuehl, T.P.; Herrmann, F.; et al. Embryonic stem cell-based cardiopatches improve cardiac function in infarcted rats. *Stem Cells Transl. Med.* **2012**, *1*, 248–260. [CrossRef]

250. Gaetani, R.; Doevendans, P.A.; Metz, C.H.G.; Alblas, J.; Messina, E.; Giacomello, A.; Sluijter, J.P.G. Cardiac tissue engineering using tissue printing technology and human cardiac progenitor cells. *Biomaterials* **2012**, *33*, 1782–1790. [CrossRef]

251. Gaetani, R.; Feyen, D.A.M.; Verhage, V.; Slaats, R.; Messina, E.; Christman, K.L.; Giacomello, A.; Doevendans, P.A.F.M.; Sluijter, J.P.G. Epicardial application of cardiac progenitor cells in a 3D-printed gelatin/hyaluronic acid patch preserves cardiac function after myocardial infarction. *Biomaterials* **2015**, *61*, 339–348. [CrossRef] [PubMed]

252. Lu, T.-Y.; Lin, B.; Kim, J.; Sullivan, M.; Tobita, K.; Salama, G.; Yang, L. Repopulation of decellularized mouse heart with human induced pluripotent stem cell-derived cardiovascular progenitor cells. *Nat. Commun.* **2013**, *4*, 2307. [CrossRef] [PubMed]

253. Huby, A.-C.; Beigi, F.; Xiang, Q.; Gobin, A.; Taylor, D. Porcine decellularized heart tissue enhance the expression of contractile proteins in human cardiomyocytes and differentiated cardiac progenitor cells. *Circ. Res.* **2016**, *119*, A29.

254. Padin-Iruegas, M.E.; Misao, Y.; Davis, M.E.; Segers, V.F.M.; Esposito, G.; Tokunou, T.; Urbanek, K.; Hosoda, T.; Rota, M.; Anversa, P.; et al. Cardiac progenitor cells and biotinylated insulin-like growth factor-1 nanofibers improve endogenous and exogenous myocardial regeneration after infarction. *Circulation* **2009**, *120*, 876–887. [CrossRef] [PubMed]

255. Tokunaga, M.; Liu, M.-L.; Nagai, T.; Iwanaga, K.; Matsuura, K.; Takahashi, T.; Kanda, M.; Kondo, N.; Wang, P.; Naito, A.T.; et al. Implantation of cardiac progenitor cells using self-assembling peptide improves cardiac function after myocardial infarction. *J. Mol. Cell. Cardiol.* **2010**, *49*, 972–983. [CrossRef] [PubMed]

256. Li, Z.; Guo, X.; Matsushita, S.; Guan, J. Differentiation of cardiosphere-derived cells into a mature cardiac lineage using biodegradable poly(*N*-isopropylacrylamide) hydrogels. *Biomaterials* **2011**, *32*, 3220–3232. [CrossRef] [PubMed]

257. Liu, Q.; Tian, S.; Zhao, C.; Chen, X.; Lei, I.; Wang, Z.; Ma, P.X. Porous nanofibrous poly(l-lactic acid) scaffolds supporting cardiovascular progenitor cells for cardiac tissue engineering. *Acta Biomater.* **2015**, *26*, 105–114. [CrossRef]

258. Ciocci, M.; Mochi, F.; Carotenuto, F.; Di Giovanni, E.; Prosposito, P.; Francini, R.; De Matteis, F.; Reshetov, I.; Casalboni, M.; Melino, S.; et al. Scaffold-in-scaffold potential to induce growth and differentiation of cardiac progenitor cells. *Stem Cells Dev.* **2017**, *26*, 1438–1447. [CrossRef]

259. Johnson, T.D.; DeQuach, J.A.; Gaetani, R.; Ungerleider, J.; Elhag, D.; Nigam, V.; Behfar, A.; Christman, K.L. Human versus porcine tissue sourcing for an injectable myocardial matrix hydrogel. *Biomater. Sci.* **2014**, *2*, 735–744. [CrossRef]

260. van Marion, M.H.; Bax, N.A.M.; van Turnhout, M.C.; Mauretti, A.; van der Schaft, D.W.J.; Goumans, M.J.T.H.; Bouten, C.V.C. Behavior of CMPCs in unidirectional constrained and stress-free 3D hydrogels. *J. Mol. Cell. Cardiol.* **2015**, *87*, 79–91. [CrossRef]

261. Gaetani, R.; Yin, C.; Srikumar, N.; Braden, R.; Doevendans, P.A.; Sluijter, J.P.G.; Christman, K.L. Cardiac-derived extracellular matrix enhances cardiogenic properties of human cardiac progenitor cells. *Cell Transplant.* **2016**, *25*, 1653–1663. [CrossRef] [PubMed]

262. French, K.M.; Boopathy, A.V.; DeQuach, J.A.; Chingozha, L.; Lu, H.; Christman, K.L.; Davis, M.E. A naturally derived cardiac extracellular matrix enhances cardiac progenitor cell behavior in vitro. *Acta Biomater.* **2012**, *8*, 4357–4364. [CrossRef] [PubMed]

263. Ng, S.L.J.; Narayanan, K.; Gao, S.; Wan, A.C.A. Lineage restricted progenitors for the repopulation of decellularized heart. *Biomaterials* **2011**, *32*, 7571–7580. [CrossRef] [PubMed]

264. Rajabi, S.; Pahlavan, S.; Ashtiani, M.K.; Ansari, H.; Abbasalizadeh, S.; Sayahpour, F.A.; Varzideh, F.; Kostin, S.; Aghdami, N.; Braun, T.; et al. Human embryonic stem cell-derived cardiovascular progenitor cells efficiently colonize in bFGF-tethered natural matrix to construct contracting humanized rat hearts. *Biomaterials* **2018**, *154*, 99–112. [CrossRef]

265. Sánchez, P.L.; Fernández-Santos, M.E.; Costanza, S.; Climent, A.M.; Moscoso, I.; Gonzalez-Nicolas, M.A.; Sanz-Ruiz, R.; Rodríguez, H.; Kren, S.M.; Garrido, G.; et al. Acellular human heart matrix: A critical step toward whole heart grafts. *Biomaterials* **2015**, *61*, 279–289. [CrossRef]

266. Bejleri, D.; Streeter, B.W.; Nachlas, A.L.Y.; Brown, M.E.; Gaetani, R.; Christman, K.L.; Davis, M.E. A bioprinted cardiac patch composed of cardiac-specific extracellular matrix and progenitor cells for heart repair. *Adv. Healthc. Mater.* **2018**, *7*, 1800672. [CrossRef]

267. Silva, A.C.; Rodrigues, S.C.; Caldeira, J.; Nunes, A.M.; Sampaio-Pinto, V.; Resende, T.P.; Oliveira, M.J.; Barbosa, M.A.; Thorsteinsdóttir, S.; Nascimento, D.S.; et al. Three-dimensional scaffolds of fetal decellularized hearts exhibit enhanced potential to support cardiac cells in comparison to the adult. *Biomaterials* **2016**, *104*, 52–64. [CrossRef]

268. Chamberland, C.; Martinez-Fernandez, A.; Beraldi, R.; Nelson, T.J. Embryonic decellularized cardiac scaffold supports embryonic stem cell differentiation to produce beating cardiac tissue. *ISRN Stem Cells* **2014**, *2014*, 1–10. [CrossRef]

269. Rajabi-Zeleti, S.; Jalili-Firoozinezhad, S.; Azarnia, M.; Khayyatan, F.; Vahdat, S.; Nikeghbalian, S.; Khademhosseini, A.; Baharvand, H.; Aghdami, N. The behavior of cardiac progenitor cells on macroporous pericardium-derived scaffolds. *Biomaterials* **2014**, *35*, 970–982. [CrossRef]

270. Chimenti, I.; Rizzitelli, G.; Gaetani, R.; Angelini, F.; Ionta, V.; Forte, E.; Frati, G.; Schussler, O.; Barbetta, A.; Messina, E.; et al. Human cardiosphere-seeded gelatin and collagen scaffolds as cardiogenic engineered bioconstructs. *Biomaterials* **2011**, *32*, 9271–9281. [CrossRef]

271. Takehara, N.; Tsutsumi, Y.; Tateishi, K.; Ogata, T.; Tanaka, H.; Ueyama, T.; Takahashi, T.; Takamatsu, T.; Fukushima, M.; Komeda, M.; et al. Controlled delivery of basic fibroblast growth factor promotes human cardiosphere-derived cell engraftment to enhance cardiac repair for chronic myocardial infarction. *J. Am. Coll. Cardiol.* **2008**, *52*, 1858–1865. [CrossRef] [PubMed]

272. Zhang, W.; Li, X.; Sun, S.; Zhang, X. Implantation of engineered conduction tissue in the rat heart. *Mol. Med. Rep.* **2019**, *19*, 2687–2697. [CrossRef] [PubMed]

273. Wang, J.; Cui, C.; Nan, H.; Yu, Y.; Xiao, Y.; Poon, E.; Yang, G.; Wang, X.; Wang, C.; Li, L.; et al. Graphene sheet-induced global maturation of cardiomyocytes derived from human induced pluripotent stem cells. *ACS Appl. Mater. Interfaces* **2017**, *9*, 25929–25940. [CrossRef] [PubMed]

274. Savchenko, A.; Cherkas, V.; Liu, C.; Braun, G.B.; Kleschevnikov, A.; Miller, Y.I.; Molokanova, E. Graphene biointerfaces for optical stimulation of cells. *Sci. Adv.* **2018**, *4*, eaat0351. [CrossRef]

275. Feiner, R.; Engel, L.; Fleischer, S.; Malki, M.; Gal, I.; Shapira, A.; Shacham-Diamand, Y.; Dvir, T. Engineered hybrid cardiac patches with multifunctional electronics for online monitoring and regulation of tissue function. *Nat. Mater.* **2016**, *15*, 679–685. [CrossRef]

276. Li, J.; Minami, I.; Shiozaki, M.; Yu, L.; Yajima, S.; Miyagawa, S.; Shiba, Y.; Morone, N.; Fukushima, S.; Yoshioka, M.; et al. Human pluripotent stem cell-derived cardiac tissue-like constructs for repairing the infarcted myocardium. *Stem Cell Rep.* **2017**, *9*, 1546–1559. [CrossRef]

277. Nunes, S.S.; Miklas, J.W.; Liu, J.; Aschar-Sobbi, R.; Xiao, Y.; Zhang, B.; Jiang, J.; Massé, S.; Gagliardi, M.; Hsieh, A.; et al. Biowire: A platform for maturation of human pluripotent stem cell–derived cardiomyocytes. *Nat. Methods* **2013**, *10*, 781–787. [CrossRef]

278. Asahi, Y.; Hamada, T.; Hattori, A.; Matsuura, K.; Odaka, M.; Nomura, F.; Kaneko, T.; Abe, Y.; Takasuna, K.; Sanbuissho, A.; et al. On-chip spatiotemporal electrophysiological analysis of human stem cell derived cardiomyocytes enables quantitative assessment of proarrhythmia in drug development. *Sci. Rep.* **2018**, *8*, 14536. [CrossRef]

279. Qian, F.; Huang, C.; Lin, Y.-D.; Ivanovskaya, A.N.; O'Hara, T.J.; Booth, R.H.; Creek, C.J.; Enright, H.A.; Soscia, D.A.; Belle, A.M.; et al. Simultaneous electrical recording of cardiac electrophysiology and contraction on chip. *Lab Chip* **2017**, *17*, 1732–1739. [CrossRef]

280. Banerjee, M.N.; Bolli, R.; Hare, J.M. Clinical studies of cell therapy in cardiovascular medicine: Recent developments and future directions. *Circ. Res.* **2018**, *123*, 266–287. [CrossRef]

281. The Lancet Editors. Expression of concern: The SCIPIO trial. *Lancet* **2014**, *383*, 1279. [CrossRef]

282. The Lancet Editors. Retraction—Cardiac stem cells in patients with ischaemic cardiomyopathy (SCIPIO): Initial results of a randomised phase 1 trial. *Lancet* **2019**, *393*, 1084. [CrossRef]

283. Makkar, R.R.; Smith, R.R.; Cheng, K.; Malliaras, K.; Thomson, L.E.; Berman, D.; Czer, L.S.; Marbán, L.; Mendizabal, A.; Johnston, P.V.; et al. Intracoronary cardiosphere-derived cells for heart regeneration after myocardial infarction (CADUCEUS): A prospective, randomised phase 1 trial. *Lancet* **2012**, *379*, 895–904. [CrossRef]

284. Yacoub, M.H.; Terrovitis, J. CADUCEUS, SCIPIO, ALCADIA: Cell therapy trials using cardiac-derived cells for patients with post myocardial infarction LV dysfunction, still evolving. *Glob. Cardiol. Sci. Pract.* **2013**, *2013*, 3. [CrossRef] [PubMed]

285. Takehara, N.; Ogata, T.; Nakata, M.; Kami, D.; Nakamura, T.; Matoba, S.; Gojo, S.; Sawada, T.; Yaku, H.; Matsubara, H. The alcadia (autologous human cardiac-derived stem cell to treat ischemic cardiomyopathy) trial. *Circulation* **2012**, *126*, 2776–2799.

286. Malliaras, K.; Zhang, Y.; Seinfeld, J.; Galang, G.; Tseliou, E.; Cheng, K.; Sun, B.; Aminzadeh, M.; Marbán, E. Cardiomyocyte proliferation and progenitor cell recruitment underlie therapeutic regeneration after myocardial infarction in the adult mouse heart. *EMBO Mol. Med.* **2013**, *5*, 191–209. [CrossRef]

287. Menasché, P.; Vanneaux, V.; Hagège, A.; Bel, A.; Cholley, B.; Parouchev, A.; Cacciapuoti, I.; Al-Daccak, R.; Benhamouda, N.; Blons, H.; et al. Transplantation of human embryonic stem cell–derived cardiovascular progenitors for severe ischemic left ventricular dysfunction. *J. Am. Coll. Cardiol.* **2018**, *71*, 429–438. [CrossRef]

288. Ishigami, S.; Ohtsuki, S.; Tarui, S.; Ousaka, D.; Eitoku, T.; Kondo, M.; Okuyama, M.; Kobayashi, J.; Baba, K.; Arai, S.; et al. Intracoronary autologous cardiac progenitor cell transfer in patients with hypoplastic left heart syndrome: The TICAP prospective phase 1 controlled trial. *Circ. Res.* **2015**, *116*, 653–664. [CrossRef]

289. Ishigami, S.; Ohtsuki, S.; Eitoku, T.; Ousaka, D.; Kondo, M.; Kurita, Y.; Hirai, K.; Fukushima, Y.; Baba, K.; Goto, T.; et al. Intracoronary cardiac progenitor cells in single ventricle physiology: The PERSEUS (cardiac progenitor cell infusion to treat univentricular heart disease) randomized phase 2 trial. *Circ. Res.* **2017**, *120*, 1162–1173. [CrossRef]

290. Tarui, S.; Ishigami, S.; Ousaka, D.; Kasahara, S.; Ohtsuki, S.; Sano, S.; Oh, H. Transcoronary infusion of cardiac progenitor cells in hypoplastic left heart syndrome: Three-year follow-up of the transcoronary infusion of cardiac progenitor cells in patients with single-ventricle physiology (TICAP) trial. *J. Thorac. Cardiovasc. Surg.* **2015**, *150*, 1198–1208. [CrossRef]

291. Cardiac Stem/Progenitor Cell Infusion in Univentricular Physiology (APOLLON Trial). Available online: https://clinicaltrials.gov/ct2/show/NCT02781922 (accessed on 9 October 2019).

292. Transcoronary Infusion of Cardiac Progenitor Cells in Pediatric Dilated Cardiomyopathy. Available online: https://clinicaltrials.gov/ct2/show/NCT03129568 (accessed on 9 October 2019).

293. Malliaras, K.; Makkar, R.R.; Smith, R.R.; Cheng, K.; Wu, E.; Bonow, R.O.; Marbán, L.; Mendizabal, A.; Cingolani, E.; Johnston, P.V.; et al. Intracoronary cardiosphere-derived cells after myocardial infarction. *J. Am. Coll. Cardiol.* **2014**, *63*, 110–122. [CrossRef] [PubMed]

294. Allogeneic Heart Stem Cells to Achieve Myocardial Regeneration. Available online: https://clinicaltrials.gov/ct2/show/NCT01458405 (accessed on 9 October 2019).

295. Sanz-Ruiz, R.; Casado Plasencia, A.; Borlado, L.R.; Fernández-Santos, M.E.; Al-Daccak, R.; Claus, P.; Palacios, I.; Sádaba, R.; Charron, D.; Bogaert, J.; et al. Rationale and design of a clinical trial to evaluate the

safety and efficacy of intracoronary infusion of allogeneic human cardiac stem cells in patients with acute myocardial infarction and left ventricular dysfunction: The randomized multicenter double-blind controlled CAREMI trial (cardiac stem cells in patients with acute myocardial infarction). *Circ. Res.* **2017**, *121*, 71–80. [PubMed]

296. Dilated CardiomYopathy iNtervention with Allogeneic MyocardIally-Regenerative Cells (DYNAMIC). Available online: https://clinicaltrials.gov/ct2/show/NCT02293603 (accessed on 9 October 2019).

297. Bolli, R.; Hare, J.M.; March, K.L.; Pepine, C.J.; Willerson, J.T.; Perin, E.C.; Yang, P.C.; Henry, T.D.; Traverse, J.H.; Mitrani, R.D.; et al. Rationale and design of the CONCERT-HF trial (combination of mesenchymal and c-kit $^+$ cardiac stem cells as regenerative therapy for heart failure). *Circ. Res.* **2018**, *122*, 1703–1715. [CrossRef] [PubMed]

298. Regression of Fibrosis & Reversal of Diastolic Dysfunction in HFPEF Patients Treated with Allogeneic CDCs. Available online: https://clinicaltrials.gov/ct2/show/NCT02941705 (accessed on 9 October 2019).

299. Sahara, M.; Santoro, F.; Chien, K.R. Programming and reprogramming a human heart cell. *EMBO J.* **2015**, *34*, 710–738. [CrossRef]

300. Amini, H.; Rezaie, J.; Vosoughi, A.; Rahbarghazi, R.; Nouri, M. Cardiac progenitor cells application in cardiovascular disease. *J. Cardiovasc. Thorac. Res.* **2017**, *9*, 127–132. [CrossRef]

301. Sanchez-Freire, V.; Lee, A.S.; Hu, S.; Abilez, O.J.; Liang, P.; Lan, F.; Huber, B.C.; Ong, S.-G.; Hong, W.X.; Huang, M.; et al. Effect of human donor cell source on differentiation and function of cardiac induced pluripotent stem cells. *J. Am. Coll. Cardiol.* **2014**, *64*, 436–448. [CrossRef]

302. Martens, T.P.; Godier, A.F.G.; Parks, J.J.; Wan, L.Q.; Koeckert, M.S.; Eng, G.M.; Hudson, B.I.; Sherman, W.; Vunjak-Novakovic, G. Percutaneous cell delivery into the heart using hydrogels polymerizing in situ. *Cell Transplant.* **2009**, *18*, 297–304. [CrossRef]

303. Beeres, S.L.M.A.; Atsma, D.E.; van Ramshorst, J.; Schalij, M.J.; Bax, J.J. Cell therapy for ischaemic heart disease. *Heart* **2008**, *94*, 1214–1226. [CrossRef]

304. Liu, Q.; Yang, R.; Huang, X.; Zhang, H.; He, L.; Zhang, L.; Tian, X.; Nie, Y.; Hu, S.; Yan, Y.; et al. Genetic lineage tracing identifies in situ Kit-expressing cardiomyocytes. *Cell Res.* **2016**, *26*, 119–130. [CrossRef]

305. He, L.; Li, Y.; Li, Y.; Pu, W.; Huang, X.; Tian, X.; Wang, Y.; Zhang, H.; Liu, Q.; Zhang, L.; et al. Enhancing the precision of genetic lineage tracing using dual recombinases. *Nat. Med.* **2017**, *23*, 1488–1498. [CrossRef]

306. Li, Y.; He, L.; Huang, X.; Bhaloo, S.I.; Zhao, H.; Zhang, S.; Pu, W.; Tian, X.; Li, Y.; Liu, Q.; et al. Genetic lineage tracing of nonmyocyte population by dual recombinases. *Circulation* **2018**, *138*, 793–805. [CrossRef] [PubMed]

307. Marino, F.; Scalise, M.; Cianflone, E.; Mancuso, T.; Aquila, I.; Agosti, V.; Torella, M.; Paolino, D.; Mollace, V.; Nadal-Ginard, B.; et al. Role of c-Kit in myocardial regeneration and aging. *Front. Endocrinol.* **2019**, *10*, 371. [CrossRef] [PubMed]

308. Cai, C.-L.; Molkentin, J.D. The elusive progenitor cell in cardiac regeneration: Slip slidin' away. *Circ. Res.* **2017**, *120*, 400–406. [CrossRef] [PubMed]

309. Eschenhagen, T.; Bolli, R.; Braun, T.; Field, L.J.; Fleischmann, B.K.; Frisén, J.; Giacca, M.; Hare, J.M.; Houser, S.; Lee, R.T.; et al. Cardiomyocyte regeneration: A consensus statement. *Circulation* **2017**, *136*, 680–686. [CrossRef] [PubMed]

310. Marks, P.W.; Witten, C.M.; Califf, R.M. Clarifying stem-cell therapy's benefits and risks. *N. Engl. J. Med.* **2017**, *376*, 1007–1009. [CrossRef] [PubMed]

311. Maliken, B.D.; Molkentin, J.D. Undeniable evidence that the adult mammalian heart lacks an endogenous regenerative stem cell. *Circulation* **2018**, *138*, 806–808. [CrossRef]

312. Writing Group Members; Roger, V.L.; Go, A.S.; Lloyd-Jones, D.M.; Benjamin, E.J.; Berry, J.D.; Borden, W.B.; Bravata, D.M.; Dai, S.; Ford, E.S.; et al. Heart disease and stroke statistics—2012 Update: A report from the American Heart Association. *Circulation* **2012**, *125*, e2–e220.

313. Cesselli, D.; Beltrami, A.P.; D'Aurizio, F.; Marcon, P.; Bergamin, N.; Toffoletto, B.; Pandolfi, M.; Puppato, E.; Marino, L.; Signore, S.; et al. Effects of age and heart failure on human cardiac stem cell function. *Am. J. Pathol.* **2011**, *179*, 349–366. [CrossRef]

314. Yao, Y.-G.; Ellison, F.M.; McCoy, J.P.; Chen, J.; Young, N.S. Age-dependent accumulation of mtDNA mutations in murine hematopoietic stem cells is modulated by the nuclear genetic background. *Hum. Mol. Genet.* **2007**, *16*, 286–294. [CrossRef]

315. Mohsin, S.; Siddiqi, S.; Collins, B.; Sussman, M.A. Empowering adult stem cells for myocardial regeneration. *Circ. Res.* **2011**, *109*, 1415–1428. [CrossRef]
316. Frati, C.; Savi, M.; Graiani, G.; Lagrasta, C.; Cavalli, S.; Prezioso, L.; Rossetti, P.; Mangiaracina, C.; Ferraro, F.; Madeddu, D.; et al. Resident cardiac stem cells. *Curr. Pharm. Des.* **2011**, *17*, 3252–3257. [PubMed]
317. Leonardini, A.; Avogaro, A. Abnormalities of the cardiac stem and progenitor cell compartment in experimental and human diabetes. *Arch. Physiol. Biochem.* **2013**, *119*, 179–187. [CrossRef] [PubMed]
318. Kurazumi, H.; Kubo, M.; Ohshima, M.; Yamamoto, Y.; Takemoto, Y.; Suzuki, R.; Ikenaga, S.; Mikamo, A.; Udo, K.; Hamano, K.; et al. The effects of mechanical stress on the growth, differentiation, and paracrine factor production of cardiac stem cells. *PLoS ONE* **2011**, *6*, e28890. [CrossRef] [PubMed]
319. Torella, D.; Rota, M.; Nurzynska, D.; Musso, E.; Monsen, A.; Shiraishi, I.; Zias, E.; Walsh, K.; Rosenzweig, A.; Sussman, M.A.; et al. Cardiac stem cell and myocyte aging, heart failure, and insulin-like growth factor-1 overexpression. *Circ. Res.* **2004**, *94*, 514–524. [CrossRef]
320. Anversa, P.; Rota, M.; Urbanek, K.; Hosoda, T.; Sonnenblick, E.H.; Leri, A.; Kajstura, J.; Bolli, R. Myocardial aging: A stem cell problem. *Basic Res. Cardiol.* **2005**, *100*, 482–493. [CrossRef]
321. Urbanek, K.; Quaini, F.; Tasca, G.; Torella, D.; Castaldo, C.; Nadal-Ginard, B.; Leri, A.; Kajstura, J.; Quaini, E.; Anversa, P. Intense myocyte formation from cardiac stem cells in human cardiac hypertrophy. *Proc. Natl. Acad. Sci. USA* **2003**, *100*, 10440–10445. [CrossRef]

Article

The Diagnostic Value of Mir-133a in ST Elevation and Non-ST Elevation Myocardial Infarction: A Meta-Analysis

Yehuda Wexler [1] and Udi Nussinovitch [2,*]

[1]   Rappaport Faculty of Medicine and Research Institute, Technion - Israel Institute of Technology, POB 9649, Haifa 3109601, Israel; yehuda.wexler@gmail.com

[2]   Applicative Cardiovascular Research Center (ACRC) and Department of Cardiology, Meir Medical Center, Kfar Saba 44281, Israel

*   Correspondence: udi.nussinovitch@gmail.com; Tel.: +972-53-526-8535

Received: 2 March 2020; Accepted: 23 March 2020; Published: 25 March 2020

Abstract: Numerous studies have reported correlations between plasma microRNA signatures and cardiovascular disease. MicroRNA-133a (Mir-133a) has been researched extensively for its diagnostic value in acute myocardial infarction (AMI). While initial results seemed promising, more recent studies cast doubt on the diagnostic utility of Mir-133a, calling its clinical prospects into question. Here, the diagnostic potential of Mir-133a was analyzed using data from multiple papers. Medline, Embase, and Web of Science were systematically searched for publications containing "Cardiovascular Disease", "MicroRNA", "Mir-133a" and their synonyms. Diagnostic performance was assessed using area under the summary receiver operator characteristic curve (AUC), while examining the impact of age, sex, final diagnosis, and time. Of the 753 identified publications, 9 were included in the quantitative analysis. The pooled AUC for Mir-133a was 0.73. Analyses performed separately on studies using healthy vs. symptomatic controls yielded pooled AUCs of 0.89 and 0.68, respectively. Age and sex were not found to significantly affect diagnostic performance. Our findings indicate that control characteristics and methodological inconsistencies are likely the causes of incongruent reports, and that Mir-133a may have limited use in distinguishing symptomatic patients from those suffering AMI. Lastly, we hypothesized that Mir-133a may find a new use as a risk stratification biomarker in patients with specific subsets of non-ST elevation myocardial infarction (NSTEMI).

Keywords: myocardial infarction; MicroRNA; Mir-133; coronary heart disease; biomarker; meta-analysis

---

1. Introduction

Cardiovascular disease (CVD) is the leading cause of death in the United States [1] and accounts for nearly \$219 billion in spending annually and 647,000 deaths [2]. Coronary artery disease (CAD) is the most prevalent form of CVD with upwards of 365,000 American mortalities each year, primarily as a result of acute myocardial infarction (AMI) [2]. With over 800,000 Americans suffering AMIs annually [3], early detection is crucial to improving clinical outcomes and decreasing mortality.

Currently used circulating biomarkers such as cardiac troponins and creatine kinase MB act as sensitive and specific tests for myocardial damage, yet, they may be negative early in the process of ischemia. Their increase in the setting of ST-elevation MI (STEMI), a process that nearly always results from coronary plaque rupture and thrombosis formation, is usually reflective of the extent of the infarct and approximates the mass of cardiomyocytes that damaged in the process of AMI. In the setting of non-ST elevation myocardial infarction (NSTEMI), increases in different biomarkers may be suggestive of a specific underlying pathophysiology, although data is limited on such associations.

For instance, it was suggested that extreme levels of cardiac troponins are suggestive of total occlusion (TO) of the culprit artery [4], but the results were not used to assess correlations with other entities of myocardial injury, and data on such a possible association is limited. At present, biomarkers are not used to differentiate specific coronary pathologies, or assess the extent of vascular occlusion, nor are they used to detect non-CAD related myocardial cell damage [5].

Early reperfusion, usually through percutaneous coronary intervention (PCI) is a primary factor in the prognosis and clinical outcome of AMI [6,7]. Although electrocardiographic signs of the ST segment elevation are a sensitive and specific sign of coronary TO in the setting of STEMI, approximately only 25.5–34% of NSTEMI patients were found to have TO [8,9]. Patients suffering from TO are commonly underdiagnosed, receive delayed intervention, and have increased rates of complications and mortality [8]. Additionally, it may be challenging to distinguish between NSTEMI resulting from coronary atherosclerosis, and other processes of myocardial damage associated with inflammation, microvascular damage, toxic injury, vasoconstriction, etc.

Thus, highly sensitive and specific circulating biomarkers capable of diagnosing AMI (and specifically patients with TO) shortly after symptoms begin are of great clinical importance and may reduce mortality as well as improve patient outcomes.

In recent years multiple circulating micro-RNAs (MiRNAs) have been identified and investigated for their possible diagnostic and prognostic utility in CVD [10–14]. Specifically, Micro-RNA 133a (Mir-133a) has been reported as a potentially powerful biomarker for AMI and CVD. Mir-133a is a short non-coding RNA molecule, which serves to regulate target genes through post-transcriptional suppression. It has been found that Mir-133a is critical for proper cardiac development, playing an important role in early differentiation and cardiogenesis, as well as mediating various cardiac processes including apoptosis, cardiac remodeling, hypertrophy, conductance, and automaticity [15]. Increased serum levels of Mir-133a have been observed in the setting of AMI and CVD. This is most likely the result of damaged myocardium releasing Mir-133a during cellular lysis, or adjacent border zone myocardium releasing Mir-133a containing vesicles in response to the cardiac insult [16]. The research on Mir-133a's diagnostic potential, however, is strongly conflicted, with some papers reporting weak correlations between circulating Mir-133a concentrations and AMI [10,17], and others reporting strong correlations with excellent sensitivity and specificity [11,18–22]. In light of these contradictory findings, this meta-analysis synthesizes data from existing literature in order to examine the true potential of Mir-133a as a biomarker in AMI. Additionally, we analyzed the time frames in which Mir-133a was quantified, and their effect on the increases in plasma concentration, to ascertain whether Mir-133a may be useful as a very early diagnostic marker. Lastly, we compared the data of STEMI patients with those of NSTEMI patients to determine if Mir-133a might be used to distinguish between these two types of AMI. We hypothesized that TO in the setting of NSTEMI, and other specific entities of myocardial injury, may be characterized by distinct Mir-133a increase patterns, and aimed to evaluate the literature in that regard. Notably, correct identification of high risk NSTEMIs has a critical impact on the course of treatment [8], and therefore Mir-133a may serve as a valuable biomarker in this respect.

2. Materials and Methods

2.1. Search Strategy

Three electronic databases (Pubmed, Embase, and Web of Science) were searched for articles written prior to December 1st, 2019 that included the terms microRNA, microRNA-133, and cardiovascular disease, as well as common synonyms for these terms. The complete search strategy for all databases can be found in the Supplementary Materials (Tables S1–S3) in accordance with PRISMA guidelines [23].

2.2. Inclusion and Exclusion Criteria

All papers retrieved in the literature search were subjected to the following criteria for inclusion:

1. STEMI or NSTEMI was the clinical diagnosis in study patients.
2. The study was either case-controlled or a cohort.
3. Mir-133a was quantified from plasma using qRT-PCR with either SYBR or TaqMan probes.
4. Sample size, area under the standard receiver operator characteristic curve (AUC), location of study, and maximum plasma sample collection time must be stated.
5. A sample size of 5 or more patients was required for each subgroup.

The following criteria were used for exclusion:

1. Papers written in languages other than English.
2. Reviews, meta-analyses, posters, and correspondence letters.
3. Experimental design based solely on animal models.

Studies meeting all the inclusion criteria and none of the exclusion criteria were used for the quantitative analysis. If two or more included papers were based on the same clinical data, only the most relevant study was included. Screening was performed in accordance with PRISMA guidelines [23]. The completed checklist can be found in the Supplementary Material (Table S4).

2.3. Data Extraction

All data used in this meta-analysis were extracted from the published versions of the included papers, their supplementary materials, and their referenced sources. No unpublished data was acquired for this analysis.

In papers where increases of Mir-133a were presented as a change in cycle threshold (ΔCT) compared to a predetermined reference, or as $\Delta\Delta$CT (ΔCT test sample–ΔCT calibrator sample), fold change was calculated using $2^{(-\Delta CT)}$ or $2^{(-\Delta\Delta CT)}$. Additionally, when fold changes were presented only for subgroups (for example STEMI and NSTEMI), the total fold change was calculated using a weighted average based on the number of patients in each subgroup.

When the demographic characteristics of subgroups were not specified in the papers (i.e., mean age, gender, etc.), it was assumed that they followed the same distribution as the larger group whose characteristics were listed.

In several papers [10,24] quantitative data was presented graphically without exact numbers being published. In these cases, the graphs and figures were digitized using GetData Graph Digitizer software Ver. 2.26 in a blinded manner, and the averaged numeric values were used in this analysis.

2.4. Statistical Analysis

The Kruskal–Wallis H test was used to compare findings between the groups. Relations between the dependent variables (fold change and AUC) and independent variables (percentage of STEMI patients (%STEMI), time from onset, and patients' age) were evaluated using linear regression analysis. Correlation analyses were estimated according to the strength and direction of a linear relationship between the two variables on a scatterplot (i.e., r). The number of patients was used as a frequency weighted variable. A p-value of <0.01 was considered statistically significant. Pooled results were expressed as mean and standard deviation for AUC and as mean and standard error of mean (SEM) for fold changes. Analyses were performed using JMP version 7.0 (SAS Institute, Cary, NC, USA) and MedCalc version 19.1.5 (MedCalc Software Ltd, Belgium). Forest plots were generated using DistillerSR Forest Plot Generator (Evidence Partners, Ottawa, ON, Canada).

3. Results

3.1. Literature Search Results

The literature selection process is shown in Figure 1. In short, the search returned 1071 results. After removal of duplicates 753 papers remained that were then assessed manually using titles and abstracts for: reviews, posters, meta-analyses, animal studies, and non-AMI related papers, which were all excluded.

We further screened the remaining 55 papers for relevance using their full texts, excluding all papers not written in English, papers, which did not publish their statistical data, and studies in which non-plasma samples were used for the quantification of Mir-133a.

The remaining 23 eligible studies [10–12,16–22,24–36] were thoroughly analyzed and subjected to the above inclusion criteria ultimately yielding 9 studies (Table 1) involving 2280 participants, with 943 AMI patients and 1337 controls that were included in the quantitative meta-analysis.

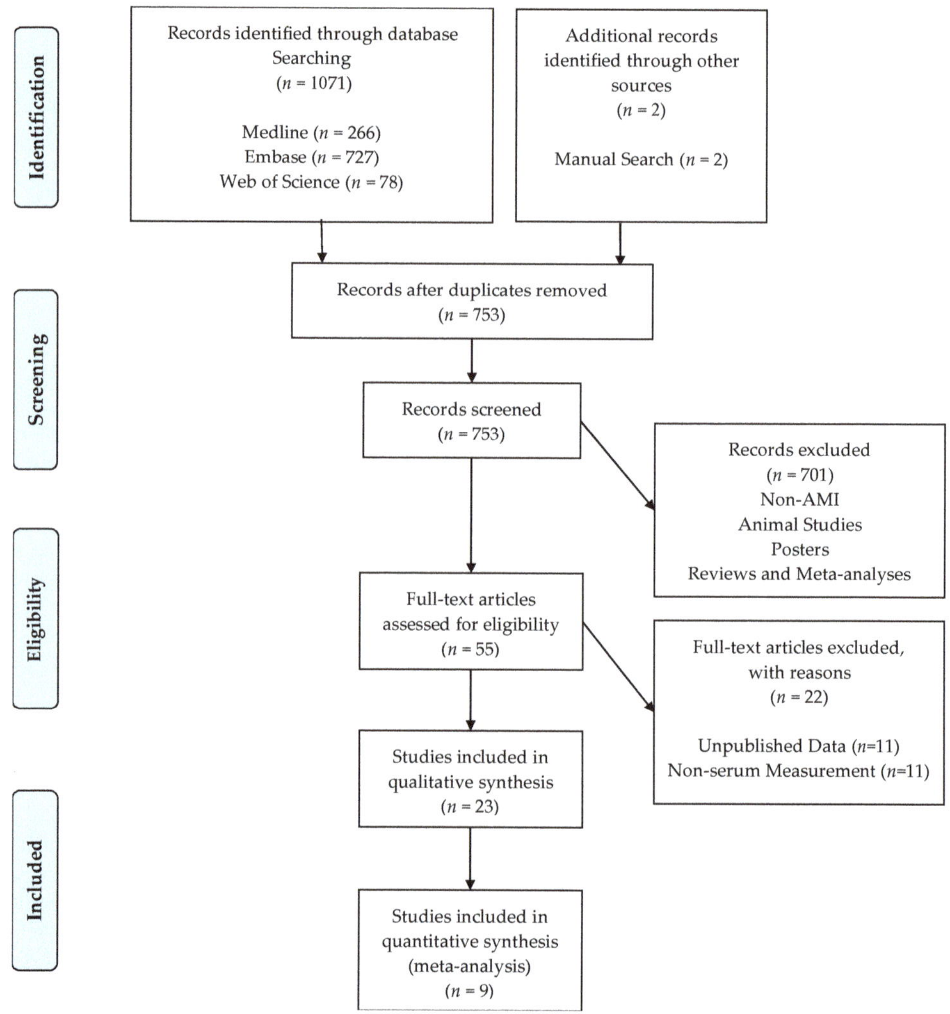

Figure 1. PRISMA flow chart of the study selection.

Table 1. Data summary from included papers.

Study (Author, Year, Reference)	Country	Number of Patients (Case/Control)	Patient Characteristics (Case)	Mean Age (Case)	Patient Characteristics (Control)	AUC	Mir Fold Increase (Total/STEMI/NSTEMI)	Max Time from Onset Until Sample Acquisition
Gidlof, O. et al. 2011 [21]	Sweden	9/11	STEMI Patients Undergoing PPCI	64.56 ± 2.7	STEMI/Healthy	0.859	70	12 h
Wang, G-K. et al. 2010 [22]	China	33/33	STEMI and NSTEMI	63.5 ± 10.1	AMI/Non-AMI ACS	0.867	——	12 h
Li, Y. et al. 2013 [19]	China	67/32	STEMI ($n = 44$) and NSTEMI ($n = 23$)	63.84 ± 11.17	AMI/Healthy	0.947	5.67	12 h
Devaux, Y. et al. 2015 [10]	Czechia, Italy, Poland, Spain, Switzerland	224/931	STEMI ($n = 45$) and NSTEMI ($n = 179$)	72	AMI/Non-AMI ACS	0.53	——	12 h
Wang, R. et al. 2011 [11]	China	58/21	STEMI and NSTEMI	60.06 ± 11.53	AMI/non-AMI ACS	0.89	4.4	24 h
Peng, L. et al. 2014 [18]	China	76/110	STEMI ($n = 25$) and NSTEMI ($n = 51$)	64.6	AMI/non-AMI ACS	0.912	7.26/7.6/7.1	——
Ji, Q. et al. 2015 [24]	China	98/23	STEMI ($n = 77$) and NSTEMI ($n = 21$)	62.33 ± 13.9	AMI/Healthy	0.787	15.26/16.65/10.9	24 h
Jia, K.-G. et al. 2016 [17]	China	233/146	STEMI ($n = 156$) and NSTEMI ($n = 77$)	62.32	AMI/Healthy and Non-AMI ACS	0.667	5.99/6.39/5.18	12 h
Liu, G. et al. 2018 [20]	China	145/30	NSTEMI Patients	67	NSTEMI/Healthy	0.927	2.4	12 h

AMI—acute myocardial infarction; STEMI—ST elevation myocardial infarction; NSTEMI—non-ST elevation myocardial infarction; AUC—area under the curve.

3.2. Meta-Analysis Results

The most consistently reported value in all the included studies was AUC. The combined frequency weighted analysis of this parameter yielded a pooled AUC of 0.73 (95% CI 0.68–0.79) for the eight studies [10,11,17,19–22,24] in which a 95% confidence interval of the AUC was provided (Figure 2).

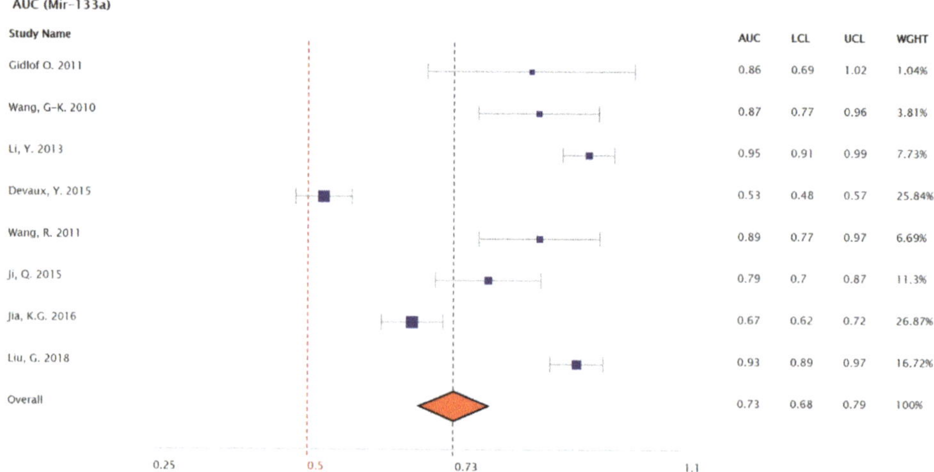

Figure 2. Forest plot of Mir-133a AUC values for the detection of AMI for each of the included studies. Pooled AUC value of 0.73 (95% CI 0.68–0.79). It is important to note that 5 out of 8 studies yielded AUC values greater than 0.86, yet, their overall weight was reduced by the relatively low number of included subjects.

Further subgroup analysis was performed in order to determine whether the distinction between STEMI and NSTEMI might account for some of the conflicting reports regarding increased Mir-133a concentrations following AMI. Relying on six papers that reported the percentage of STEMI patients in the study [17,18,20,21,24,37], and a linear regression model plotting Mir-133a fold as a function of this percentage, we found a moderate correlation ($r = 0.49$), with a trend of increasing Mir-133a concentration with higher percentages of STEMI patients (Figure 3a).

Furthermore, we compared subgroups from studies that reported data separately for STEMI or NSTEMI patients [17,18,20,21,24] and found a significantly higher value ($p < 0.001$) of the Mir-133a fold increase in STEMI patients vs. NSTEMI patients (11.6 ± 0.72 fold vs. 4.5 ± 0.14 fold, respectively; Figure 3b). Unfortunately, nearly all of the studies did not provide AUC data for these subgroups, and as such this parameter could not be analyzed.

Of the included studies, four [10,11,18,22] were controlled with non-AMI patients presenting with symptoms of acute coronary syndrome (ACS), four used healthy volunteers [19–21,24], and one used a mixed control population, with a majority of non-healthy recruits [17]. In order to assess whether choice of control partially explains the discrepancies in reported AUC values for Mir-133a in AMI, we compared these two groups using boxplots (Figure 4a). We found that the AUC was significantly greater ($p < 0.001$) in studies that recruited healthy controls as opposed to those who used non-healthy controls (pooled AUC of 0.89 ± 0.07 vs. 0.68 ± 0.14, respectively).

In the included studies there was diversity in the percentage of male participants and mean age. To examine whether these factors impacted upon the reported results we used a linear regression model and found that both age and gender had little effect upon the reported results (Supplementary Figures S1 and S2).

Lastly, we divided the studies into two groups based on the reported time from onset of symptoms until sample acquisition. Of the included studies, six were conducted such that all samples were acquired within 12 h [10,17,19–22], two reported sample acquisition within 24 h [11,24], and one did not list a maximal time for sample acquisition and, as such, was not included in this analysis [18]. A significantly higher AUC value ($p < 0.001$) was found in the 24 h group when compared to the 12 h group (pooled AUC of 0.825 ± 0.05 vs. 0.715 ± 0.16 respectively; Figure 4b).

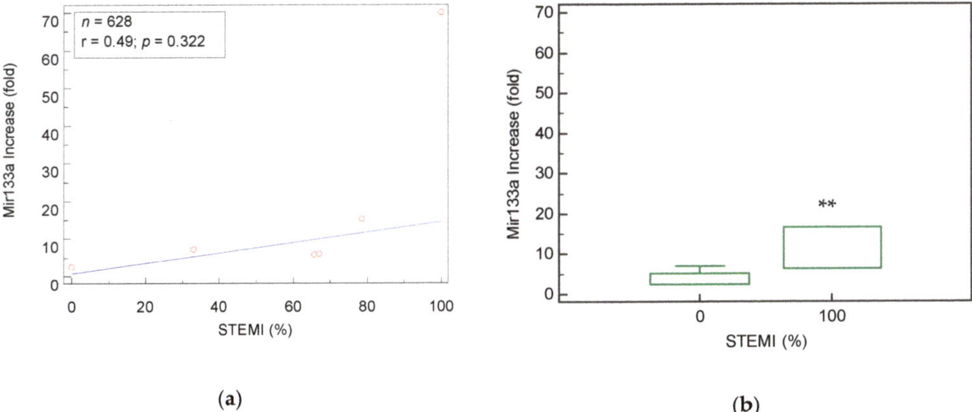

(a)

(b)

Figure 3. Linear regression analysis of (**a**) a relative increase (in fold) of Mir-133a plotted as a function of the percentage of patients in study with ST elevation myocardial infarction in composite groups. (**b**) Boxplot comparison of fold change in subgroups of 100% STEMI patients vs. 0% STEMI. ** $p < 0.001$.

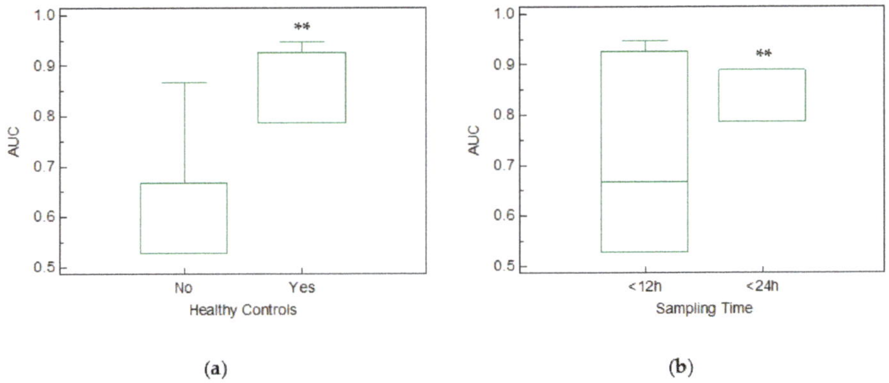

(a)

(b)

Figure 4. (**a**) Boxplot comparing AUC values based on control group characteristics (AUC was 0.89 ± 0.06 and 0.68 ± 0.14 when healthy or unhealthy controls were used, respectively), * $p < 0.001$. (**b**) Boxplot comparing AUC values based on sampling time (AUC was 0.82 ± 0.05 and 0.71 ± 0.01 for studies in which samples were acquired within 24 and 12 h, respectively). ** $p < 0.001$.

4. Discussion

AMI is a common and often deadly medical emergency. Presently, there is a great need for a quick and accurate diagnosis of AMIs, as well as improved methods for the detection of high-risk patients such as those with TO of the culprit artery, not presenting with STEMI. In the present study the diagnostic value of Mir-133a was analyzed to determine whether it may serve as a biomarker for very early detection of AMI, and to evaluate the contested claim that it may be useful in distinguishing STEMI from NSTEMI.

4.1. Mir-133a As an Early Biomarker for the Diagnosis of AMI

Historically, commonly used biomarkers for the diagnosis of acute myocardial infarction, such as cardiac troponins and creatine kinase MB, were not effective at very early diagnosis of AMI (within 0–3 h) [5]. Today, with the advent of high-sensitivity cardiac troponin tests, AMI can be diagnosed with reasonable accuracy even within the first hour [38,39]. Nevertheless, the search for additional early biomarkers, especially ones with different underlying molecular mechanisms, may allow for even greater sensitivity and specificity in shorter time frames.

To this end, early studies on Mir-133a reported high (>0.86, and as high as 0.95) AUC values [11,18,19,21,22], while some of the subsequent studies found lower sensitivities and specificities [10,17,24]. In this meta-analysis it was found that the pooled AUC for Mir-133a was 0.73 (95% CI 0.68–0.79). This value highlights its relatively weak sensitivity and specificity for the diagnosis of AMI, especially when measured against current troponin based methods that have markedly greater sensitivities and specificities [38,39]. Furthermore, it has yet to be shown whether Mir-133a, even in combination with current biomarkers, offers any diagnostic advantage and as such, its clinical value for the early diagnosis of AMI remains unclear. Notably, only two of the included papers suggested threshold values for the optimal diagnosis of AMI [17,18], highlighting the current lack of standardized values and measurements.

Additionally, we attempted to establish whether Mir-133a might be a more effective biomarker in the earliest stages of AMI, as was reported by Ji et al., 2015 [24]. Unfortunately, it is difficult in practice to determine how long after the onset of symptoms the samples are taken, and most studies did not report precise time frames. Therefore, we were only able to subdivide the studies into two main groups: (1) measurements made within 12 h [10,17,19–22] and (2) measurements made within 24 h [11,24]. The results of this analysis showed that, contrary to prior studies [11,24], measurements made within 24 h had a significantly higher pooled AUC. Regrettably, this does not settle the issue, as the distribution of individual measurements within each study is unclear, and only two studies were included in the 24 h group. Yet, the fact that measurements conducted within 12 h from symptoms yielded AUC values as high as 0.95 according to some reports [19], might be suggestive of Mir-133a's potential as a diagnostic tool in specific patient populations. Further research, with larger sample sizes, as well as careful and repeated time measurements, are necessary in order to clarify the plasma concentration dynamics of Mir-133a in various conditions associated with myocardial damage.

4.2. Mir-133a in Healthy and Unhealthy Controls

A significant methodological issue regarding research on the diagnostic potential of Mir-133a is the choice of controls. As mentioned above, choices included healthy volunteers, patients with comorbidities and acute chest pain, and mixed populations. We suspected that the reported AUCs might have been greatly affected by the choice of controls, thus further obfuscating the clinical potential of Mir-133a. After analyzing the studies separately, based on the controls used, we found, in agreement with Jia et al. 2016 [17], that Mir-133a had a significantly higher ($p < 0.001$) pooled AUC of 0.89 ± 0.06 when healthy controls were used in comparison to 0.68 ± 0.14 in unhealthy. Hence, it may be concluded that Mir-133a can more efficiently distinguish between AMI patients and healthy volunteers, but it is not nearly as effective when testing patients presenting with symptoms of AMI. This may partially explain the apparent discordance in the existing literature. As increases in Mir-133a concentration are indicative of myocardial damage [30], it is reasonable to assume that a larger overlap will exist between patients presenting with acute chest pain and AMI patients than between healthy volunteers and AMI patients. For this reason, it may be concluded that Mir-133a might have greater diagnostic potential in patients presenting without classic symptoms of cardiac distress. If true, it can be speculated that Mir-133a might be of clinical importance in detecting troponin-based false-diagnosis of AMI in certain populations [40]. Current medical literature contains little information on this topic, and this speculation needs to be further evaluated in prospective studies.

4.3. Mir-133a As a Biomarker that Distinguishes Between STEMI and NSTEMI

In addition to early diagnosis of AMI, a clinical need for the detection of NSTEMI with TO of the culprit artery, as well as other serious cardiac conditions associated with a lack of ST segment elevation, exists. Due to inherent limitations in the standard 12-lead ECG, a complete electrocardiographic picture of the heart is not obtained. As a result, patients with acute occlusion of a coronary vessel may present with NSTEMI, and, as studies have shown [8], patients suffering from complete culprit artery occlusion presenting with AMI and no ST segment elevation are at a higher risk for mortality and adverse cardiac events than their ST elevated counterparts. It is believed that this is primarily due to the delay in identification and, as a result, establishment of reperfusion. Although several factors have been suggested to aid in the identification of patients with TO (such as prolonged duration of continuous chest pain, higher levels of the creatine kinase-MB fraction [41], and higher levels of high sensitivity troponin [4]), currently a delay of more than 24 h before an invasive procedure is performed, is common according to previous reports [8]. Furthermore, although patients with TO were reported to have higher mean Global Registry of Acute Coronary Events (GRACE) scores compared to those suffering subtotal occlusion (STO; 131 (range of 120–140) vs. 117 (range of 104–126); $p = 0.032$) [4], these values are still lower than the proposed cutoff of 140, which serves as an indication for the performance of PCI within 24 h according to accepted practice [8].

Consequently, it is of great importance to develop new and improved risk stratification tools that will allow clinicians to recognize high risk NSTEMI patients, and specifically those with occult TO, as early as possible.

Our results show that there was a trend of increasing Mir-133a plasma concentrations as the percentage of STEMI patients in the study grew. Moreover, we found a significantly greater increase ($p < 0.001$) in Mir-133a concentration amongst subgroups containing only STEMI patients vs. only NSTEMI patients. These data corroborate the claims made by Devaux et al., 2015 [10] and Ji et al., 2015 [24] that levels of Mir-133a are increased to a greater degree during STEMI as opposed to NSTEMI, and contradict the results reported by Li et al., 2013 [19] who reported no significant difference between the two. This difference in results may stem from the overall smaller sample size that was used or the relatively small number of NSTEMI patients included in the report by Li et al., 2013 [19].

We hypothesize that specific entities of myocardial injury may be characterized by distinct Mir-133a increase patterns. Yet, a concise meta-analysis of the literature yielded limited data in this regard. The included studies on Mir-133a did not specifically evaluate those with TO and STO in NSTEMI. Therefore, it is unknown whether NSTEMI patients with TO of culprit artery will present with similar Mir-133a concentrations as the STEMI group, but if such a correlation can be found, it may be used to identify these higher risk patients. Moreover, the underlying cause of myocardial damage in the included studies was not reported, and so the cases likely include different entities of NSTEMI such as myocardial infarction with nonobstructive coronary arteries (MINOCA), microvascular dysfunction, and vascular anatomies with varying degree of occlusion. It is unknown whether these entities may also present with markedly different Mir-133a elevation patterns, which might have diagnostic importance prior to coronary angiography. Additionally, since the present NSTEMI groups may contain a significant number of TO patients, it is possible that subanalysis of non-TO NSTEMI groups will demonstrate an even greater relative increase in Mir-133a plasma concentration relative to the TO group.

Further studies specifically designed to answer these questions are necessary in order to fully assess Mir-133a's diagnostic potential in this respect. Future studies should also focus on the correlation between Mir-133a and other cardiac biomarkers in various populations of cardiac patients, as well as on the association with plaque vulnerability. As such, it remains to be determined whether differences in Mir-133a plasma concentration and elevation dynamics in NSTEMI may be used to identify different underlying pathophysiologies, and whether these differences may be used to accurately stratify risk groups.

4.4. Study Limitations and Methodological Issues

During the systematic literature review we encountered several methodological issues that hindered our ability to perform full comparisons between the data in each paper. It is our contention that these differences in methods and designs are the primary cause of conflicting reports as to Mir-133's diagnostic ability. A prime example of this is the "endogenous control" used for the qRT-PCR analysis of Mir-133a plasma concentration. Multiple studies used truly endogenous microRNAs such as Mir-16, Mir-17, and U6 [11,18,24,27,29], though these controls, and especially U6, have been found to vary markedly in the same patients [42,43] and therefore may be unsuitable as reference microRNAs. Other studies used arbitrary CT values or the median CT for comparison [21,36], and multiple studies used single or multiple *C. elegans* microRNA "spike ins" [10,12,17,19]. To further complicate the picture, no uniform method of fold-change calculation was used. This resulted in non-standardized data, which makes for a major limitation when attempting to perform comparisons or draw conclusions from a meta-analysis. Furthermore, no standard protocol, equipment, or probes (SYBR/TaqMan) were implemented. In this study we relied primarily on AUC values to avoid these limitations, but in the case of distinguishing STEMI from NSTEMI only fold change data was available, so our conclusions in that instance may be limited.

Another significant design issue is the selection of controls. As we showed above, the pooled AUC value for Mir-133a was significantly greater in studies that used healthy controls as opposed to unhealthy controls. This fact limits analyses conducted by combining these two groups, and likely explains, at least in part, the large variations reported in the Mir-133a's diagnostic ability.

Plasma concentration dynamics of Mir-133a in AMI are not yet fully understood. It is possible that time is a key factor in Mir-133a's sensitivity and specificity. Due to limited reporting of precise times in the published literature, we were unable to conduct a thorough investigation of the impact that time has on the reported results. Though we showed that Mir-133a's AUC is actually increased in studies with a longer duration from the onset of symptoms until sample acquisition, this conclusion is limited because the time windows are large, and the timing of individual measurements is not reported.

The studies included in this meta-analysis ranged in number of subjects from 9 in the smallest to 233 in the largest. To account for these differences the studies were weighted according to their size (n). While this is common practice, it does introduce a strong bias towards the larger studies, which also limits the conclusions drawn in this analysis.

We have posited the hypothesis that a contributing factor to the difference between Mir-133a plasma concentration in STEMI vs. NSTEMI is the degree of occlusion of the culprit artery, and therefore, that Mir-133a may be used for NSTEMI risk stratification. Since there are varied species of NSTEMI (MINOCA, TO, STO, etc.) that result from different underlying pathophysiologies that might each effect Mir-133a concentrations differently, further studies assessing the causal relation between degree of occlusion in NSTEMI and Mir-133a plasma concentration will be necessary to determine whether this hypothesis is true.

Finally, in order to rigorously test the clinical potential of Mir-133a in the setting of AMI, further studies will be needed with larger sample sizes, accurate timeline assessments, standardized methods of Mir-133a plasma concentration quantification, use of accepted reference values, and separate analyses based on subgroups.

5. Conclusions

Mir-133a has been investigated for its diagnostic potential for over a decade, yet a conclusive answer as to its clinical applicability is still lacking. In this meta-analysis we found that Mir-133a does possess a diagnostic ability (pooled AUC of 0.73), though it remains inferior to existing modalities [38,44]. Furthermore, we speculated Mir-133a may have an unrealized potential as a biomarker for the identification of high risk NSTEMI patients, and we suggest that it may be useful for detecting specific kinds of cardiac injuries and false-positive cardiac troponin increases. Further research will be needed in order to determine Mir-133a's clinical applicability in these various scenarios. Lastly, we highlighted several significant methodological issues that prevent accurate comparisons between studies in this field and may be the cause of incongruent results.

Supplementary Materials: The following are available online at http://www.mdpi.com/2073-4409/9/4/793/s1, Figure S1: Trends as a function of %males, Figure S2: Trends as a function of age, Tables S1–S3: Complete Search Strategy. Table S4: PRISMA guidelines checklist.

Author Contributions: Conceptualization, Y.W. and U.N.; methodology, Y.W. and U.N.; formal analysis, U.N.; data curation, Y.W.; writing—original draft preparation, Y.W. and U.N.; writing—review and editing, Y.W. and U.N.; visualization, Y.W. and U.N.; supervision, U.N. All authors have read and agreed to the published version of the manuscript.

Funding: This research received no external funding.

Acknowledgments: We thank the Alfred Goldschmidt Medical Science Library staff, and specifically Tal Kaminski-Rosenberg, for their assistance in designing, refining, and finalizing the meta-analysis search strategy.

Conflicts of Interest: The authors declare no conflict of interest.

References

1. Heron, M. Deaths: Leading Causes for 2015. *Natl. Vital Stat. Rep.* **2017**, *66*, 1–76. [PubMed]
2. Benjamin, E.J.; Muntner, P.; Alonso, A.; Bittencourt, M.S.; Callaway, C.W.; Carson, A.P.; Chamberlain, A.M.; Chang, A.R.; Cheng, S.; Das, S.R.; et al. Heart Disease and Stroke Statistics-2019 Update: A Report From the American Heart Association. *Circulation* **2019**, *139*, e56–e66. [CrossRef] [PubMed]
3. Fryar, C.D.; Chen, T.C.; Li, X. Prevalence of uncontrolled risk factors for cardiovascular disease: United States, 1999–2010. *NCHS Data Brief* **2012**, *103*, 1–8.
4. Baro, R.; Haseeb, S.; Ordoñez, S.; Costabel, J.P. High-sensitivity cardiac troponin T as a predictor of acute Total occlusion in patients with non-ST-segment elevation acute coronary syndrome. *Clin. Cardiol.* **2019**, *42*, 222–226. [CrossRef] [PubMed]
5. Aydin, S.; Ugur, K.; Aydin, S.; Sahin, İ.; Yardim, M. Biomarkers in acute myocardial infarction: Current perspectives. *Vasc. Health Risk Manag.* **2019**, *15*, 1–10. [CrossRef] [PubMed]
6. Cannon, C.P.; Gibson, C.M.; Lambrew, C.T.; Shoultz, D.A.; Levy, D.; French, W.J.; Gore, J.M.; Weaver, W.D.; Rogers, W.J.; Tiefenbrunn, A.J. Relationship of symptom-onset-to-balloon time and door-to-balloon time with mortality in patients undergoing angioplasty for acute myocardial infarction. *J. Am. Med. Assoc.* **2000**, *283*, 2941–2947. [CrossRef]
7. Rathore, S.S.; Curtis, J.P.; Chen, J.; Wang, Y.; Nallamothu, B.K.; Epstein, A.J.; Krumholz, H.M.; Hines, H.H. Association of door-to-balloon time and mortality in patients admitted to hospital with ST elevation myocardial infarction: National cohort study. *BMJ* **2009**, *338*, 1312–1315. [CrossRef]
8. Khan, A.R.; Golwala, H.; Tripathi, A.; Bin Abdulhak, A.A.; Bavishi, C.; Riaz, H.; Mallipedi, V.; Pandey, A.; Bhatt, D.L. Impact of total occlusion of culprit artery in acute non-ST elevation myocardial infarction: A systematic review and meta-analysis. *Eur. Heart J.* **2017**, *38*, 3082–3089. [CrossRef]
9. Hung, C.-S.; Chen, Y.-H.; Huang, C.-C.; Lin, M.-S.; Yeh, C.-F.; Li, H.-Y.; Kao, H.-L. Prevalence and outcome of patients with non-ST segment elevation myocardial infarction with occluded "culprit" artery—A systemic review and meta-analysis. *Crit. Care* **2018**, *22*, 34. [CrossRef]
10. Devaux, Y.R.; Mueller, M.R.; Haaf, P.R.; Goretti, E.R.; Twerenbold, R.R.; Zangrando, J.R.; Vausort, M.R.; Reichlin, T.R.; Wildi, K.R.; Moehring, B.R.; et al. Diagnostic and prognostic value of circulating microRNAs in patients with acute chest pain. *J. Intern. Med.* **2015**, *277*, 260–271. [CrossRef]
11. Wang, R.; Li, N.; Zhang, Y.; Ran, Y.; Pu, J. Circulating MicroRNAs are Promising Novel Biomarkers of Acute Myocardial Infarction. *Intern. Med.* **2011**, *50*, 1789–1795. [CrossRef] [PubMed]
12. Widera, C.; Gupta, S.K.; Lorenzen, J.M.; Bang, C.; Bauersachs, J.; Bethmann, K.; Kempf, T.; Wollert, K.C.; Thum, T. Diagnostic and prognostic impact of six circulating microRNAs in acute coronary syndrome. *J. Mol. Cell. Cardiol.* **2011**, *51*, 872–875. [CrossRef] [PubMed]
13. Wronska, A.; Kurkowska-Jastrzebska, I.; Santulli, G. Application of microRNAs in diagnosis and treatment of cardiovascular disease. *Acta Physiol.* **2015**, *213*, 60–83. [CrossRef] [PubMed]
14. Wojciechowska, A.; Braniewska, A.; Kozar-Kamińska, K. MicroRNA in cardiovascular biology and disease. *Adv. Clin. Exp. Med.* **2017**, *26*, 865–874. [CrossRef]

15. Xiao, Y.; Zhao, J.; Tuazon, J.P.; Borlongan, C.V.; Yu, G. MicroRNA-133a and Myocardial Infarction. *Cell Transpl.* **2019**, *28*, 831–838. [CrossRef] [PubMed]

16. Kuwabara, Y.; Ono, K.; Horie, T.; Nishi, H.; Nagao, K.; Kinoshita, M.; Watanabe, S.; Baba, O.; Kojima, Y.; Shizuta, S.; et al. Increased MicroRNA-1 and MicroRNA-133a Levels in Serum of Patients With Cardiovascular Disease Indicate Myocardial Damage. *Circ. Cardiovasc. Genet.* **2011**, *4*, 446–454. [CrossRef] [PubMed]

17. Ke-Gang, J.; Zhi-Wei, L.; Xin, Z.; Jing, W.; Ping, S.; Xue-Jing, H.; Hong-Xia, T.; Xin, T.; Xiao-Cheng, L. Evaluating Diagnostic and Prognostic Value of Plasma miRNA133a in Acute Chest Pain Patients Undergoing Coronary Angiography. *Medicine* **2016**, *95*. [CrossRef] [PubMed]

18. Peng, L.; Chun-Guang, Q.; Bei-Fang, L.; Xue-Zhi, D.; Zi-Hao, W.; Yun-Fu, L.; Yan-Ping, D.; Yang-Gui, L.; Wei-Guo, L.; Tian-Yong, H.; et al. Clinical impact of circulating miR-133, miR-1291 and miR-663b in plasma of patients with acute myocardial infarction. *Diagn. Pathol.* **2014**, *9*, 89. [CrossRef]

19. Li, Y.-Q.; Zhang, M.-F.; Wen, H.-Y.; Hu, C.-L.; Liu, R.; Wei, H.-Y.; Ai, C.-M.; Wang, G.; Liao, X.-X.; Li, X.; et al. Comparing the diagnostic values of circulating microRNAs and cardiac troponin T in patients with acute myocardial infarction. *Clinics* **2013**, *68*, 75–80. [CrossRef]

20. Liu, G.; Niu, X.; Meng, X.; Zhang, Z. Sensitive miRNA markers for the detection and management of NSTEMI acute myocardial infarction patients. *J. Thorac. Dis.* **2018**, *10*, 3206–3215. [CrossRef]

21. Gidlof, O.; Andersson, P.; van der Pals, J.; Gotberg, M.; Erlinge, D. Cardiospecific microRNA plasma levels correlate with troponin and cardiac function in patients with ST elevation myocardial infarction, are selectively dependent on renal elimination, and can be detected in urine samples. *Cardiology* **2011**, *118*, 217–226. [CrossRef] [PubMed]

22. Wang, G.-K.; Zhu, J.-Q.; Zhang, J.-T.; Li, Q.; Li, Y.; He, J.; Qin, Y.-W.; Jing, Q. Circulating microRNA: A novel potential biomarker for early diagnosis of acute myocardial infarction in humans. *Eur. Heart J.* **2010**, *31*, 659–666. [CrossRef] [PubMed]

23. Moher, D.; Liberati, A.; Tetzlaff, J.; Altman, D.G. Preferred Reporting Items for Systematic Reviews and Meta-Analyses: The PRISMA Statement. *PLoS Med.* **2009**, *6*, e1000097. [CrossRef] [PubMed]

24. Ji, Q.; Jiang, Q.; Yan, W.; Li, X.; Zhang, Y.; Meng, P.; Shao, M.; Chen, L.; Zhu, H.; Tian, N.; et al. Expression of circulating microRNAs in patients with ST segment elevation acute myocardial infarction. *Minerva Cardioangiol.* **2015**, *63*, 397–402.

25. Maciejak, A.; Kiliszek, M.; Opolski, G.; Segiet, A.; Matlak, K.; Dobrzycki, S.; Tulacz, D.; Sygitowicz, G.; Burzynska, B.; Gora, M. miR-22-5p revealed as a potential biomarker involved in the acute phase of myocardial infarction via profiling of circulating microRNAs. *Mol. Med. Rep.* **2016**, *14*, 2867–2875. [CrossRef]

26. Goldbergova, M.P.; Lipkova, J.; Fedorko, J.; Sevcikova, J.; Parenica, J.; Spinar, J.; Masarik, M.; Vasku, A. MicroRNAs in pathophysiology of acute myocardial infarction and cardiogenic shock. *Bratisl. Lek. Listy* **2018**, *119*, 341–347. [CrossRef]

27. Olivieri, F.; Antonicelli, R.; Lorenzi, M.; D'Alessandra, Y.; Lazzarini, R.; Santini, G.; Spazzafumo, L.; Lisa, R.; La Sala, L.; Galeazzi, R.; et al. Diagnostic potential of circulating miR-499-5p in elderly patients with acute non ST-elevation myocardial infarction. *Int. J. Cardiol.* **2013**, *167*, 531–536. [CrossRef]

28. Ai, J.; Zhang, R.; Li, Y.; Pu, J.; Lu, Y.; Jiao, J.; Li, K.; Yu, B.; Li, Z.; Wang, R.; et al. Circulating microRNA-1 as a potential novel biomarker for acute myocardial infarction. *Biochem. Biophys. Res. Commun.* **2010**, *391*, 73–77. [CrossRef]

29. Wang, F.; Long, G.; Zhao, C.; Li, H.; Chaugai, S.; Wang, Y.; Chen, C.; Wang, D. Plasma microRNA-133a is a new marker for both acute myocardial infarction and underlying coronary artery stenosis. *J. Transl. Med.* **2013**, *11*, 222.

30. Eitel, I.; Adams, V.; Dieterich, P.; Fuernau, G.; Waha, S.D.; Desch, S.; Schuler, G.; Thiele, H. Relation of circulating MicroRNA-133a concentrations with myocardial damage and clinical prognosis in ST-elevation myocardial infarction. *Am. Heart J.* **2012**, *164*, 706–714. [CrossRef]

31. Hromádka, M.; Černá, V.; Pešta, M.; Kučerová, A.; Jarkovský, J.; Rajdl, D.; Rokyta, R.; Moťovská, Z. Prognostic Value of MicroRNAs in Patients after Myocardial Infarction: A Substudy of PRAGUE-18. *Dis. Markers* **2019**, *2019*, 1–9. [CrossRef] [PubMed]

32. Eryilmaz, U.; Akgullu, C.; Beser, N.; Yildiz, O.; Omurlu, I.K.; Bozdogan, B. Circulating microRNAs in patients with ST-elevation myocardial infarction. *Anatol. J. Cardiol.* **2016**, *16*, 392–396. [CrossRef] [PubMed]

33. Corsten, M.F.; Dennert, R.; Jochems, S.; Kuznetsova, T.; Devaux, Y.; Hofstra, L.; Wagner, D.R.; Staessen, J.A.; Heymans, S.; Schroen, B. Circulating MicroRNA-208b and MicroRNA-499 Reflect Myocardial Damage in Cardiovascular Disease. *Circ. Cardiovasc. Genet.* **2010**, *3*, 499–506. [CrossRef] [PubMed]

34. Jaguszewski, M.; Osipova, J.; Ghadri, J.-R.; Napp, L.C.; Widera, C.; Franke, J.; Fijalkowski, M.; Nowak, R.; Fijalkowska, M.; Volkmann, I.; et al. A signature of circulating microRNAs differentiates takotsubo cardiomyopathy from acute myocardial infarction. *Eur. Heart J.* **2014**, *35*, 999–1006. [CrossRef]

35. Gacon, J.; Kablak-Ziembicka, A.; Stepien, E.; Enguita, F.J.; Karch, I.; Derlaga, B.; Zmudka, K.; Przewlocki, T. Decision-making microRNAs (MIR-124, -133a/b, -34a and -134) in patients with occluded target vessel in acute coronary syndrome. *Kardiol. Pol.* **2016**, *74*, 280–288.

36. Dalessandra, Y.; Devanna, P.; Limana, F.; Straino, S.; Carlo, A.D.; Brambilla, P.G.; Rubino, M.; Carena, M.C.; Spazzafumo, L.; Simone, M.D.; et al. Circulating microRNAs are new and sensitive biomarkers of myocardial infarction. *Eur. Heart J.* **2010**, *31*, 2765–2773. [CrossRef]

37. Li, Y.; Ouyang, M.; Shan, Z.; Ma, J.; Li, J.; Yao, C.; Zhu, Z.; Zhang, L.; Chen, L.; Chang, G.; et al. Involvement of microRNA-133a in the development of arteriosclerosis obliterans of the lower extremities via RHoA targeting. *J. Atheroscler. Thromb.* **2014**, *22*, 424–432. [CrossRef]

38. Su, Q.; Guo, Y.; Liu, H.; Qin, Y.; Zhang, J.; Yuan, X.; Zhao, X. Diagnostic role of high-sensitivity cardiac troponin T in acute myocardial infarction and cardiac noncoronary artery disease. *Arch. Med. Res.* **2015**, *46*, 193–198. [CrossRef]

39. Neumann, J.T.; Sörensen, N.A.; Ojeda, F.; Renné, T.; Schnabel, R.B.; Zeller, T.; Karakas, M.; Blankenberg, S.; Westermann, D. Early diagnosis of acute myocardial infl. Neumann, J.T.; et al. Early diagnosis of acute myocardial infarction using high-sensitivity troponin i. *PLoS ONE* **2017**, *12*. [CrossRef]

40. Wang, A.Z.; Schaffer, J.T.; Holt, D.B.; Morgan, K.L.; Hunter, B.R. Troponin Testing and Coronary Syndrome in Geriatric Patients With Nonspecific Complaints: Are We Overtesting? *Acad. Emerg. Med.* **2020**, *27*, 6–14. [CrossRef]

41. Ho Jeong, M.; Ho Jung, D.; Hun Kim, K.; Seok Lee, W.; Hong Lee, K.; Joo Yoon, H.; Sik Yoon, N.; Youn Moon, J.; Joon Hong, Y.; Wook Park, H.; et al. Predictors of total occlusion of the infarct-related artery in patients with acute Non-ST elevation myocardial infarction. *Korean J. Med.* **2008**, *74*, 271–280.

42. Xiang, M.; Zeng, Y.; Yang, R.; Xu, H.; Chen, Z.; Zhong, J.; Xie, H.; Xu, Y.; Zeng, X. U6 is not a suitable endogenous control for the quantification of circulating microRNAs. *Biochem. Biophys. Res. Commun.* **2014**, *454*, 210–214. [CrossRef]

43. Peltier, H.J.; Latham, G.J. Normalization of microRNA expression levels in quantitative RT-PCR assays: Identification of suitable reference RNA targets in normal and cancerous human solid tissues. *RNA* **2008**, *14*, 844–852. [CrossRef] [PubMed]

44. Gimenez, M.R.; Twerenbold, R.; Reichlin, T.; Wildi, K.; Haaf, P.; Schaefer, M.; Zellweger, C.; Moehring, B.; Stallone, F.; Sou, S.M.; et al. Direct comparison of high-sensitivity-cardiac troponin i vs. T for the early diagnosis of acutemyocardial infarction. *Eur. Heart J.* **2014**, *35*, 2303–2311. [CrossRef] [PubMed]

Article

Cardioprotective Potential of Human Endothelial-Colony Forming Cells from Diabetic and Nondiabetic Donors

Marcus-André Deutsch [1], Stefan Brunner [2], Ulrich Grabmaier [2], Robert David [3], Ilka Ott [4] and Bruno C. Huber [2,*

1. Department of Thoracic and Cardiovascular Surgery, Heart and Diabetes Center NRW, Ruhr-University Bochum, Georgstr. 11, D-32545 Bad Oeynhausen, Germany; mdeutsch@hdz-nrw.de
2. Department of Internal Medicine I, Ludwig-Maximilians-University, Campus Grosshadern, Marchioninistr. 15, D-81377 Munich, Germany; stefan.brunner@med.uni-muenchen.de (S.B.); ulrich.grabmaier@med.uni-muenchen.de (U.G.)
3. Reference- and Translation Center for Cardiac Stem Cell Therapy (RTC), Rostock University Medical Center, Department of Cardiac Surgery, Department Life, Light & Matter (LL&M), 18057 Rostock, Germany; Robert.David@med.uni-rostock.de
4. Department of Internal Medicine, Division of Cardiology, Helios Klinikum Pforzheim, Kanzlerstraße 2-6, D-75175 Pforzheim, Germany; ilka.ott@helios-gesundheit.de
* Correspondence: Bruno.Huber@med.uni-muenchen.de; Tel.: +49-89-44-000

Received: 14 January 2020; Accepted: 27 February 2020; Published: 2 March 2020

Abstract: Objective: The potential therapeutic role of endothelial progenitor cells (EPCs) in ischemic heart disease for myocardial repair and regeneration is subject to intense investigation. The aim of the study was to investigate the proregenerative potential of human endothelial colony-forming cells (huECFCs), a very homogenous and highly proliferative endothelial progenitor cell subpopulation, in a myocardial infarction (MI) model of severe combined immunodeficiency (SCID) mice. Methods: CD34$^+$ peripheral blood mononuclear cells were isolated from patient blood samples using immunomagnetic beads. For generating ECFCs, CD34$^+$ cells were plated on fibronectin-coated dishes and were expanded by culture in endothelial-specific cell medium. Either huECFCs (5×10^5) or control medium were injected into the peri-infarct region after surgical MI induction in SCID/beige mice. Hemodynamic function was assessed invasively by conductance micromanometry 30 days post-MI. Hearts of sacrificed animals were analyzed by immunohistochemistry to assess cell fate, infarct size, and neovascularization (huECFCs $n = 15$ vs. control $n = 10$). Flow-cytometric analysis of enzymatically digested whole heart tissue was used to analyze different subsets of migrated CD34$^+$/CD45$^+$ peripheral mononuclear cells as well as CD34$^-$/CD45$^-$ cardiac-resident stem cells two days post-MI (huECFCs $n = 10$ vs. control $n = 6$). Results: Transplantation of human ECFCs after MI improved left ventricular (LV) function at day 30 post-MI (LVEF: $30.43 \pm 1.20\%$ vs. $22.61 \pm 1.73\%$, $p < 0.001$; $\Delta P/\Delta T_{max}$ 5202.28 ± 316.68 mmHg/s vs. 3896.24 ± 534.95 mmHg/s, $p < 0.05$) when compared to controls. In addition, a significantly reduced infarct size ($50.3 \pm 4.5\%$ vs. $66.1 \pm 4.3\%$, $p < 0.05$) was seen in huECFC treated animals compared to controls. Immunohistochemistry failed to show integration and survival of transplanted cells. However, anti-CD31 immunohistochemistry demonstrated an increased vascular density within the infarct border zone (8.6 ± 0.4 CD31$^+$ capillaries per HPF vs. 6.2 ± 0.5 CD31$^+$ capillaries per HPF, $p < 0.001$). Flow cytometry at day two post-MI showed a trend towards increased myocardial homing of CD45$^+$/CD34$^+$ mononuclear cells ($1.1 \pm 0.3\%$ vs. $0.7 \pm 0.1\%$, $p = 0.2$). Interestingly, we detected a significant increase in the population of CD34$^-$/CD45$^-$/Sca1$^+$ cardiac resident stem cells ($11.7 \pm 1.7\%$ vs. $4.7 \pm 1.7\%$, $p < 0.01$). In a subgroup analysis no significant differences were seen in the cardioprotective effects of huECFCs derived from diabetic or nondiabetic patients. Conclusions: In a murine model of myocardial infarction in SCID mice, transplantation of huECFCs ameliorated myocardial function by attenuation of adverse post-MI remodeling, presumably through paracrine effects. Cardiac repair is enhanced by increasing myocardial neovascularization and the pool of Sca1$^+$ cardiac resident stem cells. The use of huECFCs for treating ischemic heart disease warrants further investigation.

Keywords: cardiovascular diseases; adult stem cells; cardiac regeneration; myocardial infraction; angiogenesis

---

1. Background

Ischemic heart disease following acute myocardial infarction (AMI) is the leading cause of morbidity and mortality in the Western world [1]. Most of the clinically approved therapeutics focus on modulating hemodynamics to reduce early mortality but do not facilitate cardiac repair, which would be needed to reduce the incidence of heart failure [2]. Therefore, the concept of cell-based therapies may have the potential to transform the treatment and prognosis of heart failure through regeneration or repair of injured cardiac tissue [3,4]. Most clinical trials have used bone marrow derived mononuclear cells (BMCd), which have demonstrated inconsistent and, overall, modest efficacy [5].

In recent years, there has been an intense investigative effort to uncover the mechanism by which transplanted stem cells preserve the function of infarcted hearts. Based on these findings, the attenuation of ischemic cardiomyopathy after cell transplantation is not attributable to cardiomyocyte repopulation or transdifferentiation. Rather, functional benefits after stem cell transplantation might be attributable to an augmentation of the natural process of myocardial healing by paracrine signaling and promoting neovascularization [6].

Endothelial progenitor cells (EPCs) are a minor population of mononuclear cells migrating from the bone marrow into the bloodstream, which are able to home in on sites of injury and promote neovascularization, which finally is associated with increased blood flow and tissue repair [7]. EPCs can be isolated from peripheral or umbilical cord blood and culturing in vitro can produce two different EPC types. These two distinct EPC subtypes have been named as early EPCs (eEPCs) and late outgrowth EPCs (LOEPCs) or endothelial colony-forming cells (ECFCs). ECFCs comply with the definition of bona fide EPCs and represent a well-characterized and homogeneous cell population of endothelial origin with high proliferative capacity and inherent vasculogenic activity [8–10]. In vivo, their potential has been investigated in different mouse models where ECFCs increased neovascularization and tissue regeneration [11]. In a rat stroke model, ECFCs treatment was associated with reduced glial scarring and increased functional recovery, which could be explained by stimulation of angiogenesis and a marked reduction in apoptosis [12]. However, there are only a few studies investigating the potential of human ECFCs to regenerate ischemic myocardium. To create a clinically relevant scenario, which aims to address the effect of cell therapy after acute MI, we transplanted ECFCs from patients with coronary artery disease (CAD) into immunodeficient SCID mice and employed functional studies, immunohistology, as well as flow cytometry to assess their potential in facilitating cardiac regeneration.

2. Methods

2.1. Isolation and Culture of ECFCs

Human adult ECFCs were collected by leukapheresis after G-CSF-induced mobilization of CD34$^+$ cells from diabetic ($n = 9$) and nondiabetic ($n = 8$) patients with coronary artery disease. The diagnosis of diabetes was made in accordance with current guidelines (mean HbA1c 7.5% ± 0.3%).

For ECFC collection, mononuclear cells from leukapheresis were isolated by density gradient centrifugation for 20 min at 1000× g (Ficoll-Hypaque, Seromed, Berlin, Germany). CD34$^+$ cells were isolated using immunomagnetic beads (Miltenyi Biotec, Bergisch Gladbach, Germany) [13]. The purity of the isolated CD34$^+$ cells ranged between 86% and 99% as assessed by flow cytometry (EPICS XL,

Couter, Hialeah, FL, USA). This study was approved by the Medical Ethics Committee of the Technical University of Munich.

CD34$^+$ cord blood (CB) and peripheral blood (PB) cells were cultured using a modified protocol as described in [14]. Briefly, CD34$^+$ cells from mobilized PB was cultured on 1% gelatin (Sigma, Hamburg, Germany) or fibronectin (10 μg/cm^2, Cellsystems, St. Katharinen, Germany) in Iscove's Modified Dulbecco's Medium (IMDM, Gibco, Paisley, UK), with 10% horse serum and 10% fetal calf serum (PAN-Biotech, Aidenbach, Germany) supplemented with penicillin/streptomycin (Gibco), 50 ng/mL recombinant human stem cell factor (SCF, R&D Systems, Abingdon, UK), 50 ng/mL vascular endothelial growth factor (VEGF, R&D Systems), 20 ng/mL basic fibroblast growth factor (FGF-2, R&D Systems), and 20 ng/mL stem cell growth factor (SCGFβ, Peprotech, London, UK). This medium (ECM) was replaced 3 times a week. After 3 weeks, cells were adapted from ECM to the low-serum EGM-2 medium (Cellsystems). To analyze EC colony-forming units (CFU-EC), CD34$^+$ cells were plated in a limiting dilution series of cell concentrations in 24-well plates and treated as above. These multiwell tissue culture plates were scored for the presence ("positive") or absence ("negative") of EC colonies between 21 and 35 days. Adherent cells were cultured to confluence in 1% gelatin-coated chamber slides (Nalge Nunc, Naperville, IL, USA). Cells were washed twice in phosphate-buffered saline (PBS), fixed, and permeabilized using Fix and Perm (Dianova, Hamburg, Germany). Samples were then incubated for 2 h with primary antibodies: antihuman specific CD31 (Sertotec, Raleigh, NC, USA), anti-CD105, anti-CD144 (VE-cadherin, Coulter-Immunotech, Krefeld, Germany), anti-CD45 and anti-vWF (Dako, Hamburg, Germany), anti VEGF-R2 (KDR-1 and KDR-2, Sigma), anti-Flt1, anti-Flt4, anti-Tie-2 (Santa Cruz Biotechnologies Inc., Heidelberg, Germany), and CD146 (Chemicon, Limburg, Germany). To visualize antibody binding (mouse and rabbit antibodies), the peroxidase-labeled avidin-biotin method (Universal Dako LSAB®-Kit, Dako, Santa Clara, CA, USA) was used according to the manufacturer's recommendations. For goat primary antibodies, donkey antigoat antibodies directly conjugated to peroxidase were used (Jackson Laboratories, Dianova, Hamburg, Germany). Isotype-matched control antibodies (Coulter-Immunotech, Jackson Laboratories, Brea, CA, USA) served as negative controls. In selected experiments, nuclear staining was performed with hematoxylin staining solution (Merck, Darmstadt, Germany).

2.2. Animal Model

For this study, we used 8–12 weeks old male severe combined immunodeficiency (SCID)beige mice (Charles River, Wilmington, MA, USA), which were kept under pathogen-free conditions. MI was induced by surgical occlusion of the left anterior descending artery (LAD) through a left anterolateral approach as described previously [15,16]. Briefly, animals were anesthetized with a mixture of 100 mg/kg ketamine (Sigma, St. Louis, MO, USA) and 5 mg/kg xylazine (Sigma) intraperitoneally. Subsequently, they were intubated and artificially ventilated with room air at 200 breaths/min using a mouse ventilator (HUGO SACHS, March-Hugstetten, Germany). A left anterolateral thoracotomy was performed, and MI was induced by surgical occlusion of the left anterior descending artery (LAD) with an 8-0 Prolene suture. Animal care and all experimental procedures were performed in strict accordance with the German and National Institutes of Health animal legislation guidelines and were approved by the local animal care committees (AZ 209.1/211-2531-117/02). The investigation conforms to the Guide for the Care and Use of Laboratory Animals published by the US National Institutes of Health (NIH Publication No. 85-23, revised 1996).

2.3. Cell Delivery

For cell delivery, a 15 μL suspension containing 5×10^5 human ECFCs or a 15 μL saline solution was administered directly after LAD ligation by two injections in the border zone of the infarcted myocardium using a 10 μL 32G Hamilton syringe (Reno, NV, USA). One injection was performed on the medial and one at the lateral side of the infarcted area.

2.4. Invasive Evaluation of Cardiac Function

For evaluation of myocardial function, mice of the previously described groups underwent impedance-micromanometer catheterization. The method as well as data analyses were performed as previously described in the literature [17,18]. Briefly, the animals were anesthetized with thiopental (100 mg/kg intraperitoneal) and ventilated using a mouse ventilator (HUGO SACHS). After that, a 1.4 French impedance-micromanometer catheter (Millar Instruments, Houston, TX, USA) was introduced into the left ventricle retrogradely via the right carotid artery, and pressure—volume loops were recorded. The method was based on measuring the time-varying electrical conductance signal of two segments of blood in the left ventricle from which total volume is calculated. Raw conductance volumes were corrected for parallel conductance by the hypertonic saline dilution method [15].

2.5. Flow Cytometry of Nonmyocyte Cardiac Cells

We previously hypothesized that EPC may not be responsible only for the formation of new vessels but may also recruit local cells [13]. Therefore, we performed flow cytometry in order to evaluate the effects of cell transplantation on proangiogenic cardiac cell populations.

Hearts of the mice were investigated by flow cytometry (FACS) as described previously [19]. Briefly, for cardiac FACS analyses, infarcted hearts of the mice were explanted at day 2 and retrogradely perfused with saline (0.9% NaCl) to wash out circulating blood cells. Thereafter, a "myocyte-depleted" cardiac cell suspension was prepared, incubating minced myocardium in 0.1% collagenase IV (Gibco, Co Dublin, Ireland) for 30 min at 37 °C, lethal to most adult mouse cardiomyocytes. Cells from peripheral blood and hearts were incubated for 40 min in the dark at 4 °C with the following fluoresceinisothiocyanate (FITC), phycoerythrin (PE), and peridininchlorophyll-protein (PerCP) conjugated monoclonal antibodies: CD45-PerCP, CD34-FITC, and CXCR4-PE (all from BD Pharmingen). A matching isotype antibody served as control. Cells were analyzed by 3-color flow cytometry using a Coulter® Epics® XL-MCLTM flow cytometer (Beckman Coulter, Brea, USA). Each analysis included 50,000 events.

2.6. Histology and Immunohistochemical Analyses

Infarct size was calculated as the average of four coronal sections sampled at 2 mm intervals from the apex to the base using the following Equation (1) developed by Pfeffer et al. [20]:

$$Infarct\ Size\ =\ \frac{Coronal\ Infarct\ Perimeter\ (Epicardal\ +\ Endocardial)}{Total\ Coronal\ Perimeter\ (Epicardal\ +\ Endocardial)}\times 100 \qquad (1)$$

infarct size $\frac{1}{4}$ [coronal infarct perimeter (epicardial plus endocardial)/total coronal perimeter (epicardial plus endocardial)] × 100. Infarct wall thickness was measured in Masson's trichrome stained sections by taking the average length of five segments along evenly spaced radii from the centre of the LV through the infarcted free LV wall [15]. To assess the incorporation and phenotype of injected EPCs in infarcted myocardium, we performed standard histological procedures (hematoxylin/eosin and Masson Trichrome) and immunostaining, which was performed as follows:

For immunohistochemical analyses, hearts were fixed in 4% phosphate-buffered formalin overnight and embedded in paraffin as described previously [15]. Before immunostaining, mounted tissue sections were deparaffinized by rinsing 3× for 5 min in Xylene followed by 2× for 5 min 100%, 2× for 5 min 96%, and 2× for 5 min 70% ethanol rinses. Endogenous peroxidases were quenched in 7.5% H_2O_2 in distilled water for 10 min. Following that, slides were rinsed in distilled water for 10 min and twice in TRISbuffer (pH 7.5) for 5 min. Finally, sections were incubated at room temperature for 60 min with either a primary antibody detecting vimentin (monoclonal mouse anti-human; Dako, Glostrup, Denmark) or class I human leukocyte antigen (HLA) (monoclonal mouse anti-human HLA-A, B, C; WAK-Chemie, Steinbach, Germany). Pretreatment was performed for 30 min (microwave 750 W) using citrate buffer (10 mM, pH 6.0) for vimentin or a target retrieval solution (Dako) for HLA-A,B,C, respectively. The detection system for vimentin and HLA-A, B, C was Dako REAL and APAAP mouse.

2.7. Statistical Analyses

Results were expressed as mean ± SEM. Multiple group comparisons were performed by one-way analyses of variance (ANOVA) followed by the Bonferroni procedure for comparison of means. Comparisons between two groups were performed using the unpaired Student's *t*-test. Data were considered statistically significant at a value of $p \leq 0.05$.

3. Results

3.1. Improvement of Cardiac Function after Intramyocardial Injection of Human ECFCS into Ischemic Myocardium

To investigate the regenerative potential of human ECFCs transplanted intramyocardially, 5×10^5 cells were injected directly after surgical occlusion of the LAD and functional parameters, and histological and immunohistochemical data were collected. To assess the functional parameters, pressure-volume relations of ECFC-treated and saline-treated hearts were measured at day 30 after MI using conductance catheters (Figure 1). Heart rates did not differ significantly between the groups, showing that experimental conditions such as anesthesia did not influence the measurements (data not shown). At day 30 after MI, we detected a significant improvement in cardiac contractility (Figure 1A; dPdt max: 5202.28 ± 316.68 mmHg/s vs. 3896.24 ± 534.95 mmHg/s, $p = 0.03$) in human ECFC-treated mice compared to saline-treated animals. Moreover, compared to saline-treated control animals, human ECFC-treated mice revealed significantly improved stroke work 30 days after MI (CO 3470.70 ± 254.44 µL/min vs. 2006.71 ± 243.18 µL/min; $p < 0.001$, Figure 1B–D). Accordingly, ECFC treatment was associated with improved LV ejection fraction (Figure 1E, LVEF 30.43 ± 1.20% vs. 22.61 ± 1.73%; $p < 0.001$).

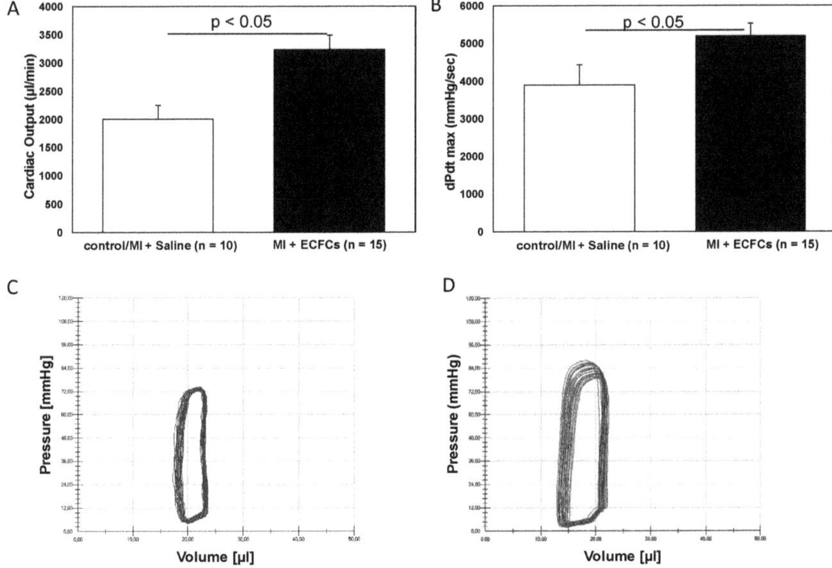

Figure 1. *Cont.*

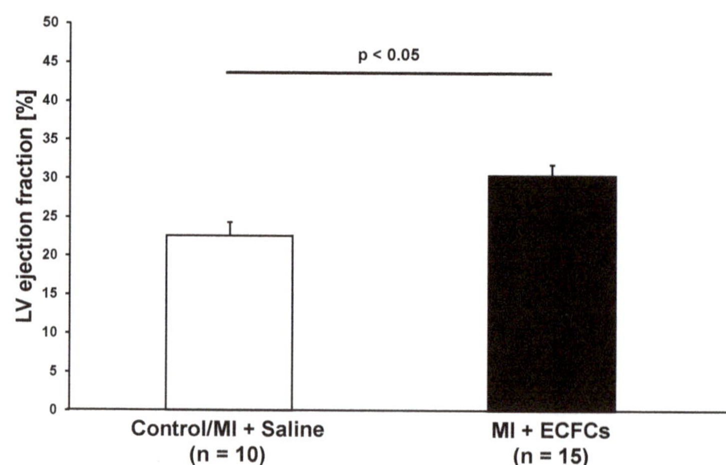

Figure 1. Improved myocardial function after human endothelial colony-forming cells (ECFC) transplantation into ischemic myocardium. Bar graphs representing cardiac output (**A**), contractility (**B**), and ejection fraction (**E**) in saline-treated animals (white bars) and ECFC-treated animals (black bars) 30 days after myocardial infarction (MI). Data represent mean ± SEM; n.s., not significant. Representative pressure volume loops of saline-treated animals (**C**) and ECFC-treated animals (**D**) 30 days after MI.

3.2. Attenuated Infarct Remodeling after Transplantation of Human ECFCS into Ischemic Myocardium

After functional profiling, we performed histological analysis of explanted hearts at day 30 following MI and cell transplantation. As reported in prior studies [15,19], permanent occlusion of the LAD artery resulted in a typical pattern of injury with transmural involvement of the myocardium in regions supplied by the main branches of the left coronary artery. Histological analyses revealed less pronounced thinning of the LV anterior wall after treatment with ECFCs (0.28 ± 0.08 mm vs. 0.20 ± 0.04 mm; not significant). However, this difference did not reach statistical significance. Infarction size was significantly diminished among human ECFC-treated animals compared to control animals 30 days after MI (50.3 ± 4.5% vs. 66.1 ± 4.3%, $p < 0.05$, Figure 2). Thirty days post-MI, as assessed by human-specific antibodies against HLA and vimentin, we were not able to detect any retained ECFCs (data not shown).

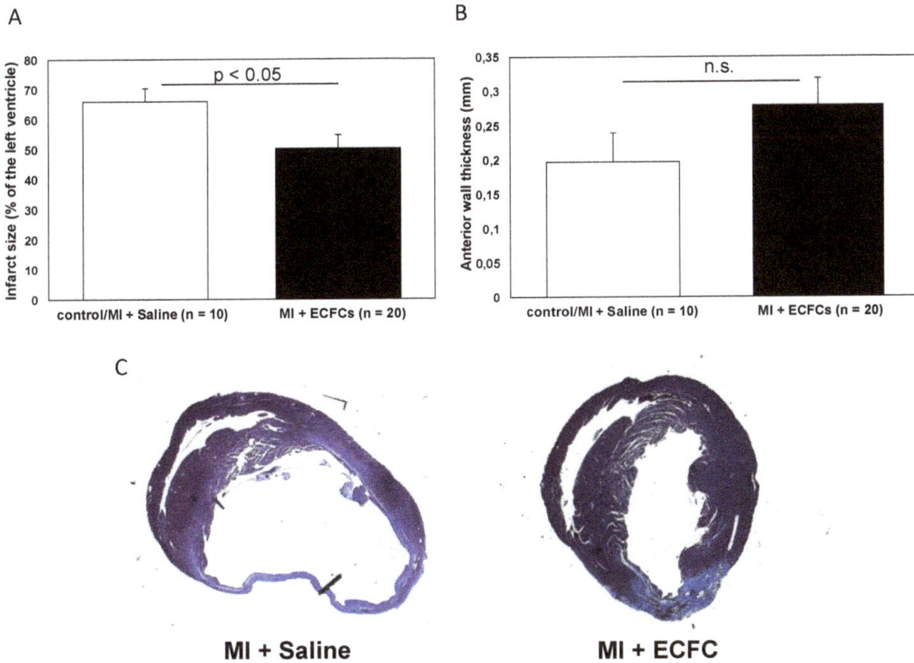

Figure 2. Attenuated infarct remodeling after human ECFC transplantation into ischemic myocardium. (**A**) Bar graphs representing the size of infarction (%) in saline-treated animals (white bar) and ECFC-treated animals (black bar) 30 days after MI. (**B**) Bar graphs representing anterior wall thickness (mm) in saline-treated animals (white bar) and ECFC-treated animals (black bar) 30 days after MI. Data represent mean ± SEM; n.s., not significant. (**C**) Representative Masson trichrome staining of infarcted hearts 30 days after MI.

3.3. Increased Neovascularization after Intramyocardial Injection of Human ECFCS into Ischemic Myocardium

We hypothesized that attenuated cardiac remodeling after transplantation of ECFCs might be a result of cell-induced enhanced neovascularization. Therefore, we performed anti-CD31 immunohistochemistry to analyze the extent of neovascularization in the border zone of animals treated with ECFCs and control animals. Consistent with the smaller infarct size after ECFC therapy, heart sections of these animals revealed a significantly increased capillary density at the infarct border zone compared to the control animals (8.6 ± 0.4 vs. 6.2 ± 0.5, $p < 0.001$, Figure 3) 30 days post-MI. In accordance to these data on enhanced neovascularization after cell transplantation in vivo, we detected increased expression of the proangiogenic transcription factors HIF-1alpha and MMP-2 in ECFCs compared to human umbilical vein endothelial cells (HUVEC) cells assessed by qPCR in vitro (Figure S1A,B).

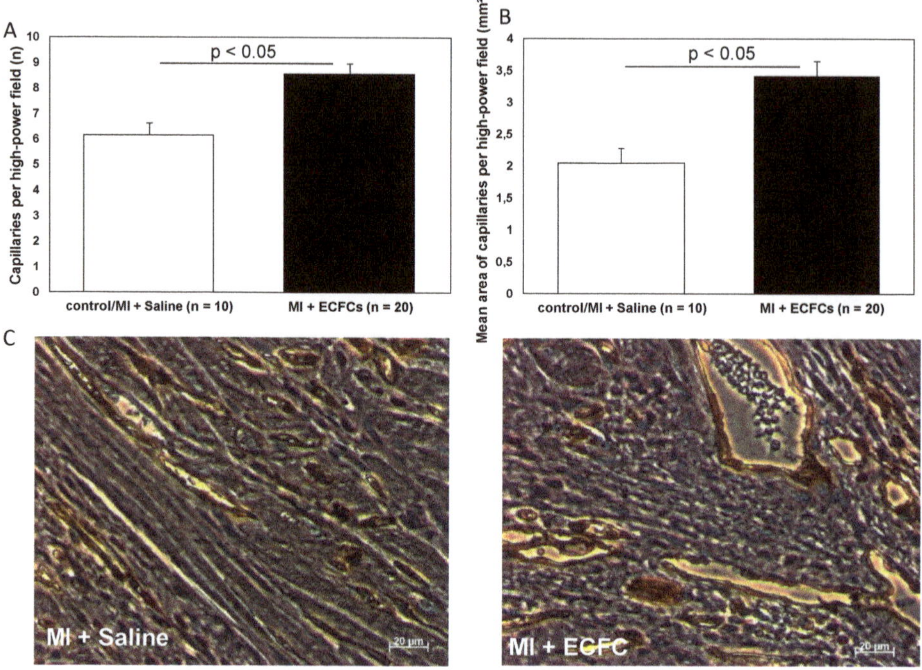

Figure 3. Increased neovascularization after transplantation of human ECFCs into ischemic myocardium. Histograms showing the numbers (**A**) and mean area (**B**) of CD31$^+$ capillaries at the infarct border zone of saline-treated control animals (white bars) and after ECFC transplantation (black bars) 30 days after MI. Data represent mean ± SEM. (**C**) Representative immunohistochemical staining of CD31 (brown) in infarcted hearts 30 days after MI.

3.4. No Enhanced Homing of BM-Derived Progenitor Cells after Human ECFC Injection into Ischemic Myocardium

Because circulating BMCs cells are known carriers of angiogenic growth factors, we sought to address whether transplantation of ECFCs is able to attract BMCs from the peripheral blood to the ischemic myocardium, thereby facilitating augmented neovascularization. To address this question, we isolated a myocyte-depleted fraction of cardiac cells and performed flow cytometry respectively. In a first step, we were interested in the amount of CD45$^+$/CD34$^+$ BMCs within the ischemic myocardium. Transplantation of ECFCs was associated with increased number of cardiac homing of CD45$^+$/CD34$^+$ cells (1.1 ± 0.3% vs. 0.7 ± 0.1%), but the values did not reach statistical significance (Figure 4A,B). Next, we further characterized CD45$^+$/CD34$^+$ cells utilizing the additional markers CD31, c-kit, Sca-1, CXCR4, Flk-1, LFA-1, and VLA-4. Compared to PBS treated controls, transplantation of ECFCs increased the number of all subfractions without reaching statistical significance (Figure 4C). Interestingly, among the different subfractions, the CD45$^+$/CD34$^+$/Sca-1 and CD45$^+$/CD34$^+$/CXCR4$^+$ showed the highest enrichment (CD45$^+$/CD34$^+$/Sca-1 + 0.8 ± 0.1% vs. 0.5 ± 0.1% and CD45$^+$/CD34$^+$/CXCR4$^+$ 0.6 ± 0.1% vs. 0.4 ± 0.1%).

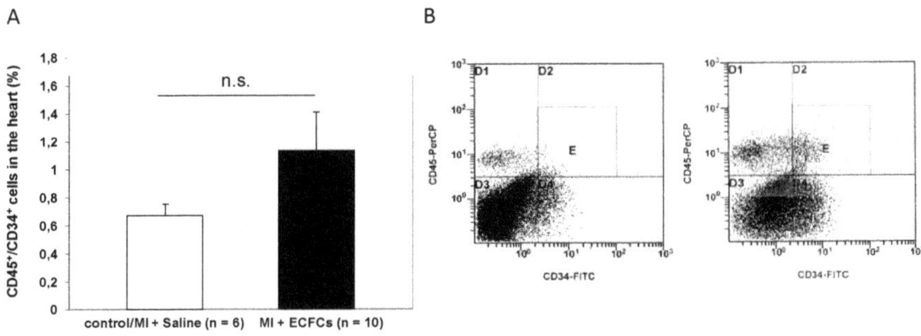

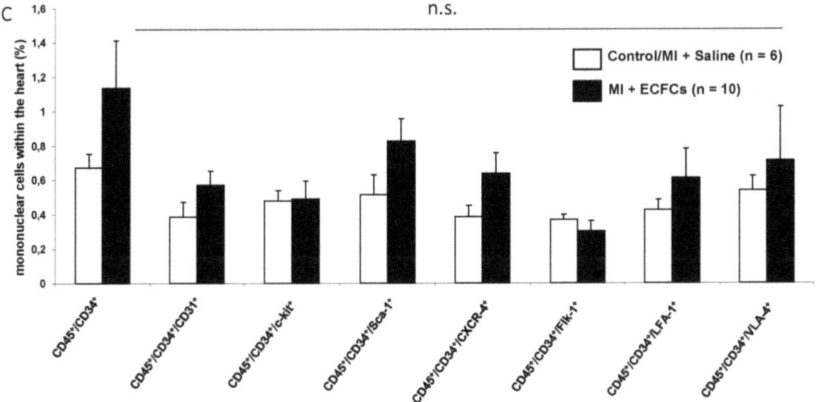

Figure 4. No enhanced homing of bone marrow (BM)-derived progenitor cells after human ECFC injection into ischemic myocardium. (**A**) Bar graphs representing the percentage of CD45$^+$/CD34$^+$ stem cells in the ischemic hearts of control animals (white bar) and after ECFC transplantation (black bar) 2 days after MI. Data represent mean ± SEM; n.s., not significant. (**B**) Representative flow cytometry (FACS) analyses of CD45$^+$/CD34$^+$ cells in the heart of control animals (left) and after ECFC transplantation (right) 2 days after MI. (**C**) Bar graphs representing the percentage of CD45$^+$/CD34$^+$ subpopulations in the ischemic hearts of control animals (white bars) and after ECFC transplantation (black bars) 2 days after MI. Data represent mean ± SEM;n.s., not significant.

3.5. Increased Numbers of Sca-1 Positive Resident Cardiac Stem Cells after ECFCS Injection into Ischemic Myocardium

In recent years, there has been emerging evidence that the heart contains a reservoir of resident cardiac progenitor cells [21]. These cells are positive for various markers, such as c-kit or Sca-1. We hypothesized that these resident cells may play a role in the repair of the injured heart, i.e., by secretion of angiogenic growth factors and contribution to improved neovascularization. To investigate these cells, we analyzed the fraction of CD45$^-$/CD34$^-$ cells within the ischemic myocardium of control and ECFC-treated animals and further characterized cells that additionally expressed c-kit or Sca-1. Transplantation of ECFCs significantly increased the number of CD45$^-$/CD34$^-$/Sca-1$^+$ progenitor cells 2 days after myocardial ischemia compared to controls (11.70 ± 1.67% vs. 4.47 ± 1.71%, $p < 0.05$). In contrast, no difference in the number of CD45$^-$/CD34$^-$/c-kit$^+$ could be observed in ECFC-treated compared to control animals (0.33 ± 0.11% vs. 0.40 ± 0.06%; not significant). Results are depicted in Figure 5.

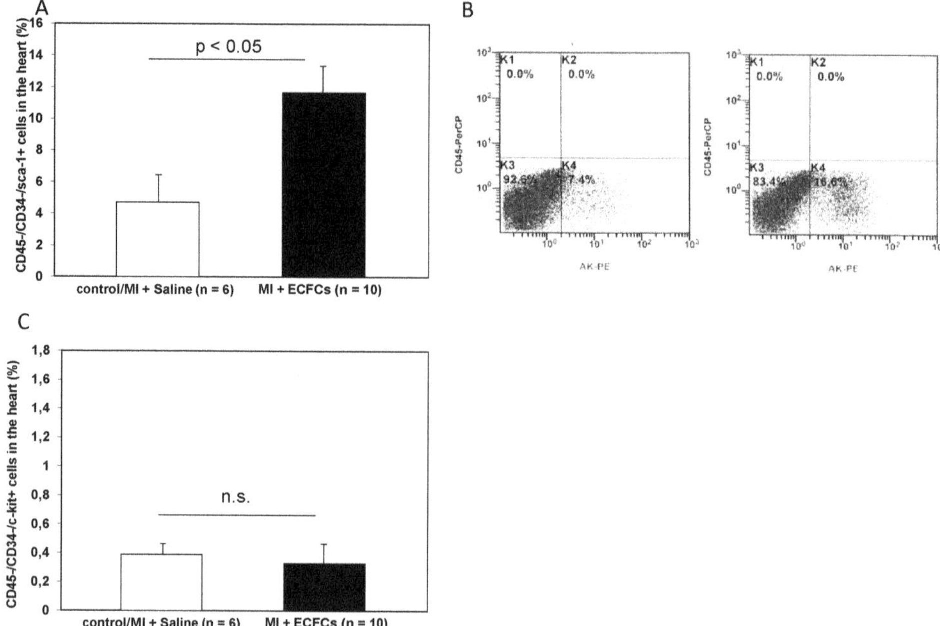

Figure 5. Increased numbers of Sca-1$^+$ resident cardiac stem cells after ECFC injection into ischemic myocardium. (**A**) Bar graphs representing the percentage of Sca-1$^+$ resident cardiac stem cells in the ischemic hearts of control animals (white bar) and after ECFC transplantation (black bar) 2 days after MI. Data represent mean ± SEM. (**B**) Representative FACS analyses of Sca-1$^+$ resident cardiac stem cells in the heart of control animals (left) and after ECFC transplantation (right) 2 days after MI. (**C**) Bar graphs representing the percentage of c-kit$^+$ resident cardiac stem cells in the ischemic hearts of control animals (white bar) and after ECFC transplantation (black bar) 2 days after MI. Data represent mean ± SEM; n.s., not significant.

3.6. No Difference in Therapeutic Effectiveness of ECFCS Derived from Diabetic or Nondiabetic Donors after Transplantation into Ischemic Myocardium

To evaluate the hypothesis that diabetes mellitus may impair the protective potential of ECFCs, we stratified cell donor patients into "diabetic" versus "nondiabetic". Interestingly, there was no statistical difference in the amount of CD45$^+$/CD34$^+$ BMCs within the ischemic myocardium between animals treated with either ECFCs from diabetic or nondiabetic patients (1.1 ± 0.4 (non DM-ECFCs) vs. 0.8 ± 0.2 (DM-ECFCs) vs. 0.7 ± 0.1% (control); for non DM-ECFCs vs. DM-ECFCs $p = 0.18$). Furthermore, LV function assessed by catheterization revealed no significant difference between animals transplanted with ECFC from diabetic or nondiabetic patients after MI. However, both ECFCs from diabetic or nondiabetic patients were able to significantly improve cardiac contractility compared to saline-treated control animals (dPdt max: 5154.65 ± 469.37 (non DM-ECFCs) vs. 5240.02 ± 447.51 (DM-ECFCs) vs. 3896.24 ± 534.95 (control) mmHg/s, for nonDM-ECFCs vs. DM-ECFCs $p = 0.28$; CO: 3821.92 ± 238.14 (non DM-ECFCs) vs. 2006.71 ± 243.18 vs. 3236.54 ± 384.78 (DM-ECFCs); for nonDM-ECFCs vs. DM-ECFCs $p = 0.28$; Figure 6).

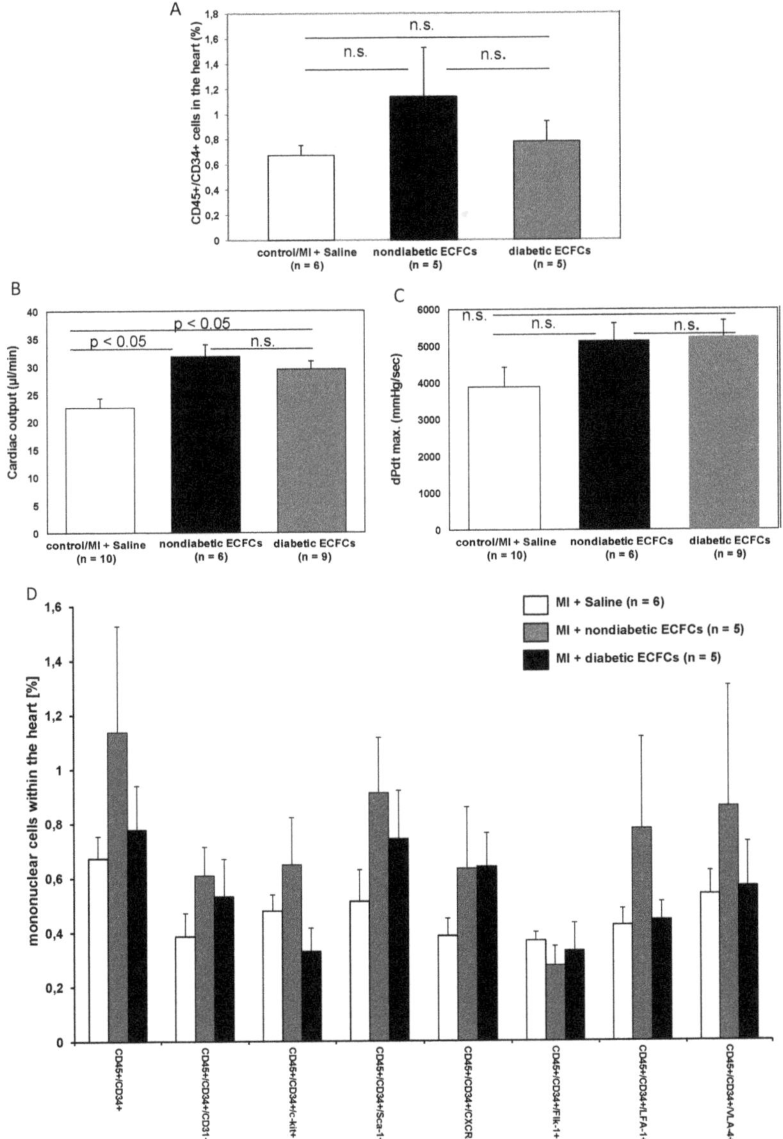

Figure 6. No difference in therapeutic effectiveness of ECFCs derived from diabetic or nondiabetic donors after transplantation into ischemic myocardium. (**A**) Bar graphs representing the percentage of CD45$^+$/CD34$^+$ stem cells in the ischemic hearts of control animals (white bar) after ECFC transplantation from nondiabetic donors (black bar) and after ECFC transplantation from diabetic donors (grey bar) 2 days after MI. Data represent mean ± SEM; n.s., not significant. (**B**) Bar graphs representing cardiac output and contractility (**C**) of control animals (white bar) after ECFC transplantation from nondiabetic donors (black bars) and after ECFC transplantation from diabetic donors (grey bars) 30 days after MI. Data represent mean ± SEM; n.s., not significant. (**D**) Bar graphs representing the percentage of CD45$^+$/CD34$^+$ subpopulations in the ischemic hearts of control animals (white bar) after ECFC transplantation from nondiabetic donors (black bars) and after ECFC transplantation from diabetic donors (grey bars) 2 days after MI. Data represent mean ± SEM; n.s., not significant.

4. Discussion

This study aimed to evaluate a proregenerative/reparative potential of human ECFCs in a murine model of MI. Our main findings were the following: (1) transplantation of human ECFCs into ischemic myocardium resulted in an improved cardiac function and reduced size of infarction 30 days after MI; (2) the attenuated remodeling after ECFC transplantation was associated with enhanced neoangiogenesis; (3) treatment with ECFCs increased the number of resident Sca-1$^+$ stem cells without significant homing of migrated BMCs; and (4) there was no difference in therapeutic effectiveness between transplanted ECFCs derived from diabetic or nondiabetic donors.

In the present study, we used immunocompromised SCID animals to prevent cell death due to cell rejection. However, 30 days after MI, we were not able to detect any transplanted cells by histological analysis. Sheikh et al. [6] investigated recently the survival kinetics and gene expression changes of transplanted BMMCs after transplantation into ischemic myocardium. Utilizing molecular-genetic bioluminescence imaging, they demonstrated short-lived survival of cells following transplant, with less than 1% of cells surviving by six weeks post-transplantation [6]. In line with these data, Higuchi et al. [22] showed in a rat study of chronic myocardial ischemia that human EPCs transduced with the sodium iodide symporter (NIS) gene for reporter gene imaging by (124)I-PET and labeled with iron oxides for visualization by MRI showed poor cell engraftment after injection. The (124)I uptake decreased on day three and was undetectable on day seven, which was confirmed by histological analysis with CD31 and CD68 antibodies [22].

However, in accordance with previous preclinical [23,24] and clinical studies, we were able to see functional improvement after transplantation of ECFCs into ischemic myocardium. These functional effects were associated with an attenuation of adverse post-MI cardiac remodeling reflected by smaller size of infarction and less pronounced thinning of the LV wall after treatment with ECFCs. Moreover, our data showed an improved neovascularization reflected by a high number of CD31$^+$ vessels at the infarct border zone. It is an ongoing matter of debate how transplanted stem and progenitor cells are able to increase neovascularization. On the one hand, it has been shown that EPCs are carriers of proangiogenic cytokines and growth factors like VEGF, HGF, angiopoietins, and IGF-1 [25–27]. In this regard, we detected increased expression of proangiogenic transcription factors HIF-1alpha and MMP-2 in ECFCs in vitro. Dubois et al. [28] investigated the paracrine activation after ECFC transplantation in a pig model of acute MI. After infusion of autologous ECFCs, they observed decreased infarct size and a greater vascular density, which was associated with higher levels of the proangiogenic placental growth factor (PLGF) [28]. Likewise, in a rat model of transient middle cerebral artery occlusion (MCAO), injection of ECFCs was associated with a reduction in the number of apoptotic cells and an increase in capillary density at the site of injury. These effects were associated with an increased expression of growth factors VEGF and IGF-1 in the infarct area 14 days after MCAO [12]. Furthermore, neovascularization also increased cell survival, and decreased apoptosis is an important contributor of cardio-protective paracrine effects after cell transplantation. In this regard, Kim et al. [29] showed a significant decrease of terminal deoxynucleotidyl transferase dUTP nick end labeling (TUNEL)-positive nuclei after transplantation of ECFCs compared to control animals in tissue sections collected from the peri-infarct zones at day seven [29].

On the other hand, there is some evidence that migration and homing of BMCs as well as expansion of resident cardiac stem cells beneficially influences cardiac remodeling after MI [3]. We used cardiac FACS analysis to quantify the number of migrated CD45$^+$/CD34$^+$ BMCs after ECFC transplantation. We were not able to see a significant increase in the absolute number of BMCs and their subpopulations. However, we found a higher number of CD45$^-$/CD34$^-$/Sca-1$^+$ cardiac-resident cells within the myocardium of human ECFC treated animals injected with ECFCs. In the group of mice treated with huECFC, we observed a more than 2-fold increase in the number of CD34$^-$/CD45$^-$/Sca1$^+$ cells 2 days after MI when compared to control animals. Sca1$^+$ cells within the mouse heart have been described as a multipotent stem cell population that are able to differentiate into cardiomyocytes, endothelial cells, and smooth muscle cells in vitro and after cardiac grafting. In fact, more recently and in contrast to previous reports, it has been shown that adult Sca-1$^+$ cardiac-resident cells appear to lack significant cardiomyogenic potential [30]. Lineage tracing of Sca-1 expressing cells in the mouse heart revealed only a very limited ability to significantly contribute new cardiomyocytes to the heart after MI in the mouse. In fact, they demonstrated a mostly endothelial expression pattern in Sca-1^{mCm}R26^{dTomato} hearts that colocalized with CD31 expression [31]. Tang et al. [32] by targeting endogenous Sca1$^+$ cells in Sca1-2A-CreER; R26-tdTomato mice through pulse labeling experiments showed that 94.25 ± 0.54% tdTomato$^+$ cells were CD31$^+$ endothelial cells, indicating that most Sca1$^+$ cells adopt an endothelial cell fate during cardiac homeostasis. By immunostaining for tdTomato and VE-cad on post-MI murine heart sections, they detected a substantial number of tdTomato$^+$ coronary endothelial cells in the injured region. Roughly, two thirds of VE-cad$^+$ cells expressed tdTomato, whereas 91.77 ± 1.12% tdTomato$^+$ cells expressed VE-cadherin in the injured myocardium. They concluded that Sca1$^+$ cardiac progenitor cells mainly differentiate into cardiac endothelial cells and fibroblasts but not cardiomyocytes during cardiac homeostasis and after injuries. In a different experimental approach, Vagnozzi et al. [33] generated Ly6a gene-targeted mice containing either a constitutive or an inducible Cre recombinase to perform genetic lineage tracing of Sca-1$^+$ cells in vivo. Similarly, they observed that the contribution of endogenous Sca-1$^+$ cells to the cardiomyocyte population in the heart was < 0.005% throughout all of cardiac development, with aging, or after MI. Moreover, pulse labeling of Sca-1$^+$ cells with an inducible Ly6a-MerCreMer allele revealed that Sca-1$^+$ cells rather represent a subset of vascular endothelial cells that expand postnatally with enhanced responsiveness to pathological stress in vivo [33]. In our study, huECFCs treatment were a potent stimulus for Sca-1$^+$ endogenous angiogenic cells. Anti-CD31 immunohistochemistry confirmed improved neovascularization within the infarct border zone, thereby contributing to protection of nonischemic areas. Sca1$^+$ cells also have been described as differentiating into fibroblasts that may positively act during the early phase of infarct healing, thereby mediating positive effects on post-MI adverse remodeling.

In our study, we did not use specific cardiac stromal cell markers to define the phenotype of CD45$^-$/CD34$^-$/Sca-1$^+$ cells found in the myocardium (along with absence of CD31 expression). It must be the goal of future studies to further characterize our population of Sca-1$^+$ cells and to completely rule out that they are not vessel-related cells or part of the mobilized pool from a noncardiac origin.

After we had established the cardioprotective potential of ECFCs, we further evaluated our ECFC patient collective for cardiovascular risk factors and discovered that a fraction of these patients had diabetes mellitus Typ II (DM Typ II) according to current guidelines. Interestingly, we were not able to detect functional or cellular differences within the ischemic myocardium between animals treated with either ECFCs from diabetic or nondiabetic patients. In this context, Tan et al. [34] injected ECFCs of diabetic and healthy rabbits intramyocardially in diabetic rabbits. They found that injection of transplantation of diabetic ECFCs was not able to rescue the ischemic myocardium of diabetic rabbits. In comparison, transplantation of healthy ECFCs resulted in increased neovascularization and paracrine activation. However, methodological differences might explain the different findings. Tan et al. [34] used recipient animals with severe, untreated diabetes, suggesting that the recipient environment plays an important role for the magnitude of therapeutical effects of cell-based therapies. Likewise, others have reported that the dysfunction of diabetic ECFCs can be restored by improved glycemic control [35,36].

A sole mechanism of action for ECFC beneficial effects remains to be elucidated. In some of these studies, ECFCs facilitate vascular repair by directly integrating within the host vasculature, while in others, vascular engraftment of these cells is low or completely absent [11]. Herein, we could show that ECFC transplantation is associated with a secondary increase in Sca-1$^+$ cardiac resident progenitor cells and propose a novel proreparative mechanism of ECFC action.

In the present study, we were not able to perform noninvasive and repetitive analyses of the animals because mice were sacrificed after invasive hemodynamic measurements and for histological analyses. Therefore, it will be the goal of future studies to use noninvasive imaging techniques such as cardiac MRI to facilitate serial assessments of cardiac function [37].

The rapid donor cell death 30 days after transplantation seen in our study is consistent with previous studies showing poor donor cell survival after transplantation of different cell types [6]. However, reporter gene PET imaging could be employed in future studies to monitor stem cell survival in vivo. Likewise, molecular imaging will possibly help to unravel the mechanistic puzzle about the contributory roles of inflammation, ischemia, apoptosis, anoikis, and autophagy after cell transplantation [38].

5. Conclusions

In conclusion, in a murine model of myocardial infarction in SCID mice, transplantation of huECFCs ameliorated myocardial function by attenuation of adverse post-MI remodeling, presumably through paracrine effects. Cardiac repair was enhanced by increasing myocardial neovascularization and the pool of Sca1$^+$ cardiac resident stem cells without significant differences in the cardioprotective effects of huECFCs derived from diabetic or nondiabetic patients. The use of huECFCs for treating ischemic heart disease warrants further investigation.

Supplementary Materials: The following are available online at http://www.mdpi.com/2073-4409/9/3/588/s1, Figure S1: ECFC expression of proangiogenic transcription factors in vitro assessed by RT-PCR.

Author Contributions: M.-A.D. designed and performed experiments, analyzed data, and contributed to the writing of the manuscript; S.B., U.G., R.D., and I.O. performed experiments; B.C.H. contributed to design of research, data analysis, and interpretation and the writing of the manuscript. All authors have read and agreed to the published version of the manuscript.

Funding: This work was funded by the FöFoLe program of the Ludwig-Maximilians-University [Reg.-Nr. 634].

Acknowledgments: We are grateful to Judith Arcifa for her excellent technical assistance.

Conflicts of Interest: The authors declare no conflict of interest.

References

1. Lloyd-Jones, D.; Adams, R.J.; Brown, T.M.; Carnethon, M.; Dai, S.; De Simone, G.; Ferguson, T.B.; Ford, E.; Furie, K.; Gillespie, C.; et al. Executive summary: Heart Disease and Stroke Statistics—2010 Update: A report from the American Hear Association. *Circulation* **2010**, *121*, 948–954. [PubMed]
2. Velagaleti, R.S.; Pencina, M.J.; Murabito, J.M.; Wang, T.J.; Parikh, N.I.; D'Agostino, R.B.; Levy, D.; Kannel, W.B.; Vasan, R.S. Long-term trends in the incidence of heart failure after myocardial infarction. *Circulation* **2008**, *118*, 2057–2062. [CrossRef] [PubMed]
3. Dimmeler, S.; Zeiher, A.M.; Schneider, M.D. Unchain my heart: The scientific foundations of cardiac repair. *J. Clin. Investig.* **2005**, *115*, 572–583. [CrossRef] [PubMed]
4. Wollert, K.C.; Drexler, H. Cell therapy for the treatment of coronary heart disease: A critical appraisal. *Nat. Rev. Cardiol.* **2010**, *7*, 204–215. [CrossRef]
5. Gyöngyösi, M.; Haller, P.M.; Blake, D.J.; Martin-Rendon, E. Meta-Analysis of Cell Therapy Studies in Heart Failure and Acute Myocardial Infarction. *Circ. Res.* **2018**, *123*, 301–308. [CrossRef]
6. Sheikh, A.Y.; Huber, B.C.; Narsinh, K.H.; Spin, J.M.; Van Der Bogt, K.; De Almeida, P.E.; Ransohoff, K.J.; Kraft, D.L.; Fajardo, G.; Ardigo, D.; et al. In Vivo Functional and Transcriptional Profiling of Bone Marrow Stem Cells after Transplantation into Ischemic Myocardium. *Arterioscler. Thromb. Vasc. Boil.* **2012**, *32*, 92–102. [CrossRef]
7. Fadini, G.P.; Losordo, U.; Dimmeler, S. Critical reevaluation of endothelial progenitor cell phenotypes for therapeutic and diagnostic use. *Circ. Res.* **2012**, *110*, 624–637. [CrossRef]
8. Yoder, M.C.; Mead, L.E.; Prater, D.; Krier, T.R.; Mroueh, K.N.; Li, F.; Krasich, R.; Temm, C.J.; Prchal, J.T.; Ingram, D.A. Redefining endothelial progenitor cells via clonal analysis and hematopoietic stem/progenitor cell principals. *Blood* **2007**, *109*, 1801–1809. [CrossRef]
9. Hirschi, K.K.; Ingram, D.A.; Yoder, M.C. Assessing identity, phenotype, and fate of endothelial progenitor cells. *Arterioscler. Thromb. Vasc. Boil.* **2008**, *28*, 1584–1595. [CrossRef]

10. Medina, R.J.; Barber, C.L.; Sabatier, F.; Dignat-George, F.; Melero-Martin, J.M.; Khosrotehrani, K.; Ohneda, O.; Randi, A.M.; Chan, J.K.; Yamaguchi, T.; et al. Endothelial Progenitors: A Consensus Statement on Nomenclature. *Stem Cells Transl. Med.* **2017**, *6*, 1316–1320. [CrossRef]

11. O'Neill, C.L.; McLoughlin, K.J.; Chambers, S.E.J.; Guduric-Fuchs, J.; Stitt, A.W.; Medina, R.J. The Vasoreparative Potential of Endothelial Colony Forming Cells: A Journey Through Pre-clinical Studies. *Front. Med.* **2018**, *5*, 273. [CrossRef] [PubMed]

12. Moubarik, C.; Guillet, B.; Youssef, B.; Codaccioni, J.-L.; Piercecchi, M.-D.; Sabatier, F.; Lionel, P.; Dou, L.; Foucault-Bertaud, A.; Velly, L.; et al. Transplanted Late Outgrowth Endothelial Progenitor Cells as Cell Therapy Product for Stroke. *Stem Cell Rev. Rep.* **2011**, *7*, 208–220. [CrossRef] [PubMed]

13. Ott, I.; Keller, U.; Knoedler, M.; Götze, K.S.; Doss, K.; Fischer, P.; Urlbauer, K.; Debus, G.; Von Bubnoff, N.; Rudelius, M.; et al. Endothelial-like cells expanded from CD34+ blood cells improve left ventricular function after experimental myocardial infarction. *FASEB J.* **2005**, *19*, 992–994. [CrossRef] [PubMed]

14. Gehling, U.M.; Ergun, S.; Schumacher, U.; Wagener, C.; Pantel, K.; Otte, M.; Schuch, G.; Schafhausen, P.; Mende, T.; Kilic, N.; et al. In vitro differentiation of endothelial cells from AC133-positive progenitor cells. *Blood* **2000**, *95*, 3106–3112. [CrossRef]

15. Deindl, E.; Zaruba, M.-M.; Brunner, S.; Huber, B.; Mehl, U.; Assmann, G.; E Hoefer, I.; Mueller-Hoecker, J.; Franz, W.-M. G-CSF administration after myocardial infarction in mice attenuates late ischemic cardiomyopathy by enhanced arteriogenesis. *FASEB J.* **2006**, *20*, 956–958. [CrossRef] [PubMed]

16. Huber, B.C.; Fischer, R.; Brunner, S.; Groebner, M.; Rischpler, C.; Segeth, A.; Zaruba, M.M.; Wollenweber, T.; Hacker, M.; Franz, W.-M. Comparison of parathyroid hormone and G-CSF treatment after myocardial infarction on perfusion and stem cell homing. *Am. J. Physiol. Circ. Physiol.* **2010**, *298*, H1466–H1471. [CrossRef]

17. Pacher, P.; Nagayama, T.; Mukhopadhyay, P.; Bátkai, S.; Kass, D.A. Measurement of cardiac function using pressure-volume conductance catheter technique in mice and rats. *Nat. Protoc.* **2008**, *3*, 1422–1434. [CrossRef]

18. Lips, D.J.; Nagel, T.V.D.; Steendijk, P.; Palmen, M.; Janssen, B.J.; Dantzig, J.-M.V.; Windt, L.; Doevendans, P. Left ventricular pressure?volume measurements in mice: Comparison of closed-chest versus open-chest approach. *Basic Res. Cardiol.* **2004**, *99*, 351–359. [CrossRef]

19. Huber, B.C.; Brunner, S.; Segeth, A.; Nathan, P.; Fischer, R.; Zaruba, M.M.; Vallaster, M.; Theiss, H.D.; David, R.; Gerbitz, A.; et al. Parathyroid hormone is a DPP-IV inhibitor and increases SDF-1-driven homing of CXCR4+ stem cells into the ischaemic heart. *Cardiovasc. Res.* **2011**, *90*, 529–537. [CrossRef]

20. Pfeffer, M.A.; Braunwald, E. Ventricular enlargement following infarction is a modifiable process. *Am. J. Cardiol.* **1991**, *68*, 127–131. [CrossRef]

21. Noseda, M.; Paiva, M.S.A.; Schneider, M. The Quest for the Adult Cardiac Stem Cell. *Circ. J.* **2015**, *79*, 1422–1430. [CrossRef] [PubMed]

22. Higuchi, T.; Anton, M.; Dumler, K.; Seidl, S.; Pelisek, J.; Saraste, A.; Welling, A.; Hofmann, F.; Oostendorp, R.A.; Gansbacher, B.; et al. Combined Reporter Gene PET and Iron Oxide MRI for Monitoring Survival and Localization of Transplanted Cells in the Rat Heart. *J. Nucl. Med.* **2009**, *50*, 1088–1094. [CrossRef] [PubMed]

23. Kalka, C.; Masuda, H.; Takahashi, T.; Kalka-Moll, W.M.; Silver, M.; Kearney, M.; Li, T.; Isner, J.M.; Asahara, T. Transplantation of ex vivo expanded endothelial progenitor cells for therapeutic neovascularization. *Proc. Natl. Acad. Sci. USA* **2000**, *97*, 3422–3427. [CrossRef] [PubMed]

24. Kawamoto, A.; Tkebuchava, T.; Yamaguchi, J.-I.; Nishimura, H.; Yoon, Y.; Milliken, C.; Uchida, S.; Masuo, O.; Iwaguro, H.; Ma, H.; et al. Intramyocardial transplantation of autologous endothelial progenitor cells for therapeutic neovascularization of myocardial ischemia. *Circulation* **2003**, *107*, 461–468. [CrossRef] [PubMed]

25. Urbich, C.; Aicher, A.; Heeschen, C.; Dernbach, E.; Hofmann, W.K.; Zeiher, A.M.; Dimmeler, S. Soluble factors released by endothelial progenitor cells promote migration of endothelial cells and cardiac resident progenitor cells. *J. Mol. Cell. Cardiol.* **2005**, *39*, 733–742. [CrossRef]

26. Cho, H.-J.; Lee, N.; Lee, J.Y.; Choi, Y.J.; Ii, M.; Wecker, A.; Jeong, J.-O.; Curry, C.; Qin, G.; Yoon, Y. Role of host tissues for sustained humoral effects after endothelial progenitor cell transplantation into the ischemic heart. *J. Exp. Med.* **2007**, *204*, 3257–3269. [CrossRef]

27. Yang, Y.; Min, J.-Y.; Rana, J.S.; Ke, Q.; Cai, J.; Chen, Y.; Morgan, J.P.; Xiao, Y.-F. VEGF enhances functional improvement of postinfarcted hearts by transplantation of ESC-differentiated cells. *J. Appl. Physiol.* **2002**, *93*, 1140–1151. [CrossRef]

28. Dubois, C.; Liu, X.; Claus, P.; Marsboom, G.; Pokreisz, P.; Vandenwijngaert, S.; Dépelteau, H.; Streb, W.; Chaothawee, L.; Maes, F.; et al. Differential Effects of Progenitor Cell Populations on Left Ventricular Remodeling and Myocardial Neovascularization After Myocardial Infarction. *J. Am. Coll. Cardiol.* **2010**, *55*, 2232–2243. [CrossRef]

29. Kim, S.-W.; Jin, H.L.; Kang, S.-M.; Kim, S.; Yoo, K.-J.; Jang, Y.; Kim, H.O.; Yoon, Y. Therapeutic effects of late outgrowth endothelial progenitor cells or mesenchymal stem cells derived from human umbilical cord blood on infarct repair. *Int. J. Cardiol.* **2016**, *203*, 498–507. [CrossRef]

30. Soonpaa, M.H.; Lafontant, P.J.; Reuter, S.; Scherschel, J.A.; Srour, E.F.; Zaruba, M.-M.; Der Lohe, M.R.-V.; Field, L.J. Absence of Cardiomyocyte Differentiation Following Transplantation of Adult Cardiac-Resident Sca-1 + Cells Into Infarcted Mouse Hearts. *Circulation* **2018**, *138*, 2963–2966. [CrossRef]

31. Neidig, L.E.; Weinberger, F.; Palpant, N.J.; Mignone, J.; Martinson, A.M.; Sorensen, D.W.; Bender, I.; Nemoto, N.; Reinecke, H.; Pabon, L.; et al. Evidence for Minimal Cardiogenic Potential of Stem Cell Antigen 1–Positive Cells in the Adult Mouse Heart. *Circulation* **2018**, *138*, 2960–2962. [CrossRef] [PubMed]

32. Tang, J.; Li, Y.; Huang, X.; He, L.; Zhang, L.; Wang, H.; Yu, W.; Pu, W.; Tian, X.; Nie, Y.; et al. Fate Mapping of Sca1+ Cardiac Progenitor Cells in the Adult Mouse Heart. *Circulation* **2018**, *138*, 2967–2969. [CrossRef] [PubMed]

33. Vagnozzi, R.J.; Sargent, M.A.; Lin, S.-C.J.; Palpant, N.J.; Murry, C.E.; Molkentin, J.D. Genetic Lineage Tracing of Sca-1+ Cells Reveals Endothelial but Not Myogenic Contribution to the Murine Heart. *Circulation* **2018**, *138*, 2931–2939. [CrossRef] [PubMed]

34. Tan, Q.; Qiu, L.; Li, G.; Li, C.; Zheng, C.; Meng, H.; Yang, W. Transplantation of healthy but not diabetic outgrowth endothelial cells could rescue ischemic myocardium in diabetic rabbits. *Scand. J. Clin. Lab. Investig.* **2010**, *70*, 313–321. [CrossRef] [PubMed]

35. Tanaka, R.; Vaynrub, M.; Masuda, H.; Ito, R.; Kobori, M.; Miyasaka, M.; Mizuno, H.; Warren, S.M.; Asahara, T. Quality-Control Culture System Restores Diabetic Endothelial Progenitor Cell Vasculogenesis and Accelerates Wound Closure. *Diabetes* **2013**, *62*, 3207–3217. [CrossRef] [PubMed]

36. Langford-Smith, A.W.W.; Hasan, A.; Weston, R.; Edwards, N.; Jones, A.M.; Boulton, A.J.M.; Bowling, F.L.; Rashid, S.T.; Wilkinson, F.L.; Alexander, M.Y. Diabetic endothelial colony forming cells have the potential for restoration with glycomimetics. *Sci. Rep.* **2019**, *9*, 1–12. [CrossRef]

37. Huber, B.C.; Ransohoff, J.D.; Ransohoff, K.J.; Riegler, J.; Ebert, A.; Kodo, K.; Gong, Y.; Sanchez-Freire, V.; Dey, D.; Kooreman, N.G.; et al. Costimulation-adhesion blockade is superior to cyclosporine A and prednisone immunosuppressive therapy for preventing rejection of differentiated human embryonic stem cells following transplantation. *Stem Cells.* **2013**, *31*, 2354–2363. [CrossRef]

38. Kooreman, N.G.; Ransohoff, J.D.; Wu, J.C. Tracking gene and cell fate for therapeutic gain. *Nat. Mater.* **2014**, *13*, 106–109. [CrossRef]

Communication

Angiogenic Potential of Bone Marrow Derived CD133$^+$ and CD271$^+$ Intramyocardial Stem Cell Trans-Plantation Post MI

Sarah Sasse [1,†], Anna Skorska [1,2,†], Cornelia Aquilina Lux [1], Gustav Steinhoff [1,2], Robert David [1,2,*] and Ralf Gaebel [1,2]

[1] Department of Cardiac Surgery, Rostock University Medical Center, 18059 Rostock, Germany; sarah-sasse@t-online.de (S.S.); anna.skorska@med.uni-rostock.de (A.S.); cornelia.lux@gmx.de (C.A.L.); gustav.steinhoff@med.uni-rostock.de (G.S.); ralf.gaebel@med.uni-rostock.de (R.G.)

[2] Department Life, Light & Matter (LL&M), University of Rostock, A.-Einstein-Str. 25, 18057 Rostock, Germany

[*] Correspondence: robert.david@med.uni-rostock.de; Tel.: +49-381-4988973; Fax: +49-381-4988970

[†] Authors contributed equally to this work.

Received: 6 October 2019; Accepted: 24 December 2019; Published: 27 December 2019

Abstract: Background: Bone marrow (BM)-derived stem cells with their various functions and characteristics have become a well-recognized source for the cell-based therapies. However, knowledge on their therapeutic potential and the shortage for a cross-link between distinct BM-derived stem cells, primed after the onset of myocardial infarction (MI), seems to be still rudimentary. Therefore, the post-examination of the therapeutic characteristics of such primed hematopoietic CD133$^+$ and mesenchymal CD271$^+$ stem cells was the object of the present study. Methods and Results: The effects of respective CD133$^+$ and CD271$^+$ mononuclear cells alone as well as in the co-culture model have been explored with focus on their angiogenic potential. The phenotypic analysis revealed a small percentage of isolated cells expressing both surface markers. Moreover, target stem cells isolated with our standardized immunomagnetic isolation procedure did not show any negative alterations following BM storage in regard to cell numbers and/or quality. In vitro network formation relied predominantly on CD271$^+$ stem cells when compared with single CD133$^+$ culture. Interestingly, CD133$^+$ cells contributed in the tube formation, only if they were cultivated in combination with CD271$^+$ cells. Additional to the in vitro examination, therapeutic effects of the primed stem cells were investigated 48 h post MI in a murine model. Hence, we have found a lower expression of transforming growth factor βeta 3 (TGFβ3) as well as an increase of the proangiogenic factors after CD133$^+$ cell treatment in contrast to CD271$^+$ cell treatment. On the other hand, the CD271$^+$ cell therapy led to a lower expression of the inflammatory cytokines. Conclusion: The interactions between CD271$^+$ and CD133$^+$ subpopulations the extent to which the combination may enhance cardiac regeneration has still not been investigated so far. We expect that the multiple characteristics and various regenerative effects of CD271$^+$ cells alone as well as in combination with CD133$^+$ will result in an improved therapeutic impact on ischemic heart disease.

Keywords: bone marrow stem cells; angiogenesis; myocardial infarction; cardiac regeneration

---

1. Introduction

In order to cover up the deficit of regenerative potential, new treatment approaches for cardiac diseases are needed. Several studies suggest that approaches based on the application of stem cells may be the future strategy in order to compensate the regenerative deficiency of the heart. As myocardial infarction (MI) occurs due to the absent blood flow in the coronary arteries, the formation of new blood vessels would be an essential prerequisite for the regeneration. Yet, in the search for suitable stem cell subtypes, various other properties are of interest as well. This includes cells that interfere with pathological immune mechanisms as well as cellular mediators and chemokines that cause migration and homing processes. Besides, it is of the tremendous importance whether the cells are proofed to be safe in application. Furthermore, the cells can also be treated in advance in order to develop desirable properties. As the CD133$^+$ stem cell therapies and other approaches with hematopoietic cells provide promising results, research has to examine more cell types in order to achieve additional effects and thus to further improve the therapeutic measures. Until now, no stem cell-related cardiac complications following intramyocardial transplantation of bone marrow (BM) derived stem cells have been reported so that the therapy is considered to be safe [1]. In addition, the BM also contains a population named after their ability to differentiate into various cell types of mesenchymal origin. These multipotent stem cells are expandable and modifiable in cell culture and offer several distinctive characteristics, which make them a promising therapeutic approach for damaged heart tissue [2]. Especially mesenchymal stem cells (MSC) are able to decrease infarct size and myocardial fibrosis [3]. Thereby, cardiac remodeling in total may be attenuated as well as the heart's contractility improves [3]. Furthermore, MSC have paracrine effects that inhibit apoptosis of endothelial cells in hypoxic condition, increase their survival, and thus stimulate angiogenesis processes [4]. As another advantage, MSC do not express any antigens of histocompatibility and thus can be used in allogenic transplantation as well. Quite the contrary, these cells even have immunomodulatory paracrine effects that counteract inflammatory responses [5,6]. As inflammatory processes play a decisive role in the development of chronic heart diseases and have a negative influence on the heart function, these effects may also improve the therapy significantly. After the transplantation, MSC may be able to stimulate the production of reparative growth factors, inhibit local inflammatory response, and replace damaged cells [7]. Apart from these positive effects, difficulties with the practical application arise from the large heterogeneity of MSC as well. Thus, it is uncertain to predict which exact cell type arises from the MSC population after the application. Indeed, a study by Yoon et al. showed that the direct transplantation of unselected BM cells into the infarcted myocardium induces significant intramyocardial calcification [8]. This clearly demonstrates the research demands in order to guarantee the safety of MSC as therapeutic option. The development of methods for the identification and isolation of MSC subpopulations is crucial for a determined specific function and a successful clinical application. A currently much examined MSC subtype is characterized by the expression of the surface marker CD271. Several studies showed that this marker is one of the most specific concerning BM derived MSC [9]. Thus, there may be a particular suitability of CD271$^+$ MSC for preselected cell products. CD271 enables prospective cell isolation of MSC from BM derived mononuclear cells (MNC) [10] and have been shown to be essential in the formation of the heart as they influence the survival of endothelial cells, vascular smooth muscle cells, and cardiac myocytes. Furthermore, these proteins regulate angiogenesis and vasculogenesis by autocrine and paracrine mechanisms [10–12].

Knowledge of the therapeutic potential attributable to distinct BM derived stem cell subtypes primed due to MI of the donors is still greatly lacking. Therefore, this study intends to examine the benefits of CD271$^+$ and CD133$^+$ stem cells. In this context, we have explored the effects of CD271$^+$ cells alone as well as in combination with CD133$^+$ with the main focus on their angiogenic potential (in vitro examination). Moreover, the therapeutic capacities of both primed BM stem cells were investigated following cardiac ischemia/reperfusion in mice.

2. Materials and Methods

2.1. Bone Marrow Aspiration

Informed donors gave written consent to the aspiration of their BM according to the Declaration of Helsinki. The ethical committee of the University of Rostock has approved the presented study (registered as no. A201023) as of 29 April, 2010. BM samples were obtained by sternal aspiration from patients undergoing coronary artery bypass grafting (CABG) surgery at Rostock University,

Germany. Anticoagulation was achieved by heparinization with 250 i.E./mL sodium heparine (B. Braun Melsungen AG, Melsungen Germany).

2.2. Cell Isolation (Single and Dual Method)

Mononuclear cells (MNC) were isolated by density gradient centrifugation using Lymphocyte Separation Medium (LSM; 1.077 g/L, PAA Laboratories GmbH, Pasching, Austria). $CD133^+$ and $CD271^+$ stem cells were enriched by positive magnetic selection using the magnet activated cell sorting (MACS) system (Miltenyi Biotec, Bergisch Gladbach, Germany). In case of $CD133^+$ direct labeling was applied and indirect labeling for $CD271^+$ cells (CD271-APC/anti-APC-microbeads; all Miltenyi Biotec). Purified stem cells were appropriately employed in further in vitro and in vivo experiments. For the dual method, the $CD133^+$ population was isolated from the $CD271^-$ fraction and vice versa.

2.3. Phenotypical Characterization by Flow Cytometry

Cell fractions were suspended in cold MACS-buffer containing PBS, 2 mM EDTA and 0.5% BSA. To reduce unspecific binding, FcR blocking reagent (Miltenyi Biotec) was added to all samples. Cells were stained using the mouse anti-human-CD133-phycoerythrin (PE; 293C2), -CD34-fluorescein isothiocyanate (FITC) and -CD271-allophycocyanine (all Miltenyi Biotec), -CD45-allophycocyanin-H7 as well as -CD45-Horizon-V500 (both BD Biosciences, Heidelberg, Germany) following incubation for 10 min in the dark at 4 °C. For optimal multicolor setting and correction of the spectral overlap single stained mouse isotype antibodies were considered as controls. The gating strategy was performed with matched isotype/fluorescence minus one control. After performing antibody staining 15 µM 4′,6-Diamidino-2-phenylindole (DAPI, Thermo Fisher Scientific, Waltham, MA, USA.) was added, cells were incubated for 2 min and then immediately acquired by BD™ LSRII flow cytometer and data were analyzed using FACS-Diva software, version 6.1.2 (both Becton Dickinson, Franklin Lakes, NJ, USA). Purity and viability of all cell isolations were analyzed using near-IR live dead stain (Thermo Fisher).

2.4. Angiogenesis Assay

Freshly isolated stem cells were prepared at 4 °C in an extracellular growth matrix using 100 µL BD Matrigel™ Matrix (BD Biosciences) and 100 µL complete endothelial growth medium (EGM-2, Lonza, Basel, Switzerland) per well of a 24-well-plate (Greiner Bio-One, Kremsmünster, Austria). For the assay 100,000 $CD133^+$ or $CD271^+$ cells (single model) or both cell types in a ratio of 50,000 cells each (co-culture model) were seeded per well. After 20 min polymerization of the matrix at 37 °C the matrix 200 µL EGM-2was added carefully onto each well. The medium was changed every two days for another 14 days.

2.5. Three Dimensional (3D) Microscopy Analysis

After two weeks the tube formation assays were conducted according to the criteria of network length and count of nodal points. For this, z-stack images were acquired using the Zeiss high-resolution microscope ELYRA PS.1 LSM 780 confocal imaging system and corresponding Zen2011 software (both Carl Zeiss AG, Oberkochen, Germany). Z-Stack images are transversal slice images (two-dimensional) of the 3D assay and thus allow representative analysis of their structures. For each well, five z-stacks with ten images were taken and included into a final interpretation. The evaluation was performed using the ImageJ free software (NIH, Bethesda, MD, USA). For better illustration, the different levels of the z-stack were marked in different colors. Likewise, the amount of nodal points was counted, respectively. In total, 900 images ($n = 6$) were analyzed and measured toward network length and count of nodal points.

2.6. Cell Tracking within Matrigel Matrix

In order to further investigate the cell networks accomplished in Matrigel matrix, immunofluorescence staining was carried out on angiogenesis assay. For better discrimination and alterations within the matrix, freshly isolated CD133$^+$ cells were stained with the lipophilic cell permeable dye CFDA-SE as well as CD271$^+$ cells with the red fluorescent lipophilic tracer PKH26 (both Sigma-Aldrich, Saint Louis, MO, USA). Additionally, both cell types were stained for nuclei discrimination with Hoechst 33324 (Thermo Fisher). Acquisition and analyzes were performed using the Axiovert 40 CFL fluorescence microscope with Axio Cam MRm ZEN software (both Carl Zeiss AG).

2.7. Immunofluorescence Staining within 3D Matrix

Mouse anti-human-CD29 allophycocyanin as well as -CD73-phycoerythrin antibodies (both BD Biosciences) were diluted with EGM-2 in 1:10 ratio and incubated with the cells for 30 min. Afterwards, the assays were washed with EGM-2. For each marker an isotype control was applied in the same way in order to obtain a negative control. Additionally, both cell types were stained with Hoechst 33324. The analysis was performed by means of the Zeiss high-resolution microscope ELYRA PS.1 LSM 780 confocal imaging system and corresponding Zen2011 software Z-stack images were used for 3D reconstructions.

2.8. Gene Expression Analysis by Quantitative Real-Time-PCR

Cells derived from the single and co-culture models were collected after the termination of the angiogenesis assay and have undergone lysis in TRIzoL$^®$ reagent (Thermo Fisher). RNA was extracted following the manufacturer's instructions. For reverse transcription of total RNA amount (2 µg) and cDNA synthesis, SuperScript$^®$ III Reverse Transcriptase (Thermo Fisher) and oligo (dT)$_{15}$ Primers (Promega, Fitchburg, WI, USA) were used. Quantitative real time-PCR was performed with StepOnePlus^{TM} Real-Time PCR System (Applied Biosystems, Waltham, MA, USA) in TaqMan$^®$ Universal Master Mix (Thermo Fisher) according to the instructions of the manufacturer. The expression of the housekeeping gene ribosomal protein, large, P0 (human RPLP0, TaqMan^{TM} VIC$^®$ Endogenous Control 4310879E) was used on each cell type. Similarly, human *ACTA* (alpha smooth muscle actin) (TaqMan$^®$ Assay ID: Hs00909449_m1, FAM-MGB), *NGFR* (nerve growth factor receptor) (TaqMan$^®$ Assay ID: Hs00609976_m1, FAM-MGB) and *vWF* (von Willebrand factor) (TaqMan$^®$ Assay ID: Hs01109446_m1, FAM-MGB, all Thermo Fisher) were analyzed in duplicates and normalized to RPLP0. Negative controls were included in each assay. Cycle thresholds (C_T) for single reactions were determined with StepOne^{TM} Software 2.0 (formula: $\Delta C_{T\ mean} = C_{T\ mean} - C_{T\ mean\ RPLP0}$).

2.9. Animals

All animal procedures were in conformity with the guidelines from Directive 2010/63/EU of the European Parliament on the protection of animals used for scientific purposes. The federal animal care committee of LALLF Mecklenburg-Vorpommern (Germany) approved the study protocol (approval number LALLF M-V/TSD/7221.3-1.1-088/11). Severe Combined Immunodeficiency beige mice (SCID *beige*; strain CB17.Cg-*Prkdc^{scid}Lyst^{bg-1}*/Crl, female, 22 ± 2 g, Charles River, Sulzfeld, Germany) were randomly assigned to 4 groups: Healthy control (SHAM, $n = 3$), two MI groups with implanted human stem cells of the respective source (MI133, MI271 each $n = 3$) and untreated MI control group (MIC $n = 3$).

2.10. Generation of Reperfused MI in Mice and Stem Cell Implantation

Mice were anesthetized with 50mg/kg Pentobarbital (Vetmedica GmbH, Ingelheim, Germany) intraperitoneal injection. After thoracotomy and preparation, the left anterior descending coronary artery (LAD) was ligated. After 45 min each mouse received an intramyocardial cell injection. For cell treatment, 100,000 stem cells were suspended in 10 µL PBS containing 0.5% BSA and 2 mM EDTA and mixed with an equal amount of reduced growth-factor BD Matrigel^{TM} Matrix. The same experimental set-up was applied to control groups using cell suspension buffer and Matrigel^{TM} for application. Four injections, 5 μL each were given along the border of the blanched myocardium. Subsequently, the ligation was removed. Healthy control (SHAM)-operated mice underwent identical surgical procedures without left anterior descending coronary artery (LAD)-ligation.

2.11. Organ Harvesting

Mice underwent euthanization 48 h after intervention by cervical dislocation. Each heart was removed, embedded in O.C.T.^{TM} Compound (Sakura Finetek, Alphen aan den Rijn, Nederlands) and snap-frozen in liquid nitrogen. For further examinations of the infarction area [13] and RT-PCR, the heart tissue was divided into four horizontal levels from the apex to the base and cut into slices.

2.12. Immunofluorescence Staining within Tissue Sections

For immunofluorescence staining slices of cryosectioned hearts were fixed with 4% paraformaldehyde (Sigma Aldrich) for 20 min at room temperature following treatment with M.O.M.^{TM} Mouse Ig Blocking Reagent and M.O.M.^{TM} protein concentrate according to the instructions of Vector® M.O.M.^{TM} Immunodetection Kit (LINARIS GmbH, Mannheim, Germany). To identify human cells, monoclonal anti-human-nuclei (Merck Millipore, Darmstadt, Germany) primary antibody as first and anti-mouse Alexa-Fluor® 594 as secondary antibody was used. Nuclei were stained with DAPI and the tissue was mounted using coverslip and Dako mounting medium (Dako, Santa Clara, CA, USA).

2.13. Gene Expression Analysis by RT^2 Profiler PCR Array

Interlayers of the cryosectioned hearts were used for RNA-isolation, which was performed with TRIzol® reagent. Reverse transcription as well as PCR were performed using RT^2 Profiler PCR Array Mouse Angiogenesis® kit (Qiagen, Germantown, MD, USA). According to the instructions of the manufacturer an RT^2 SYBR Green ROX qPCR Mastermix (Qiagen) was used. The expression of the housekeeping genes *ActB*, *B2M*, *GAPDH*, *GUSB*, and *HSP90AB1* as well as the gene panel were analyzed in triplicates and normalized to the mean of all the housekeeping genes. No genomic DNA was found in the samples. Moreover, we observed no apparent inhibition by any impurities that affect the reverse transcription as well as the amplification of the gene-specific products.

2.14. Statistical Analysis

Statistical analysis was performed using SigmaStat (Version 3.5, Systat Software, San Jose, CA, USA) and IBM SPSS Statistics for Windows (Version 22.0., IBM Corp., Armonk, NY, USA). Comparisons of two experimental groups were performed using Mann–Whitney U test, when data were not normally distributed (Shapiro–Wilk's test) otherwise Student t-test was utilized. p values ≤ 0.05 were considered as statistically significant.

3. Results

3.1. Immunomagnetic Separation of CD133$^+$ and CD271$^+$ Cells from BM

In order to optimize the procedure of cell isolation it had to be determined which impact the dual isolation strategy has on the cell product quality. Therefore, the CD133$^+$ population was isolated from the CD271$^-$ fraction as well as vice versa. Thus, the quality of the cell populations resulting from the isolation was compared with each other. The average volume achieved by BM aspiration was approximately 53.8 mL. By the use of the manual immunomagnetic cell separation 150×10^6 MNC were isolated on average. From these a mean of 0.42×10^6 viable CD133$^+$ and 0.29×10^6 viable CD271$^+$ cells were separated (Figure 1A). Furthermore, analysis revealed that the amount of CD133$^+$ as well as CD271$^+$ cells showed a linear dependence on the initial amount of MNC that has been used for isolation (Figure 1B).

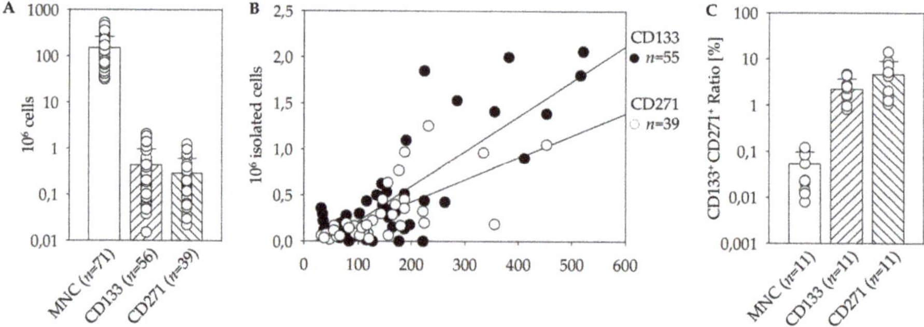

Figure 1. Quality of the immunomagnetic cell separation of CD271$^+$ and CD133$^+$ cells from fresh bone marrow (BM). Mean cell numbers of mononuclear cells (MNC), CD133$^+$, and CD271$^+$ cells achieved by magnet activated cell sorting (MACS) direct cell isolation (**A**). Linear dependence of CD133$^+$ as well as CD271$^+$ cell numbers on the initial amount of MNC (linear regression plots: R_{CD133} = 0.821; R_{CD271} = 0.689; **B**). The percentage of the CD133/CD271 double-positive cells determined within the different populations MNC, CD133$^+$, as well as CD271$^+$ cells using flow cytometry (**C**).

3.2. Flow Cytometric Characterization of Stem Cell Populations, Quality, and Storage

Furthermore, we have examined whether CD133$^+$ and CD271$^+$ cells form separate populations within BM or if the surface proteins are also expressed simultaneously by an overlap population, which is an important aspect in order to evaluate the feasibility of a combined therapeutic approach. Taking into account that the majority of CD133$^+$ cells express CD34$^+$, MACS-purified CD133$^+$ cells were co-stained with CD34 antibody [14,15]. Moreover, the CD271$^+$ population was co-stained for CD45 for further distinguishing of the MSC subpopulation lacking the CD45 surface marker [16]. As demonstrated below, 0.04 ± 0.01% of BM-MNC co-expressed CD133 and CD271. This subpopulation represents 2.21 ± 0.43% of total CD133$^+$ stem cells and 4.65 ± 1.68% of total CD271$^+$ stem cells (Figure 1C). As another aspect of cell isolation, the impact of the dual isolation strategy was investigated. Therefore, the CD133$^+$ population was isolated from the CD271$^-$ fraction and vice versa. The respective comparisons of the qualities between the obtained subpopulations were conducted. We found that 1.14 ± 0.50% of the isolated CD133$^+$ population were positive for CD271, and 3.00 ± 0.99% of the isolated CD271$^+$ cells were also positive for CD133. At this point, no indication that the order of isolation has a negative or significant influence on the quality or cell number was detected.

Additionally, the impact of the storage on the cell quality versus the quality of the fresh samples was examined. In the clinical scenario, the BM samples reach the laboratory depending on the time the heart surgery is performed, which cannot always be predicted precisely. Considering the fact that one single MACS-isolation takes around five hours, the possibility of sample storage would make experimental work much easier and effective. Furthermore, it is unavoidable that the cells undergo storage for at least several hours due to processing and transportation. This applies for storage of BM intended for mono-cell type isolation as well as for negative fractions assigned for dual isolations. We could find that overnight storage of the initial material at 4 °C had no significant negative influence on numbers or quality of the target cells resulting from the isolation (Table 1).

Table 1. Influence of the storage (4 °C) and order of the isolation during the dual isolation of CD133$^+$ or CD271$^+$ cells on their cell numbers and quality. Single MACS isolations of CD133$^+$ or CD271$^+$ cells were utilized on fresh (I) or stored (II) BM samples. For dual isolations, respective negative fractions (III or IV) were stored and respective target isolations (CD133$^+$ or CD271$^+$) were conducted after overnight storage.

	CD133$^+$ Stem Cells						CD271$^+$ Stem Cells					
	Count		Purity		Viability		Count		Purity		Viability	
	$\times 10^6$	n	[%]	n	[%]	n	$\times 10^6$	n	[%]	n	[%]	n
Single isolation												
I. Fresh BM	0.299	19	83.422	10	94.123	10	0.112	8	76.435	7	58.753	7
II. Storage BM	0.506	8	85.401	6	93.272	6	0.482	4	82.148	4	89.509	4
Dual isolation												
III. Storage CD133$^-$ fraction	0.338	5	73.213	3	82.895	3	0.203	4	47.500	2	77.290	2
IV. Storage CD271$^-$ fraction	0.137	7	88.292	3	96.182	3	0.170	7	88.430	6	89.824	6

3.3. In Vitro Network Formation

In order to investigate to what extend the stem cells build networks of endothelial cells and thus to evaluate their angiogenic capabilities, the isolated fractions of CD133$^+$ and CD271$^+$ cells were cultivated in Matrigel and EGM-2. Matrigel was used to create a complex extracellular environment with structural proteins like laminin, entactin, and collagen which resembles natural conditions. For the purpose of stimulating the cell growth and endothelial differentiation EGM-2 was used as an endothelial growth medium (Figure 2A). Each patient's CD133$^+$ and CD271$^+$ cells were seeded separately in order to compare their tendency to build network structures. Additionally, one assay consisting of both cell types in equal amounts was created with the aim of investigating whether they influence each other and alter the network formation. All assays contained cells of one patient only. The angiogenesis assays were analyzed according to the criteria of network length and count of nodal points using two-dimensional transversal slice images of the three-dimensional assay (Figure 2B) as well as in regard with levels of the gene expression for *ActA*, *NGFR*, and *vWF*. This allowed representative analysis of the structures (Figure 2C). The CD271$^+$ cells formed well-defined, 3D networks within Matrigel matrix, whereas CD133$^+$ cells only have formed minimal networks in three out of six experiments at all. The assays consisting of CD133$^+$/CD271$^+$ cells in co-culture also showed well-defined, 3D networks. CD271$^+$ cells generated networks with mean lengths of 81.20 ± 86.02 mm, combined CD133$^+$/CD271$^+$ cell culture assays formed networks with 60.35 ± 44.81 mm in length, and CD133$^+$ cells only formed 1.5 ± 4.74 mm of such network structure. Thus, in terms of network length CD271$^+$ alone and assays of the co-culture consisting of CD133$^+$/CD271$^+$ cells were both significant compared to CD133$^+$ cells alone. Moreover, networks formed by CD133$^+$/CD271$^+$ cells in co-culture contained the highest number of nodal points (893.2 ± 838.4), which revealed as significant when compared to CD133$^+$ cells alone (18.1 ± 75.4). The amount of nodal points built in CD271$^+$ cell networks regarded also as significant in comparison with those formed in CD133$^+$ cell assay (639.9 ± 901 vs. 18.1 ± 75.4, respectively). A quantitative mRNA-expression analysis demonstrated a 10-fold higher level of the Actin assembly-inducing protein ActA in CD133$^+$ cell monoculture compared to CD271$^+$ cell influenced assays (Figure 2D). No differences were observed between CD271$^+$ and CD133$^+$/CD271$^+$ cell assays for *ActA*, the *NGFR* as well as the angiogenic marker *vWF* genes expression (Figure 2E).

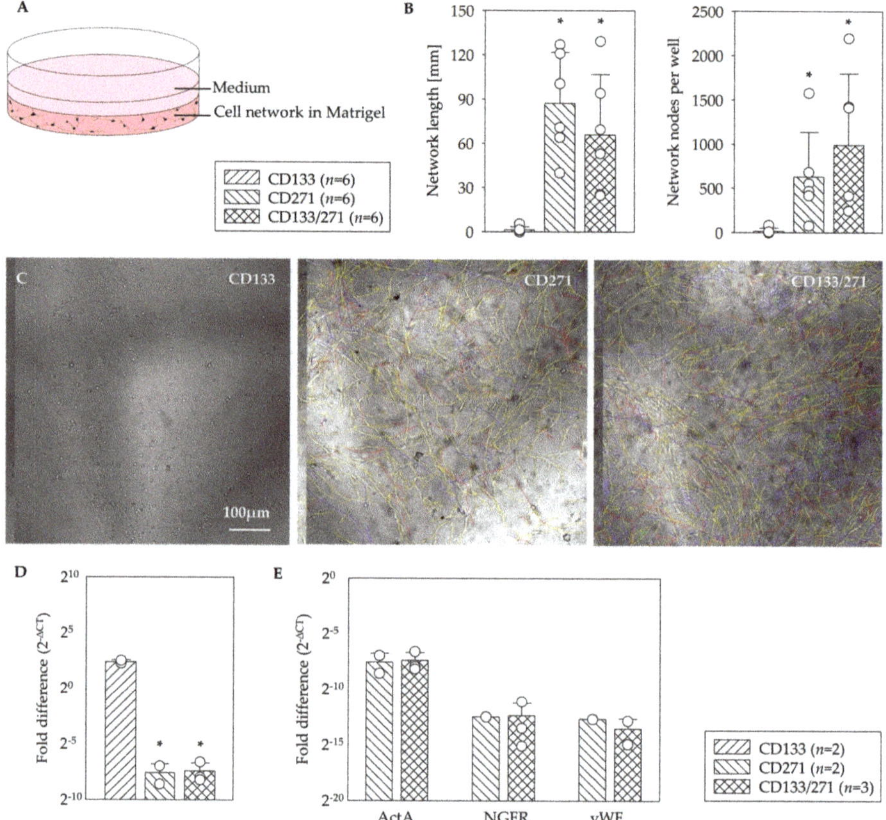

Figure 2. In vitro angiogenesis assay. Schematic illustration of the network formation assay within the Matrigel for testing of the angiogenesis potential (**A**). In contrast to CD133$^+$ cells, a significant formation of network length and highest number of nodal points in the network of CD271$^+$ cells as well as CD133$^+$/CD271$^+$ cell co-culture were observed (**B**). Representative phase contrast z-stack images of the mono- and co-culture assays (**C**). CD133$^+$ cell culture in Matrigel led to a 10-fold higher level of *ActA* gene in comparison to CD133$^+$/271$^+$ cell co-culture (**D**). No differences were observed between CD271$^+$ and CD133$^+$/271$^+$ cell culture assays for *ActA*, *NGFR*, and *vWF* gene expression (**E**). Mean ± SD; * $p \leq 0.015$ vs. CD133, Mann–Whitney *U* test.

3.4. Identification of Stem Cell Phenotype

In order to further investigate the cell networks accomplished in Matrigel, respective immunohistochemical analysis was carried out. Therefor cells were labelled with different fluorescent dyes, allowing the structures to become more perceptible and differentiable. In this regard, the assays containing both cell types in co-culture were of particular interest. By using different dyes for the stem cell subpopulations the staining should reveal whether both cell types take part in building the network structures and to which extent. The CD133$^+$ and CD271$^+$ cells have been labelled with CFDA-SE and PKH 26, respectively. Both fluorescent dyes could be detected within the network structures. Figure 3A shows a colocalization of CFDA labelled CD133$^+$ cells and PKH26 labelled CD271$^+$ cells at nodal points of the CD133$^+$/CD271$^+$ cell co-culture angiogenesis assay but no connections were observed within the CD133$^+$ mono cell assay. Of note, the true CFDA-SE fluorescent signals come from inside of the cellular extensions (Figure 3B) although CD133$^+$ cell monocultures built minimal networks. In turn, it demonstrates that CD133$^+$ cells form networks when cultured with CD271$^+$ cells.

To obtain information on the stem cell differentiation degree and phenotypic fate within the network, differences in the extent and distributions of the subpopulations in those assays have been tested. At this, the cells were positive for CD29 (Figure 3C). The same applies to the staining of CD271[+] cell culture with CD73 (Figure 3D). Conversely, the staining with CD271 antibody and its isotype control were both negative. Based on these results, the CD133[+]/CD271[+] cell co-culture network showed CD73 expression (Figure 3E). Additionally, no signals for CD73 were detectable in assays lacking network formation.

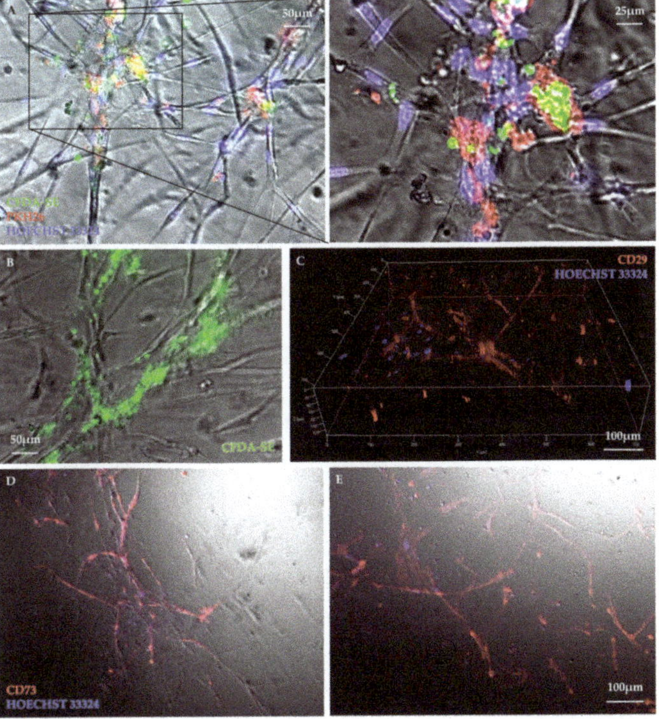

Figure 3. Representative images illustrate cellular interactions within the network structure. Colocalization of CFDA-SE labelled CD133[+] cells (green) and PKH26-labelled CD271[+] cells (red) at nodal points of the CD133[+]/CD271[+] cell co-culture angiogenesis assay but with very rare or no connections of CD133[+] cell culture were observed (**A**). Although CD133[+] cell cultures built minimal networks and true CFDA-SE fluorescent signal originate inside the network structures (**B**). 3D-image of a vital mesenchymal cell network was taken 2 weeks after seeding (**C**). CD271[+] cell (**D**) as well as CD133[+]/CD271[+] cell co-culture (**E**) angiogenesis assays were stained with CD73-PE antibody (red) and counterstained for nuclei with Hoechst dye (blue).

3.5. In Vivo Angiogenic Benefit

Currently, animal experiments are necessary to evaluate the effects of stem cell therapies on the infarcted hearts. The complex pathophysiological processes have to be taken into account, as the intact circulation, vegetative nervous system, and cellular interactions of the entire organism are essential to guarantee reliable scientific results. By the use of a small animal model, applied stem cells were investigated regarding their capability to differentiate as well as which gene expression patterns were associated. For quantitative assessment of the angiogenic benefit we performed Real-Time-PCR assays using the RT^2 Profiler PCR Array Mouse Angiogenesis® kit. The $\Delta\Delta C_T$ method was used to calculate fold-differences in target gens compared to the SHAM operation. Following CD133[+] cell treatment (MI133) we observed a significant lower expression of the inflammatory cytokine TIE1 (Tyrosine kinase with immunoglobulin-like and EGF-like domains) and fibrotic marker TGFβ3 and a significant improvement of the pro-angiogenic factors such as CXCL1 and CXCL2 compared to MIC as well as MI271 group (Figure 4A). After CD271$^+$ cell treatment we found a lower expression of the inflammatory cytokines IL1β, IL6, CCL11, and TIMP1 in contrast to a significant increase after CD133$^+$ stem cell treatment versus MIC and the angiogenic factor VEGFC (vascular endothelial growth factor) was significantly improved compared to MI133 (Figure 4B). We were able to find the transplanted human cells 48 h after transplantation at the border of the infarct (Figure 4C).

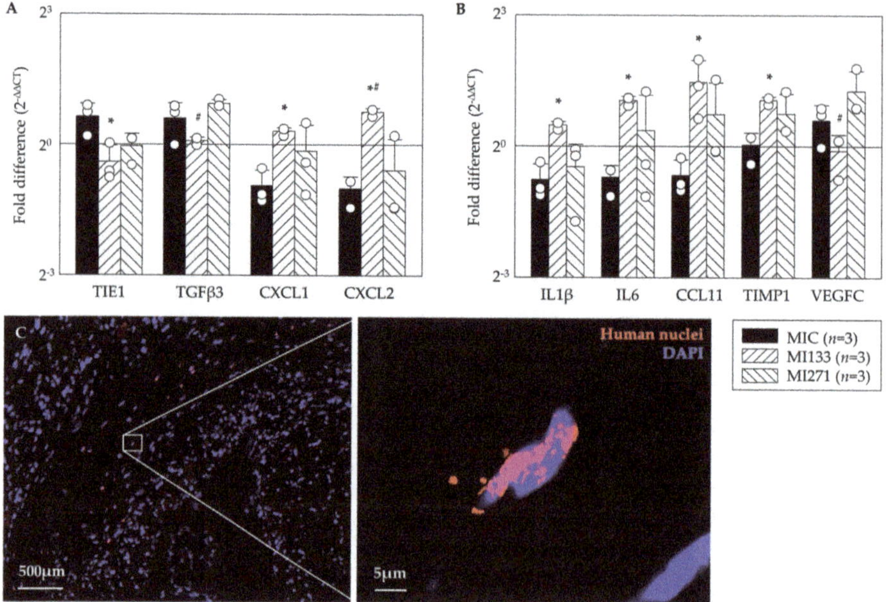

Figure 4. Angiogenic benefit 48 h after myocardial infarction (MI). Data show a significant lower expression of fibrotic markers as well as significant improvement of pro angiogenic factors after CD133$^+$ cell treatment compared to MIC as well as MI271 group (**A**). Lower expression of inflammatory cytokines after CD271$^+$ stem cell treatment in contrast to significantly increase after CD133$^+$ stem cell treatment as well as significantly improvement of the pro angiogenic factor VEGFC (vascular endothelial growth factor) after CD271$^+$ cell treatment. (**B**). Mean ± SD normalized to SHAM operation; * $p \leq 0.05$ vs. MIC; # $p \leq 0.05$ vs. MI271 (*t*-test). Representative images of CD271$^+$ cell derivatives illustrate the engrafted human stem cells 48 h post transplantation (**C**).

4. Discussion

Our present study predominantly relied on investigation of angiogenic capabilities of primed human CD133$^+$ and CD271$^+$ BM derived stem cells alone and in a co-culture in vitro as well as in mice using combined cell transplantation post MI. Recently, we already have demonstrated the beneficial therapy when a single administration of stem cells was applied [13].

First, we were able to achieve suitable isolation protocols and storage parameters. The results from the manual immuno-magnetic cell separation in combination with flow cytometric analysis revealed relatively few CD133$^+$ and CD271$^+$ cells within BM. Moreover, a linear dependence between target stem cells and the initial amount of MNC has been found (Figure 1B). However, the initial cell numbers and percentages differed eminently within the patient's cohort [17,18].

Whether CD133 and CD271 are separate subsets or co-expression exists, is an important aspect in order to evaluate the feasibility of a combined therapeutic approach. In conclusion, CD133$^+$ CD271$^+$

double-positive cells are represented at a higher proportion within the CD133$^+$ as well as the CD271$^+$ cells than within the total MNC fraction (Figure 1C). Nevertheless, only a small percentage of isolated cells express both surface markers. During magnetic cell separation no selective enrichment or depletion of double or single positive cells took place. Therefore, the separation of the two populations is considered to be sufficient for this study and the investigation of their separate and combined effects.

In clinical setting, autologous stem cells may be transplanted 24 to 30 h after BM aspiration [19]. A recent study by our laboratory closely investigated the impact of short-term storage on human CD133$^+$ cells. At this, we used standardized non-freezing storage conditions even for up to 72 h. We have shown that cell number as well as metabolic activity decreased after 30 h, although no significant alterations were observed in cell viability at this point over the time. In line with that, also present results do not indicate a negative influence of storage on cell numbers or quality (Table 1). This applies to storage of BM as well as for negative fractions intended for dual isolation.

In the present study, we have characterized the phenotype of both stem cells types. The flow cytometric strategy relied on our previous work [13,15]. Cultivated MSC are often characterized by the expression of several surface marker including CD73, CD105, and CD44 as well as the absence of others, including CD45 and CD34 [2,13]. In our analysis the expression of CD73 and CD105 was significantly increased on CD271$^+$ cells compared to remaining BM subtypes. We have already detected a favorable survival pattern, improved healing performance, and a more robust preservation of cardiac function for CD105$^+$ cells in infarcted hearts [20]. Therefore, the increased expression of CD105 on the CD271$^+$ cell population can be expected to have a positive influence on the angiogenic potential. Whereas most of cultivated MSC do not express CD45, the largest proportion of freshly isolated CD271$^+$ cells expressed CD45^{dim} - a surface marker that has been correlated to mesenchymal colony formation by Cuthbert and coworkers [16]. The percentage of CD45$^-$ cells was small and showed considerable fluctuations between patients. In conclusion, the most frequent population in purified CD271$^+$ cells presumably consists of mesenchymal progenitors [13]. Of note, as our recent flow cytometric observation revealed, the same CD271$^+$ MSC used for the present study have undergone phenotypic alterations toward a pericytic phenotype [13]. This is in agreement with the observation by Kutcher and coworkers and Bellagamba and associates [21,22].

Proliferative capacity of MSC is associated with their clonogenic properties. In accordance with Quirici and coworkers [10], colony-formation was only detected in MNC and CD271$^+$ fraction, but neither in CD271$^-$ nor in CD133$^+$ population [13]. The ability to proliferate is a major contributor to the impacts and effectiveness of stem cells in clinical applications. This applies to various functions such as differentiation into specialized mature cells as well as self-renewal or angiogenic processes. Although CD133$^+$ HSC do not show colony formation in CFU-F assays, they have been demonstrated to proliferate and generate offspring with endothelial characteristics in CFU-EC assays [19]. CFU-EC acquire the functional phenotype of EC and also show typical gene expression including vWF [19] and angiogenic factors such as FGF3, PDGFB, and others [23]. Furthermore, we detected a significant augmentation of EC-colonies after hypoxic preconditioning of CD133$^+$ cells and thus assumed that hypoxia may lead to the differentiation of CD133$^+$ cells towards endothelial lineage [23]. Our results indicate that in vitro network formation relies on CD271$^+$ cells. Since CD133$^+$ cells alone rarely build network structures at all, it is remarkable that they will nevertheless contribute to the network, if they are cultivated in combination with CD271$^+$ cells (Figure 2). The fluorescence images provided evidence for this contribution as both fluorescent signals were detected from within the networks (Figure 3). Noteworthy, we previously observed similar network formation also in co-culture when primed cardiac CD45$^-$ CD44$^+$ DDR2$^+$ MSC from MI-induced rats were setup together with endothelial cells [24]. Immunoflorescence staining indicated a strong expression of CD73 marker within the networks built by CD271$^+$ as well as CD133$^+$/CD271$^+$ cells of the angiogenesis assays. CD73 is often referred to us the MSC phenotype [25,26]. CD133$^+$ cell networks also expressed CD73 whereas the maker was not detected on CD133$^+$ cells that did not form network structures. For this study the involvement of the marker in angiogenic processes is of particular importance. At this, CD73 does not only seem to be associated with cells that are involved in angiogenesis, but appears to play a crucial role in the process itself. A study by Wang and associates found that capillary-like structures were formed more in $CD73^{+/+}$ pulmonary microvascular endothelial cells (PMEC) than in $CD73^{-/-}$ PMEC in vitro. In accordance with that they also observed that the extent of tumor angiogenesis as well as the tumor size was greater in $CD73^{+/+}$ mice compared to $CD73^{-/-}$ mice in vivo. Furthermore, CD73 expression decreased the adhesion of EC to collagen IV and promoted migration. Thus, they concluded that CD73 contributes to EC forming new vessels especially in cancer conditions [27]. A study by Allard and associates demonstrated that both tumor and host-derived CD73 are involved in these processes. In fact, tumor-derived CD73 enhances the production of vascular endothelial growth factor (VEGF) by tumor cells whereas host-derived CD73 is required for in vivo angiogenic responses. Furthermore, they also confirmed that EC require CD73 expression for tube formation and migration [28]. Consequently, measures against CD73 have been shown to impair angiogenesis in tumors [28,29]. However, Böring and associates stated that CD73 deficiency has no effect on angiogenesis in their mice model of hind limb ischemia [30]. Thus, it is not certain if the angiogenic effect applies on tumor tissue only and might be different in non-tumor environment. The increased expression of CD73 especially on $CD271^{+}$ cells may have positive impact on the angiogenic potential of the stem cell treatment. We were able to show that CD271 network structures express ActA and thus a marker for muscular differentiation. This expression was 10 fold higher in assays consisting of $CD133^{+}$ cell cultures (Figure 2D). To simulate network formation in vitro, we used Matrigel as a 3D framework. Interestingly the $CD133^{+}$ cell culture in Matrigel led to a 10-fold higher level of *ActA* (*α-SMA*) gene in comparison to the $CD133^{+}/271^{+}$ cell co-culture. However, Lu and associates observed that following culture in endothelial cell-promoting environment $CD133^{+}$ cells did not produce endothelial-like cells that expressed α-SMA, but only HUVEC and $CD34^{+}$ progenitor cells [31]. As we have already shown recently, flow cytometric analysis of the $CD133^{+}$ cell product isolated from sternal BM (similar patient's disease baseline as in the present study) was characterized by the co-expression of approx. 80% of CD34 marker [15]. Therefore, detected *ActA* mRNA levels in our $CD133^{+}$ cell angiogenesis assay seem to be reasonable. Interestingly, $CD133^{+}$ $αSMA^{+}$ phenotype was found only in association with blood vessels of cancer-associated fibroblasts [32], confirming the linkage of α-SMA expression to endothelial-like origin and cell activation. Furthermore, the possible reason of less or comparable *ActaA* mRNA levels with a $CD271^{+}$ cell culture for the $CD133^{+}/CD271^{+}$ cell co-culture might be explained by the phenotypical alterations of $CD133^{+}$ hematopoietic stem cells when applied together with MSC and/or due to the maintained undifferentiated state of MSC in co-culture model [33,34]. Hence, these in vitro co-culture conditions may have a reducing effect on the *α-SMA* mRNA levels in total.

Moreover, we examined mRNA expression profiles using the infarcted area of heart cross sections. In line with recent observations of the proliferative capacities and induction of regenerative pathways, demonstrated by our group [24], the present study focused on testing of possible angiogenic potential in the early phase post MI. On the one hand in a murine model, the switch between the early inflammatory and the proliferative phase eventuate 2-4 days after MI [35]. At the same time, neovascularization of surrounding, viable myocardium in the infarct border zone is also crucial during this process of tissue remodeling, although angiogenesis occurs in the granulation tissue that will ultimately form the infarct scar [36]. On the other hand, a complete engraftment and no significant differences in the retention of both cell subsets were evident in the hearts 48 h following MI as shown in our own previous work [13]. Indeed, in a recent long term study using the identical human stem cell types as well as the identical mouse model we did observe functional improvement [13].

Recently, we could show the co-localization of $CD133^{+}$ cell derivatives with functional blood vessels 3 weeks post transplantation. This direct impact of $CD133^{+}$ stem cells on angiogenesis was not detected for $CD271^{+}$ cells, suggesting different cellular mechanisms for cardiac regeneration [13]. Coherently, indications of the present study 48 h after MI reveal positive effects of the stem cell therapy regarding angiogenesis. The expression of the pro-angiogenic cytokines Cxcl1 and Cxcl2 has been significantly increased in the $CD133^{+}$ cell treated animal group, whereas the MI271 group showed less expression (Figure 4A). However, the same applies to the expression of IL1B and IL6 as well as Timp1 [37] (Figure 4B). Al-Amran and co-workers demonstrated that down-regulation of Cxcl2 results in decrement in myocardial neutrophil infiltration, which is associated with less reactive oxygen species and reactive nitrogen species formation after global ischemia reperfusion apoptosis [38]. Therefore, we assume, that CD133$^+$ cell therapy may have a valuable impact on a moderate course of the neutrophil-mediated tissue injury [39], although, we only could observe a significant lower *TGFβ3* mRNA level of this fibrotic marker in MI133 therapy group when compared with MI271 [40]. Our long term follow up experiments revealed unanticipated data. Subsequently, both cell therapy groups led to a decreased collagen deposition at the infarction border as well as improved vessel infiltration of scar tissue compared to the MIC group [13,40–42]. Again, these data show possible different cellular mechanisms for angiogenic effects mediated via CD133$^+$ vs. CD271$^+$. Furthermore, many existing in vitro studies confirmed the involvement of TGFβ in the activation of cardiac fibroblasts. Nevertheless, the in vivo functioning has not completely been explained because of the complexity and the context-dependent signaling cascades [40]. Interestingly, mRNA levels for pro-inflammatory interleukins (see above) within MI133 therapy group were significantly enhanced, while TGFβ3 levels were rather reduced. Since IL-6 bears pleiotropic properties, a possible switch from pro- into anti-inflammatory or even immunomodulatory signaling cascades, would be in favor for a respective therapy. Importantly, it has been shown, that both IL-6 with TGFβ play a crucial role in the shift between inhibitory Tregs and pro-inflammatory Th17 cells and this is associated with their ratio at all [43].

Finally, we have observed significantly increased mRNA levels of the pro angiogenic factor VEGFC after CD271$^+$ stem cell treatment compared to MI133 group (Figure 4B). It is known that MSC promotes angiogenesis by releasing VEGF [44–46]. Accordingly, Liu and associates observed enhanced cardiac function, increased perfusion and angiogenesis in vivo using MSC as a VEGF delivery system [47]. Collectively, analysis of hearts of the MI271 therapy in the early ischemic phase revealed rather immunomodulatory fate of CD271$^+$ stem cells.

5. Limitations

It must be stated that the present study and the generated data rely on a relatively small sample size. Thereby, sternal bone marrow for the present study was obtained from patients with CABG surgery. In general, we were able to perform standard quality control experiments at a high sample size (Sections 3.1 and 3.2). However, downstream use of these samples served to address various related projects, resulting in a limited sample size for the respective specific experiments.

Author Contributions: S.S., A.S., C.A.L., and R.G. conducted the experiments and/or acquired the data. C.A.L. and G.S. designed the research study. S.S. and A.S. contributed to the analysis and interpretation of the results. R.G. and R.D. supervised the research study and were involved in writing/correction of the manuscript, which all authors reviewed and approved. All authors have read and agreed to the published version of the manuscript.

Funding: This work was supported by the Federal Ministry of Education and Research Germany (FKZ 0312138A and FKZ 316159), the State Mecklenburg-Western Pomerania with EU Structural Funds (ESF/IVWM-B34-0030/10 and ESF/IVBM-B35-0010/12) and the DFG (DA1296/6-1) and the German Heart Foundation (F/01/12). In addition, RD is supported by the EU Structural Fund (ESF/14-BM-A55-0024/18), the FORUN Program of Rostock University Medical Centre (889001 and 889003), the Josef and Käthe Klinz Foundation (T319/29737/2017), the DAMP Foundation and the BMBF (VIP+ 00240).

Conflicts of Interest: The authors declare that they have no conflict of interest.

References

1. Donndorf, P.; Steinhoff, G. CD133-Positive Cells for Cardiac Stem Cell Therapy: Current Status and Outlook. *Adv. Exp. Med. Biol.* **2013**, *777*, 215–227. [CrossRef] [PubMed]
2. Pittenger, M.F.; Mackay, A.M.; Beck, S.C.; Jaiswal, R.K.; Douglas, R.; Mosca, J.D.; Moorman, M.A.; Simonetti, D.W.; Craig, S.; Marshak, D.R. Multilineage Potential of Adult Human Mesenchymal Stem Cells. *Science* **1999**, *284*, 143–147. [CrossRef] [PubMed]

3. Elnakish, M.T.; Kuppusamy, P.; Khan, M. Stem Cell Transplantation as a Therapy for Cardiac Fibrosis. *J. Pathol.* **2013**, *229*, 347–354. [CrossRef] [PubMed]

4. Jain, M.; Pfister, O.; Hajjar, R.J.; Liao, R. Mesenchymal Stem Cells in the Infarcted Heart. *Coron. Artery Dis.* **2005**, *16*, 93–97. [CrossRef]

5. Sánchez-Guijo, F.; Caballero-Velázquez, T.; López-Villar, O.; Redondo, A.; Parody, R.; Martinez, C.; Olavarria, E.; Andreu, E.; Prosper, F.; Diez-Campelo, M.; et al. Sequential Third-Party Mesenchymal Stromal Cell Therapy for Refractory Acute Graft-Versus-Host Disease. *Biol. Blood Marrow Transplant.* **2014**, *20*, 1580–1585. [CrossRef]

6. Delorme, B.; Ringe, J.; Gallay, N.; Le Vern, Y.; Kerboeuf, D.; Jorgensen, C.; Rosset, P.; Sensebe, L.; Layrolle, P.; Häupl, T.; et al. Specific Plasma Membrane Protein Phenotype of Culture-Amplified and Native Human Bone Marrow Mesenchymal Stem Cells. *Blood* **2008**, *111*, 2631–2635. [CrossRef]

7. Nesselmann, C.; Ma, N.; Bieback, K.; Wagner, W.; Ho, A.; Konttinen, Y.T.; Zhang, H.; Hinescu, M.E.; Steinhoff, G. Mesenchymal Stem Cells and Cardiac Repair. *J. Cell. Mol. Med.* **2008**, *12*, 1795–1810. [CrossRef]

8. Yoon, Y.S.; Park, J.S.; Tkebuchava, T.; Luedeman, C.; Losordo, D.W. Unexpected Severe Calcification after Transplantation of Bone Marrow Cells in Acute Myocardial Infarction. *Circulation* **2004**, *109*, 3154–3157. [CrossRef]

9. Bühring, H.J.; Battula, V.L.; Treml, S.; Schewe, B.; Kanz, L.; Vogel, W. Novel Markers for the Prospective Isolation of Human MSC. *Ann. N. Y. Acad. Sci.* **2007**, *1106*, 262–271. [CrossRef]

10. Quirici, N.; Soligo, D.; Bossolasco, P.; Servida, F.; Lumini, C.; Deliliers, G.L. Isolation of Bone Marrow Mesenchymal Stem Cells by Anti-Nerve Growth Factor Receptor Antibodies. *Exp. Hematol.* **2002**, *30*, 783–791. [CrossRef]

11. Meloni, M.; Caporali, A.; Graiani, G.; Lagrasta, C.; Katare, R.; Linthout, S.V.; Spillmann, F.; Campesi, I.; Madeddu, P.; Quaini, F.; et al. Nerve Growth Factor Promotes Cardiac Repair Following Myocardial Infarction. *Circ. Res.* **2010**, *106*, 1275–1284. [CrossRef] [PubMed]

12. Lecht, S.; Foerster, C.; Arien-Zakay, H.; Marcinkiewicz, C.; Lazarovici, P.; Lelkes, P.I. Cardiac microvascular endothelial cells express and release nerve growth factor but not fibroblast growth factor-2. *In Vitro Cell. Dev. Biol. Anim.* **2010**, *46*, 469–476. [CrossRef] [PubMed]

13. Lemcke, H.; Gaebel, R.; Skorska, A.; Voronina, N.; Lux, C.A.; Petters, J.; Sasse, S.; Zarniko, N.; Steinhoff, G.; David, R. Mechanisms of Stem Cell Based Cardiac Repair-Gap Junctional Signaling Promotes the Cardiac Lineage Specification of Mesenchymal Stem Cells. *Sci. Rep.* **2017**, *7*, 9755. [CrossRef] [PubMed]

14. Sutherland, D.R.; Anderson, L.; Keeney, M.; Nayar, R.; Chin-Yee, I. The ISHAGE Guidelines for CD34+ Cell Determination by Flow Cytometry. International Society of Hematotherapy and Graft Engineering. *J. Hematotherapy* **1996**, *5*, 213–226. [CrossRef]

15. Skorska, A.; Müller, P.; Gaebel, R.; Große, J.; Lemcke, H.; Lux, C.A.; Bastian, M.; Hausburg, F.; Zarniko, N.; Bubritzki, S.; et al. GMP-conformant on-site manufacturing of a CD133+ stem cell product for cardiovascular regeneration. *Stem Cell Res. Ther.* **2017**, *8*, 33. [CrossRef]

16. Cuthbert, R.; Boxall, S.A.; Tan, H.B.; Giannoudis, P.V.; McGonagle, D.; Jones, E. Single-Platform Quality Control Assay to Quantify Multipotential Stromal Cells in Bone Marrow Aspirates Prior to Bulk Manufacture or Direct Therapeutic Use. *Cytotherapy* **2012**, *14*, 431–440. [CrossRef]

17. Skorska, A.; von Haehling, S.; Ludwig, M.; Lux, C.A.; Gaebel, R.; Kleiner, G.; Klopsch, C.; Dong, J.; Curato, C.; Altarche-Xifró, W.; et al. The CD 4+ AT 2R+ T cell subpopulation improves post-infarction remodelling and restores cardiac function. *J. Cell. Mol. Med.* **2015**, *19*, 1975–1985. [CrossRef]

18. Müller, P.; Gaebel, R.; Lemcke, H.; Wiekhorst, F.; Hausburg, F.; Lang, C.; Zarniko, N.; Westphal, B.; Steinhoff, G.; David, R. Intramyocardial fate and effect of iron nanoparticles co-injected with MACS®purified stem cell products. *Biomaterials* **2017**, *135*, 74–84. [CrossRef]

19. Lux, C.A.; Mark, P.; Klopsch, C.; Laupheimer, M.; Tu-Rapp, H.; Li, W.; Ma, N.; Steinhoff, G.; David, R. Impact of Short-Term Liquid Storage on Human CD133+ Stem Cells. *Cell Transplant.* **2015**, *24*, 2409–2422. [CrossRef]

20. Gaebel, R.; Furlani, D.; Sorg, H.; Polchow, B.; Frank, J.; Bieback, K.; Wang, W.; Klopsch, C.; Ong, L.L.; Li, W.; et al. Cell Origin of Human Mesenchymal Stem Cells Determines a Different Healing Performance in Cardiac Regeneration. *PLoS ONE* **2011**, *6*, e15652. [CrossRef]

21. Kutcher, M.E.; Herman, I.M. The pericyte: Cellular regulator of microvascular blood flow. *Microvasc. Res.* **2009**, *77*, 235–246. [CrossRef] [PubMed]

22. Bellagamba, B.C.; Grudzinski, P.B.; Ely, P.B.; de Jesus Hartmann Nader, P.; Nardi, N.B.; da SilvaMeirelles, L. Induction of Expression of CD271 and CD34 in Mesenchymal Stromal CellsCultured as Spheroids. *Stem Cells Int.* **2018**, *2018*, 7357213. [CrossRef] [PubMed]

23. Ong, L.L.; Li, W.; Oldigs, J.K.; Kaminski, A.; Gerstmayer, B.; Piechaczek, C.; Wagner, W.; Li, R.K.; Ma, N.; Steinhoff, G. Hypoxic/Normoxic Preconditioning Increases Endothelial Differentiation Potential of Human Bone Marrow CD133+ cells. *Tissue Eng. Part C Methods* **2010**, *16*, 1069–1081. [CrossRef] [PubMed]

24. Klopsch, C.; Skorska, A.; Ludwig, M.; Lemcke, H.; Maass, G.; Gaebel, R.; Beyer, M.; Lux, C.; Toelk, A.; Müller, K.; et al. Intramyocardial angiogenetic stem cells and epicardial erythropoietin save the acute ischemic heart. *Dis. Models Mech.* **2018**, *11*, dmm033282. [CrossRef]

25. Mark, P.; Kleinsorge, M.; Gaebel, R.; Lux, C.A.; Toelk, A.; Pittermann, E.; David, R.; Steinhoff, G.; Ma, N. Human Mesenchymal Stem Cells Display Reduced Expression of CD105 after Culture in Serum-Free Medium. *Stem Cells Int.* **2013**, *2013*, 698076. [CrossRef]

26. Cashman, T.J.; Gouon-Evans, V.; Costa, K.D. Mesenchymal Stem Cells for Cardiac Therapy: Practical Challenges and Potential Mechanisms. *Stem Cell Rev. Rep.* **2013**, *9*, 254–265. [CrossRef]

27. Wang, L.; Tang, S.; Wang, Y.; Xu, S.; Yu, J.; Zhi, X.; Ou, Z.; Yang, J.; Zhou, P.; Shao, Z. Ecto-5′-Nucleotidase (CD73) Promotes Tumor Angiogenesis. *Clin. Exp. Metastasis* **2013**, *30*, 671–680. [CrossRef]

28. Allard, B.; Turcotte, M.; Spring, K.; Pommey, S.; Royal, I.; Stagg, J. Anti-CD73 Therapy Impairs Tumor Angiogenesis. *Int. J. Cancer* **2014**, *134*, 1466–1473. [CrossRef]

29. Antonioli, L.; Blandizzi, C.; Malavasi, F.; Ferrari, D.; Haskó, G. Anti-CD73 Immunotherapy: A Viable Way to Reprogram the Tumor Microenvironment. *Oncoimmunology* **2016**, *5*, e1216292. [CrossRef]

30. Böring, Y.C.; Flögel, U.; Jacoby, C.; Heil, M.; Schaper, W.; Schrader, J. Lack of Ecto-5′-Nucleotidase (CD73) Promotes Arteriogenesis. *Cardiovasc. Res.* **2013**, *97*, 88–96. [CrossRef]

31. Lu, X.; Dunn, J.; Dickinson, A.M.; Gillespie, J.I.; Baudouin, S.V. Smooth muscle α-actin expression in endothelial cells derived from CD34+ human cord blood cells. *Stem Cells Dev.* **2004**, *13*, 521–527. [CrossRef]

32. Vered, M.; Shnaiderman-Shapiro, A.; Zlotogorski-Hurvitz, A.; Salo, T.; Yahalom, R. Cancer-associated fibroblasts in the tumor microenvironment of tongue carcinoma is a heterogeneous cell population. *Acta Histochem.* **2019**, *121*, 151446. [CrossRef]

33. Kinner, B.; Zaleskas, J.M.; Spector, M. Regulation of smooth muscle actin expression and contraction in adult human mesenchymal stem cells. *Exp. Cell Res.* **2002**, *278*, 72–83. [CrossRef]

34. Hung, S.C.; Kuo, P.Y.; Chang, C.F.; Chen, T.H.; Ho, L.L.T. Alpha-smooth muscle actin expression and structure integrity in chondrogenesis of human mesenchymal stem cells. *Cell Tissue Res.* **2006**, *324*, 457–466. [CrossRef]

35. Yang, F.; Liu, Y.H.; Yang, X.P.; Xu, J.; Kapke, A.; Carretero, O.A. Myocardial Infarction and Cardiac Remodelling in Mice. *Exp. Physiol.* **2002**, *87*, 547–555. [CrossRef]

36. Cochain, C.; Channon, K.M.; Silvestre, J.S. Angiogenesis in the infarcted myocardium. *Antioxid. Redox Signal.* **2013**, *18*, 1100–1113. [CrossRef]

37. Wang, M.; Hu, Y.; Shima, I.; Stearns, M.E. IL-10/IL-10 Receptor Signaling Regulates TIMP-1 Expression. *Cancer Biol. Ther.* **2002**, *1*, 556–563. [CrossRef]

38. Al-Amran, F.F.; Shahkolahi, M. Oxytocin Ameliorates the Immediate Myocardial Injury in Heart Transplant through Down Regulation of the Neutrophil Dependent Myocardial Apoptosis. *Heart Views* **2014**, *15*, 37–45. [CrossRef]

39. De Filippo, K.; Dudeck, A.; Hasenberg, M.; Nye, E.; van Rooijen, N.; Hartmann, K.; Gunzer, M.; Roers, A.; Hogg, N. Mast Cell and Macrophage Chemokines CXCL1/CXCL2 Control the Early Stage of Neutrophil Recruitment during Tissue Inflammation. *Blood* **2013**, *121*, 4930–4937. [CrossRef]

40. Frangogiannis, N.G. The Role of Transforming Growth Factor (TGF)-beta in the Infarcted Myocardium. *J. Thorac. Dis.* **2017**, *9*, S52–S63. [CrossRef]

41. English, K.; Barry, F.P.; Field-Corbett, C.P.; Mahon, B.P. IFN-γ and TNF-α differentially regulate immunomodulation by murine mesenchymal stem cells. *Immunol. Lett.* **2007**, *110*, 91–100. [CrossRef]

42. Zhou, L.; Lopes, J.E.; Chong, M.M.; Ivanov, I.I.; Min, R.; Victora, G.D.; Shen, Y.; Du, J.; Rubtsov, Y.P.; Rudensky, A.Y.; et al. TGF-β-induced Foxp3 inhibits T H 17 cell differentiation by antagonizing RORγt function. *Nature* **2008**, *453*, 236. [CrossRef] [PubMed]

43. Holan, V.; Hermankova, B.; Krulova, M.; Zajicova, A. Cytokine interplay among the diseased retina, inflammatory cells and mesenchymal stem cells-a clue to stem cell-based therapy. *World J. Stem Cells* **2019**, *11*, 957. [CrossRef]

44. Chen, L.; Tredget, E.E.; Wu, P.Y.; Wu, Y. Paracrine factors of mesenchymal stem cells recruit macrophages and endothelial lineage cells and enhance wound healing. *PLoS ONE* **2008**, *3*, e1886. [CrossRef]
45. Kinnaird, T.; Stabile, E.; Burnett, M.S.; Shou, M.; Lee, C.W.; Barr, S.; Fuchs, S.; Epstein, S.E. Local delivery of mar-row-derived stromal cells augments collateral perfusion through paracrine mechanisms. *Circulation* **2004**, *109*, 1543–1549. [CrossRef]
46. Ferrara, N.; Gerber, H.P.; LeCouter, J. The biology of VEGF and its receptors. *Nat. Med.* **2003**, *9*, 669. [CrossRef]
47. Liu, G.; Li, L.; Huo, D.; Li, Y.; Wu, Y.; Zeng, L.; Cheng, P.; Xing, M.; Zeng, W.; Zhu, C. A VEGF delivery system targeting MI improves angiogenesis and cardiac function based on the tropism of MSCs and layer-by-layer self-assembly. *Biomaterials* **2017**, *127*, 117–131. [CrossRef]

 *cells*

Review

Engineered Maturation Approaches of Human Pluripotent Stem Cell-Derived Ventricular Cardiomyocytes

Feixiang Ge [1], Zetian Wang [2] and Jianzhong Jeff Xi [1],*

[1] State Key Laboratory of Natural and Biomimetic Drugs, Department of Biomedical Engineering, College of Engineering, Peking University, Beijing 100871, China; ge_fei_xiang@163.com

[2] Institute of Microelectronics, Peking University, Beijing 100871, China; zt.wang@pku.edu.cn

*   Correspondence: jzxi@pku.edu.cn

Received: 14 November 2019; Accepted: 16 December 2019; Published: 18 December 2019

Abstract: Heart diseases such as myocardial infarction and myocardial ischemia are paroxysmal and fatal in clinical practice. Cardiomyocytes (CMs) differentiated from human pluripotent stem cells provide a promising approach to myocardium regeneration therapy. Identifying the maturity level of human pluripotent stem cell-derived cardiomyocytes (hPSC-CMs) is currently the main challenge for pathophysiology and therapeutics. In this review, we describe current maturity indicators for cardiac microtissue and microdevice cultivation technologies that accelerate cardiac maturation. It may provide insights into regenerative medicine, drug cardiotoxicity testing, and preclinical safety testing.

Keywords: human pluripotent stem cell; cardiomyocytes; ventricular; cardiac tissue engineering; maturation

---

1. Introduction

It is well-acknowledged that significant differences in hearts exist between human and model organisms. These morphological and physiological differences can lead to complex problems, such as low pathological reproducibility in clinical practice [1]. On the other hand, human pluripotent stem cells (hPSCs), including human embryonic stem cells (hESCs) and human-induced pluripotent stem cells (hiPSCs), benefitting from the property of indefinite proliferation in vitro and the capacity to differentiate into different types of somatic cells, are a promising tool in biomedical applications. Compared with the transdifferentiation of human somatic cells, the differentiation of hPSCs seems to be more efficient in terms of productivity, safety, and cost. The rapid development of hPSC research in the past few decades has made it possible to utilize hPSC-derived cardiomyocytes in large-scale cardiac tissue engineering directly.

Several effective protocols have been successfully developed to induce hPSCs to become cardiomyocytes. Embryoid body (EB) was first used in the differentiation of cardiomyocytes from hESCs and hiPSCs; however, its effectiveness and reproducibility were found to be problematic because of serum quality instabilities and heterogeneous EB sizes [2]. Differentiation protocols developed using serum-free and compound-defined media were subsequently used and improved the efficiency and reproducibility of cardiomyocytes generated from EBs. Engineering approaches to the production of homogeneous EBs emerged several years ago [3,4]. This method produced more homogenously sized EBs compared with conventional methods that used 96-well plates, and it was also appropriate to scale-up. Because of the limitations of EB protocols, monolayer three-dimensional (3D) approaches have drawn increasing attention over the past several years. Uniform hESC colonies were plated on Matrigel via a microcontact approach and the size range was optimized for maximizing mesoderm formation and cardiac induction. The method of activation of canonical Wnt signaling by the glycogen synthase kinase-3 (GSK-3) inhibitor (CHIR99021) followed by inhibition of Wnt signaling via Inhibitor of Wnt Production 2 (IWP2) or Inhibitor of Wnt Production 4 (IWP4) were found to be sufficient to produce numerous functional cardiomyocytes from multiple human pluripotent stem cell lines in two weeks without exogenous growth factors or genetic manipulation in adherent culture or suspension culture system [5–7]. Paul et al. designed a cardiomyocytes differentiation strategy by using a medium including three components: RPMI-1640, L-ascorbic acid 2-phosphate, and rice-derived recombinant human albumin [8]. It is essential for cardiac development in vitro through an appropriate addition of different growth factors, including fibroblast growth factor-2 (FGF-2), transforming growth factor-β (TGF-β), superfamily growth factors activin A, bone morphogenetic protein-4 (BMP-4), vascular endothelial growth factor (VEGF), and dickkopf WNT signaling pathway inhibitor 1 (DKK-1). All of these factors were found to assist human pluripotent stem cells generate myocardial precursor cells and cardiomyocytes when added in order [9].

In clinical practice, patient-derived hiPSC-CMs are an optimal disease model for personalized medicine involving inherited cardiac diseases and stem cell therapies to repair or replace injured heart tissues. hiPSC-CMs can be used to model several heart diseases, including Duchenne muscular dystrophy [10], Leopard syndrome [11], long QT syndrome [12], Timothy syndrome [13], Fabry disease [14], Danon disease [15], and familial hypertrophic cardiomyopathy [16]. In addition, hiPSC-CMs from mitochondrial cardiomyopathy of Barth syndrome (BTHS) have been used to generate a platform for pathogenesis and medical therapeutics. This cardiomyopathy model shows irregular sarcomeres, abnormal myocardial contraction, and defective heart function; more importantly, it mimics mitochondrial functional impairment caused by a mature cardiolipin defect [17]. Masahide et al. constructed a Torsade de Pointes (TdP) arrhythmias model from hiPSCs to mimic a patient's disease condition and provide a chance to study the mechanisms of TdP generation and develop an anti-arrhythmias drug test [18]. Overexpression of CDK1, CDK4, cyclin B1, and cyclin D1 efficiently induced cell cycle progress in at least 15% of post-mitotic murine and human cardiomyocytes [19]. Nutlin-3a can selectively activate the p53 signaling pathway and induce cell apoptosis of DNA-damaged iPSCs except for DNA-damage-free cells. These iPSC-CMs were engrafted into an ischemic mouse heart to enhance mouse cardiac beating [20]. This technology may bring about potential benefits for patients with a cardiac disease in clinical medicine.

However, evidence indicates that cardiomyocytes differentiated by these approaches are not as mature as an adult phenotype, thus they may not be able to reflect the physiological response of the adult heart accurately. In addition, with respect to cardiac tissue engineering, cardiomyocytes that more closely resemble those of the native myocardium would contribute more to myocardial repair. For example, as cardiovascular diseases predominantly occur in elderly humans, immature hPSC-CMs may cause modeling to be imprecise and futile [21]. In this review, we discuss the state of current approaches to obtaining more mature cardiomyocytes.

2. Characteristics of Mature and Immature Cardiomyocytes

The maturation level of human cardiomyocytes is crucial to clinical therapy. Immature cardiomyocytes fail to maintain full cardiac function and may lead to an aberrant remodeling of cardiac wall and cardiomyocyte hypertrophy because of differences in cell size, myofibrillar switch, conduction velocity, metabolism, and calcium handling between the two statuses. Table 1 briefly summarizes some typical physiological and chemical differences between mature human cardiomyocytes and immature hPSC-CMs and provides criteria for defining mature cardiomyocytes [22–25]. For a review of the details, we refer the reader to the work of Xiulang Yang et al. [23].

Table 1. Distinctions between mature human cardiomyocytes (CMs) and immature CMs.

		Mature CMs	**Immature CMs**
Structure	Structure	Rod-shaped	Round and irregular
	Alignment	Orderly	Disorderly
	Nucleation	20–30% binuclear or polynuclear	Slightly binuclear
	Beating	Quiescent	Spontaneous
	Length–width ratio	5–10:1	1–3:1
	Sarcomere banding	Z discs, I band, H band, A band, M band	Z discs, I band
	Sarcomere length	2.2 μm	1.6 μm
	Troponin	cTnT, high β-MHC/α-MHC, high MLC2v/MLC2a, high cTnI/fetal ssTnI, Titin isoform N2B, ADRA1A	cTnT, low β-MHC/α-MHC, nondeterministic MLC2v/MLC2a, low cTnI/fetal ssTnI, Titin isoform N2BA
	SRP	High CSQ, PLN, RYR2, SERCA/ATP2A2	Low CSQ, PLN, RYR2, SERCA/ATP2A2
	T-tubules	Present	Not present
	Mitochondria	Regularly distributed; 20–40% of cell volume	Irregularly distributed; paucity
	LGJ	Intercalated discs	Circumferential
Biochemistry	Metabolism	Fatty acid β-oxidative	Glycolysis and lactate
Biophysical	Force	40–80 mN/mm^2 for muscle lines μN range for a single cell	0.08-4 mN/mm^2 for 3D cultivation nN range for a single cell
Electrophysiology	Capacitance	150 pF	10–30 pF
	RMP	−80 to −90 mV	−20 to −60 mV
	Upstroke velocity	100–300 V/s	10–50 V/s
	Conduction velocity	60 cm/s	10–20 cm/s
	APA	100–110 mV	70–120 mV

cTnT, Cardiac troponin T2; β-MHC, Myosin heavy β chain; MLC2v, Myosin light chain 2 ventricular isoform; MLC2a, Myosin light chain 2 atrial isoform; ADRA1A, Adrenoceptor α1A; cTnI, Cardiac troponin I3; CSQ, Calsequestrin; PLN, Phospholamban; RYR2, Ryanodine receptor 2; SERCA/ATP2A2, Sarco/endoplasmic reticulum calcium transport ATPases; SRP, Sarcoplasmic Reticulum Proteins; T-tubule, Transverse tubule; RMP, Resting Membrane Potential; LGJ, Location of Gap Junctions; APA, Action Potential Amplitude.

3. Approaches to Obtaining Mature Cardiomyocytes

In order to obtain more mature and functional hPSC-CMs, the provision of a similar physiological microenvironment in the process of cardiomyocyte development may be a feasible adult direction. In recent years, academics have performed various experiments to stimulate cardiomyocyte maturity, including biophysical, biochemical, electrophysiological, and mechanical experiments.

3.1. Biophysical and Biochemical Factors

Several practicable methods have been used to promote the maturation of cardiomyocytes, including long-term cultivation, a specified material, a three-dimensional culture, a microfluidic system, a co-culture with other cells following transplantation to model organisms, a dynamic sustainability system, drugs, and metabolic regulation.

Long-term cultivation and stiff matrix have been shown to enhance human pluripotent stem cell-derived cardiomyocyte sarcomere formation, calcium handling, and ion channel protein expression [26,27]. Collagen-coated polyacrylamide gels with an elastic moduli of 10 kPa have been shown to lead to aligned sarcomeres in comparison with a stiffer substrate [28]. Mihic et al. generated human-engineered cardiac tissues from hESC-CMs in a large gelatin. Human-engineered cardiac tissues were subjected to a cyclic stretch and their cell size increased, their Z discs were organized, and the Connexin-43 expression increased significantly [29]. Biohybrids of collagen and pristine graphene increased the metabolic activity of human pluripotent stem cell-derived cardiomyocytes and enhanced sarcomere structures [30].

Three-dimensional (3D) culture systems can mimic the native cardiomyocyte microenvironment in vivo to support the maturation of cardiomyocytes. Tulloch et al. generated 3D human-engineered cardiac tissues from hPSC-CMs in collagen that was seeded into a channel with a silicon floor plus nylon mesh anchors. After seven days, myofibril and Z-disc alignment increased [31]. Lee et al. used 3D bio-print collagen to obtain a human heart tissue model that possessed synchronized contraction and directional action potential propagation [32]. hESC-CMs in 3D patches exhibited more mature characteristics, including significantly faster conduction velocities and longer sarcomeres as compared with two-dimensional (2D) monolayers. The conduction velocities of these cardiac patches increased significantly as the purity of the cardiomyocytes increased [33]. Human cardiac muscle patches transplanted into swine were shown to prominently improve left ventricular function and myocardial stress, promote myocardial hypertrophy, and reduce myocardial apoptosis [34]. It was shown that a 3D culture suppressed smooth muscle -actin content and increased the expression of several cardiac markers [35].

Microfluidic systems can be used to study disease and organoid models. In recent years, combinations of microfluidic systems and functional human myocardium have been developed for drug cardiotoxicity testing [36]. Flow culture systems provide continuous gas and nutrient exchange to induce cardiomyocyte maturation. Dynamic cultures result in an enhancement in sarcomeric protein expression, an increase in size, augmentation of the contraction force, and a higher conduction velocity [37].

In addition, mixtures of human primary or human-induced pluripotent stem cell-derived cardiomyocytes (hiPSC-CMs), fibroblasts, and endothelial cells have been used to obtain vascularized functional myocardium. This improvement has allowed us to introduce blood flow into cardiomyocyte cultivation systems and paves the way to cardiomyocyte metabolism and maturation [38]. Human cardiomyocyte patches, through several types of cells derived from hPSCs, were shown to enhance the capacity for excitation contraction coupling, calcium handling, and force generation. Moreover, Johannes and co-workers found that co-transplantation of hESC-derived epicardial cells and cardiomyocytes could double the cardiomyocyte proliferation and augmented angiogenesis between the graft and the host simultaneously [39]. Triiodothyronine is essential to myosin heavy chains (MHCs) and Titin isoform switchover in normal cardiac development. Addition of triiodothyronine was shown to compel immature cardiomyocytes to show several maturation characteristics [40]. The co-inhibition of HIF1 (hypoxia-inducible factor 1) and lactate dehydrogenase A promoted the function maturation of hPSC-CM as mitochondria prefer to conduct oxidative phosphorylation rather than aerobic glycolysis and resulted in sarcomere length increase and contraction stress enhance [41].

These biophysical and biomedical approaches promote the growth and proliferation of immature cardiomyocytes and the formation of adult-like cardiac tissue with an organized ultrastructure, longer sarcomeres, more intensively developed mitochondria, more T-tubules, a more mature oxidative metabolism, and more rapid calcium handling.

3.2. Electrophysiological Stimulation

The spontaneous beating of cardiomyocytes is directly regulated by the cells of atrio-ventricular nodes in vivo. A combination of 3D cultivation with 6 Hz of electrophysiological stimulation was shown to markedly increase myofibril ultrastructural organization and cardiomyocyte size, elevate the conduction velocity, and improve both electrophysiological and calcium ion handling in hPSC-CMs [42]. Electrophysiological stimulation of 2 Hz was used to culture hESC-CMs in a 3D matrix, and significant improvements in contraction and calcium handling were obtained [43]. Chiu et al. found that an electrical field with a symmetric biphasic square and strengths of 2–5 V/cm at a frequency of 1 Hz could enhance the hallmarks of cardiomyocyte maturation in vitro [44]. Thus, the maturation process of cardiomyocytes in vitro progresses more quickly when accompanied by optimized electrophysiological stimulation.

More and more researchers are becoming aware of the fact that electrophysiological stimulation can promote the maturation of hPSC-CMs. However, it remains hard to compare the results from different assays as a universal and gold standard is currently lacking. Besides this, it is obvious that diverse electric field intensities and stimulation frequencies, hPSC cell types, and cultivation conditions can lead to differences in the maturation of cardiomyocytes. Thus, there may be benefits to establish a compatible platform to assess the maturation process, and make comparisons in different stimulation models.

3.3. Mechanical Stress

Functional cardiomyocytes are linked to varieties of cells and structure in vivo. These structures provide the cells with anchors to contract and contribute to physiological hypertrophy. Mechanical loading may be the efficient factor with the most potential when considering the explosion of research in this area in recent years. Mechanical stress increases cells' size and improves the contraction force that is associated with hypertrophic growth. Jianzhong et al. devised a system for assembling muscle-powered microdevices based on precise manipulation of materials to monitor muscle tissue function [45]. Furthermore, periodic stretching of hPSC-CMs in a 3D structure mixture was shown to cause faster force production, higher calcium influxes, an increased expression of β-MHC and cTnT [35]. Schmelter et al. demonstrated that cyclic mechanical stretching activated the Reactive Oxygen Species (ROS) signaling pathway and enhanced the differentiation of ESCs into cardiomyocytes [46]. Ronaldson et al. formed cardiac grafts from early stage iPSC-derived cardiomyocytes and trained them via cyclic mechanical stress for several weeks. After one month, the grafted cardiomyocytes showed adult-like gene expression profiles, increased sarcomere length, enhanced density of mitochondria, the presence of T-tubules, metabolism switch, and functional calcium handling [47]. Table 2 summarizes some typical engineered approaches to the maturation of human and rodent cardiomyocytes.

Table 2. Different methods for maturing cardiomyocytes.

Stimuluses	Cultured Cell Types	Maturation Conditions	Reference
Electric stimulation	Hes3 hESCs	After 4 days of culturing in the presence of electric field stimulation (a 6.6 V/cm, 1 Hz, and 2 ms pulse), hESC-CM elongation and troponin-T enhancement.	[48]
	Hes2 and Hes3 hESCs and CDI-MRB HR-I-2Cr-2R hiPSCs	Biowires increased myofibril ultrastructural organization, elevated conduction velocity, improved Calcium handling properties, and produced better electrophysiological performance.	[42]
	C25 hiPSCs	2 Hz in the first week and 1.5 Hz thereafter, developed 1.5-fold contractile forces.	[49]
	hiPSC-CMs (ReproCardio 2)	Efficient electrical stimulations were formed by a hydrogel-based microchamber with organic electrodes. The large interfacial capacitance of the electrodes eliminated cytotoxic bubbles.	[50]
Electric stimulation and mechanical strain	Neonatal Rat Heart Cells	Engineered heart muscle was subjected to electric stimulation at 0, 2, 4, or 6 Hz for 5 days and engineered flexible poles facilitated auxotonic contractions by straining. Force–frequency relationships of 2 and 4 Hz stimulation were divergent.	[51]
	C2A, WTC-11, IMR90, and BS2 hiPSCs	After the first contraction was observed, tissue was immediately subjected to 21 days of increasingly intense electromechanical strain. Cell properties were then evaluated by a multiplex test.	[52]

Table 2. *Cont.*

Stimuluses	Cultured Cell Types	Maturation Conditions	Reference
Mechanical loading	HES2 hESCs	Cyclical stretching produced Cardiac troponin T elevation, cell elongation, and an increase in gap junction.	[29]
	IMR90 ESCs and IBJ hiPSCs	Compared to a 2D culture, a 3D environment increased the number of cardiomyocytes and decreased the number of smooth muscles. With cyclic stress, expression of several cardiac markers increased, including β-myosin heavy chains and cardiac troponin T.	[35]
Mechanical loading and vascular co-culture	H7 hESCs	Cyclic stress enhanced cardiomyocyte hypertrophy and proliferation rates significantly and endothelial cells showed the formation of vessel-like structures.	[31]
Textile based-culturing	UTA.04602 hiPSC	Gelatin-coated polyethylene terephthalate-based textiles were used as the culturing surface. hiPSC-CMs showed improved structural properties.	[53]
Substrate stiffness	Neonatal Rat Ventricular Myocytes	Substrates of varying elastic moduli were fabricated. Cardiomyocytes matured on 10 kPa gels were similar to the native myocardium and generated a greater mechanical force and the largest calcium transients.	[28]

Mechanical loading can improve the rate of maturation of hPSC-CMs and contractile properties. All these characteristics reflect the state of maturity of these cells. However, there is little published data about the real-time monitoring of cardiomyocyte development; to date, the shortage of clinical feedback has slowed its application.

4. Conclusions

In this review, we summarized the approaches that have been adopted to improve the maturity of hPSC-CMs under different conditions. Human embryonic cardiac development and postnatal physiological hypertrophy processes are difficult to study because of species specificity and the lack of availability of human heart tissues. Mature hPSC-CMs may reflect the pathological state in adults more accurately and serve as preferential disease models for clinical use.

With the rapid development in this multidisciplinary field, our understanding of the maturation of human cardiomyocytes has been growing in recent years. Many studies on the maturation of cardiomyocytes have been published in multiple journals in the last decade. Some methods have been applied to successfully produce adult-like cardiomyocytes with respect to the biochemical indicators; however, the remaining methods still need improvement. Current hurdles include achieving adult-like cardiomyocytes with angiogenesis and organized, mixed assemblies of multi-layer 3D heart tissues. A real-time assessment system is required to compare different approaches and obtain an optimized maturation status in hPSC-CMs. Besides this, in order to monitor feedback from cell and tissue signals, we may need an electrophysiological surveillance system that can regulate the differentiation and maturation of cardiomyocytes in real time in clinical practices. A mature and functional human cardiac model in vitro could play a role in myocardial tissue development research, cardiotoxicity drug screening, and clinical therapies.

Author Contributions: F.G. wrote original draft. F.G., Z.W. and J.J.X. reviewed and edited manuscript. All authors have read and agreed to the published version of the manuscript.

Funding: This research received no external funding.

Acknowledgments: Special thanks are given to Smina Mukhtar and Martyn Rittman for excellent scientific secretary assistance and English editing service.

Conflicts of Interest: The authors declare no conflict of interest.

References

1. Milani-Nejad, N.; Janssen, P.M. Small and large animal models in cardiac contraction research: Advantages and disadvantages. *Pharmacol. Ther.* **2014**, *141*, 235–249. [CrossRef] [PubMed]
2. Boheler, K.R.; Czyz, J.; Tweedie, D.; Yang, H.T.; Anisimov, S.V.; Wobus, A.M. Differentiation of pluripotent embryonic stem cells into cardiomyocytes. *Circ. Res.* **2002**, *91*, 189–201. [CrossRef]
3. Bauwens, C.L.; Peerani, R.; Niebruegge, S.; Woodhouse, K.A.; Kumacheva, E.; Husain, M.; Zandstra, P.W. Control of human embryonic stem cell colony and aggregate size heterogeneity influences differentiation trajectories. *Stem Cells* **2008**, *26*, 2300–2310. [CrossRef] [PubMed]
4. Peerani, R.; Rao, B.M.; Bauwens, C.; Yin, T.; Wood, G.A.; Nagy, A.; Kumacheva, E.; Zandstra, P.W. Niche-mediated control of human embryonic stem cell self-renewal and differentiation. *EMBO J.* **2007**, *26*, 4744–4755. [CrossRef] [PubMed]
5. Lian, X.; Zhang, J.; Azarin, S.M.; Zhu, K.; Hazeltine, L.B.; Bao, X.; Hsiao, C.; Kamp, T.J.; Palecek, S.P. Directed cardiomyocyte differentiation from human pluripotent stem cells by modulating Wnt/β-catenin signaling under fully defined conditions. *Nat. Protoc.* **2012**, *8*, 162–175. [CrossRef] [PubMed]
6. Lian, X.; Hsiao, C.; Wilson, G.; Zhu, K.; Hazeltine, L.B.; Azarin, S.M.; Raval, K.K.; Zhang, J.; Kamp, T.J.; Palecek, S.P. Robust cardiomyocyte differentiation from human pluripotent stem cells via temporal modulation of canonical Wnt signaling. *Proc. Natl. Acad. Sci. USA* **2012**, *109*, E1848–E1857. [CrossRef]
7. Halloin, C.; Schwanke, K.; Lobel, W.; Franke, A.; Szepes, M.; Biswanath, S.; Wunderlich, S.; Merkert, S.; Weber, N.; Osten, F.; et al. Continuous WNT Control Enables Advanced hPSC Cardiac Processing and Prognostic Surface Marker Identification in Chemically Defined Suspension Culture. *Stem Cell Rep.* **2019**, *13*, 366–379. [CrossRef]
8. Burridge, P.W.; Matsa, E.; Shukla, P.; Lin, Z.C.; Churko, J.M.; Ebert, A.D.; Lan, F.; Diecke, S.; Huber, B.; Mordwinkin, N.M.; et al. Chemically defined generation of human cardiomyocytes. *Nat. Methods* **2014**, *11*, 855–860. [CrossRef]
9. Yang, L.; Soonpaa, M.H.; Adler, E.D.; Roepke, T.K.; Kattman, S.J.; Kennedy, M.; Henckaerts, E.; Bonham, K.; Abbott, G.W.; Linden, R.M.; et al. Human cardiovascular progenitor cells develop from a KDR+ embryonic-stem-cell-derived population. *Nature* **2008**, *453*, 524–528. [CrossRef]
10. Macadangdang, J.; Guan, X.; Smith, A.S.; Lucero, R.; Czerniecki, S.; Childers, M.K.; Mack, D.L.; Kim, D.H. Nanopatterned Human iPSC-based Model of a Dystrophin-Null Cardiomyopathic Phenotype. *Cell. Mol. Bioeng.* **2015**, *8*, 320–332. [CrossRef]
11. Carvajal-Vergara, X.; Sevilla, A.; D'Souza, S.L.; Ang, Y.S.; Schaniel, C.; Lee, D.F.; Yang, L.; Kaplan, A.D.; Adler, E.D.; Rozov, R.; et al. Patient-specific induced pluripotent stem-cell-derived models of LEOPARD syndrome. *Nature* **2010**, *465*, 808–812. [CrossRef] [PubMed]
12. Moretti, A.; Bellin, M.; Welling, A.; Jung, C.B.; Lam, J.T.; Bott-Flugel, L.; Dorn, T.; Goedel, A.; Hohnke, C.; Hofmann, F.; et al. Patient-specific induced pluripotent stem-cell models for long-QT syndrome. *N. Engl. J. Med.* **2010**, *363*, 1397–1409. [CrossRef] [PubMed]
13. Yazawa, M.; Hsueh, B.; Jia, X.; Pasca, A.M.; Bernstein, J.A.; Hallmayer, J.; Dolmetsch, R.E. Using induced pluripotent stem cells to investigate cardiac phenotypes in Timothy syndrome. *Nature* **2011**, *471*, 230–234. [CrossRef] [PubMed]
14. Itier, J.M.; Ret, G.; Viale, S.; Sweet, L.; Bangari, D.; Caron, A.; Le-Gall, F.; Benichou, B.; Leonard, J.; Deleuze, J.F.; et al. Effective clearance of GL-3 in a human iPSC-derived cardiomyocyte model of Fabry disease. *J. Inherit. Metab. Dis.* **2014**, *37*, 1013–1022. [CrossRef]
15. Nishino, I.; Fu, J.; Tanji, K.; Yamada, T.; Shimojo, S.; Koori, T.; Mora, M.; Riggs, J.E.; Oh, S.J.; Koga, Y.; et al. Primary LAMP-2 deficiency causes X-linked vacuolar cardiomyopathy and myopathy (Danon disease). *Nature* **2000**, *406*, 906–910. [CrossRef]
16. Giacomelli, E.; Mummery, C.L.; Bellin, M. Human heart disease: Lessons from human pluripotent stem cell-derived cardiomyocytes. *Cell. Mol. Life Sci.* **2017**, *74*, 3711–3739. [CrossRef]
17. Wang, G.; McCain, M.L.; Yang, L.; He, A.; Pasqualini, F.S.; Agarwal, A.; Yuan, H.; Jiang, D.; Zhang, D.; Zangi, L.; et al. Modeling the mitochondrial cardiomyopathy of Barth syndrome with induced pluripotent stem cell and heart-on-chip technologies. *Nat. Med.* **2014**, *20*, 616–623. [CrossRef]

18. Kawatou, M.; Masumoto, H.; Fukushima, H.; Morinaga, G.; Sakata, R.; Ashihara, T.; Yamashita, J.K. Modelling Torsade de Pointes arrhythmias in vitro in 3D human iPS cell-engineered heart tissue. *Nat. Commun.* **2017**, *8*, 1078. [CrossRef]

19. Mohamed, T.M.A.; Ang, Y.S.; Radzinsky, E.; Zhou, P.; Huang, Y.; Elfenbein, A.; Foley, A.; Magnitsky, S.; Srivastava, D. Regulation of Cell Cycle to Stimulate Adult Cardiomyocyte Proliferation and Cardiac Regeneration. *Cell* **2018**, *173*, 104–116.e112. [CrossRef]

20. Kannappan, R.; Turner, J.F.; Miller, J.M.; Fan, C.; Rushdi, A.G.; Rajasekaran, N.S.; Zhang, J. Functionally Competent DNA Damage-Free Induced Pluripotent Stem Cell-Derived Cardiomyocytes for Myocardial Repair. *Circulation* **2019**, *140*, 520–522. [CrossRef]

21. Feric, N.T.; Radisic, M. Maturing human pluripotent stem cell-derived cardiomyocytes in human engineered cardiac tissues. *Adv. Drug Deliv. Rev.* **2016**, *96*, 110–134. [CrossRef] [PubMed]

22. Denning, C.; Borgdorff, V.; Crutchley, J.; Firth, K.S.; George, V.; Kalra, S.; Kondrashov, A.; Hoang, M.D.; Mosqueira, D.; Patel, A.; et al. Cardiomyocytes from human pluripotent stem cells: From laboratory curiosity to industrial biomedical platform. *Biochim. Biophys. Acta* **2016**, *1863*, 1728–1748. [CrossRef] [PubMed]

23. Yang, X.; Pabon, L.; Murry, C.E. Engineering adolescence: Maturation of human pluripotent stem cell-derived cardiomyocytes. *Circ. Res.* **2014**, *114*, 511–523. [CrossRef] [PubMed]

24. Veerman, C.C.; Kosmidis, G.; Mummery, C.L.; Casini, S.; Verkerk, A.O.; Bellin, M. Immaturity of human stem-cell-derived cardiomyocytes in culture: Fatal flaw or soluble problem? *Stem Cells Dev.* **2015**, *24*, 1035–1052. [CrossRef] [PubMed]

25. Jonsson, M.K.; Vos, M.A.; Mirams, G.R.; Duker, G.; Sartipy, P.; de Boer, T.P.; van Veen, T.A. Application of human stem cell-derived cardiomyocytes in safety pharmacology requires caution beyond hERG. *J. Mol. Cell. Cardiol.* **2012**, *52*, 998–1008. [CrossRef] [PubMed]

26. Rajamohan, D.; Matsa, E.; Kalra, S.; Crutchley, J.; Patel, A.; George, V.; Denning, C. Current status of drug screening and disease modelling in human pluripotent stem cells. *BioEssays* **2013**, *35*, 281–298. [CrossRef] [PubMed]

27. Weber, N.; Schwanke, K.; Greten, S.; Wendland, M.; Iorga, B.; Fischer, M.; Geers-Knorr, C.; Hegermann, J.; Wrede, C.; Fiedler, J.; et al. Stiff matrix induces switch to pure beta-cardiac myosin heavy chain expression in human ESC-derived cardiomyocytes. *Basic Res. Cardiol.* **2016**, *111*, 68. [CrossRef]

28. Jacot, J.G.; McCulloch, A.D.; Omens, J.H. Substrate stiffness affects the functional maturation of neonatal rat ventricular myocytes. *Biophys. J.* **2008**, *95*, 3479–3487. [CrossRef] [PubMed]

29. Mihic, A.; Li, J.; Miyagi, Y.; Gagliardi, M.; Li, S.H.; Zu, J.; Weisel, R.D.; Keller, G.; Li, R.K. The effect of cyclic stretch on maturation and 3D tissue formation of human embryonic stem cell-derived cardiomyocytes. *Biomaterials* **2014**, *35*, 2798–2808. [CrossRef]

30. Ryan, A.J.; Kearney, C.J.; Shen, N.; Khan, U.; Kelly, A.G.; Probst, C.; Brauchle, E.; Biccai, S.; Garciarena, C.D.; Vega-Mayoral, V.; et al. Electroconductive Biohybrid Collagen/Pristine Graphene Composite Biomaterials with Enhanced Biological Activity. *Adv. Mater.* **2018**, *30*. [CrossRef]

31. Tulloch, N.L.; Muskheli, V.; Razumova, M.V.; Korte, F.S.; Regnier, M.; Hauch, K.D.; Pabon, L.; Reinecke, H.; Murry, C.E. Growth of engineered human myocardium with mechanical loading and vascular coculture. *Circ. Res.* **2011**, *109*, 47–59. [CrossRef] [PubMed]

32. Lee, A.; Hudson, A.R.; Shiwarski, D.J.; Tashman, J.W.; Hinton, T.J.; Yerneni, S.; Bliley, J.M.; Campbell, P.G.; Feinberg, A.W. 3D bioprinting of collagen to rebuild components of the human heart. *Science* **2019**, *365*, 482–487. [CrossRef] [PubMed]

33. Zhang, D.; Shadrin, I.Y.; Lam, J.; Xian, H.Q.; Snodgrass, H.R.; Bursac, N. Tissue-engineered cardiac patch for advanced functional maturation of human ESC-derived cardiomyocytes. *Biomaterials* **2013**, *34*, 5813–5820. [CrossRef] [PubMed]

34. Gao, L.; Gregorich, Z.R.; Zhu, W.; Mattapally, S.; Oduk, Y.; Lou, X.; Kannappan, R.; Borovjagin, A.V.; Walcott, G.P.; Pollard, A.E.; et al. Large Cardiac Muscle Patches Engineered From Human Induced-Pluripotent Stem Cell-Derived Cardiac Cells Improve Recovery From Myocardial Infarction in Swine. *Circulation* **2018**, *137*, 1712–1730. [CrossRef] [PubMed]

35. Ruan, J.L.; Tulloch, N.L.; Saiget, M.; Paige, S.L.; Razumova, M.V.; Regnier, M.; Tung, K.C.; Keller, G.; Pabon, L.; Reinecke, H.; et al. Mechanical Stress Promotes Maturation of Human Myocardium From Pluripotent Stem Cell-Derived Progenitors. *Stem Cells* **2015**, *33*, 2148–2157. [CrossRef]

36. Mathur, A.; Ma, Z.; Loskill, P.; Jeeawoody, S.; Healy, K.E. In vitro cardiac tissue models: Current status and future prospects. *Adv. Drug Deliv. Rev.* **2016**, *96*, 203–213. [CrossRef]

37. Jackman, C.P.; Carlson, A.L.; Bursac, N. Dynamic culture yields engineered myocardium with near-adult functional output. *Biomaterials* **2016**, *111*, 66–79. [CrossRef]

38. Richards, D.J.; Coyle, R.C.; Tan, Y.; Jia, J.; Wong, K.; Toomer, K.; Menick, D.R.; Mei, Y. Inspiration from heart development: Biomimetic development of functional human cardiac organoids. *Biomaterials* **2017**, *142*, 112–123. [CrossRef]

39. Bargehr, J.; Ong, L.P.; Colzani, M.; Davaapil, H.; Hofsteen, P.; Bhandari, S.; Gambardella, L.; Le Novere, N.; Iyer, D.; Sampaziotis, F.; et al. Epicardial cells derived from human embryonic stem cells augment cardiomyocyte-driven heart regeneration. *Nat. Biotechnol.* **2019**, *37*, 895–906. [CrossRef]

40. Yang, X.; Rodriguez, M.; Pabon, L.; Fischer, K.A.; Reinecke, H.; Regnier, M.; Sniadecki, N.J.; Ruohola-Baker, H.; Murry, C.E. Tri-iodo-l-thyronine promotes the maturation of human cardiomyocytes-derived from induced pluripotent stem cells. *J. Mol. Cell. Cardiol.* **2014**, *72*, 296–304. [CrossRef]

41. Hu, D.; Linders, A.; Yamak, A.; Correia, C.; Kijlstra, J.D.; Garakani, A.; Xiao, L.; Milan, D.J.; van der Meer, P.; Serra, M.; et al. Metabolic Maturation of Human Pluripotent Stem Cell-Derived Cardiomyocytes by Inhibition of HIF1alpha and LDHA. *Circ. Res.* **2018**, *123*, 1066–1079. [CrossRef] [PubMed]

42. Nunes, S.S.; Miklas, J.W.; Liu, J.; Aschar-Sobbi, R.; Xiao, Y.; Zhang, B.; Jiang, J.; Masse, S.; Gagliardi, M.; Hsieh, A.; et al. Biowire: A platform for maturation of human pluripotent stem cell-derived cardiomyocytes. *Nat. Methods* **2013**, *10*, 781–787. [CrossRef] [PubMed]

43. Ruan, J.L.; Tulloch, N.L.; Razumova, M.V.; Saiget, M.; Muskheli, V.; Pabon, L.; Reinecke, H.; Regnier, M.; Murry, C.E. Mechanical Stress Conditioning and Electrical Stimulation Promote Contractility and Force Maturation of Induced Pluripotent Stem Cell-Derived Human Cardiac Tissue. *Circulation* **2016**, *134*, 1557–1567. [CrossRef] [PubMed]

44. Chiu, L.L.; Iyer, R.K.; King, J.P.; Radisic, M. Biphasic electrical field stimulation aids in tissue engineering of multicell-type cardiac organoids. *Tissue Eng. Part A* **2011**, *17*, 1465–1477. [CrossRef] [PubMed]

45. Xi, J.; Schmidt, J.J.; Montemagno, C.D. Self-assembled microdevices driven by muscle. *Nat. Mater* **2005**, *4*, 180–184. [CrossRef]

46. Schmelter, M.; Ateghang, B.; Helmig, S.; Wartenberg, M.; Sauer, H. Embryonic stem cells utilize reactive oxygen species as transducers of mechanical strain-induced cardiovascular differentiation. *FASEB J.* **2006**, *20*, 1182–1184. [CrossRef]

47. Ronaldson-Bouchard, K.; Ma, S.P.; Yeager, K.; Chen, T.; Song, L.; Sirabella, D.; Morikawa, K.; Teles, D.; Yazawa, M.; Vunjak-Novakovic, G. Advanced maturation of human cardiac tissue grown from pluripotent stem cells. *Nature* **2018**, *556*, 239–243. [CrossRef]

48. Chan, Y.C.; Ting, S.; Lee, Y.K.; Ng, K.M.; Zhang, J.; Chen, Z.; Siu, C.W.; Oh, S.K.; Tse, H.F. Electrical stimulation promotes maturation of cardiomyocytes derived from human embryonic stem cells. *J. Cardiovasc. Transl. Res.* **2013**, *6*, 989–999. [CrossRef]

49. Hirt, M.N.; Boeddinghaus, J.; Mitchell, A.; Schaaf, S.; Bornchen, C.; Muller, C.; Schulz, H.; Hubner, N.; Stenzig, J.; Stoehr, A.; et al. Functional improvement and maturation of rat and human engineered heart tissue by chronic electrical stimulation. *J. Mol. Cell. Cardiol.* **2014**, *74*, 151–161. [CrossRef]

50. Yoshida, S.; Sumomozawa, K.; Nagamine, K.; Nishizawa, M. Hydrogel Microchambers Integrated with Organic Electrodes for Efficient Electrical Stimulation of Human iPSC-Derived Cardiomyocytes. *Macromol. Biosci.* **2019**, *19*. [CrossRef]

51. Godier-Furnemont, A.F.; Tiburcy, M.; Wagner, E.; Dewenter, M.; Lammle, S.; El-Armouche, A.; Lehnart, S.E.; Vunjak-Novakovic, G.; Zimmermann, W.H. Physiologic force-frequency response in engineered heart muscle by electromechanical stimulation. *Biomaterials* **2015**, *60*, 82–91. [CrossRef] [PubMed]

52. Ronaldson-Bouchard, K.; Yeager, K.; Teles, D.; Chen, T.; Ma, S.; Song, L.; Morikawa, K.; Wobma, H.M.; Vasciaveo, A.; Ruiz, E.C.; et al. Engineering of human cardiac muscle electromechanically matured to an adult-like phenotype. *Nat. Protoc.* **2019**, *14*, 2781–2817. [CrossRef] [PubMed]

53. Pekkanen-Mattila, M.; Hakli, M.; Polonen, R.P.; Mansikkala, T.; Junnila, A.; Talvitie, E.; Koivisto, J.T.; Kellomaki, M.; Aalto-Setala, K. Polyethylene Terephthalate Textiles Enhance the Structural Maturation of Human Induced Pluripotent Stem Cell-Derived Cardiomyocytes. *Materials* **2019**, *12*, 1805. [CrossRef] [PubMed]

Article

RNA-Based Strategies for Cardiac Reprogramming of Human Mesenchymal Stromal Cells

Paula Mueller [1,2], Markus Wolfien [3], Katharina Ekat [4], Cajetan Immanuel Lang [5], Dirk Koczan [6], Olaf Wolkenhauer [3,7], Olga Hahn [4], Kirsten Peters [4], Hermann Lang [8], Robert David [1,2,*] and Heiko Lemcke [1,2]

1   Department of Cardiac Surgery, Reference and Translation Center for Cardiac Stem Cell Therapy (RTC), Rostock University Medical Center, 18057 Rostock, Germany; Paula.Mueller@uni-rostock.de (P.M.); Heiko.Lemcke@med.uni-rostock.de (H.L.)
2   Faculty of Interdisciplinary Research, Department Life, Light & Matter, University Rostock, 18059 Rostock, Germany
3   Institute of Computer Science, Department of Systems Biology and Bioinformatics, University of Rostock, 18057 Rostock, Germany; Markus.Wolfien@uni-rostock.de (M.W.); Olaf.wolkenhauer@uni-rostock.de (O.W.)
4   Department of Cell Biology, Rostock University Medical Center, 18057 Rostock, Germany; Katharina.Ekat@med.uni-rostock.de (K.E.); olga.hahn@med.uni-rostock.de (O.H.); Kirsten.Peters@med.uni-rostock.de (K.P.)
5   Department of Cardiology, Rostock University Medical Center, 18057 Rostock, Germany; Cajetan.lang@med.uni-rostock.de
6   Institute of Immunology, Rostock University Medical Center, 18057 Rostock, Germany; Dirk.Koczan@med.uni-rostock.de
7   Stellenbosch Institute of Advanced Study, Wallenberg Research Centre, Stellenbosch University, 7602 Stellenbosch, South Africa
8   Department of Operative Dentistry and Periodontology, Rostock University Medical Center, 18057 Rostock, Germany; Herman.Lang@med.uni-rostock.de
*   Correspondence: Robert.David@med.uni-rostock.de; Tel.: +49-381-498-8973

Received: 6 December 2019; Accepted: 17 February 2020; Published: 22 February 2020

Abstract: Multipotent adult mesenchymal stromal cells (MSCs) could represent an elegant source for the generation of patient-specific cardiomyocytes needed for regenerative medicine, cardiovascular research, and pharmacological studies. However, the differentiation of adult MSC into a cardiac lineage is challenging compared to embryonic stem cells or induced pluripotent stem cells. Here we used non-integrative methods, including microRNA and mRNA, for cardiac reprogramming of adult MSC derived from bone marrow, dental follicle, and adipose tissue. We found that MSC derived from adipose tissue can partly be reprogrammed into the cardiac lineage by transient overexpression of GATA4, TBX5, MEF2C, and MESP1, while cells isolated from bone marrow, and dental follicle exhibit only weak reprogramming efficiency. qRT-PCR and transcriptomic analysis revealed activation of a cardiac-specific gene program and up-regulation of genes known to promote cardiac development. Although we did not observe the formation of fully mature cardiomyocytes, our data suggests that adult MSC have the capability to acquire a cardiac-like phenotype when treated with mRNA coding for transcription factors that regulate heart development. Yet, further optimization of the reprogramming process is mandatory to increase the reprogramming efficiency.

Keywords: mesenchymal stromal cells (MSC); mRNA; miRNA; cardiac reprogramming; cardiac differentiation

---

1. Introduction

Mesenchymal stromal cells (MSC) represent a multipotent cell population capable to differentiate into different cell types [1]. They are an easily-accessible cell source as they can be isolated at high yields from various kinds of human tissue, such as umbilical cord, bone marrow, dental pulp, adipose tissue, placenta, etc. [1]. The common mesenchymal cell types that emanate from MSC are osteocytes, chondrocytes, and adipocytes [2]. Due to their plasticity, MSC are considered as one of the most important cell types for the application in regenerative medicine as demonstrated by a huge number of pre-clinical studies and several clinical trials [3,4]. In addition, MSC mediate immunomodulatory and immunosuppressive effects that promote wound healing and tissue repair, while showing no teratoma formation post transplantation [5]. Nowadays, it is commonly accepted that the observed therapeutic impact induced by MSCs is mainly based on the secretion of paracrine factors rather than on the differentiation into cardiomyocytes.

In recent years, MSC have also been utilized for the generation of mesenchymal as well as non-mesenchymal cell lineages, including neuron-like, hepatocyte-like, and cardiac-like cells [6–10]. Despite these promising results, the differentiation of human MSC into fully mature cardiomyocytes bearing all their respective phenotypical and functional characteristics is difficult [11–15]. As MSC are located in various tissues, they represent a heterogeneous progenitor cell population dependent on the tissue source and the individual donor [16]. This heterogeneity could explain the variety in differentiation characteristics [17–19]. Therefore, it remains to be investigated which type of MSC favorably undergoes cardiac trans-differentiation, thus, is a suitable candidate for cardiac reprogramming strategies. Detailed knowledge about the cardiac differentiation potential of specific MSC populations is even more important as some studies showed enhanced therapeutic effects following cardiovascular lineage commitment of MSC [12].

The development of an approach to efficiently control the cardiac differentiation of MSC would be a crucial step for the production of patient-derived cardiomyocytes without any ethical concerns. As such, they can also serve as a model system, beneficial for basic cardiovascular research, drug screening, and translational applications. Currently, several re/programming strategies exist to guide the mesenchymal and non-mesenchymal differentiation of MSC, such as treatment with small molecules and cytokines, exposure to metabolic stress, co-culture experiments, or overexpression of regulatory proteins [20–24]. For the potential clinical use, transient, non-integrative reprogramming approaches are preferred to prevent permanent alterations of the genome and to reduce tumorigenic risk. Small non-coding RNAs, like microRNAs (miRNA) and chemically modified messenger RNA (mRNAs) allow the manipulation of cell behavior for a limited period of time, e.g., triggering (trans)-differentiation by activation of lineage-specific molecular pathways. Some studies have already shown that alteration of gene expression using selected miRNAs can induce cardiac differentiation of MSC to a small extent [15,25,26], while data about mRNA-based cardiac reprogramming is still lacking.

Unlike multipotent MSCs, pluripotent stem cells (PSCs) have been demonstrated to efficiently differentiate into cardiomyocytes, characterized by a profound sarcomere organization and spontaneous beating behavior [27]. Yet, these PSC-derived cardiomyocytes typically still represent an immature cell type, resembling a neonatal cell stage rather than an adult phenotype [28,29]. The common cardiac programming approaches used to guide cardiac differentiation of PSCs mainly relies on the application of cytokines and small molecules [30,31]. However, PSCs bear tumorigenic risk due to genome modification (induced pluripotent stem cells, iPSC) and provoke ethical concerns (embryonic stem cells, ESC). Therefore, increasing the efficiency of cardiac programming of MSC would be beneficial for cardiovascular research, including their therapeutic use.

Here, we examined whether MSC derived from different sources, including bone marrow (BM), dental follicle and subcutaneous adipose tissue can be driven towards a cardiac lineage using a transient reprogramming strategy based on miRNA and mRNA transfection. According to our results, adipose tissue-derived MSC (adMSC) were found to be the most susceptible cell type for this reprogramming approach, as shown by enhanced expression of cardiac markers. At the same time, we observed the activation of transcriptome pathways involved in cardiac development following mRNA treatment.

2. Material and Methods

2.1. Cell Culture

BM-derived MSC (BM MSC) were obtained by sternal aspiration from donors undergone coronary bypass graft surgery. Anticoagulation was achieved by heparinization with 250 i.E./mL sodium heparin (Ratiopharm, Ulm, Germany). Mononuclear cells were isolated by density gradient centrifugation on 1077 Lymphocyte Separation Medium (LSM; PAA Laboratories, Pasching, Germany). MSC were enriched by plastic adherence and sub-cultured in MSC basal medium supplemented with SingleQuot (all Lonza, Cologne, Germany) and 1% Zellshield (Biochrom, Berlin, Germany).

Isolation of adMSC was performed by liposuction of healthy individuals. The extracted tissue was treated with collagenase for 30 min, followed by several filtrations and washing steps. The detailed process of adMSC isolation has been already described previously [32].

Dental follicle stem cells (DFSCs) were isolated from dental follicles of extracted wisdom teeth before tooth eruption. Following tooth removal, the follicle was removed and subjected to enzymatic treatment as presented earlier [33]. Upon tissue digestion, cells were seeded on tissue flasks and obtained by plastic adherence. DFSCs were maintained in DMEM-F12 (Thermo Fisher, Waltham, USA) supplemented with 10% FCS and 1% Zellshield.

All three types of stromal cells were maintained at 37 °C and 5% CO_2 humidified atmosphere. Medium was changed every 2–3 days. Sub-cultivation was performed when cells reached a confluency of ~80–90%.

All donors have given their written consent for the donation of their tissue according to the Declaration of Helsinki. The study was approved by the ethical committee of the Medical Faculty of the University of Rostock (registration number: bone marrow A2010-23; renewal in 2015; adipose tissue: A2013-0112, renewal in 2019, dental tissue: A 2017-0158).

2.2. Fluorescence-Activated Cell Sorting

The expression of cell surface markers was quantified by flow cytometric analysis. Stromal cells were labelled with antibodies CD29-APC, CD44-PerCP-Cy5.5, CD45-V500, CD73-PE, CD117-PE-Cy7, PerCP-Cy5.5 CD90 (BD Biosciences, San Jose, USA), and CD105-AlexaFluor488 (AbD Serotec, Oxford, UK). Respective isotype antibodies served as negative controls. A measurement of 3×10^4 events was carried out using BD FACS LSRII flow cytometer (BD Biosciences).

To evaluate miRNA and mRNA uptake efficiency, cells were treated with different amounts of Cy3-labeled Pre-miRNA Negative Control #1 (AM17120, Thermo Fisher) or GFP-mRNA (Trilink, San Diego, USA) and analyzed by flow cytometry 24 h post transfection. To detect cytotoxicity, cells were labelled with Near-IR LIVE/DEAD fixable dead cell stain kit (Molecular Probes, Eugene, USA). Analysis of flow cytometry data, including gating, was conducted with the FACSDiva software, Version 8. (Becton Dickinson).

2.3. Cardiac Reprogramming

For cardiac reprogramming, 1×10^5 cells/well were seeded on 0.1% gelatin-coated 6 well plates and cultured to 80% confluency. We transfected 40 pmol of each miRNA (Pre-miR™ hsa-miR-1, Pre-miR™hsa-miR-499a-5p, Pre-miR™hsa-miR-208a-3p, Pre-miR™hsa-miR-133a-3p, all Thermo Fisher) with Lipofectamine® 2000 according to the manufacturer's instructions (Thermo Fisher). Transfection of custom-made mRNA (Trilink) was performed with Viromer Red® transfection reagent (Lipocalyx, Halle, Germany). Cells were either transfected with 2 µg MESP1 or with a combination of 1 µg GATA4, 1 µg MEF2C and 1 µg TBX5. One day after transfection of miRNA or mRNA, cells were subjected to two different medium conditions. For cardiac induction medium I (card ind. I), cells were incubated in RPMI, supplemented with B27 without insulin (Thermo fisher) for 7 days, followed by incubation in RPMI containing B27 +insulin/- vitamin A (Thermo Fisher) for another 21 days. Additionally, culture medium was supplemented with ascorbic acid (Sigma Aldrich, St. Louis, USA) and Wnt pathway targeting small molecules, including 6 μM CHIR99021 (days 1–2), and 5 μM IWP-2 (days 4–5) (both Stemcell Technologies). For cardiac induction II (card ind II), a commercially available cardiomyocyte differentiation kit was used according to the instructions given by the manufacturer (Thermo Fisher, A2921201).

2.4. IF Staining and Calcium Imaging

To verify multipotency, BM-MSC, DFSC and adMSC were subjected to in vitro differentiation towards osteogenic, chondrogenic and adipogenic lineages using the Mesenchymal Stem Cell Functional Identification Kit (R & D). Differentiation was induced by maintaining cells under different culture conditions according to the manufacturer instructions for 20 days. Subsequently, cells were fluorescently labelled to detect fatty acid-binding protein 4 (FABP4), Aggrecan and Osteocalcin to visualize successful differentiation into adipocytes, chondrocytes, and osteocytes.

For labelling of cardiac markers, cells were seeded on coverslips and fixed with 4% PFA. Antibody staining was performed as described elsewhere [34]. Cells were labelled with anti sarcomeric α-actinin (abcam, ab9465), anti-NKX2.5 (Santa Cruz, sc-8697), anti-TBX5 (abcam, ab137833) and anti-MEF2C (Santa Cruz, sc-313).

To visualize intracellular calcium, cells were cultured on 8 well chamberslides (Ibidi). Three days after seeding, cells were incubated with the calcium sensitive dye Cal520 (AATBioquest) for one hour at 37 °C and subjected to fluorescence microscopy. All fluorescence images were acquired using Zeiss ELYRA LSM 780 (Zeiss, Oberkochen, Germany).

2.5. RNA Isolation and Quantitative Real-Time Polymerase Chain Reaction

Isolation of cellular RNA was performed using the NucleoSpin® RNA isolation kit (Macherey-Nagel, Düren, Germany) according to the manufacturer instructions. The concentration and purity of isolated RNA was assessed with NanoDrop 1000 Spectrophotometer (Thermo Fisher Scientific). Subsequently, cDNA synthesis was performed with a High-Capacity cDNA Reverse Transcription Kit (Thermo Fisher Scientific). The reverse transcription reaction was conducted using the MJ Mini™ thermal cycler (Bio-Rad).

Quantitative real-time PCR for cardiac marker genes was carried out using the StepOnePlus™ Real-Time PCR System (Applied Biosystems, Foster City, USA) with following reaction parameters (StepOne™ Software Version 2, Applied Biosystems, Germany): start at 50 °C for 2 min, initial denaturation at 95 °C for 10 min, denaturation at 95 °C for 15 s and annealing/elongation at 60 °C for 1 min with 40 cycles. A qPCR reaction contained: TaqMan® Universal PCR Master Mix (Thermo Fisher), respective TaqMan® Gene Expression Assay, UltraPure™ DNase/RNase-Free Distilled Water (Thermo Fisher), and 30 ng of the respective cDNA. The following target gene assays were used: ACTN2 (Hs00153809_m1); MYH6 (Hs01101425_m1) TBX5 (Hs00361155_m1); TNNI3 (Hs00165957_m1), GJA1 (Hs00748445_s1); HPRT (HS01003267_m1) (all Thermo Fisher). Obtained CT values were normalized to HPRT and data were calculated as fold-change expression, related to untreated control cells.

2.6. Microarray Analysis

RNA integrity was analyzed using the Agilent Bioanalyzer 2100 with the RNA Pico chip kit (Agilent Technologies). 200 ng of isolated RNAs were subjected to microarray hybridization as described in [35]. Hybridization was performed on Affymetrix Clariom™ D Arrays according to the manufacturer's instructions (Thermo Fisher).

Analysis of the microarray data was conducted with the provided Transcriptome Analysis Console Software from Thermo Fisher (Version 4.0.1, Waltham, USA). The analysis included quality control, data normalization, and statistical testing for differential expression (Limma). Transcripts were considered as significantly differentially expressed with a fold change (FC) higher than 2 or smaller −2, false discovery rate (FDR) < 0.05, and $p < 0.05$. The pathway analyses were conducted based on a gene set enrichment analysis using Fisher's Exact Test (GSEA) on the Wiki-Pathways database. Only significant pathways have been selected.

2.7. Statistical Analysis

Data are presented as mean ± SEM, obtained from three patients for each MSC type. Preparation of graphs and statistical analysis was performed using SIGMA Plot software (Systat Software GmbH, Erkrath, Germany). Statistical significance was considered as * $p \leq 0.5$, ** $p \leq 0.05$, *** $p \leq 0.001$.

3. Results

3.1. Characterization of Isolated MSC

Initially, we performed flow cytometric analysis to investigate the presence of common mesenchymal surface markers in isolated MSC. The obtained data indicated a high expression of CD29, CD44, CD73, CD105 and CD90, while very low levels were detected for CD117 and CD45, indicating that stem cells possess properties of MSC (Figure 1A,B).

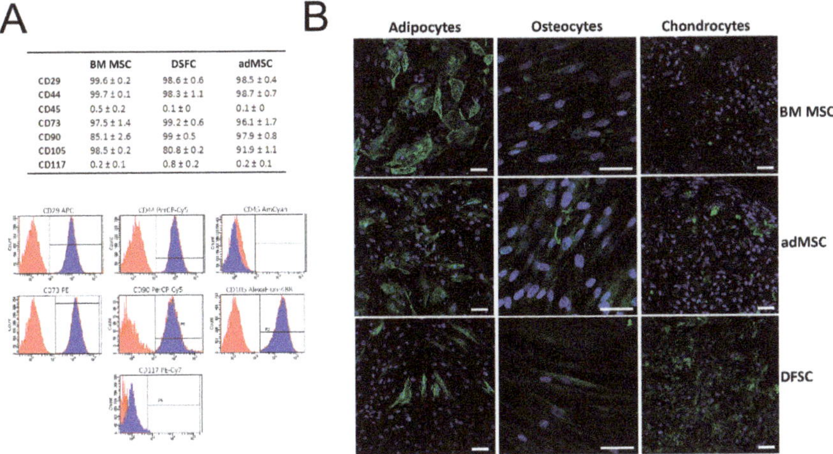

	BM MSC	DSFC	adMSC
CD29	99.6 ± 0.2	98.6 ± 0.6	98.5 ± 0.4
CD44	99.7 ± 0.1	98.3 ± 1.1	98.7 ± 0.7
CD45	0.5 ± 0.2	0.1 ± 0	0.1 ± 0
CD73	97.5 ± 1.4	99.2 ± 0.6	96.1 ± 1.7
CD90	85.1 ± 2.6	99 ± 0.5	97.9 ± 0.8
CD105	98.5 ± 0.2	80.8 ± 0.2	91.9 ± 1.1
CD117	0.2 ± 0.1	0.8 ± 0.2	0.2 ± 0.1

Figure 1. Phenotype-related and functional characterization of mesenchymal stromal cells (MSC): (**A**) Flow cytometric measurements revealed a high expression of common MSC surface markers (CD29, CD44, CD73, CD90, CD105), while very low levels were found for hematopoietic surface markers (CD45 and CD117). Representative flow cytometry charts of adipose tissue-derived MSC (adMSC) demonstrate the expression level of surface markers. Blue histograms represent measurement of CD surface marker with corresponding isotype control, shown in red. (**B**) Tri-lineage differentiation assay indicated adipogenic, osteogenic, and chondrogenic differentiation of MSC. Detection of adipocytes was performed by labelling of FABP4, while osteocytes and chondrocytes were identified by fluorescence staining of osteocalcein and aggrecan, respectively. Scale bar: 50 μm. Results in (**A**) are shown as mean ± SEM, obtained by analysis of three different donors for each MSC cell type.

MSC characteristics were further confirmed by a functional assay that demonstrated the multilineage differentiation capability of all three cell types. Upon incubation in lineage-specific induction medium, the cells were capable to differentiate into adipocytes, chondrocytes, and osteocytes, as shown by fluorescence labelling of specific differentiation markers (Figure 1B). As expected, adMSC were found to profoundly express FABP4, if compared to osteocalcin and aggrecan labelling. In contrast, DFSCs favored chondrogenic differentiation indicated by strong fluorescence intensity of aggrecan staining.

Next, we compared the different MSC by analyzing their gene expression profiles using a microarray platform. The obtained data allowed us to compare the transcription profile among both, individual donors and MSC derived from different tissue. Boxplots of signal intensity distributions for each performed microarray are shown in Figure 2, indicating good data quality prior (blue) and after (red) normalization of the gene expression data (Figure 2A). A principal component analysis (PCA) was performed to show the common clustering of the triplicates (Figure 2B, blue, red and purple) as well as the differences of tested cell types, each represented by three different donors. We found that stromal cells from BM, adipose, and dental tissue are clearly distinct with respect to their transcriptomic profile. Interestingly, we detected a high donor-dependent variety of the gene expression for MSC derived from human BM (Figure 2B), suggesting a potential donor-specific impact on the efficacy of cardiac programming. A total of 1685 differentially expressed genes were detected, while 13 genes were shared by all MSC populations (Figure 2C). Most differentially expressed transcripts (679) have been found between MSCs obtained from BM and adipose tissue, suggesting a higher gene profile related diversity within these two MSC populations (Figure 2D). A list of differentially expressed genes between all MSC types is given in Table S1.

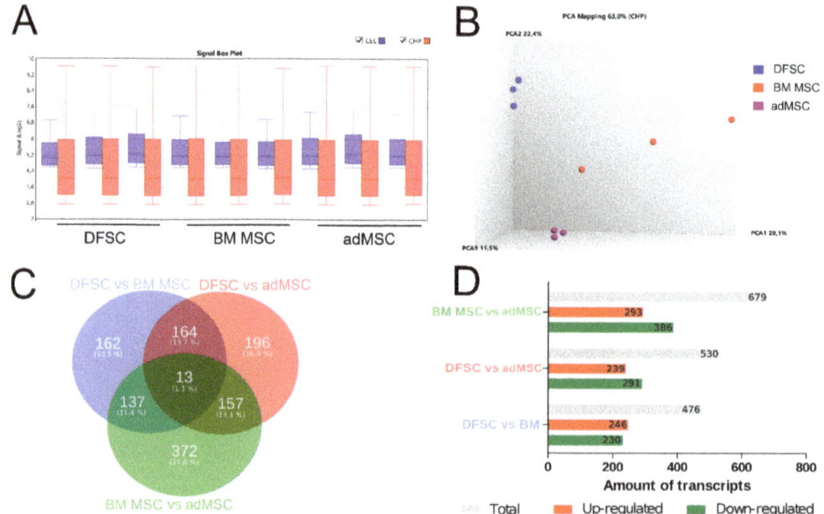

Figure 2. Comparative microarray analysis of undifferentiated dental follicle stem cells (DFSCs), bone marrow (BM) MSC, and adMSC. (**A**) Comparison of signal intensity for .cel files (blue) and .chp files (red) after normalization demonstrates sufficient data quality. (**B**) MSC from different sources are clearly distinct in regard to their transcription profile. A high patient-dependent variety was found for BM MSC, while adMSC and DFSCs demonstrate a more homogenous distribution. (**C**) Venn diagram visualizes expressed genes overlapping between different MSC cell types. (**D**) The numbers of up- and down-regulated transcripts is significantly differentially expressed in all three cell types.

3.2. Reprogramming of MSC Using miRNA and Cardiac Induction Cell Culture Conditions

In order to induce cardiac reprogramming, cells were cultured under two different medium conditions (see Section 2.3), separately or in combination with myocardial miRNAs (myo-miRNAs), that have been previously shown to induce cardiac differentiation in fibroblasts (miR-1, miR-499a, miR-208a, and miR-133a) [36]. As the efficiency of miRNA-based reprogramming largely depends on proper intracellular miRNA uptake, we evaluated miRNA transfection conditions using Cy3-labelled miRNA. Depending on the amount of transfected miRNA, uptake efficiencies of ~80–95% were achieved in all three cell types tested (Figure 3A). Importantly, only minimal cytotoxic effects were observed following transfection of miRNA (Figure 3B).

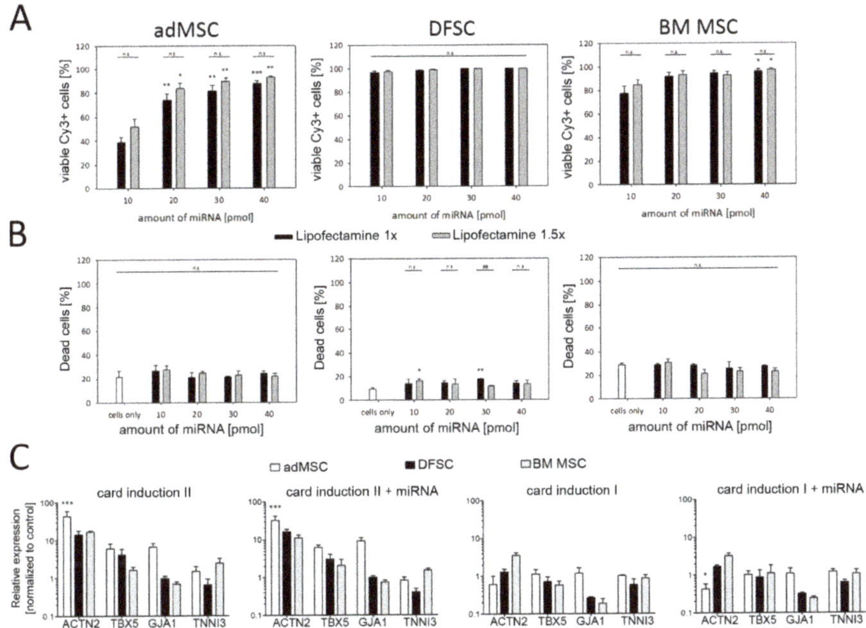

Figure 3. miRNA transfection and programming efficiency in MSC. (**A**) Uptake of miRNA was determined using Cy3-labelled miRNA and flow cytometry. (**B**) Detection of dead cells revealed low cytotoxicity induced by miRNA transfection. (**C**) Relative expression of cardiac marker genes among all tested cell types, four weeks after transfection and cultivation under different culture conditions. Reprogramming efficiency with cardiac induction medium I, II and myo-miRNAs (miR-1, miR-499, miR-208, miR-133) resulted in an up-regulation of cardiac specific markers in all types of MSC, while most profound up-regulation was found for cardiac induction medium II. Among tested MSC, the strongest increase of cardiac gene expression was observed for adMSC. Note, no beneficial effects on cardiac programming were observed following myo-miRNA transfection. Data are shown as mean ± SEM, obtained from three donors for each MSC type. Statistical analysis was performed using ANOVA test, followed by Bonferroni post-hoc analysis. * $p \leq 0.5$, ** $p \leq 0.05$, *** $p \leq 0.001$.

The success of myo-miRNA-based cardiac reprogramming was determined by qRT-PCR analysis of cardiac specific marker genes four weeks post transfection. Compared to control cells, cardiac induction medium II was found to be the most effective treatment leading to an induction of α-actinin, TBX5, GJA1, and cardiac Troponin I. While the level of α-actinin mRNA was strongly increased in all three cell types, a less pronounced effect was observed for cardiac Troponin I (Figure 3C). Notably, adMSCs showed the highest expression levels of cardiac marker genes after the treatment with cardiac induction medium II, when compared to MSCs obtained from dental follicle as well as BM, and therefore have been identified as the preferred candidate for our cardiac programming approach.

Surprisingly, our data also revealed that transfection with myo-miRNAs did not provoke an additional, beneficial effect on the expression of cardiac markers. Likewise, the cardiac induction medium containing RPMI and small molecules (Figure 3C, card induction I) did not promote the cardiac differentiation of MSC.

3.3. mRNA-Based Reprogramming of adMSC

As the transfection of miRNA did not further improve cardiac differentiation, we asked whether the application of modified mRNAs might boost the reprogramming efficiency in adMSC, which had been found to be the most promising cell type for the differentiation towards the cardiac lineage (Figure 3).

For mRNA-based programming of adMSC, cells were either transfected with single MESP1 mRNA or with a combination of GATA4, MEF2C, and TBX5 mRNA (GMT). First, mRNA transfection and translation efficiency were determined with mRNA encoding GFP to evaluate the optimal amount of mRNA showing strong expression while causing minimal cytotoxic effects. As demonstrated by flow cytometry and fluorescence microscopy, approximately 80% of cells express the GFP protein 24 h post transfection with 1 µg of mRNA (Figure 4A–C). Considering the increasing cytotoxicity when higher amounts of mRNA are transfected, reprogramming experiments were performed with 1–2 µg of individual mRNA (Figure 4D).

Analysis by qRT-PCR showed that both MESP1 and GMT transfection resulted in elevated levels of selected cardiac marker genes, compared to untreated control cells (Figure 4E). The most prominent incline of gene expression was observed for α-actinin, which was confirmed on the protein level by immunostaining showing a faint signal in cells treated with MESP1 and GMT mRNAs (Figure 4F). Additional antibody staining of early cardiac transcription factors demonstrated the expression of MEF2C and NKX2.5 on the protein level in GMT treated cells (Figure 4G and Figure S1). Interestingly, a profound increase of the expression level was also found for TBX5 that has been used for mRNA transfection in the GMT-treated group, verified by fluorescence microscopy (Figure S1).

Moreover, we observed differences of the intracellular Ca^{2+} concentration between treated groups. Following labelling of intracellular Ca^{2+}, GMT transfected cells demonstrated a more intensive fluorescence signal than observed for MESP1 treated cells and the control group (Figure S2).

To obtain a deeper understanding of the mRNA-induced effects on the gene expression profile of treated adMSC, we conducted a microarray analysis of cells that underwent cardiac reprogramming. The signal intensity values detected on each microarray had a similar spread after normalization, indicating a well-suited data quality for further downstream data analysis (Figure 5A). The PCA plot visualizes the differences in gene expression among treated groups, showing that control cells (blue) share a high similarity regarding their transcription profile (Figure 5B). In contrast, reprogramming with cardiac induction medium II (red), MESP1 (green), and GMT (purple) mRNA induced a strong donor-dependent alteration of gene levels, however, the treatment specific groups remain distinguishable from each other.

The numbers of significant total up-regulated and down-regulated transcripts are represented in Figure 5C, indicating a distinct change of gene expression following cardiac reprogramming. The highest number of genes differentially expressed was found in MESP1 (6669 transcripts) and GMT (5649) treated cells. Interestingly, more transcripts are down-regulated than up-regulated in most of the comparisons.

The corresponding Venn diagram (Figure 5D) compares the significantly expressed genes of the three different reprogramming approaches related to untreated control cells. The largest amount of transcripts (2828 transcripts, 33.6%) was found to be commonly regulated by all three treatments. The second largest proportion of differentially expressed genes is shared by GMT vs. Control and MESP1 vs. Control (1816 transcripts, 21.6%). Notably, the largest unique set of transcripts was found in cells transfected with MESP1 mRNA (1660 transcripts, 19.7%). A detailed comparison of up-regulated (Figure 5E, red) and down-regulated (Figure 5E, green) genes among these three reprogrammed groups indicates that the differences between MESP1 and GMT treatment vs. cardiac induction medium II are more profound (189 up-regulated, 276 down-regulated transcripts), while MESP1 and GMT only showed one differentially up-regulated transcript that was not previously up-regulated in other comparisons (Figure 5E). A detailed list of differentially expressed genes found in all reprogrammed groups is shown in in Table S1.

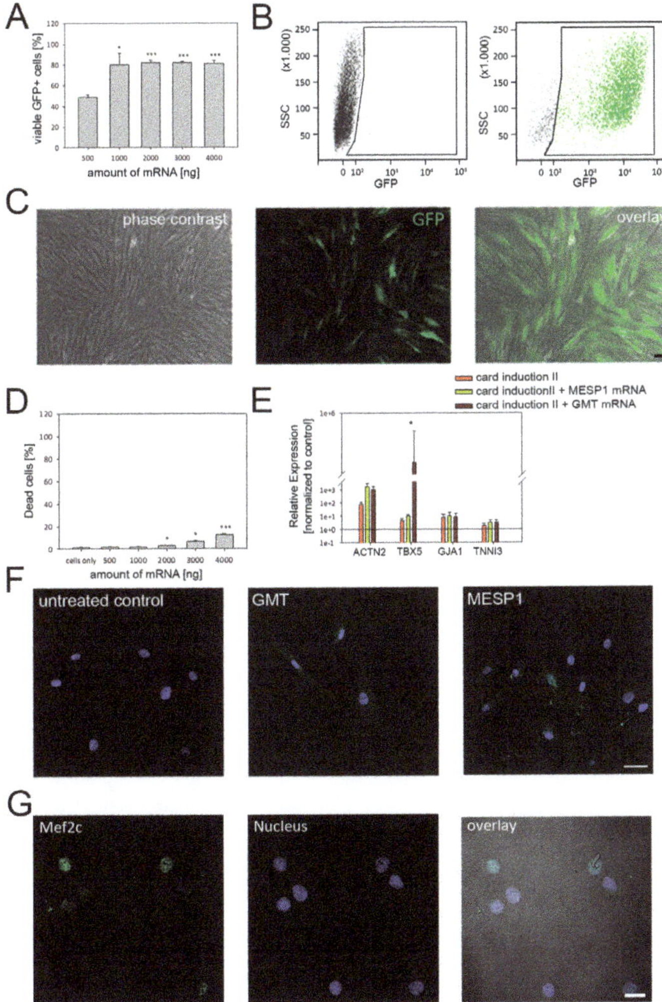

Figure 4. mRNA-based cardiac programming of adMSC. (**A**) Concentration-dependent expression of transfected mRNAs was evaluated with mRNA coding GFP. The quantative flow cytometry analysis demonstrated maximum transfection efficiency of ~80% when ≤ 1000 ng mRNA were applied. (**B**) Representative scatterplots of control cells (left) and cells transfected with GFP mRNA (right). (**C**) Corresponding microscopy images of cells expressing GFP following mRNA treatment. (**D**) Cytotoxic effects were only induced when mRNA amounts higher than 1000 ng were used for transfection. (**E**) Compared to untreated control cells, higher gene expression levels of selected cardiac markers were detected for all reprogramming conditions, in particular for α-actinin. (**F**) Immunolabeling of cells using anti α-actinin antibody results in a faint fluorescence signal in cells transfected with MESP1 and GATA4, MEF2C, and TBX5 (GMT) mRNAs, Scale Bar: 25 μm. (**G**) Moreover, GMT treated cells also demonstrated protein expression of MEF2C, an early cardiac transcription factor. Flow cytometry and qRT-PCR data are shown as mean ± SEM, obtained from three different donors. Statistical analysis was performed using one-way ANOVA. * $p \leq 0.5$, ** $p \leq 0.05$, *** $p \leq 0.001$.

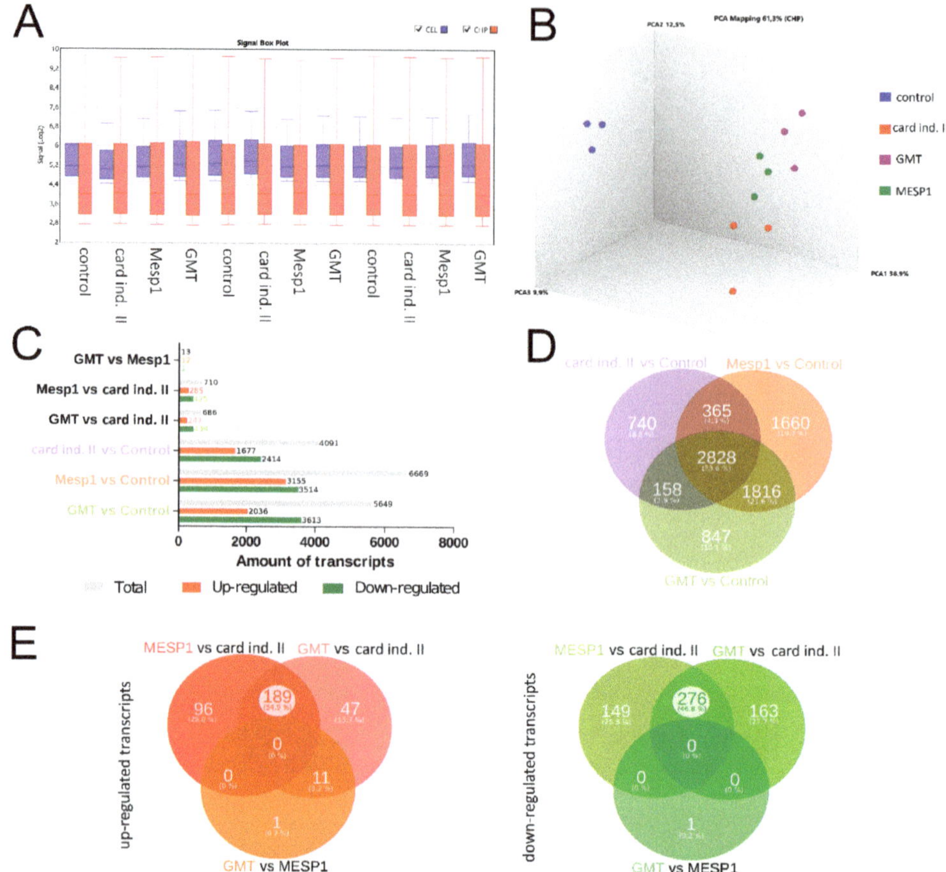

Figure 5. Transcriptome based comparison of reprogrammed adMSC. (**A**) Quality control of microarray data. Box plot of signal intensity of performed microarrays on .cel (blue) and .chp files normalization (red) confirm good data quality. (**B**) Principal component analysis (PCA) demonstrates clustering of treated groups, clearly showing the impact of respective reprogramming conditions on the transcriptomic profile compared to control cells (blue). Yet, cells subjected to MESP1 (green), GMT (purple) or cardiac induction medium II solely (red) remain distinguishable. (**C**) Up-and down-regulated transcripts and corresponding Venn diagram (**D**) showing the impact of reprogrammed cells compared to control. Most differentially expressed transcripts were regulated by all three reprogramming treatments (2828 genes), while 1816 transcripts are shared by GMT vs. control and MESP1 vs. control. (**E**) Detailed comparison of common and distinct up-regulated (red) and down-regulated (green) transcripts among the three reprogrammed groups. The differences found for optimized medium vs. MESP1 and GMT transfections are much more prominent than the differences between MESP1 and GMT.

These data indicate a strong change of gene expression when cells are subjected to cardiac induction medium II, with more distinct effects induced by mRNA transfections.

To evaluate the influence of the differentially expressed genes on important cardiac development pathways, we integrated our microarray gene expression data into the WikiPathways database and identified significantly enriched pathways for "heart development" (Figure 6A) and "cardiac progenitor differentiation" (Figure 6B). The pathway visualization indicates proteins mainly involved in cardiac development, while up-regulated and down-regulated transcripts of respective programming treatments are labelled in red and green, respectively. As shown in Figure 6, cardiac induction medium II as well as mRNA programming by MESP1 and GMT influence the gene expression profile of several key transcription factors and signaling molecules involved in cardiac differentiation, such as IGF, VEGF, TBX5, GATA4 and HAND2 (Figure 6A,B). Most changes on pathway genes were induced by GMT treatment (92%), followed by MESP1 (60%) and cardiac induction medium (52%). Additional immunofluorescence labelling of GMT treated cells, confirmed the expression of early cardiac transcription factors, including NKX2.5, TBX5 and MEF2C (Figure 4G and Figure S1)

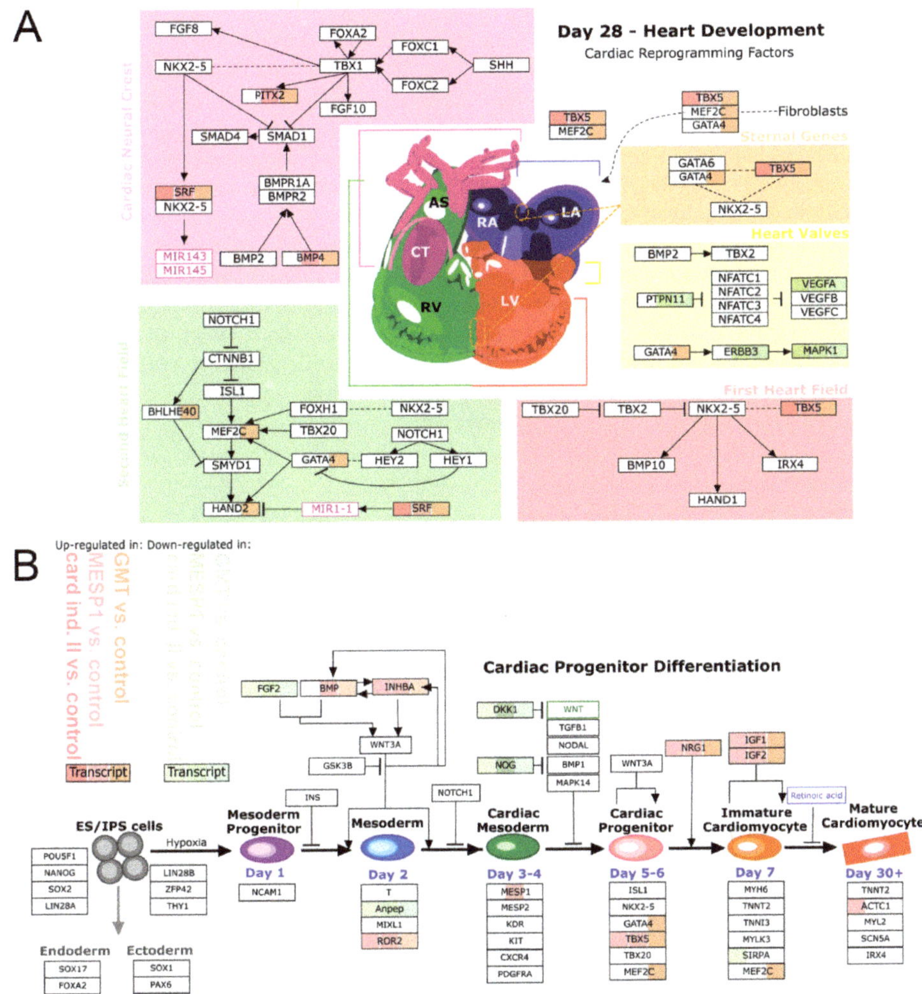

Figure 6. The impact of reprogramming on cardiac-differentiation pathways. Up-regulated and down-regulated transcripts of respective programming conditions are labelled in red or green color. (**A**,**B**) Strongest up-regulation of transcripts involved in cardiac development ((**A**) heart development, (**B**) cardiac progenitor differentiation) was mainly found in GMT reprogrammed cells, followed by MESP1 treatment and cardiac induction medium II. Key cardiac transcription factors and signaling molecules were significantly up-regulated, including TBX5, GATA4, MEF2C, HAND2, BMP4, and IGF.

Taken together, the results obtained by microarray analysis clearly indicate that reprogramming with cardiac induction medium II and mRNA induced a strong alteration of the transcription patterns with high similarity in mRNA transfected cells compared to cells cultured in cardiac induction medium solely.

4. Discussion

In vitro generated cardiomyocytes are an important tool for cardiovascular research, as they can be utilized for disease modelling or for the development of drug screening assays to assess the cardiac toxic risk of established or newly synthesized drugs [37–39]. Moreover, promising preclinical data suggests the therapeutic potential of generated cardiomyocytes for the treatment of cardiac diseases to overall improve heart regeneration and function [40,41]. Although several stem cell types are available to produce cardiac cells, the ideal source of stem cells remains elusive as each has its own advantages and drawbacks. Adult MSC can be easily isolated from human donors in large quantities, possess immunomodulatory properties and can be propagated in vitro [12]. Further, they can overcome certain limitations that have been attributed to PSCs, including ESC and iPSC. In contrast to ESC, MSC do not provoke any ethical concerns [12,37,38]. Moreover, pre-clinical studies demonstrated a tumorigenic potential of ESC and iPSC-derived cell products that has not been observed for MSC to date [42–45]. However, other pre-clinical and clinical trial data showed that the transplantation of iPSCs-derived cardiomyocytes did not result in teratoma formation [46–48]. These different outcomes might be associated with the transplantation of residual undifferentiated cells along with the PSC product that increases the possibility of tumorigenesis. In this regard, the therapeutic use of PSC requires the establishment of differentiation protocols allowing the generation of highly pure PSC-derived cell types, e.g., cardiomyocytes [49]. The major advantage in comparison to adult stem cells is the cardiac differentiation potential of ESCs and iPSCs. So far, PSC have been found to be the only stem cell type capable to differentiate into functional, premature cardiomyocytes showing pronounced sarcomere organization, contraction capacity, and subtype specific ion channel composition [50,51]. Thus, for the generation of cardiomyocytes applied in regenerative medicine PSC are currently superior to MSC as no efficient cardiac reprogramming strategies have been developed for adult stem cells yet.

The successful cardiac differentiation of human MSC into fully mature cardiomyocytes is by far more challenging. Adult cardiomyocytes are characterized by a specific cell shape, structural organization, ion channel composition and mechanical properties; important features that need to be addressed when generating stem cell-derived cardiac cells [52]. Former reports led to contradictory results about the programming efficiency of MSC. While some reports described spontaneous beating associated with the formation of sarcomeric protein structures, other studies failed to generate cardiac-like cells from adult MSC [53–57].

One reason for this might be attributed to the fact that MSC may represent a heterogeneous stem cell population with different functional and phenotype-related properties as well as varying therapeutic potential [58]. A notion that is supported by our microarray data, indicating a high diversity of the expressed transcripts among MSC obtained from BM, dental pulp and adipose tissue (Figure 2). Likewise, our functional data revealed cell type-dependent differentiation capacity of tested MSC (Figure 1). Previous studies have also reported distinct characteristics between MSC from different sources regarding surface marker expression, proliferation rate, and differentiation potency [17,19,58,59]. For example, adMSC were observed to favor osteogenic differentiation and demonstrate higher proliferation when compared with DSFCs [18,60]. Moreover, our results suggest that these different biological characteristics of MSC could have an impact on the selected strategy and efficiency of cardiac programming as adMSC demonstrated a more pronounced incline of cardiac marker expression than BM MSC and DFSCs (Figure 3). In line with these data, Kakkar et al. recently described human adMSC to be a better choice for cardiac programming using a combination of small molecules and cytokines. Compared to BM MSC, adMSC exhibited a higher expression of α-actinin, troponin and connexin43 following cardiac induction with 5-Azacytidine and TGF-β1 [61].

Similarly, a comparative study revealed that adMSC expressed significantly more cardiomyocyte specific biomarkers as DFSCs following cardiac programming with cytokine supplemented culture medium [11]. The impact of MSC origin on programming capability was also shown for non-cardiac cell lineages like hepatocytes and smooth muscle cells [59,62].

Myo-miRNA based programming has been successfully applied for the conversion of cardiac fibroblasts, into cardiomyocytes [36]. For MSCs, cardiac induction by miRNA is less efficient as shown by different groups [25,63,64]. For example, it was demonstrated that transfection with miRNA-1-2 promote the expression of GATA4, NKX2.5 and cardiac Troponin in BM MSCs [15]. Similarly, miR-149 and miR-1 were found to slightly trigger myocardial differentiation, albeit without formation of sarcomere structures or beating activity [25,65]. We did not observe any additional effects on the expression of selected cardiac marker genes following miRNA treatment. This might be attributed to the fact that the miRNA concentrations used in this study are not sufficient to significantly increase the expression level of cardiac-specific genes, although uptake efficiency for miRNA was about 80%. In this regard, some studies have used viral vectors to ensure constitutive overexpression of miRNA [25,64]. Given that miRNAs have a very short half live, transient transfection approaches, as used in our study, might be less effective.

Proper cardiac development requires the activation and inhibition of many different pathways modulated by several transcription factors [66]. MESP1 was shown to drive cardiovascular fate of stem cells during embryonic development, while the combination of GATA4, MEF2C and TBX5 was described to induce the cardiac differentiation of murine and human fibroblasts, leading to spontaneously contracting cells with cardiomyocyte-like expression profile [67–70]. Therefore, we have concluded that this approach might be applicable to reprogram human adMSC. Using an mRNA-based setting we induced the overexpression of GATA4, MEF2C, and TBX5 as well as MESP1, which provoked an incline of genes involved in cardiac differentiation (Figure 4). To our knowledge this combination of transcription factors has not been applied before to induce cardiac differentiation of human adMSC. In contrast to our strategy, most of the previous studies performed overexpression of transcription factors by application of retro- or lentiviral systems. For example, in a study by Wystrychowski et al., adMSC from cardiac tissue were treated with seven transcription factors, including GATA4, MEF2C, MESP1, and TBX5, that resulted in an elevated number of cells positive for α-actinin and troponin [71]. However, no clear sarcomere structures have been observed, suggesting a premature cardiac progenitor state. Similarly, forced expression of another factor of the T-box family, TBX20, provokes an up-regulation of sarcomeric proteins, without cardiomyocyte specific sarcomere organization [72]. These data are in line with our observations as we could also detect a moderate signal for α-actinin, albeit without the presence of sarcomere structures (Figure 4).

Yet, our programming approach leads to a strong induction of the key cardiac transcription factors GATA4, MEF2C, MESP1 and TBX5, which corresponds to the transfected mRNAs used for programming. However, it is known that mRNAs underlie fast turnover, suggesting that mRNA transfection activated the expression of its endogenous counterparts [73,74]. At the same time, the current study demonstrates that mRNA transfection boosts the cardiac programming effects induced by culture conditions targeting important signaling pathways such as the WNT cascade.

The manipulation of signaling pathways by cytokines and small molecules is the most common methodology to generate large amounts of PSC-derived functional cardiomyocytes [30,31]. In addition, the overexpression of transcription factors, like Tbx3 and MESP1, can influence cell fate decision in PSCs [75,76]. While these techniques allow highly efficient programming of ESCs and iPSCs, we observed significantly less programming efficiency for MSCs in the current study. However, the comparison of programming protocols used for PSCs and multipotent stem cells is difficult due to their different developmental stages and resulting culture conditions prerequisites. Yet, it was shown that cytokines like BMP4, IL and TGF improve cardiac development of human and non-human MSCs [57,77]. However, the cardiomyocyte-like cells derived from these programmed MSCs lack profound sarcomere formation, beating activity and physiological maturation [78,79]. This is in accordance to our data indicating that mRNA transfection could promote the expression of early cardiac proteins, while differentiation efficiency and elaboration of a terminal cardiac phenotype is profoundly limited when compared to PSC differentiation protocols [27,31].

Together with previous studies of adMSC overexpressing transcription factors, our results demonstrate the feasibility of mRNA-based cardiac reprogramming of MSC. However, the absence of sarcomere structures and spontaneous cell beating suggests a yet quite incomplete reprogramming, leading to an immature cardiac cell type. Hence, there is an urgent need for further optimization. Since mRNAs are degraded over time, multiple transfection steps might increase the reprogramming efficiency, a strategy that is already applied for the generation of iPSCs from adult cells [74,80]. Moreover, proportions of GATA4, MEF2C, and TBX5 protein expression has been described to play a crucial role for the quality of cardiac reprogramming [81], thus, different ratios of transfected mRNA could positively influence the outcome of reprogrammed adMSC. This will have to be addressed in future studies as the impact of mRNA ratios and mRNA concentration on cardiac programming might be affected in a donor specific manner. Former data already demonstrated donor-to-donor variability of MSC functional potential, including differentiation capacity [82,83]. Beside age and gender, underlying diseases are known to influence cellular properties of MSCs [82]. This is supported by our microarray results, showing a large variety of the transcription profile of BM MSCs that have been obtained from patients suffering from cardiovascular diseases. On the contrary, adMSCs and DFSCs derived from healthy donors shared similar transcription patterns, suggesting same programming conditions required to induce cardiac development. Nevertheless, it is recommended to adapt mRNA conditions for each individual patient to obtain maximum programming efficiency.

In addition, more comparative studies are required to identify and characterize MSC subtypes most susceptible for specific transdifferentiation towards the respective desired target cells, including non-mesodermal and mesodermal cell types such as cardiomyocytes.

Supplementary Materials: The following are available online at http://www.mdpi.com/2073-4409/9/2/504/s1.

Author Contributions: P.M., H.L. (Hermann Lang) and R.D. performed the study design. P.M. carried out cell culture experiments, RNA isolation, flow cytometry, qRT-PCR and respective data analysis. M.W. supported analysis of microarray data, subfigure preparation and corrected the manuscript. K.E. isolated and pre-cultured the DFSC. K.P. and O.H. isolated, characterized and pre-cultured the adMSC. D.K. carried out microarray experiments, including RNA quality measurement. K.P., H.L. (Heiko Lemcke), C.I.L., O.W., and R.D. proofread and revised the manuscript. H.L. (Hermann Lang) collected microscopy data, conceptualized and wrote the manuscript with contribution from P.M. and R.D. All authors have read and agreed to the published version of the manuscript.

Funding: This study was supported by the EU structural Fund (ESF/14-BM-A55-0024/18). In addition, R.D and P.M. are supported by the DFG (DA1296/6-1). R.D. is further supported by the DAMP foundation, the German Heart Foundation (F/01/12) and the BMBF (VIP+ 00240). In addition, H.L. is supported by the FORUN Program of Rostock University Medical Centre (889001 and 889003) and the Josef and Käthe Klinz Foundation (T319/29737/2017).

Conflicts of Interest: The authors declare no conflict of interest. The funders were not involved in study design, data collection and interpretation, and manuscript preparation.

References

1. Rajabzadeh, N.; Fathi, E.; Farahzadi, R. Stem cell-based regenerative medicine. *Stem Cell Investig.* **2019**, *6*, 19. [CrossRef] [PubMed]

2. Samsonraj, R.M.; Raghunath, M.; Nurcombe, V.; Hui, J.H.; van Wijnen, A.J.; Cool, S.M. Concise Review: Multifaceted Characterization of Human Mesenchymal Stem Cells for Use in Regenerative Medicine. *Stem Cells Transl. Med.* **2017**, *6*, 2173–2185. [CrossRef] [PubMed]

3. Squillaro, T.; Peluso, G.; Galderisi, U. Clinical trials with mesenchymal stem cells: An update. *Cell Transplant.* **2016**, *25*, 829–848. [CrossRef] [PubMed]

4. Collichia, M.; Jones, D.A.; Beirne, A.-M.; Hussain, M.; Weeraman, D.; Rathod, K.; Veerapen, J.; Lowdell, M.; Mathur, A. Umbilical cord-derived mesenchymal stromal cells in cardiovascular disease: review of preclinical and clinical data. *Cytotherapy* **2019**, *21*, 1007–1018. [CrossRef]

5. Guerrouahen, B.S.; Sidahmed, H.; Al Sulaiti, A.; Al Khulaifi, M.; Cugno, C. Enhancing Mesenchymal Stromal Cell Immunomodulation for Treating Conditions Influenced by the Immune System. *Stem Cells Int.* **2019**, *2019*, 7219297. [CrossRef]

6. Aguilera-Castrejon, A.; Pasantes-Morales, H.; Montesinos, J.J.; Cortés-Medina, L.V.; Castro-Manrreza, M.E.; Mayani, H.; Ramos-Mandujano, G. Improved Proliferative Capacity of NP-Like Cells Derived from Human Mesenchymal Stromal Cells and Neuronal Transdifferentiation by Small Molecules. *Neurochem. Res.* **2017**, *42*, 415–427. [CrossRef]

7. Tsai, W.-L.; Yeh, P.-H.; Tsai, C.-Y.; Ting, C.-T.; Chiu, Y.-H.; Tao, M.-H.; Li, W.-C.; Hung, S.-C. Efficient programming of human mesenchymal stem cell-derived hepatocytes by epigenetic regulations. *J. Gastroenterol. Hepatol.* **2017**, *32*, 261–269. [CrossRef]

8. Papadimou, E.; Morigi, M.; Iatropoulos, P.; Xinaris, C.; Tomasoni, S.; Benedetti, V.; Longaretti, L.; Rota, C.; Todeschini, M.; Rizzo, P.; et al. Direct Reprogramming of Human Bone Marrow Stromal Cells into Functional Renal Cells Using Cell-free Extracts. *Stem Cell Reports* **2015**, *4*, 685–698. [CrossRef]

9. Cai, B.; Li, J.; Wang, J.; Luo, X.; Ai, J.; Liu, Y.; Wang, N.; Liang, H.; Zhang, M.; Chen, N.; et al. microRNA-124 Regulates Cardiomyocyte Differentiation of Bone Marrow-Derived Mesenchymal Stem Cells Via Targeting STAT3 Signaling. *Stem Cells* **2012**, *30*, 1746–1755. [CrossRef]

10. Li, J.; Zhu, K.; Wang, Y.; Zheng, J.; Guo, C.; Lai, H.; Wang, C. Combination of IGF-1 gene manipulation and 5-AZA treatment promotes differentiation of mesenchymal stem cells into cardiomyocyte-like cells. *Mol. Med. Rep.* **2015**, *11*, 815–820. [CrossRef]

11. Loo, Z.X.; Kunasekaran, W.; Govindasamy, V.; Musa, S.; Abu Kasim, N.H. Comparative analysis of cardiovascular development related genes in stem cells isolated from deciduous pulp and adipose tissue. *Sci. World J.* **2014**, *2014*, 186508. [CrossRef] [PubMed]

12. Müller, P.; Lemcke, H.; David, R. Stem Cell Therapy in Heart Diseases—Cell Types, Mechanisms and Improvement Strategies. *Cell. Physiol. Biochem.* **2018**, *48*, 2607–2655. [CrossRef] [PubMed]

13. Szaraz, P.; Gratch, Y.S.; Iqbal, F.; Librach, C.L. In Vitro Differentiation of Human Mesenchymal Stem Cells into Functional Cardiomyocyte-like Cells. *J. Vis. Exp.* **2017**, *9*, 55757. [CrossRef] [PubMed]

14. Markmee, R.; Aungsuchawan, S.; Narakornsak, S.; Tancharoen, W.; Bumrungkit, K.; Pangchaidee, N.; Pothacharoen, P.; Puaninta, C. Differentiation of mesenchymal stem cells from human amniotic fluid to cardiomyocyte-like cells. *Mol. Med. Rep.* **2017**, *16*, 6068–6076. [CrossRef]

15. Shen, X.; Pan, B.; Zhou, H.; Liu, L.; Lv, T.; Zhu, J.; Huang, X.; Tian, J. Differentiation of mesenchymal stem cells into cardiomyocytes is regulated by miRNA-1-2 via WNT signaling pathway. *J. Biomed. Sci.* **2017**, *24*, 29. [CrossRef]

16. O'Connor, K.C. Molecular Profiles of Cell-to-Cell Variation in the Regenerative Potential of Mesenchymal Stromal Cells. *Stem Cells Int.* **2019**, *2019*, 1–14. [CrossRef]

17. Elahi, K.C.; Klein, G.; Avci-Adali, M.; Sievert, K.D.; MacNeil, S.; Aicher, W.K. Human Mesenchymal Stromal Cells from Different Sources Diverge in Their Expression of Cell Surface Proteins and Display Distinct Differentiation Patterns. *Stem Cells Int.* **2016**, *2016*, 1–9. [CrossRef]

18. D'Alimonte, I.; Mastrangelo, F.; Giuliani, P.; Pierdomenico, L.; Marchisio, M.; Zuccarini, M.; Di Iorio, P.; Quaresima, R.; Caciagli, F.; Ciccarelli, R. Osteogenic Differentiation of Mesenchymal Stromal Cells: A Comparative Analysis Between Human Subcutaneous Adipose Tissue and Dental Pulp. *Stem Cells Dev.* **2017**, *26*, 843–855. [CrossRef]

19. Kwon, A.; Kim, Y.; Kim, M.; Kim, J.; Choi, H.; Jekarl, D.W.; Lee, S.; Kim, J.M.; Shin, J.-C.; Park, I.Y. Tissue-specific Differentiation Potency of Mesenchymal Stromal Cells from Perinatal Tissues. *Sci. Rep.* **2016**, *6*, 23544. [CrossRef]

20. Leijten, J.; Georgi, N.; Moreira Teixeira, L.; van Blitterswijk, C.A.; Post, J.N.; Karperien, M. Metabolic programming of mesenchymal stromal cells by oxygen tension directs chondrogenic cell fate. *Proc. Natl. Acad. Sci. USA* **2014**, *111*, 13954–13959. [CrossRef]

21. Occhetta, P.; Pigeot, S.; Rasponi, M.; Dasen, B.; Mehrkens, A.; Ullrich, T.; Kramer, I.; Guth-Gundel, S.; Barbero, A.; Martin, I. Developmentally inspired programming of adult human mesenchymal stromal cells toward stable chondrogenesis. *Proc. Natl. Acad. Sci. USA* **2018**, *115*, 4625–4630. [CrossRef] [PubMed]

22. Yannarelli, G.; Pacienza, N.; Montanari, S.; Santa-Cruz, D.; Viswanathan, S.; Keating, A. OCT4 expression mediates partial cardiomyocyte reprogramming of mesenchymal stromal cells. *PLoS ONE* **2017**, *12*, e0189131. [CrossRef] [PubMed]

23. Lemcke, H.; Gaebel, R.; Skorska, A.; Voronina, N.; Lux, C.A.; Petters, J.; Sasse, S.; Zarniko, N.; Steinhoff, G.; David, R. Mechanisms of stem cell based cardiac repair-gap junctional signaling promotes the cardiac lineage specification of mesenchymal stem cells. *Sci. Rep.* **2017**, *7*, 1–17. [CrossRef] [PubMed]

24. Li, L.; Xia, Y. Study of adipose tissue-derived mesenchymal stem cells transplantation for rats with dilated cardiomyopathy. *Ann. Thorac. Cardiovasc. Surg.* **2014**, *20*, 398–406. [CrossRef]

25. Zhao, X.-L.; Yang, B.; Ma, L.-N.; Dong, Y.-H. MicroRNA-1 effectively induces differentiation of myocardial cells from mouse bone marrow mesenchymal stem cells. *Artif. Cells Nanomed. Biotechnol.* **2015**, *44*, 1665–1670. [CrossRef]

26. Dai, F.; Du, P.; Chang, Y.; Ji, E.; Xu, Y.; Wei, C.; Li, J. Downregulation of MiR-199b-5p inducing differentiation of bone-marrow mesenchymal stem cells (BMSCs) toward cardiomyocyte-like cells via HSF1/HSP70 pathway. *Med. Sci. Monit.* **2018**, *24*, 2700–2710. [CrossRef]

27. Burridge, P.W.; Matsa, E.; Shukla, P.; Lin, Z.C.; Churko, J.M.; Ebert, A.D.; Lan, F.; Diecke, S.; Huber, B.; Mordwinkin, N.M.; et al. Chemically defned generation of human cardiomyocytes. *Nat. Methods* **2014**, *11*, 855–860. [CrossRef]

28. Jiang, Y.; Park, P.; Hong, S.M.; Ban, K. Maturation of cardiomyocytes derived from human pluripotent stem cells: Current strategies and limitations. *Mol. Cells* **2018**, *41*, 613–621.

29. Chen, R.; He, J.; Wang, Y.; Guo, Y.; Zhang, J.; Peng, L.; Wang, D.; Lin, Q.; Zhang, J.; Guo, Z.; et al. Qualitative transcriptional signatures for evaluating the maturity degree of pluripotent stem cell-derived cardiomyocytes. *Stem Cell Res. Ther.* **2019**, *10*, 113. [CrossRef]

30. D'Antonio-Chronowska, A.; Donovan, M.K.R.; Young Greenwald, W.W.; Nguyen, J.P.; Fujita, K.; Hashem, S.; Matsui, H.; Soncin, F.; Parast, M.; Ward, M.C.; et al. Association of Human iPSC Gene Signatures and X Chromosome Dosage with Two Distinct Cardiac Differentiation Trajectories. *Stem Cell Rep.* **2019**, *13*, 924–938. [CrossRef]

31. Lian, X.; Zhang, J.; Azarin, S.M.; Zhu, K.; Hazeltine, L.B.; Bao, X.; Hsiao, C.; Kamp, T.J.; Palecek, S.P. Directed cardiomyocyte differentiation from human pluripotent stem cells by modulating Wnt/β-catenin signaling under fully defined conditions. *Nat. Protoc.* **2013**, *8*, 162–175. [CrossRef] [PubMed]

32. Meyer, J.; Salamon, A.; Herzmann, N.; Adam, S.; Kleine, H.-D.; Matthiesen, I.; Ueberreiter, K.; Peters, K. Isolation and Differentiation Potential of Human Mesenchymal Stem Cells From Adipose Tissue Harvested by Water Jet-Assisted Liposuction. *Aesthetic Surg. J.* **2015**, *35*, 1030–1039. [CrossRef] [PubMed]

33. Müller, P.; Ekat, K.; Brosemann, A.; Köntges, A.; David, R.; Lang, H. Isolation, characterization and microRNA-based genetic modification of human dental follicle stem cells. *J. Vis. Exp.* **2018**, *2018*, e58089.

34. Thiele, F.; Voelkner, C.; Krebs, V.; Müller, P.; Jung, J.J.; Rimmbach, C.; Steinhoff, G.; Noack, T.; David, R.; Lemcke, H. Nkx2.5 Based Ventricular Programming of Murine ESC-Derived Cardiomyocytes. *Cell. Physiol. Biochem.* **2019**, *53*, 337–354.

35. Koczan, D.; Fitzner, B.; Zettl, U.K.; Hecker, M. Microarray data of transcriptome shifts in blood cell subsets during S1P receptor modulator therapy. *Sci. Data* **2018**, *5*, 180145. [CrossRef]

36. Jayawardena, T.M.; Egemnazarov, B.; Finch, E.A.; Zhang, L.; Payne, J.A.; Pandya, K.; Zhang, Z.; Rosenberg, P.; Mirotsou, M.; Dzau, V.J. MicroRNA-mediated in vitro and in vivo direct reprogramming of cardiac fibroblasts to cardiomyocytes. *Circ. Res.* **2012**, *110*, 1465–1473. [CrossRef]

37. Sala, L.; Gnecchi, M.; Schwartz, P.J. Long QT Syndrome Modelling with Cardiomyocytes Derived from Human-induced Pluripotent Stem Cells. *Arrhythmia Electrophysiol. Rev.* **2019**, *8*, 105. [CrossRef]

38. Brodehl, A.; Ebbinghaus, H.; Deutsch, M.-A.; Gummert, J.; Gärtner, A.; Ratnavadivel, S.; Milting, H. Human Induced Pluripotent Stem-Cell-Derived Cardiomyocytes as Models for Genetic Cardiomyopathies. *Int. J. Mol. Sci.* **2019**, *20*, 4381. [CrossRef]

39. Protze, S.I.; Lee, J.H.; Keller, G.M. Human Pluripotent Stem Cell-Derived Cardiovascular Cells: From Developmental Biology to Therapeutic Applications. *Cell Stem Cell* **2019**, *25*, 311–327. [CrossRef]

40. Rikhtegar, R.; Pezeshkian, M.; Dolati, S.; Safaie, N.; Afrasiabi Rad, A.; Mahdipour, M.; Nouri, M.; Jodati, A.R.; Yousefi, M. Stem cells as therapy for heart disease: iPSCs, ESCs, CSCs, and skeletal myoblasts. *Biomed. Pharmacother.* **2019**, *109*, 304–313. [CrossRef]

41. Jackson, A.O.; Tang, H.; Yin, K. HiPS-Cardiac Trilineage Cell Generation and Transplantation: a Novel Therapy for Myocardial Infarction. *J. Cardiovasc. Transl. Res.* **2019**, *13*, 110–119. [CrossRef] [PubMed]

42. Hentze, H.; Soong, P.L.; Wang, S.T.; Phillips, B.W.; Putti, T.C.; Dunn, N.R. Teratoma formation by human embryonic stem cells: Evaluation of essential parameters for future safety studies. *Stem Cell Res.* **2009**, *2*, 198–210. [CrossRef] [PubMed]

43. Yong, K.W.; Choi, J.R.; Dolbashid, A.S.; Wan Safwani, W.K.Z. Biosafety and bioefficacy assessment of human mesenchymal stem cells: What do we know so far? *Regen. Med.* **2018**, *13*, 219–232. [CrossRef]

44. Duinsbergen, D.; Salvatori, D.; Eriksson, M.; Mikkers, H. Tumors Originating from Induced Pluripotent Stem Cells and Methods for Their Prevention. *Ann. N. Y. Acad. Sci.* **2009**, *1176*, 197–204. [CrossRef] [PubMed]

45. Seminatore, C.; Polentes, J.; Ellman, D.; Kozubenko, N.; Itier, V.; Tine, S.; Tritschler, L.; Brenot, M.; Guidou, E.; Blondeau, J.; et al. The postischemic environment differentially impacts teratoma or tumor formation after transplantation of human embryonic stem cell-derived neural progenitors. *Stroke* **2010**, *41*, 153–159. [CrossRef] [PubMed]

46. Menasché, P.; Vanneaux, V.; Hagège, A.; Bel, A.; Cholley, B.; Parouchev, A.; Cacciapuoti, I.; Al-Daccak, R.; Benhamouda, N.; Blons, H.; et al. Transplantation of Human Embryonic Stem Cell–Derived Cardiovascular Progenitors for Severe Ischemic Left Ventricular Dysfunction. *J. Am. Coll. Cardiol.* **2018**, *71*, 429–438. [CrossRef] [PubMed]

47. Funakoshi, S.; Miki, K.; Takaki, T.; Okubo, C.; Hatani, T.; Chonabayashi, K.; Nishikawa, M.; Takei, I.; Oishi, A.; Narita, M.; et al. Enhanced engraftment, proliferation, and therapeutic potential in heart using optimized human iPSC-derived cardiomyocytes. *Sci. Rep.* **2016**, *6*, 1–14. [CrossRef]

48. Liu, Y.W.; Chen, B.; Yang, X.; Fugate, J.A.; Kalucki, F.A.; Futakuchi-Tsuchida, A.; Couture, L.; Vogel, K.W.; Astley, C.A.; Baldessari, A.; et al. Human embryonic stem cell-derived cardiomyocytes restore function in infarcted hearts of non-human primates. *Nat. Biotechnol.* **2018**, *36*, 597–605. [CrossRef]

49. Ito, E.; Miyagawa, S.; Takeda, M.; Kawamura, A.; Harada, A.; Iseoka, H.; Yajima, S.; Sougawa, N.; Mochizuki-Oda, N.; Yasuda, S.; et al. Tumorigenicity assay essential for facilitating safety studies of hiPSC-derived cardiomyocytes for clinical application. *Sci. Rep.* **2019**, *9*, 1–10. [CrossRef]

50. Oikonomopoulos, A.; Kitani, T.; Wu, J.C. Pluripotent Stem Cell-Derived Cardiomyocytes as a Platform for Cell Therapy Applications: Progress and Hurdles for Clinical Translation. *Mol. Ther.* **2018**, *26*, 1624–1634. [CrossRef] [PubMed]

51. Tan, S.H.; Ye, L. Maturation of Pluripotent Stem Cell-Derived Cardiomyocytes: A Critical Step for Drug Development and Cell Therapy. *J. Cardiovasc. Transl. Res.* **2018**, *11*, 375–392. [CrossRef] [PubMed]

52. Scuderi, G.J.; Butcher, J. Naturally Engineered Maturation of Cardiomyocytes. *Front. Cell Dev. Biol.* **2017**, *5*, 50. [CrossRef] [PubMed]

53. Rose, R.A.; Jiang, H.; Wang, X.; Helke, S.; Tsoporis, J.N.; Gong, N.; Keating, S.C.J.; Parker, T.G.; Backx, P.H.; Keating, A. Bone marrow-derived mesenchymal stromal cells express cardiac-specific markers, retain the stromal phenotype, and do not become functional cardiomyocytes in vitro. *Stem Cells* **2008**, *26*, 2884–2892. [CrossRef] [PubMed]

54. Shim, W.S.N.; Jiang, S.; Wong, P.; Tan, J.; Chua, Y.L.; Seng Tan, Y.; Sin, Y.K.; Lim, C.H.; Chua, T.; Teh, M.; et al. Ex vivo differentiation of human adult bone marrow stem cells into cardiomyocyte-like cells. *Biochem. Biophys. Res. Commun.* **2004**, *324*, 481–488. [CrossRef] [PubMed]

55. Martin-Rendon, E.; Sweeney, D.; Lu, F.; Girdlestone, J.; Navarrete, C.; Watt, S.M. 5-Azacytidine-treated human mesenchymal stem/progenitor cells derived from umbilical cord, cord blood and bone marrow do not generate cardiomyocytes in vitro at high frequencies. *Vox Sang.* **2008**, *95*, 137–148. [CrossRef]

56. Ramkisoensing, A.A.; Pijnappels, D.A.; Askar, S.F.A.; Passier, R.; Swildens, J.; Goumans, M.J.; Schutte, C.I.; de Vries, A.A.F.; Scherjon, S.; Mummery, C.L.; et al. Human embryonic and fetal Mesenchymal stem cells differentiate toward three different cardiac lineages in contrast to their adult counterparts. *PLoS ONE* **2011**, *6*, e24164. [CrossRef]

57. Shi, S.; Wu, X.; Wang, X.; Hao, W.; Miao, H.; Zhen, L.; Nie, S. Differentiation of Bone Marrow Mesenchymal Stem Cells to Cardiomyocyte-Like Cells Is Regulated by the Combined Low Dose Treatment of Transforming Growth Factor-? 1 and 5-Azacytidine. *Stem Cells Int.* **2016**, *2016*, 11. [CrossRef]

58. Jin, H.J.; Bae, Y.K.; Kim, M.; Kwon, S.J.; Jeon, H.B.; Choi, S.J.; Kim, S.W.; Yang, Y.S.; Oh, W.; Chang, J.W. Comparative analysis of human mesenchymal stem cells from bone marrow, adipose tissue, and umbilical cord blood as sources of cell therapy. *Int. J. Mol. Sci.* **2013**, *14*, 17986–18001. [CrossRef]

59. Li, J.; Xu, S.; Zhao, Y.; Yu, S.; Ge, L.; Xu, B.; Yu, S.; Yu, S.; Ge, L.; Ge, L.; et al. Comparison of the biological characteristics of human mesenchymal stem cells derived from exfoliated deciduous teeth, bone marrow, gingival tissue, and umbilical cord. *Mol. Med. Rep.* **2018**, *18*, 4969–4977. [CrossRef]

60. Mohamed-Ahmed, S.; Fristad, I.; Lie, S.A.; Suliman, S.; Mustafa, K.; Vindenes, H.; Idris, S.B. Adipose-derived and bone marrow mesenchymal stem cells: a donor-matched comparison. *Stem Cell Res. Ther.* **2018**, *9*, 168. [CrossRef]

61. Kakkar, A.; Nandy, S.B.; Gupta, S.; Bharagava, B.; Airan, B.; Mohanty, S. Adipose tissue derived mesenchymal stem cells are better respondents to TGFβ1 for in vitro generation of cardiomyocyte-like cells. *Mol. Cell. Biochem.* **2019**, *460*, 53–66. [CrossRef] [PubMed]

62. Bajek, A.; Olkowska, J.; Walentowicz-Sadłecka, M.; Sadłecki, P.; Grabiec, M.; Porowińska, D.; Drewa, T.; Roszkowski, K. Human adipose-derived and amniotic fluid-derived stem cells: A preliminary in vitro study comparing myogenic differentiation capability. *Med. Sci. Monit.* **2018**, *24*, 1733–1741. [CrossRef] [PubMed]

63. Guo, X.; Bai, Y.; Zhang, L.; Zhang, B.; Zagidullin, N.; Carvalho, K.; Du, Z.; Cai, B. Cardiomyocyte differentiation of mesenchymal stem cells from bone marrow: New regulators and its implications. *Stem Cell Res. Ther.* **2018**, *9*, 44. [CrossRef] [PubMed]

64. Neshati, V.; Mollazadeh, S.; Fazly Bazzaz, B.S.; de Vries, A.A.F.; Mojarrad, M.; Naderi-Meshkin, H.; Neshati, Z.; Mirahmadi, M.; Kerachian, M.A. MicroRNA-499a-5p Promotes Differentiation of Human Bone Marrow-Derived Mesenchymal Stem Cells to Cardiomyocytes. *Appl. Biochem. Biotechnol.* **2018**, *186*, 245–255. [CrossRef] [PubMed]

65. Lu, M.; Xu, L.; Wang, M.; Guo, T.; Luo, F.; Su, N.; Yi, S.; Chen, T. MiR-149 promotes the myocardial differentiation of mouse bone marrow stem cells by targeting Dab2. *Mol. Med. Rep.* **2018**, *17*, 8502–8509. [CrossRef] [PubMed]

66. Fujita, J.; Tohyama, S.; Kishino, Y.; Okada, M.; Morita, Y. Concise Review: Genetic and Epigenetic Regulation of Cardiac Differentiation from Human Pluripotent Stem Cells. *Stem Cells* **2019**, *37*, 992–1002. [CrossRef] [PubMed]

67. Ieda, M.; Fu, J.-D.; Delgado-Olguin, P.; Vedantham, V.; Hayashi, Y.; Bruneau, B.G.; Srivastava, D. Direct reprogramming of fibroblasts into functional cardiomyocytes by defined factors. *Cell* **2010**, *142*, 375–386. [CrossRef]

68. Chen, J.X.; Krane, M.; Deutsch, M.A.; Wang, L.; Rav-Acha, M.; Gregoire, S.; Engels, M.C.; Rajarajan, K.; Karra, R.; Abel, E.D.; et al. Inefficient reprogramming of fibroblasts into cardiomyocytes using Gata4, Mef2c, and Tbx5. *Circ. Res.* **2012**, *111*, 50–55. [CrossRef]

69. Fu, J.D.; Stone, N.R.; Liu, L.; Spencer, C.I.; Qian, L.; Hayashi, Y.; Delgado-Olguin, P.; Ding, S.; Bruneau, B.G.; Srivastava, D. Direct reprogramming of human fibroblasts toward a cardiomyocyte-like state. *Stem Cell Reports* **2013**, *1*, 235–247. [CrossRef]

70. David, R.; Brenner, C.; Stieber, J.; Schwarz, F.; Brunner, S.; Vollmer, M.; Mentele, E.; Müller-Höcker, J.; Kitajima, S.; Lickert, H.; et al. MesP1 drives vertebrate cardiovascular differentiation through Dkk-1-mediated blockade of Wnt-signalling. *Nat. Cell Biol.* **2008**, *10*, 338–345. [CrossRef]

71. Wystrychowski, W.; Patlolla, B.; Zhuge, Y.; Neofytou, E.; Robbins, R.C.; Beygui, R.E. Multipotency and cardiomyogenic potential of human adipose-derived stem cells from epicardium, pericardium, and omentum. *Stem Cell Res. Ther.* **2016**, *7*, 84. [CrossRef] [PubMed]

72. Neshati, V.; Mollazadeh, S.; Fazly Bazzaz, B.S.; de Vries, A.A.; Mojarrad, M.; Naderi-Meshkin, H.; Neshati, Z.; Kerachian, M.A. Cardiomyogenic differentiation of human adipose-derived mesenchymal stem cells transduced with Tbx20-encoding lentiviral vectors. *J. Cell. Biochem.* **2018**, *119*, 6146–6153. [CrossRef] [PubMed]

73. Chen, Y.-H.; Coller, J. A Universal Code for mRNA Stability? *Trends Genet.* **2016**, *32*, 687–688. [CrossRef] [PubMed]

74. Warren, L.; Lin, C. mRNA-Based Genetic Reprogramming. *Mol. Ther.* **2019**, *27*, 729–734. [CrossRef] [PubMed]

75. Weidgang, C.E.; Russell, R.; Tata, P.R.; Kühl, S.J.; Illing, A.; Müller, M.; Lin, Q.; Brunner, C.; Boeckers, T.M.; Bauer, K.; et al. TBX3 directs cell-fate decision toward mesendoderm. *Stem Cell Reports* **2013**, *1*, 248–265. [CrossRef] [PubMed]

76. Chan, S.S.K.; Shi, X.; Toyama, A.; Arpke, R.W.; Dandapat, A.; Iacovino, M.; Kang, J.; Le, G.; Hagen, H.R.; Garry, D.J.; et al. Mesp1 patterns mesoderm into cardiac, hematopoietic, or skeletal myogenic progenitors in a context-dependent manner. *Cell Stem Cell* **2013**, *12*, 587–601. [CrossRef]

77. Lv, Y.; Gao, C.-W.; Liu, B.; Wang, H.-Y.; Wang, H.-P. BMP-2 combined with salvianolic acid B promotes cardiomyocyte differentiation of rat bone marrow mesenchymal stem cells. *Kaohsiung J. Med. Sci.* **2017**, *33*, 477–485. [CrossRef]

78. Bhuvanalakshmi, G.; Arfuso, F.; Kumar, A.P.; Dharmarajan, A.; Warrier, S. Epigenetic reprogramming converts human Wharton's jelly mesenchymal stem cells into functional cardiomyocytes by differential regulation of Wnt mediators. *Stem Cell Res. Ther.* **2017**, *8*, 185. [CrossRef]

79. Ibarra-Ibarra, B.R.; Franco, M.; Paez, A.; López, E.V.; Massó, F. Improved efficiency of cardiomyocyte-like cell differentiation from rat adipose tissue-derived mesenchymal stem cells with a directed differentiation protocol. *Stem Cells Int.* **2019**, *2019*, 8940365. [CrossRef]

80. Steinle, H.; Weber, M.; Behring, A.; Mau-Holzmann, U.; Schlensak, C.; Wendel, H.P.; Avci-Adali, M. Generation of iPSCs by Nonintegrative RNA-Based Reprogramming Techniques: Benefits of Self-Replicating RNA versus Synthetic mRNA. *Stem Cells Int.* **2019**, *2019*, 1–16. [CrossRef]

81. Wang, L.; Liu, Z.; Yin, C.; Asfour, H.; Chen, O.; Li, Y.; Bursac, N.; Liu, J.; Qian, L. Stoichiometry of Gata4, Mef2c, and Tbx5 influences the efficiency and quality of induced cardiac myocyte reprogramming. *Circ. Res.* **2015**, *116*, 237–244. [CrossRef] [PubMed]

82. Qayed, M.; Copland, I.; Galipeau, J. Allogeneic Versus Autologous Mesenchymal Stromal Cells and Donor-to-Donor Variability. In *Mesenchymal Stromal Cells*; Elsevier: Amsterdam, The Netherlands, 2017; pp. 97–120.

83. McLeod, C.M.; Mauck, R.L. On the origin and impact of mesenchymal stem cell heterogeneity: new insights and emerging tools for single cell analysis. *Eur. Cell. Mater.* **2017**, *34*, 217–231. [CrossRef] [PubMed]

Article

Generation of *GLA*-Knockout Human Embryonic Stem Cell Lines to Model Autophagic Dysfunction and Exosome Secretion in Fabry Disease-Associated Hypertrophic Cardiomyopathy

Hui-Yung Song [1,2], **Chian-Shiu Chien** [1,2], **Aliaksandr A. Yarmishyn** [1,2], **Shih-Jie Chou** [1,2], **Yi-Ping Yang** [2,3,4], **Mong-Lien Wang** [1,2,3,4], **Chien-Ying Wang** [3,5], **Hsin-Bang Leu** [2,3,6], **Wen-Chung Yu** [3,6], **Yuh-Lih Chang** [1,3,4,7,*] and **Shih-Hwa Chiou** [1,2,3,8,*]

[1] Institute of Pharmacology, National Yang-Ming University, Taipei 11221, Taiwan; shy770307@gmail.com (H.-Y.S.); cschien6688@gmail.com (C.-S.C.); yarmishyn@gmail.com (A.A.Y.); ohyeahchou@gmail.com (S.-J.C.); monglien@gmail.com (M.-L.W.)

[2] Department of Medical Research, Taipei Veterans General Hospital, Taipei 11217, Taiwan; molly0103@gmail.com (Y.-P.Y.); hbleu@vghtpe.gov.tw (H.-B.L.)

[3] School of Medicine, National Yang-Ming University, Taipei 11221, Taiwan; wangcy@vghtpe.gov.tw (C.-Y.W.); wcyu@vghtpe.gov.tw (W.-C.Y.)

[4] School of Pharmaceutical Sciences, National Yang-Ming University, Taipei 11221, Taiwan

[5] Department of Emergent Medicine, Taipei Veterans General Hospital, Taipei 11217, Taiwan

[6] Division of Cardiology & Department of Medicine, Taipei Veterans General Hospital, Taipei 11217, Taiwan

[7] Department of Pharmacy, Taipei Veterans General Hospital, Taipei 11217, Taiwan

[8] Genomics Research Center, Academia Sinica, Taipei 11574, Taiwan

[*] Correspondence: ylchang@vghtpe.gov.tw (Y.-L.C.); shchiou@vghtpe.gov.tw (S.-H.C.); Tel.: +886-2-2875-7394; Fax: +886-2-2875-7435

Received: 25 February 2019; Accepted: 4 April 2019; Published: 8 April 2019

Abstract: Fabry disease (FD) is a rare inherited disorder characterized by a wide range of systemic symptoms; it is particularly associated with cardiovascular and renal problems. Enzyme replacement therapy and pharmacological chaperone migalastat are the only approved and effective treatment strategies for FD patients. It is well documented that alpha-galactosidase A (GLA) enzyme activity deficiency causes globotriaosylceramide (Gb3) accumulation, which plays a crucial role in the etiology of FD. However, the detailed mechanisms remain unclear, and the lack of a reliable and powerful disease model is an obstacle. In this study, we created such a model by using CRISPR/Cas9-mediated editing of *GLA* gene to knockout its expression in human embryonic stem cells (hESCs). The cardiomyocytes differentiated from these hESCs (GLA-null CMs) were characterized by the accumulation of Gb3 and significant increases of cell surface area, the landmarks of FD-associated cardiomyopathy. Furthermore, we used mass spectrometry to compare the proteomes of GLA-null CMs and parental wild type CMs and found that the Rab GTPases involved in exocytotic vesicle release were significantly downregulated. This caused impairment of autophagic flux and protein turnover, resulting in an increase of reactive oxygen species and apoptosis. To summarize, we established a FD model which can be used as a promising tool to study human hypertrophic cardiomyopathy in a physiologically and pathologically relevant manner and to develop new therapies by targeting Rab GTPases signaling-related exosomal vesicles transportation.

Keywords: Fabry disease; human embryonic stem cells; CRISPR/Cas9 genomic editing; Mass spectrometry proteomic analysis; hypertrophic cardiomyopathy; disease model

1. Introduction

Approximately 3–5% of inherited hypertrophic cardiomyopathies result from lysosome storage disorders [1]. One of these cardiomyopathies is Fabry disease (FD), which is caused by alpha-galactosidase A (GLA) deficiency leading to accumulation of globotriaosylceramide (Gb3) in several tissues. FD is particularly manifested in renal and cardiovascular dysfunctions [2]. A number of in vivo and in vitro studies revealed that the loss of GLA results in left ventricular hypertrophy and develops into heart failure, myocardial infarction and life-threatening arrhythmias due to Gb3 deposition. The levels of Gb3 and its deacylated derivative, globotriaosylsphingosine (lyso-Gb3) in plasma and tissues are used as diagnostic biomarkers of FD, and are applied for the screening/monitoring of FD patients by liquid chromatography-tandem mass spectrometry (LC-MS/MS) [3–6]. However, the limited sample volume and variable levels of lyso-Gb3 in different matrices makes the quantitation very challenging [7]. Furthermore, the detailed mechanism of how Gb3 accumulation results in hypertrophic cardiomyopathy still needs to be elucidated. In order to develop new therapeutic strategies for FD-associated vasculopathy, it is essential to understand the underlying pathogenesis mechanisms, as well as to discover the potential prognostic biomarkers.

Previously, we generated CRISPR/Cas9-edited GLA-null human HEK293 cells; however, this cell line may not be an appropriate model with which to study pathological events occurring in cardiomyocytes [8]. The pluripotent stem cells, including embryonic stem cells (ESCs) and induced pluripotent cells (iPSCs), offer a great potential for modelling human diseases as they can be differentiated into the tissue affected by the pathology. Although several FD-specific iPSC lines have been obtained from patients carrying different *GLA* gene mutations, which claimed to be useful for FD cardiomyopathy research [9,10], one major limitation of such approach is the influence of variable genetic background, which can be significant even for monogenic, dominant and highly penetrant disease in FD [11,12]. Currently, CRISPR/Cas9 emerges as a powerful genome editing technique, providing the opportunity to efficiently delete genes to establish isogenic cells [13–15]. Therefore, our strategy in this study was to generate *GLA* knockout in human pluripotent stem cells by CRISPR/Cas9-mediated gene editing, and compare them with the parental cells of the same genetic background to study the mechanisms of FD-associated cardiomyopathy.

Recently, it became increasingly clear that lysosomal storage disorders have an impact on autophagic dysfunction [16]. Dysregulated ceramide metabolism can trigger cytotoxic signaling cascades, including apoptosis and necroptosis, missorting and accumulation of these sphingolipids in the membrane subdomains may destabilize lipid bilayer and cause their permeabilization [17]. Exosomes are small vesicles secreted upon fusion of multivesicular endolysosomal compartments with the plasma membrane and are derived from the intraluminal vesicles (ILVs) of those organelles. Exosomes may participate in the control of cellular homeostasis by promoting the release of intracellular harmful components, including proteins, lipids or nucleic acids. Emerging evidence from the studies of normal development, as well as multiple disease studies, is beginning to reveal a coordinated exosome–autophagy response that functions to maintain homeostasis through lysosomal degradation and release of cellular cargo [18–20]. However, little effort has been made to investigate the impact of autophagic dysfunction in FD on biogenesis and secretion of exosomes.

In the current study, we applied CRISPR/Cas9-mediated genomic editing to deplete *GLA* expression in ESC-derived cardiomyocytes to recapitulate FD cardiac hypertrophy in vitro and performed proteomic analysis by LC-MS/MS. We identified that Rab GTPase signaling-related vesicle secretion is the factor that may initiate or exacerbate the development of FD-associated cardiomyopathy. Such information will be extremely important for potential application in the prevention of and in interventions for the adverse effects of the cardiac hypertrophy in FD patients.

2. Materials and Methods

2.1. CRISPR/Cas9 Plasmid Construction and Transfection

The CRISPR/Cas9 with T2A-eGFP co-expression vector pSpCas9(BB)–2A-GFP (PX458) was a gift from Feng Zhang (Addgene plasmid). The exon 1 of *GLA* was selected for guiding RNA design and the sequence (5′-AGGAACCCAGAACTACATCT-3′) was cloned into PX458 (abbreviated as GLA-Cas9-GFP) as previously described [8]. The *GLA*-specific targeting plasmid was transfected into hESC line (WA09) by electroporation using Nucleofector System (Lonza, Basel, Switzerland) following the manufacturer's protocol. Briefly, hESC cells were cultured to 80–90% confluence, then harvested and washed with PBS without Ca^{2+} and Mg^{2+}. Approximately 4×10^5 cells were resuspended in the pre-mixture solution with 2 µg GLA-Cas9-GFP plasmid, and the optimized protocol (program B016) was used for electroporation. The cells were plated on Matrigel-coated 6-well plate in mTeSR1 medium containing 10 µM Y27632. 48 h later, the proportion of cells expressing EGFP was enriched by flow cytometry using FACSCalibur (BD Biosciences, San Jose, CA, USA). Three days later, cells were detached with TrypLE (Invitrogen, Thermo Fisher Scientific, Waltham, MA, USA), separated into single cells and seeded with a density of 1 cell/well of a 96-well dish.

2.2. Analysis of CRISPR/Cas9-Induced Mutations in GLA Gene

To identify the presence of indels in *GLA* gene, the genomic DNA was extracted and used for PCR amplification of the target site with the primer pair 5′-CACACACCAACCTCTAACGATACC-3′ (forward) and 5′-CCAGGAAAGGTCACACAGAGAAAG-3′ (reverse). PCR products were TA-cloned into pGEM-T Easy vector (Promega, Madison, WI, USA). Subsequently, DNA from the clones #19 and #27 was sequenced using T7 forward and Sp6 Reverse primer. Vector NTI software was used to align the results of sequencing and determine the indel spectra in *GLA* target site.

2.3. hESC Culture and Differentiation to Cardiomyocytes

The hESCs were cultured on tissue culture dishes coated with Geltrex (Life Technologies, Thermo Fisher Scientific) in mTeSR1 culture medium (STEMCELL Technologies, Vancouver, BC, Canada) with daily media changes. The cells were passaged every 3–4 days using Accutase (STEMCELL Technologies). The undifferentiated phenotype of the hESCs was checked daily using a light microscope. In order to differentiate hESCs to cardiomyocytes, they were dissociated by Versene (Life Technologies, Thermo Fisher Scientific), resuspended in mTeSR1 + 5 µM Y27632 and seeded onto Geltrex-coated plates at a density of 3×10^5 cells/cm^2 and grown for the next four days with daily medium change. Following that, the cells were treated with 6 µM CHIR99021 (Selleckchem, Houston, TX, USA) in insulin-free RPMI/B27 medium (Life Technologies) for 24 h. The medium was replaced with basal medium for another 2 days. At day 3, the culture medium was subsequently replaced with 5 µM IWP-2 (Tocris Bioscience, Minneapolis, MN, USA) in insulin-free RPMI/B27 for 48 h. On day 7, the culture medium was changed to RPMI/B27 containing insulin (Life Technologies, Thermo Fisher Scientific), and the culture medium was refreshed every 3 days thereafter.

2.4. Alkaline Phosphatase Staining

Cells were washed in PBS twice, fixed with 80% alcohol for at least 2 h at 4 °C, followed by soaking in ddH$_2$O for 2–3 min, and 100 mM Tris-HCl (pH 8.2–8.5) for 5 min. Alkaline phosphatase substrate working solution (Vector Laboratories, Burlingame, CA, USA) was added for 1 h and stained colonies were visualized under light microscope.

2.5. Reverse Transcription-Polymerase Chain Reaction (RT-PCR)

Total RNA was isolated with TRIzol reagent (Invitrogen, Thermo Fisher Scientific) and quantified by spectrophotometry at 260 nm. 3 µg of total RNA was reverse transcribed using SuperScript III

Reverse Transcriptase (Invitrogen, Thermo Fisher Scientific) at 55 °C for 1 h into complementary DNA, which was then used as the template for subsequent PCR reactions. The PCR reactions were run with the following cycling conditions: 94 °C for 5 min, followed by 25 or 30 cycles at 94 °C (denaturation) for 30 s, 58–62 °C for 30 s (annealing), 72 °C for 45 s (synthesis). The primer sequences are shown in Supplementary Tables S1 and S2. Amplified RT-PCR products were analyzed on 2% agarose gels and visualized using ethidium bromide staining and SPOT camera system (Diagnostic Instruments, Sterling Height, MI, USA).

2.6. Immunofluorescence Staining

First, the cells were rinsed in PBS and fixed with 1% (*v/v*) paraformaldehyde for 10 min followed by treatment with 70% ethanol (*v/v*) for 10 min at room temperature. The cells were permeabilized with 0.1% NP-40 (Sigma-Aldrich, St. Louis, MO, USA) for 20 min, then washed twice with PBS. To block cells, blocking solution (0.3% BSA and 5% serum in PBS) was applied for 30 min. Cells were incubated with primary antibodies in the blocking solution overnight at 4 °C, washed three times in PBS, then stained with secondary antibodies at 1:200 in PBS for 1 h at room temperature. Cells were washed three times in PBS, and nuclei were stained with Hoechst 33,342 (Life Technologies, Thermo Fisher Scientific) at 1:5000 in PBS for 5 min at room temperature. Prior to imaging, cells mounted with SlowFade Gold Antifade Mountant (Millipore, Sigma, Burlington, MA, USA). Images were obtained using fluorescent microscopy and a digital camera. Antibody for characterization of pluripotency is listed in Supplementary Table S3.

2.7. Measurement of Cardiomyocyte Size

The size of iPSC-derived cardiomyocytes was evaluated by measuring the cell area. Twenty days after cardiac induction, the spontaneously beating embryoid bodies were dissociated into single cells using Accutase™ (STEMCELL Technologies). These cells were then plated onto gelatin-coated dishes for further experiments and analysis. Subsequently, the cellular images of cTnT-positive cells were recorded at days 30, 40 and 60 post-induction of differentiation using FV10i confocal microscope (Olympus, Tokyo, Japan). The pixel area of cTnT-positive cells was measured and analyzed using ImageJ software package (NIH). About total two hundred cells were analyzed in five independent experiments.

2.8. Western Blotting

Cells were lysed in RIPA lysis buffer (0.5M Tris-HCl, pH 7.4, 1.5M NaCl, 2.5% deoxycholic acid, 10% NP-40, 10 mM EDTA, protease inhibitor), and the protein lysates were subjected to SDS-PAGE followed by electroblotting onto a PVDF membrane. Membranes were probed with the following primary antibodies: α-galactosidase A, TSG101 (GeneTex, Irvine, CA, USA), CD63 (Santa Cruz Biotechnology, Dallas, TX, USA), Calnexin (Abcam, Cambridge, MA, USA) GDRID2, VPS36, VTI1A (Proteintech Group, Wuhan, Hubei, P.R.C), LC3 (Novus Biologicals, Centennial, CO, USA), Rab11 and GAPDH (Cell Signaling Technology, Denver, MA, USA). The bands were visualized by chemiluminescence detection reagents.

2.9. Lipid Extraction

The cells were grown to confluency on 150 mm cell culture dishes and harvested by scraping into 700 µL PBS. The cell suspension was transferred into 16 mm × 100 mm glass tubes, and 1 mL chloroform and 2.4 mL methanol were added. After water bath sonication, protein was precipitated by centrifuging at 2400× *g* for 30 min. The supernatant was transferred to a new glass tube and 4.5 mL chloroform and 1.2 mL 0.9% NaCl were added. The sample was centrifuged at 900× *g* for 5 min. The upper aqueous phase was discarded, and the lower organic phase was washed twice with 2 mL methanol and 0.8 mL 0.9% NaCl. The lower phase was extracted using a 1 mL glass syringe (Hamilton, Reno, NV, USA) and transferred to a new glass tube. Lipids were dried under a stream of nitrogen.

2.10. Quantification of Gb3 by Thin Layer Chromatography (TLC)

First, 100 nmole of total phospholipid was applied to a silica high performance TLC plate (Sigma-Aldrich, St. Louis, MO, USA). The plate was first developed in a solvent system consisting of chloroform/methanol (98:2), and air-dried. The plate was then developed in a solvent system consisting of chloroform/methanol/acetic acid/water (61/31/5/3) and air-dried. Plates were submerged in 8% cupric sulfate pentahydrate in water/methanol/H_3PO_4 (60:32:8), and charred for 10 min at 150 °C, or were sprayed with 1% orcinol in 11% H_2SO_4 and charred at 130 °C for 5 min. Plates were scanned and densinometry measured using ImageJ software. Lipids were quantified by running Gb3 standards on a plate (Matreya LLC, State College, PA, USA).

2.11. Liquid Chromatography-Tandem Mass Spectrometry (LC-MS/MS) Analysis

LC-MS/MS analysis was performed using Orbitrap Mass Analyzer (Thermo Fisher Scientific), according to the manufacturer's protocol. Briefly, each sample of digested peptides was reconstituted in 20 µL of 0.1% formic acid. Peptides were first separated by the nanoflow HPLC on Agilent 1100 (Agilent Technologies, Santa Clara, CA, USA) using C18 column (Agilent Technologies) with a flow rate of 0.4 µL/min, and were ionized after passing through the nanospray tip (New Objective, Woburn, MA, USA). LC gradient for the LC-MS/MS system ramped from 2–40% ACN in 120 min, and the system was set up for automated data-dependent acquisition, with a mode of 200–2000 m/z full scan for the maximum three most intense peaks from each Orbitrap MS scan. Peptides with +2 or +3 charge were further subjected to CID. Spectra were obtained in raw data files with Xcalibur (version 2.0 SR2). Protein identification was accomplished by TurboSEQUEST (Thermo Fisher Scientific) using the UniProt database. A protein was confirmed once three peptides with Xcorr >2.5 were matched in sequencing.

2.12. Transmission Electron Microscopy

The morphology of differentiated cardiomyocytes was characterized using JEM-2000 EX II transmission electron microscope (JEOL, Tokyo, Japan). The cardiomyocytes were covered with 400 mesh carbon-coated copper TEM grid. After 15 min, the grid was tapped with filter paper to remove the excess water followed by staining with 1% phosphotungstic acid (Sigma-Aldrich) for 20 min. The samples were allowed to air-dry for 24 h and then observed under TEM.

2.13. Array-Based Comparative Genomic Hybridization (CGH-Array

Genomic DNA was isolated and intermittently sonicated using a Digital Sonifier 450 sonicator probe (Branson Ultrasonics, Danbury, CT, USA). DNA samples were amplified using the GenomePlex WGA kit (Thermo Fisher Scientific). Genomic DNA ULS Labeling Kit (Agilent) was used to label the amplified DNA with either Cy3 or Cy5. As recommended by Agilent, 2.0–2.5 µg of amplified DNA was used as the input starting material for each labeling reaction. Scanning and image analysis were conducted according to Agilent Oligonucleotide Array-based CGH for Genomic DNA analysis Protocol (version 4.0). Microarrays were scanned using an Agilent G2565BA DNA Microarray Scanner (Agilent). Agilent Feature Extraction software (v9.1.3) was used to extract data from raw microarray image files. Agilent CGH Analytics software (v3.4) was used to visualize, detect and analyze the aberration patterns from CGH microarray profiles.

2.14. Exosome Isolation and Characterization

Exosomes were isolated from cell culture supernatants using Total Exosome Isolation Reagent (Thermo Fisher Scientific) following the manufacturer's protocol. Culture media samples were centrifuged at 2000× *g* for 30 min to remove cells and debris. The supernatant was transferred to sterile tubes and an exosome precipitation solution was added at a 2:1 ratio. Samples were mixed and left overnight at 4 °C. Samples were then centrifuged at 10,000× *g* for 60 min and supernatant carefully removed. The precipitated exosome pellets were re-suspended with PBS and either used immediately or stored at −80 °C until required. Immunoaffinity capture assay was used to characterize the purity of exosomes using CD63 antibody-conjugated dynabeads. Exosomes were incubated with antibody-conjugated dynabeads overnight and washed by PBS containing 0.1% BSA two times. Further, CD63 PE-conjugated antibody was used to stain the exosome-bound dynabeads. All samples were examined by flow cytometry using FACSCanto System (BD Biosciences) and FACSDIVA software was used to analyze the population of exosomes.

2.15. Quantification of Isolated Exosomes

Acetylcholinesterase activity assay was performed using EXOCET Exosome Quantitation Kit (System Biosciences, Palo Alto, CA, USA). Isolated exosomes were re-suspended with PBS and lysed with lysis buffer. Each sample was incubated at 37 °C for 5 min and mixed with reaction buffer in 96-well plates. Mixed samples were incubated for 20 min at room temperature and quantified by spectrophotometry at OD 405 using standard samples containing known numbers of exosomes. The final number of exosomes was converted to micrograms, per the manufacturer's guideline. To characterize exosomes, isolated exosome preparations were re-suspended with PBS and analyzed using Zetasizer Nano dynamic light scattering system (Malvern Instruments, Malvern, UK).

2.16. Mitochondrial Superoxide Stress Quantification

Mitochondrial superoxide production was quantified using MitoSOX Red mitochondrial superoxide indicator (Thermo Fisher Scientific) according to the manufacturer's protocol. Briefly, the cells were incubated with 5 µM MitoSOX Red in the culture medium in the dark for 60 min at 37 °C. Stained cells were counterstained as desired and mounted and analyzed by fluorescent microscopy.

2.17. Annexin V Staining

Cells were washed once with PBS and analyzed using FITC Annexin V Apoptosis Detection Kit (BD Biosciences) according to the manufacturer's protocol.

2.18. GLA Enzyme Activity Assay

Cells were washed twice with 1X PBS and were lysed in 60 µL lysis buffer (27 mM sodium citrate, 46 mM sodium phosphate dibasic, 0.5% Triton X-100). 10 µL of cell lysate was added to 50 µL assay buffer containing 6 mM 4-methylumbelliferyl-α-d-galactopyranoside (Sigma-Aldrich) and 117 mM N-acetyl-D-galactosamine (Sigma-Aldrich) and incubated at 37 °C for 1 hr. The 4-methylumbelliferone (Sigma-Aldrich) dissolved in methanol was used as standard ranging from 0.15 µM to 5000 µM. Thereafter, 70 µL glycine-carbonate solution (pH 10.8) was then added to stop the reaction and fluorescence was detected by microplate reader (em/ex = 365/448 nm). The enzyme activity was normalized by protein concentration of cell lysate.

2.19. Statistical Analysis

The quantifiable data are presented as the means ± standard deviation (SD) and compared with Student's *t*-test by GraphPad Prism 6 (GraphPad Prism Software). * $p < 0.05$, ** $p < 0.01$, *** $p < 0.005$ and **** $p < 0.001$.

3. Results

3.1. CRISPR/Cas9-Mediated Knockout of Expression of GLA in hESCs

In order to investigate the functions of *GLA* and its therapeutic potential for FD, we performed CRISPR/Cas9-mediated gene editing to knock out expression of this gene in human embryonic stem cells (hESCs). The procedure was similar to that described in our previous work, where we established the GLA-null HEK293 cell line [8]. Briefly, the *GLA*-specific single-guide RNA (sgRNA) was designed by using Optimized CRISPR Design tool (http://crispr.mit.edu/) to introduce the frameshift mutations in exon 1 of *GLA* gene as shown in Figure 1A. The parental male hESC line H9 (WA09) was transfected with Cas9-encoding plasmid with EGFP reporter, along with the sgRNA. Following the transfection, the proportion of the successfully transfected EGFP-expressing cells was enriched by FACS, and these cells were seeded to 96-well dishes to obtain pure colonies. Out of 36 wells where the single cells were seeded, the clones #20 to #24 and #28 did not grow up. The remaining 30 colonies were screened for GLA protein expression by western blotting (Figure 1B). The GLA protein expression was completely ablated in four clones (#3, #19, #26, and #27) and partially suppressed in two (#25 and #31) (Figure 1B). By performing preliminary differentiation of these clones into cardiomyocytes (CMs), we found that clone #26-derived CMs displayed slight GLA expression, which could be due to contamination with parental non-edited cells, therefore, this clone was excluded from further analysis (Supplementary Figure S1A). *GLA* mRNA levels in these CM clones were not significantly decreased as compared to the control, only in clone #27-derived CMs mRNA levels were downregulated by approximately 50% (Supplementary Figure S1B). On the other hand, GLA enzyme activity was completely absent in clones #3, #19, and #27, which was consistent with the complete absence of GLA protein in these cells (Supplementary Figure S1C). Subsequently, the mutations introduced by CRISPR/Cas9 were identified by Sanger sequencing in the remaining *GLA*-knockout hESC clones #3, #19, and #27. Clone #3 displayed mixed genetic background at the target site, due to putative non-homogeneous population, therefore, it was also excluded from the analysis. (Supplementary Figure S1D). Two of GLA-null hESC clones were selected for the following study because they were hemizygous for 2 bp deletion (clone #19) and 1 bp insertion (clone #27) (Figure 1C). T7 Endonuclease I (T7E1) was used to confirm the presence CRISPR/Cas9-introduced mutations in the *GLA* gene, whereby the cleaved mismatch products were detected in clones #19 and #27 (Figure 1D). On the other hand, T7E1 assay revealed that there were no CRISPR/Cas9-introduced mutations in the potential off-target genes predicted by Optimized CRISPR Design tool (Suppl. 2). Therefore, by conducting the CRISPR/Cas9-mediated genome editing, we successfully established two GLA-null hESC clones.

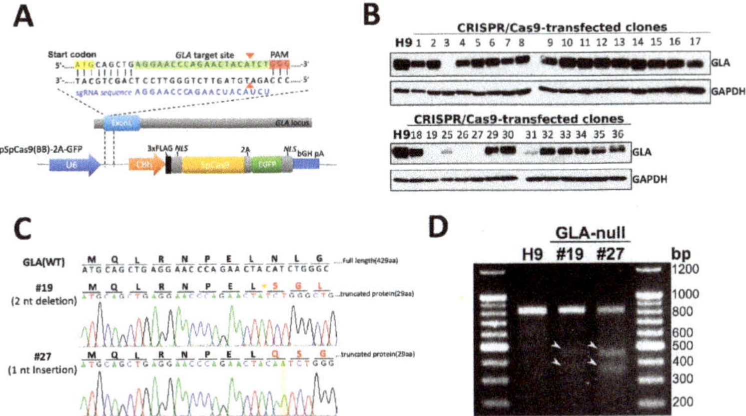

Figure 1. CRISPR/Cas9-mediated knockout of expression of *GLA* in hESCs. (**A**) Schematic depiction of sgRNA-guided Cas9 target site within exon 1 of *GLA* gene. The sgRNA sequence (5'-AGGAACCCAGAACUACAUCU-3') is labeled in blue font and PAM recognition sequence highlighted in red. The gRNA targeting site in the *GLA* exon 1 region is highlighted in green and the double-strand breaking site is indicated by the red arrowheads. The start codon is highlighted in yellow. (**B**) Western blot screening of 30 CRISPR/Cas9-corrected clones for expression of GLA protein. H9 cells were used as wild type parental control. GAPDH used as a loading control. (**C**) Sanger sequencing analysis confirming two nucleotides deletion and one nucleotide insertion in CRSIPR/Cas9-edited hESC clones #19 and #27, respectively. (**D**) T7E1 digestion assay of the mutants at the target site in *GLA* gene. The mismatch T7E1 cleavage products are marked with white arrowheads.

3.2. Characterization of GLA-Null hESC Clones

To confirm the pluripotency status of the established GLA-null hESC clones #19 and #27, several stemness markers were detected by RT-PCR, thus confirming that CRISPR/Cas9-mediated *GLA* knockout did not significantly influence the pluripotency gene expression in hESCs (Supplementary Figure S3A). Both GLA-null hESC clones displayed typical pluripotent morphology features, including small and tightly packed cells, high nucleus to cytoplasm ratio, and positive alkaline phosphatase activity (Supplementary Figure S3B). The expression of pluripotency markers OCT4, NANOG, TRA-1-60 and TRA-1-81 was verified by immunofluorescence staining (Figure 2A). Both null-GLA hESC clones were able to spontaneously differentiate in vitro into embryoid bodies expressing markers of three germ layers: HNF3β of endoderm, α-SMA of mesoderm, and nestin of ectoderm (Figure 2B). Both clones exhibited normal karyotypes (Figure 2C). To summarize, both established GLA-null hESC clones exhibited normal stem cell properties, and therefore, were suitable for differentiation into FD-specific cardiomyocytes to investigate the mechanisms of involvement of GLA deficiency in the pathology of FD cardiomyopathy.

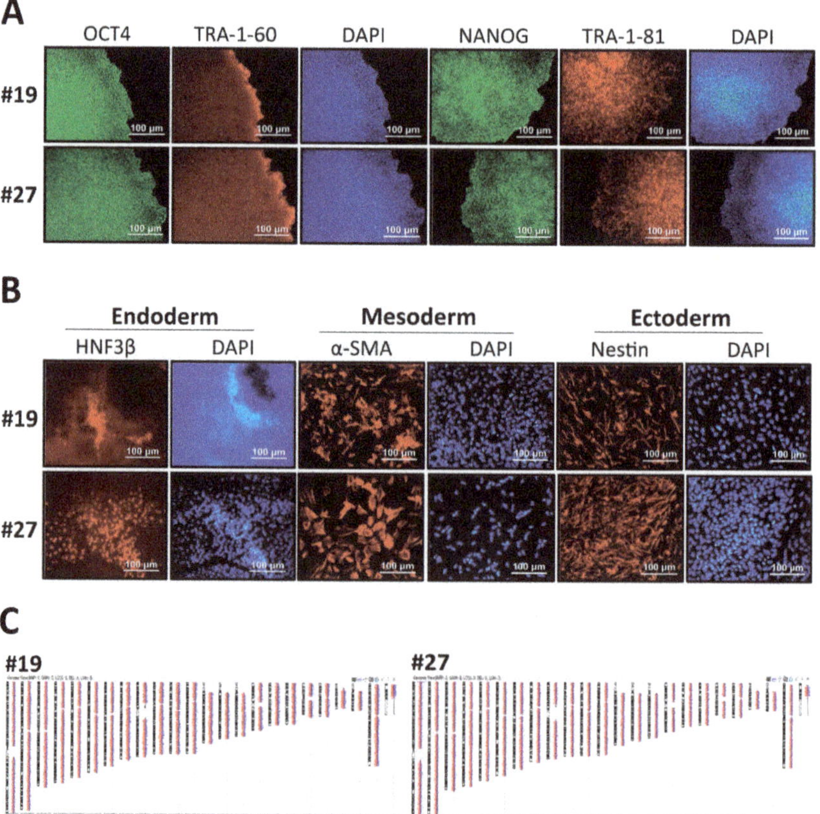

Figure 2. Characterization of CRISPR/Cas9-edited GLA-null hESC clones. (**A**) Immunofluorescence staining demonstrating the protein expression of pluripotency markers OCT4, TRA-1-60, NANOG and TRA-1-81 in GLA-null hESC clones #19 and #27. Nuclei stained with DAPI. (**B**) Embryoid body formation assay showing differentiation of GLA-null hESC clones #19 and #27 into three germ layers: endoderm immunoreactive for HNF3β, mesoderm immunoreactive for α-SMA, and ectoderm immunoreactive for nestin. Nuclei stained with DAPI. (**C**) Representative karyograms GLA-null hESC clones #19 and #27. Visualization was performed using Agilent CytoGenomics software.

3.3. Recapitulation of FD-Specific Cardiac Abnormalities in GLA-Null Cardiomyocytes

Given the fact that CM hypertrophy is one of the symptoms of FD-associated cardiomyopathy, we examined whether GLA-null CMs could recapitulate such features. hESCs were differentiated into CMs according to the conventional CM differentiation protocol [21]. Both GLA-null CMs and CMs derived from the parental H9 hESCs (H9 CMs) exhibited spontaneous contractions as early as twelve days following initiation of cardiac differentiation. Noticeably, no significant differences in the efficiency of cardiac differentiation were observed between GLA-null CMs and H9 CMs, which suggested that the GLA deficiency did not significantly affect the differentiation of hESCs to CMs. The accumulation of Gb3 in CMs is the most prominent hallmark of FD-associated cardiomyopathy. Compared with H9 CMs, GLA-null CMs accumulated more Gb3, as was demonstrated by transmission electron microscopy (Figure 3A) and thin layer chromatography (Figure 3B). GLA protein was not expressed in #19 and #27 GLA-null CM clones at day 60 of differentiation (Figure 3C). Cardiac hypertrophy is often accompanied by reactivation of fetal genes which are active during fetal cardiac development and quiescent in adult hearts [22]. We measured the expression level of several of such genes by qRT-PCR and showed that atrial natriuretic peptide (ANP), B-type natriuretic peptide (BNP), but not troponin T (TnT) were upregulated in GLA-null CMs compared to H9 CMs. In addition, GLA-null CMs were characterized by shifted balance between α- and β-cardiac myosin heavy chain (MYH6/MYH7 ratio) expression, which is a common response to cardiac injury and a hallmark of cardiac hypertrophy [23] (Figure 3D). Immunofluorescent staining of cardiac troponin T (cTnT) revealed that GLA-null CMs were of significantly larger size (26% ± 9.7%, $p < 0.001$) as compared to the control H9 CMs (Figure 3E,F), which was consistent with the phenotype of FD-associated cardiomyopathy. In summary, we have shown that CRISPR/Cas9-edited hESC-derived GLA-null CMs recapitulated the typical phenotype features of FD-affected CMs and, therefore, could be a useful model to study the FD-associated cardiomyopathy.

3.4. Proteomic Analysis of GLA-Null CMs

To further investigate the potential mechanism of FD-related cardiomyopathy, we used proteomic analysis. We focused our analysis to the genes that where downregulated by *GLA* knockout, i.e., whose function was negatively affected. By applying LC-MS/MS, we identified 60 proteins downregulated in GLA-null CMs as compared to H9 CMs with a false discovery rate (FDR) below 0.01. By performing gene ontology (GO) analysis using FatiGO software [24], we found that these 60 proteins were significantly enriched in cellular component GO terms related to extracellular vesicle transportation, secretion, and exocytosis (Figure 4A,B). Among the downregulated proteins involved in regulated exosome release were Ras-Related Protein Rab-11 (RAB11), Rho GDP-dissociation inhibitor 2 (GDIR2, ARHGDIB), VPS36 and VTI1A (Supplementary Table S4). The levels of these proteins were validated and quantified by western blotting; they were shown to be downregulated in GLA-null CMs (Figure 4C,D).

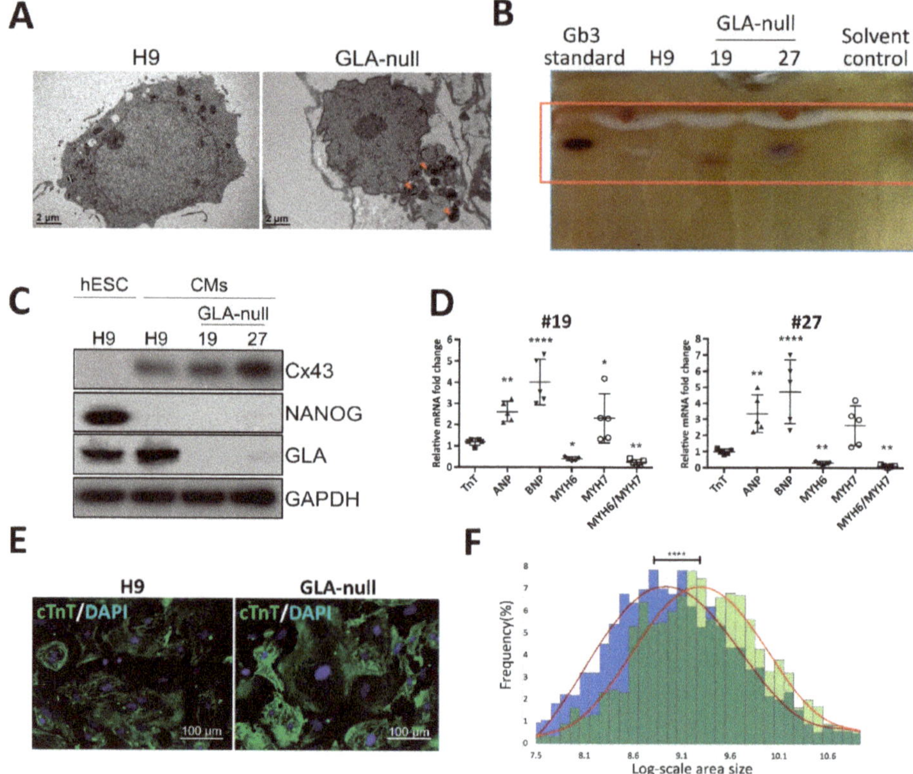

Figure 3. Recapitulation of FD-specific cardiac abnormalities in GLA-null cardiomyocytes. (**A**) TEM images showing the ultrastructure of parental type (H9) and GLA-null CMs. The red arrowheads indicate the multilayered lysosomal structure. (**B**) Gb3 content in parental (H9) and GLA-null CMs (clones #19 and #27) analyzed by TLC. (**C**) Western blot showing lack of expression of GLA in clones #19 and #27 of GLA-null CMs as compared to the parental wild type CMs and hESCs (H9). Connexin 43 (Cx43) and NANOG served as cardiomyocyte and pluripotency markers, respectively. GAPDH used as a loading control. (**D**) qRT-PCR analysis of expression several fetal heart markers in CM clones #19 and #27. The results are expressed as fold change relative to H9 CMs. (**E**) Immunostaining of cTnT showing significantly enlarged size of GLA-null CMs compared to H9 CMs. (**F**) Quantification of area size of GLA-null CMs (green columns) and H9 CMs (blue columns). At least 200 cells were analyzed individually and statistical difference is $p < 0.001$.

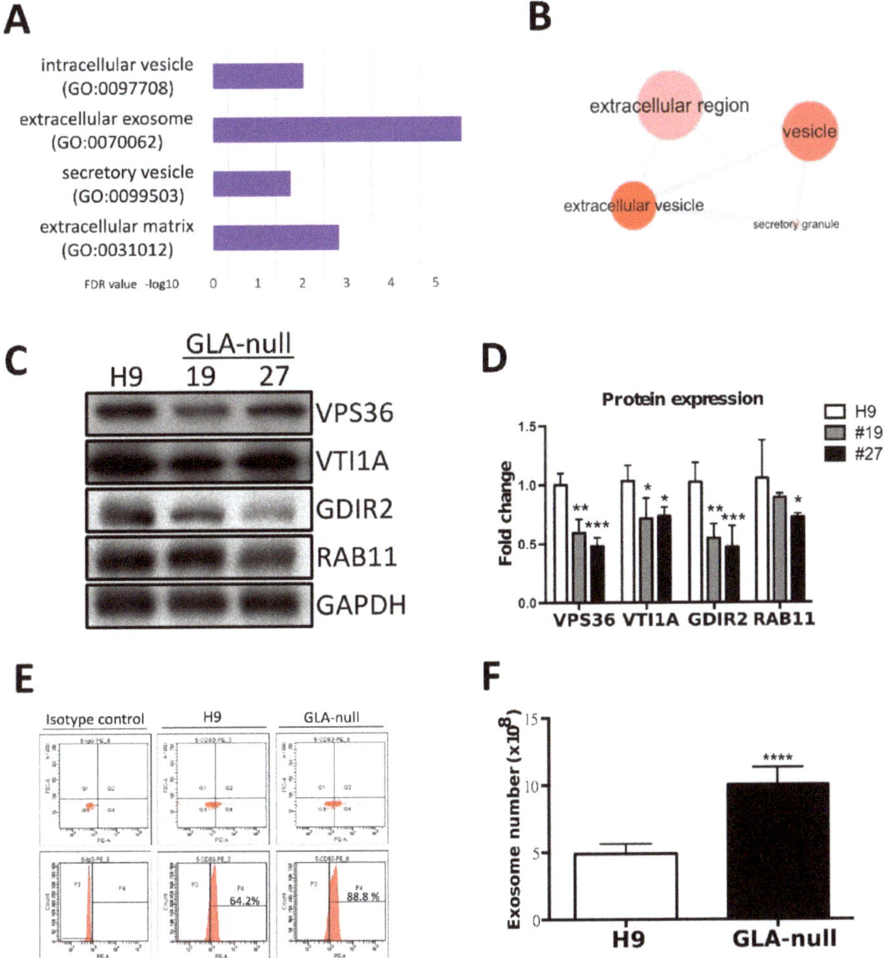

Figure 4. Proteomic analysis of GLA-null CMs. (**A**) The most enriched cellular component GO terms in the list of 60 genes downregulated in GLA-null CMs. (**B**) Visualization of the most enriched cellular component GO terms using REVIGO software. (**C**) Western blot showing expression of the mediators of vesicular trafficking in GLA-null CMs #19 and #27 as compared to wild type control (H9). GAPDH served as a loading control. (**D**) Quantitation of expression of proteins annotated as the mediators of vesicular trafficking. The data are presented as mean ± standard deviation error bars from three independent experiments. (**D**) Flow cytometric analysis of CD63 expression in H9- and GLA-null CM-derived exosome isolated with magnetic beads directly from cell culture medium. (**E**) Quantification of exosome numbers in the supernatant from H9 and GLA-null CMs.

3.5. GLA-Null CMs Secrete More Exosomes Than H9 CMs

Given the fact that the proteome analysis revealed changes in proteins involved in vesicle secretion, we further investigated whether the exosome biogenesis was affected in GLA-null CMs. The exosomes were isolated from the culture medium, and their identity was confirmed by electron microscopy. Multivesicular bodies were observed and exhibited the typical characteristic cup-shaped morphology (Supplementary Figure S4A) and size (diameter between 50 and 150 nm). Exosome size was measured by the Nanosight tracking system (Supplementary Figure S4B) and their identity was confirmed by detecting expression of exosomal markers TSG101 and CD63 (Supplementary Figure S4C). Quantification of exosome though CD63 PE-conjugated fluorescence by flow cytometry analysis revealed that at higher CD63 positive levels in exosomes derived from GLA-null CMs compared to H9 CMs (Figure 4D). Next, we investigated the number of isolated exosomes from H9 and GLA-null CMs. Interestingly, GLA-null CMs produced significantly larger number of exosomes than H9 CMs (10.8×10^8 vs. 5.1×10^8, $p < 0.001$) (Figure 4E). These results indicate that GLA-null CMs increased the production and secretion of exosomes.

3.6. Vesicle Turnover Impairment Induces Cardiotoxicity in GLA-Null CMs

Several studies have shown that the molecular machinery and regulatory mechanisms are shared between exosome biogenesis and autophagy [25,26], suggesting that these two processes are intimately linked. Since the proteomic analysis revealed the downregulation of components of vesicular trafficking machinery in GLA-null CMs, we hypothesized that dysfunctional autophagy pathway may underlie the impairment of cellular homeostasis in FD-related cardiomyopathy. To investigate whether autophagy flux impairment plays a role in cardiac phenotype of FD, we used western blotting to analyze the expression of autophagy markers LC3-I and LC3-II. The levels of these proteins were monitored in a time course of four hours after induction of autophagy with HBSS medium, and it was revealed that LC3-II accumulated by two hours after autophagy induction in both H9 and GLA-null CMs indicating to initiation of autophagy flux (Figure 5A,B). However, after four hours of autophagy induction, LC3-II levels dropped significantly in H9, but remained high in GLA-null CMs, signifying the digestion of autophagosome by lysosomes in the former and the block of autophagy flux in the latter (Figure 5A,B). Previous studies have indicated that undigested material within the autophagosome, including dysfunctional mitochondria, may be a source of free radicals, which in turn can result in cellular dysfunction and apoptosis [27,28]. Mitochondrial integrity and turnover play an important role in CM bioenergetics and function. MitoSOX Red staining demonstrated significant activation of mitochondrial superoxide production in GLA-null CMs (Figure 5C). The quantification of the MitoSOX Red staining intensity indicated that mitochondrial superoxide production in GLA-null CMs was 3.4-fold higher than that in H9 CMs (Figure 5D). To assess CM apoptosis, annexin V assay was used. GLA-null CMs displayed 4-fold increase in staining with annexin V in comparison with H9 CMs (Figure 5E,F). Furthermore, we observed significant 2-fold increase of lactate dehydrogenase (LDH), a marker of necrosis, in the supernatant derived from GLA-null CMs (Figure 5G). Collectively, our results demonstrate that GLA-null CMs were characterized by autophagy impairment and active mitochondrial ROS production that caused apoptosis and necrosis.

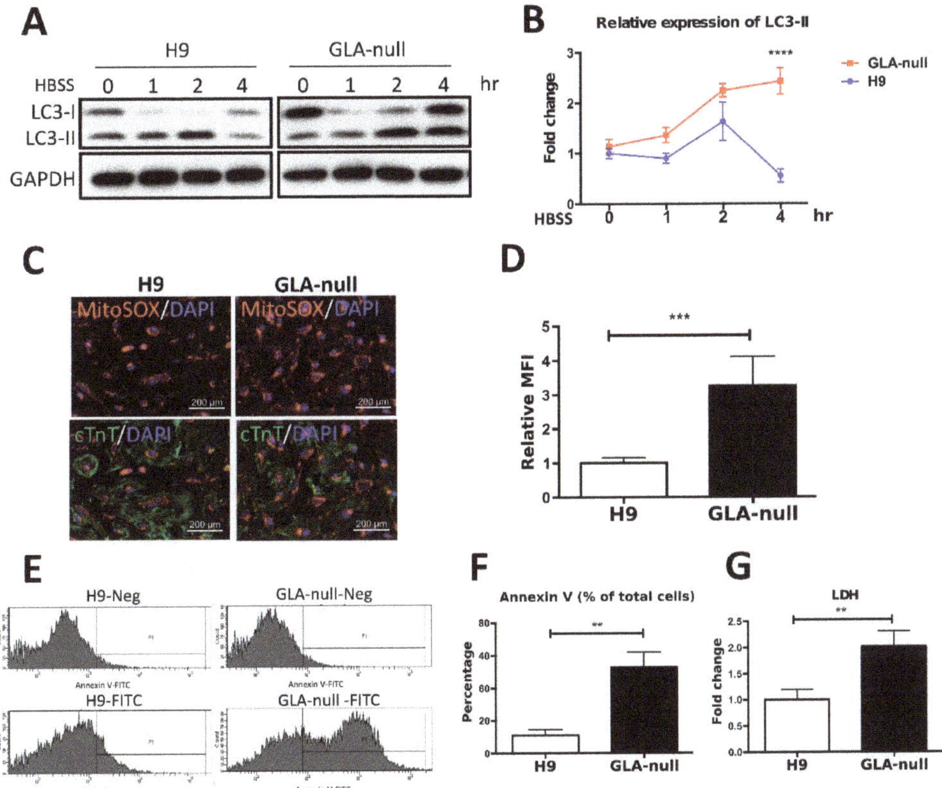

Figure 5. Vesicle turnover impairment induces cardiotoxicity in GLA-null CMs. (**A**) Western blot showing expression of LC3 protein isoforms (LC3-I and LC3-II) in a time course of induction of autophagy with HBSS medium. GAPDH was used as a loading control. (**B**) The expression levels of LC3-II were measured by using ImageJ and the quantification results presented as mean ± standard deviation error bars from three independent experiments. (**C**) Staining of mitochondrial ROS in GLA-null CMs and H9 control CMs with MitoSOX Red. (**D**) MitoSOX Red fluorescence intensity quantified by flow cytometry presented as mean ± standard deviation from three independent experiments. (**E**) Flow cytometry analysis of annexin V-positive cells in populations of GLA-null and H9 CMs. (**F**) Quantitation of annexin V-positive cells. (**G**) The level of LDH secreted by H9 and GLA-null CMs. The data are presented as a fold change relative to H9 control. These data are presented as mean ± standard deviation error bars from three independent experiments.

4. Discussion

The lysosome storage dysfunction (LSD) is the major factor in etiology of inherited hypertrophic cardiomyopathy, including FD; however, the underlying mechanisms of FD-associated cardiomyopathy are not fully understood. In order to expand the therapeutic strategies for FD, the construction of in vitro disease models using primary human cells is essential and unavoidable. Whereas the patients' cells can be used to directly model the effects of drugs on humans, their availability and capacity for expansion are limited and finite compared to in vitro derived cell lines, especially for vital organs such as heart and brain. These drawbacks restrict the capacity of these models to faithfully simulate human disease. By comparison, ESCs can sidestep these limitations and thus provide a powerful and versatile tool for disease therapy, as well as basic research. Combined with the advancements in genome editing technology though CRISPR/Cas9, it is now possible to model human diseases in a physiologically, pathologically, and genetically relevant manner.

Mass spectrometry-based proteomics has been recognized as a powerful tool with a potential to uncover detailed changes in protein expression [29]. To date, most of the proteomics studies performed on FD patients examined FD-affected renal tissue or plasma; however, few studies of protein expression have used FD-affected human heart tissue [30,31]. Although it has been revealed that Gb3 accumulation induces endothelial KCa3.1 degradation in *Gla*-knockout mice through clathrin and Rab5C, which are the critical components of endosome maturation machinery [32], the proteomic profiling performed in our study revealed that *GLA* knockout led to the downregulation of Rab GTPases, including RAB11 subfamily, which are involved in recycling from an endosomal compartment to the plasma membrane, and was shown to contribute to exosome secretion in neuronal cells [33], although the molecular mechanism of RAB11 function in exosome secretion has yet to be deciphered, especially regarding its downstream effectors. On the other hand, our proteomic profile also revealed downregulation of Rho GDP-dissociation inhibitor 2 (GDIR2) in GLA-null CMs. This observation is consistent with previously published clinic proteomic profiles, where PBMCs isolated from FD patients were analyzed, and among the downregulated proteins were calnexin, Rho GDP-dissociation inhibitor 1 (GDIR1), Rho GDP-dissociation inhibitor 2 (GDIR2), chloride intracellular channel protein [34]. Rho GDIs play an important role in regulating Rho GTPases, which are members of the Ras superfamily of GTP binding proteins that participate in the regulation of cytoskeleton and other cellular functions including proliferation, differentiation, and apoptosis [35,36]. Rho GDI is ubiquitously expressed and binds to all Rho family proteins, including RhoA [37]. The small G-protein RhoA regulates the actin cytoskeleton, and its involvement in cell proliferation has also been established. In cardiomyocytes grown in vitro, RhoA induces hypertrophic cell growth and gene expression [38,39]. In vivo, however, cardiac-specific overexpression of RhoA leads to development of heart failure [40,41]. These correlated evidences suggest that exosome secretion regulated by Rab GTPase/RhoGDI signaling pathway may utilize as a target for the potential therapeutic strategy for FD-associated cardiomyopathy.

FD is characterized by failures of cellular autophagy associated with accumulation of glycogen granules and intracytoplasmic vacuoles that contain autophagic material. Impairment of autophagic flux in FD, which was manifested as defects of autophagosome maturation in renal endothelial cells and mesangial cells [42,43]. Autophagy is an evolutionary conserved process of self-degradation of cellular components by autophagosomes, which are delivered to the lysosomal machinery. Several studies have shown that starvation-induced autophagy reduces exosome secretion due to the fusion of multivesicular bodies with autophagic vacuoles [44]. In contrast, cellular stresses, such as senescence and ER stress, increase exosome secretion [45,46]. It is not clear why cells respond to stress by releasing more exosomes, but this could be an alternative way of eliminating waste products. The secreted exosomes may be targeted to and degraded by phagocytes, but they may also have other destinations. Exosomes secreted as waste are likely to affect neighboring cells and possibly induce pathological conditions. Another possibility is that cells may communicate with neighboring cells about intracellular stress by increasing exosome release. Therefore, preventing waste accumulation and rescuing the autophagic ability in FD-affected CMs may be utilized as another therapy approach for FD cardiomyopathy. It has been observed that exosomes derived from CMs harbor a variety of mRNAs, miRNAs and proteins, which may be transferred to the adjacent endothelial cells and modulate their function [47]. Interestingly, exosomes derived from ESCs/iPSCs were shown to possess regenerative power on CMs by augmenting and modulating endogenous repair mechanisms [48,49]. Emerging evidence from the studies of normal development, as well as multiple disease studies, revealed that exosome secretion and autophagy act in a coordinated manner to maintain homeostasis through lysosomal degradation and/or release of cellular cargo [50]. Therefore, considering the role of exosomes in physiological and pathological conditions, strategies that interfere with the release of exosomes and impair exosome-mediated cell-to-cell communication could potentially be exploited therapeutically in FD cardiomyopathy.

The objective of the present study was to explore the influence of Gb3, a lysosomal glycolipid accumulating in FD-affected cells, on modulation of cellular vesicle cycling and the possible mechanism underlying cardiomyopathy. We demonstrated that disruption of *GLA* with CRISPR/Cas9 resulted in the complete ablation of GLA protein expression in hESCs. Results from this study may provide mechanistic insights into how Gb3 accumulation modulates vesicles formation, particularly the autophagy flux in CMs. Such information may be extremely important for potential application in prevention and intervention of adverse effects of FD-associated cardiomyopathy.

5. Conclusions

To summarize, in the present study we show that CRISPR/Cas9-mediated *GLA* knockout of hESC-derived CMs can serve as an in vitro FD model for studying hypertrophic cardiomyopathy. Here, we adopted CRISPR/Cas9-mediated genomic editing to successfully generate GLA-deficient hESC clones (Figure 1). These GLA-deficient hESC clones displayed the properties of pluripotency, and were differentiated into CMs, which exhibited the typical biochemical and pathological abnormalities of FD including ablated GLA expression, enlarged cellular size, increased expression of cardiac hypertrophy genes and Gb3 accumulation. Therefore, these GLA-null CMs clones recapitulated the typical characteristics of FD-associated cardiomyopathy (Figures 2 and 3). To determine the global protein expression changes involved in the hypertrophic response to GLA deficiency, we used proteomic analysis and demonstrated markedly reduced expression of proteins involved in cytoskeleton dynamics and secretion of extracellular vesicles (Figure 4). Since exosome biogenesis and autophagy are highly interrelated processes, we have shown that *GLA* knockout led to autophagic impairment, which resulted in increase of mitochondrial ROS production and cell death (Figure 5). Additional investigation of the role of autophagy in FD is highly promising for the development of novel chemical modulation-based therapeutic approaches that may be more economically viable than traditionally-used ERT.

Supplementary Materials: Supplementary materials are available online at http://www.mdpi.com/2073-4409/8/4/327/s1.

Author Contributions: Conceptualization: H.-Y.S., and C.-S.C.; Methodology: H.-Y.S., C.-S.C. and S.-J.C. writing—original draft preparation: H.-Y.S. and A.A.Y.; writing—review and editing: A.A.Y., M.-L.W. and Y.-P.Y.; Supervision: C.-Y.W., H.-B.L., W.-C.Y.; project administration and funding acquisition Y.-L.C. and S.-H.C. All authors read and approved the final manuscript.

Funding: This work was financially supported by the Ministry of Science and Technology in Taiwan (MOST 106-2911-I-010-502) New Partnership Program for the Connection to Top Labs in the World; MOST 107-2633-B-009-003−107-0210-01-19-01−and 106-2319-B-001-003; Research Foundation of Cardiovascular Medicine, the Department of Health Cancer Center Research of Excellence (MOHW105-TDUB-211-134003, MOHW105-TDU-B-211-133017, MOHW106-TDU-B-211-113001, and MOHW107-TDU-B-211-123001); VGH, TSGH, NDMC, AS Joint Research Program (VTA105-V1-5-1, and VTA107-V1-5-1), VGH, NTUH Joint Research Program (VN106-02, VN107-16); National Health Research Institutes in Taiwan (NHRIEX106-10621BI, and NHRI-EX107-10621BI); "Development and Construction Plan(107F-M01)" of the School of Medicine, National Yang-Ming University; "Aiming for the SPROUT Project-Center for Intelligent Drug Systems and Smart Bio-devices (IDS2B)" of National Chiao Tung University from The Featured Areas Research Center Program within the framework of the Higher Education Sprout Project by the Ministry of Education (MOE) in Taiwan.

Acknowledgments: pSpCas9(BB)-2A-GFP (PX458) was a gift from Feng Zhang (Addgene plasmid # 48138; http://n2t.net/addgene:48138; RRID: Addgene_48138). We thank Ms. Yi-Ching Tsai, Yu-Ling Kuo, Hsiao-Yun Tai, Yu-Hsuan Yang, Fu-Ting Tsai from Taipei Veterans General Hospital for isolation and maintain the hESCs, assisted the experiments, data analysis, and literature review.

Conflicts of Interest: The authors declare that they have no potential conflicts of interest in the present study.

Abbreviation

LSD	lysosomal storage disease
FD	Fabry disease
Gb3	globotriaosylceramide
α-Gal A	α-galactosidase A
ERT	enzyme replacement therapy
CRISPR	Clustered regularly interspaced short palindromic repeats
sgRNA	single-guide RNA
Cas9	CRISPR-associated protein
KO	knockout
hESCs	human embryonic stem cells
CMs	cardiomyocytes
MVEs	multivesicular endosomes

References

1. Roos, J.C.; Cox, T.M. Glycogen storage diseases and cardiomyopathy. *N. Engl. J. Med.* **2005**, *352*, 2553. [PubMed]
2. von Scheidt, W.; Eng, C.M.; Fitzmaurice, T.F.; Erdmann, E.; Hubner, G.; Olsen, E.G.; Christomanou, H.; Kandolf, R.; Bishop, D.F.; Desnick, R.J. An atypical variant of Fabry's disease with manifestations confined to the myocardium. *N. Engl. J. Med.* **1991**, *324*, 395–399. [CrossRef]
3. Brady, R.O.; Tallman, J.F.; Johnson, W.G.; Gal, A.E.; Leahy, W.R.; Quirk, J.M.; Dekaban, A.S. Replacement therapy for inherited enzyme deficiency. Use of purified ceramidetrihexosidase in Fabry's disease. *N. Engl. J. Med.* **1973**, *289*, 9–14. [CrossRef] [PubMed]
4. Stepien, K.M.; Hendriksz, C.J.; Roberts, M.; Sharma, R. Observational clinical study of 22 adult-onset Pompe disease patients undergoing enzyme replacement therapy over 5years. *Mol. Genet. Metab.* **2016**, *117*, 413–418. [CrossRef] [PubMed]
5. Nowak, A.; Mechtler, T.; Kasper, D.C.; Desnick, R.J. Correlation of Lyso-Gb3 levels in dried blood spots and sera from patients with classic and Later-Onset Fabry disease. *Mol. Genet. Metab.* **2017**, *121*, 320–324. [CrossRef]
6. Roddy, T.P.; Nelson, B.C.; Sung, C.C.; Araghi, S.; Wilkens, D.; Zhang, X.K.; Thomas, J.J.; Richards, S.M. Liquid chromatography-tandem mass spectrometry quantification of globotriaosylceramide in plasma for long-term monitoring of Fabry patients treated with enzyme replacement therapy. *Clin. Chem.* **2005**, *51*, 237–240. [CrossRef] [PubMed]
7. Mitobe, S.; Togawa, T.; Tsukimura, T.; Kodama, T.; Tanaka, T.; Doi, K.; Noiri, E.; Akai, Y.; Saito, Y.; Yoshino, M.; et al. Mutant alpha-galactosidase A with M296I does not cause elevation of the plasma globotriaosylsphingosine level. *Mol. Genet. Metab.* **2012**, *107*, 623–626. [CrossRef] [PubMed]
8. Song, H.Y.; Chiang, H.C.; Tseng, W.L.; Wu, P.; Chien, C.S.; Leu, H.B.; Yang, Y.P.; Wang, M.L.; Jong, Y.J.; Chen, C.H.; et al. Using CRISPR/Cas9-Mediated GLA Gene Knockout as an In Vitro Drug Screening Model for Fabry Disease. *Int. J. Mol. Sci.* **2016**, *17*, 2089. [CrossRef]
9. Kuramoto, Y.; Naito, A.T.; Tojo, H.; Sakai, T.; Ito, M.; Shibamoto, M.; Nakagawa, A.; Higo, T.; Okada, K.; Yamaguchi, T.; et al. Generation of Fabry cardiomyopathy model for drug screening using induced pluripotent stem cell-derived cardiomyocytes from a female Fabry patient. *J. Mol. Cell. Cardiol.* **2018**, *121*, 256–265. [CrossRef] [PubMed]
10. Kawagoe, S.; Higuchi, T.; Otaka, M.; Shimada, Y.; Kobayashi, H.; Ida, H.; Ohashi, T.; Okano, H.J.; Nakanishi, M.; Eto, Y. Morphological features of iPS cells generated from Fabry disease skin fibroblasts using Sendai virus vector (SeVdp). *Mol. Genet. Metab.* **2013**, *109*, 386–389. [CrossRef]
11. Kajiwara, M.; Aoi, T.; Okita, K.; Takahashi, R.; Inoue, H.; Takayama, N.; Endo, H.; Eto, K.; Toguchida, J.; Uemoto, S.; et al. Donor-dependent variations in hepatic differentiation from human-induced pluripotent stem cells. *Proc. Natl. Acad. Sci. USA* **2012**, *109*, 12538–12543. [CrossRef]
12. Pan, X.; Ouyang, Y.; Wang, Z.; Ren, H.; Shen, P.; Wang, W.; Xu, Y.; Ni, L.; Yu, X.; Chen, X.; et al. Genotype: A Crucial but Not Unique Factor Affecting the Clinical Phenotypes in Fabry Disease. *PLoS ONE* **2016**, *11*, e0161330. [CrossRef]
13. Larson, M.H.; Gilbert, L.A.; Wang, X.; Lim, W.A.; Weissman, J.S.; Qi, L.S. CRISPR interference (CRISPRi) for sequence-specific control of gene expression. *Nat. Protoc.* **2013**, *8*, 2180–2196. [CrossRef]

14. Qi, L.S.; Larson, M.H.; Gilbert, L.A.; Doudna, J.A.; Weissman, J.S.; Arkin, A.P.; Lim, W.A. Repurposing CRISPR as an RNA-guided platform for sequence-specific control of gene expression. *Cell* **2013**, *152*, 1173–1183. [CrossRef]

15. Cong, L.; Ran, F.A.; Cox, D.; Lin, S.; Barretto, R.; Habib, N.; Hsu, P.D.; Wu, X.; Jiang, W.; Marraffini, L.A.; et al. Multiplex genome engineering using CRISPR/Cas systems. *Science* **2013**, *339*, 819–823. [CrossRef]

16. Sillence, D.J. Glucosylceramide modulates endolysosomal pH in Gaucher disease. *Mol. Genet. Metab.* **2013**, *109*, 194–200. [CrossRef]

17. de la Mata, M.; Cotan, D.; Villanueva-Paz, M.; de Lavera, I.; Alvarez-Cordoba, M.; Luzon-Hidalgo, R.; Suarez-Rivero, J.M.; Tiscornia, G.; Oropesa-Avila, M. Mitochondrial Dysfunction in Lysosomal Storage Disorders. *Diseases* **2016**, *4*, 31. [CrossRef]

18. Miranda, A.M.; Lasiecka, Z.M.; Xu, Y.; Neufeld, J.; Shahriar, S.; Simoes, S.; Chan, R.B.; Oliveira, T.G.; Small, S.A.; Di Paolo, G. Neuronal lysosomal dysfunction releases exosomes harboring APP C-terminal fragments and unique lipid signatures. *Nat. Commun.* **2018**, *9*, 291. [CrossRef]

19. De Toro, J.; Herschlik, L.; Waldner, C.; Mongini, C. Emerging roles of exosomes in normal and pathological conditions: New insights for diagnosis and therapeutic applications. *Front. Immunol.* **2015**, *6*, 203. [CrossRef] [PubMed]

20. Abdulrahman, B.A.; Abdelaziz, D.H.; Schatzl, H.M. Autophagy regulates exosomal release of prions in neuronal cells. *J. Biol. Chem.* **2018**, *293*, 8956–8968. [CrossRef] [PubMed]

21. Jha, R.; Xu, R.H.; Xu, C. Efficient differentiation of cardiomyocytes from human pluripotent stem cells with growth factors. *Methods Mol. Biol.* **2015**, *1299*, 115–131. [PubMed]

22. Chien, K.R.; Zhu, H.; Knowlton, K.U.; Miller-Hance, W.; van-Bilsen, M.; O'Brien, T.X.; Evans, S.M. Transcriptional regulation during cardiac growth and development. *Annu. Rev. Physiol.* **1993**, *55*, 77–95. [CrossRef] [PubMed]

23. Hamdani, N.; Kooij, V.; van Dijk, S.; Merkus, D.; Paulus, W.J.; Remedios, C.D.; Duncker, D.J.; Stienen, G.J.; van der Velden, J. Sarcomeric dysfunction in heart failure. *Cardiovasc. Res.* **2008**, *77*, 649–658. [CrossRef]

24. Al-Shahrour, F.; Diaz-Uriarte, R.; Dopazo, J. FatiGO: A web tool for finding significant associations of Gene Ontology terms with groups of genes. *Bioinformatics* **2004**, *20*, 578–580. [CrossRef]

25. Fader, C.M.; Colombo, M.I. Multivesicular bodies and autophagy in erythrocyte maturation. *Autophagy* **2006**, *2*, 122–125. [CrossRef] [PubMed]

26. Huotari, J.; Helenius, A. Endosome maturation. *EMBO J.* **2011**, *30*, 3481–3500. [CrossRef] [PubMed]

27. Song, S.B.; Hwang, E.S. A Rise in ATP, ROS, and Mitochondrial Content upon Glucose Withdrawal Correlates with a Dysregulated Mitochondria Turnover Mediated by the Activation of the Protein Deacetylase SIRT1. *Cells* **2018**, *8*, 11. [CrossRef] [PubMed]

28. Kuno, A.; Hosoda, R.; Sebori, R.; Hayashi, T.; Sakuragi, H.; Tanabe, M.; Horio, Y. Resveratrol Ameliorates Mitophagy Disturbance and Improves Cardiac Pathophysiology of Dystrophin-deficient mdx Mice. *Sci. Rep.* **2018**, *8*, 15555. [CrossRef] [PubMed]

29. Hawkridge, A.M.; Muddiman, D.C. Mass spectrometry-based biomarker discovery: Toward a global proteome index of individuality. *Annu. Rev. Anal. Chem.* **2009**, *2*, 265–277. [CrossRef]

30. Provencal, P.; Boutin, M.; Dworski, S.; Au, B.; Medin, J.A.; Auray-Blais, C. Relative distribution of Gb3 isoforms/analogs in NOD/SCID/Fabry mice tissues determined by tandem mass spectrometry. *Bioanalysis* **2016**, *8*, 1793–1807. [CrossRef]

31. Kuchar, L.; Faltyskova, H.; Krasny, L.; Dobrovolny, R.; Hulkova, H.; Ledvinova, J.; Volny, M.; Strohalm, M.; Lemr, K.; Kryspinova, L.; et al. Fabry disease: Renal sphingolipid distribution in the alpha-Gal A knockout mouse model by mass spectrometric and immunohistochemical imaging. *Anal. Bioanal. Chem.* **2015**, *407*, 2283–2291. [CrossRef]

32. Choi, S.; Kim, J.A.; Na, H.Y.; Cho, S.E.; Park, S.; Jung, S.C.; Suh, S.H. Globotriaosylceramide induces lysosomal degradation of endothelial KCa3.1 in fabry disease. *Arterioscler. Thromb. Vasc. Biol.* **2014**, *34*, 81–89. [CrossRef]

33. Escudero, C.A.; Lazo, O.M.; Galleguillos, C.; Parraguez, J.I.; Lopez-Verrilli, M.A.; Cabeza, C.; Leon, L.; Saeed, U.; Retamal, C.; Gonzalez, A.; et al. The p75 neurotrophin receptor evades the endolysosomal route in neuronal cells, favouring multivesicular bodies specialised for exosomal release. *J. Cell Sci.* **2014**, *127*, 1966–1979. [CrossRef]

34. Cigna, D.; D'Anna, C.; Zizzo, C.; Francofonte, D.; Sorrentino, I.; Colomba, P.; Albeggiani, G.; Armini, A.; Bianchi, L.; Bini, L.; et al. Alteration of proteomic profiles in PBMC isolated from patients with Fabry disease: Preliminary findings. *Mol. Biosyst.* **2013**, *9*, 1162–1168. [CrossRef]

35. Wei, L.; Imanaka-Yoshida, K.; Wang, L.; Zhan, S.; Schneider, M.D.; DeMayo, F.J.; Schwartz, R.J. Inhibition of Rho family GTPases by Rho GDP dissociation inhibitor disrupts cardiac morphogenesis and inhibits cardiomyocyte proliferation. *Development* **2002**, *129*, 1705–1714.

36. Wei, L.; Taffet, G.E.; Khoury, D.S.; Bo, J.; Li, Y.; Yatani, A.; Delaughter, M.C.; Klevitsky, R.; Hewett, T.E.; Robbins, J.; et al. Disruption of Rho signaling results in progressive atrioventricular conduction defects while ventricular function remains preserved. *FASEB J.* **2004**, *18*, 857–859. [CrossRef]

37. Fukumoto, Y.; Kaibuchi, K.; Hori, Y.; Fujioka, H.; Araki, S.; Ueda, T.; Kikuchi, A.; Takai, Y. Molecular cloning and characterization of a novel type of regulatory protein (GDI) for the rho proteins, ras p21-like small GTP-binding proteins. *Oncogene* **1990**, *5*, 1321–1328.

38. Aikawa, R.; Komuro, I.; Nagai, R.; Yazaki, Y. Rho plays an important role in angiotensin II-induced hypertrophic responses in cardiac myocytes. *Mol. Cell. Biochem.* **2000**, *212*, 177–182. [CrossRef]

39. Aikawa, R.; Komuro, I.; Yamazaki, T.; Zou, Y.; Kudoh, S.; Zhu, W.; Kadowaki, T.; Yazaki, Y. Rho family small G proteins play critical roles in mechanical stress-induced hypertrophic responses in cardiac myocytes. *Circ. Res.* **1999**, *84*, 458–466. [CrossRef]

40. Del Re, D.P.; Miyamoto, S.; Brown, J.H. RhoA/Rho kinase up-regulate Bax to activate a mitochondrial death pathway and induce cardiomyocyte apoptosis. *J. Biol. Chem.* **2007**, *282*, 8069–8078. [CrossRef]

41. Sah, V.P.; Minamisawa, S.; Tam, S.P.; Wu, T.H.; Dorn, G.W., 2nd; Ross, J., Jr.; Chien, K.R.; Brown, J.H. Cardiac-specific overexpression of RhoA results in sinus and atrioventricular nodal dysfunction and contractile failure. *J. Clin. Investig.* **1999**, *103*, 1627–1634. [CrossRef]

42. Liebau, M.C.; Braun, F.; Hopker, K.; Weitbrecht, C.; Bartels, V.; Muller, R.U.; Brodesser, S.; Saleem, M.A.; Benzing, T.; Schermer, B.; et al. Dysregulated autophagy contributes to podocyte damage in Fabry's disease. *PLoS ONE* **2013**, *8*, e63506. [CrossRef]

43. Chevrier, M.; Brakch, N.; Celine, L.; Genty, D.; Ramdani, Y.; Moll, S.; Djavaheri-Mergny, M.; Brasse-Lagnel, C.; Annie Laquerriere, A.L.; Barbey, F.; et al. Autophagosome maturation is impaired in Fabry disease. *Autophagy* **2010**, *6*, 589–599. [CrossRef]

44. Fader, C.M.; Sanchez, D.; Furlan, M.; Colombo, M.I. Induction of autophagy promotes fusion of multivesicular bodies with autophagic vacuoles in k562 cells. *Traffic* **2008**, *9*, 230–250. [CrossRef]

45. Beer, L.; Zimmermann, M.; Mitterbauer, A.; Ellinger, A.; Gruber, F.; Narzt, M.S.; Zellner, M.; Gyongyosi, M.; Madlener, S.; Simader, E.; et al. Analysis of the Secretome of Apoptotic Peripheral Blood Mononuclear Cells: Impact of Released Proteins and Exosomes for Tissue Regeneration. *Sci. Rep.* **2015**, *5*, 16662. [CrossRef]

46. Kanemoto, S.; Nitani, R.; Murakami, T.; Kaneko, M.; Asada, R.; Matsuhisa, K.; Saito, A.; Imaizumi, K. Multivesicular body formation enhancement and exosome release during endoplasmic reticulum stress. *Biochem. Biophys. Res. Commun.* **2016**, *480*, 166–172. [CrossRef]

47. Stoorvogel, W. Functional transfer of microRNA by exosomes. *Blood* **2012**, *119*, 646–648. [CrossRef]

48. Khan, M.; Nickoloff, E.; Abramova, T.; Johnson, J.; Verma, S.K.; Krishnamurthy, P.; Mackie, A.R.; Vaughan, E.; Garikipati, V.N.; Benedict, C.; et al. Embryonic Stem Cell-Derived Exosomes Promote Endogenous Repair Mechanisms and Enhance Cardiac Function Following Myocardial Infarction. *Circ. Res.* **2015**, *117*, 52–64. [CrossRef]

49. Wang, Y.; Zhang, L.; Li, Y.; Chen, L.; Wang, X.; Guo, W.; Zhang, X.; Qin, G.; He, S.H.; Zimmerman, A.; et al. Exosomes/microvesicles from induced pluripotent stem cells deliver cardioprotective miRNAs and prevent cardiomyocyte apoptosis in the ischemic myocardium. *Int. J. Cardiol.* **2015**, *192*, 61–69. [CrossRef]

50. Minakaki, G.; Menges, S.; Kittel, A.; Emmanouilidou, E.; Schaeffner, I.; Barkovits, K.; Bergmann, A.; Rockenstein, E.; Adame, A.; Marxreiter, F.; et al. Autophagy inhibition promotes SNCA/alpha-synuclein release and transfer via extracellular vesicles with a hybrid autophagosome-exosome-like phenotype. *Autophagy* **2018**, *14*, 98–119. [CrossRef]

Communication

Blocking LFA-1 Aggravates Cardiac Inflammation in Experimental Autoimmune Myocarditis

Ludwig T. Weckbach [1,2,3,4,*], Andreas Uhl [1,2,3], Felicitas Boehm [1,2,3], Valentina Seitelberger [1], Bruno C. Huber [1], Gabriela Kania [5], Stefan Brunner [1,†] and Ulrich Grabmaier [1,4,†]

[1] Medizinische Klinik und Poliklinik I, Klinikum der Universität, LMU Munich, 81377 Munich, Germany; andreas.uhl@med.uni-muenchen.de (A.U.); felicitas.boehm@lrz.uni-muenchen.de (F.B.); valentina.seitelberger@med.uni-muenchen.de (V.S.); bruno.huber@med.uni-muenchen.de (B.C.H.); stefan.brunner@med.uni-muenchen.de (S.B.); ulrich.grabmaier@med.uni-muenchen.de (U.G.)

[2] Walter Brendel Centre of Experimental Medicine, University Hospital, LMU Munich, 82152 Planegg-Martinsried, Germany

[3] Institute of Cardiovascular Physiology and Pathophysiology, Biomedical Center, LMU Munich, 82152 Planegg-Martinsried, Germany

[4] German Center for Cardiovascular Research, Partner Site Munich Heart Alliance, 80802 Munich, Germany

[5] Center of Experimental Rheumatology, Department of Rheumatology, University Hospital Zurich, 8952 Schlieren, Switzerland; gabriela.kania@uzh.ch

* Correspondence: ludwig.weckbach@med.lmu.de

† Contributed equally.

Received: 17 September 2019; Accepted: 9 October 2019; Published: 17 October 2019

Abstract: The lymphocyte function-associated antigen 1 (LFA-1) is a member of the beta2-integrin family and plays a pivotal role for T cell activation and leukocyte trafficking under inflammatory conditions. Blocking LFA-1 has reduced or aggravated inflammation depending on the inflammation model. To investigate the effect of LFA-1 in myocarditis, mice with experimental autoimmune myocarditis (EAM) were treated with a function blocking anti-LFA-1 antibody from day 1 of disease until day 21, the peak of inflammation. Cardiac inflammation was evaluated by measuring infiltration of leukocytes into the inflamed cardiac tissue using histology and flow cytometry and was assessed by analysis of the heart weight/body weight ratio. LFA-1 antibody treatment severely enhanced leukocyte infiltration, in particular infiltration of CD11b+ monocytes, F4/80+ macrophages, CD4+ T cells, Ly6G+ neutrophils, and CD133+ progenitor cells at peak of inflammation which was accompanied by an increased heart weight/body weight ratio. Thus, blocking LFA-1 starting at the time of immunization severely aggravated acute cardiac inflammation in the EAM model.

Keywords: myocarditis; inflammation; leukocytes

1. Introduction

Myocarditis is a major cause of heart failure in young adults. Infections with cardiotropic viruses represent the most common etiology of myocarditis in the Western World. Release of cardiac self-antigens can subsequently lead to breakdown of heart-specific tolerance tissue and can evolve into autoimmune-mediated inflammation sustaining the disease upon eradication of the virus. Sustained cardiac inflammation may eventually lead to cardiac remodeling and end-stage heart failure with dilation of the ventricles and deteriorating contractility of the cardiac muscle, a condition called inflammatory dilated cardiomyopathy (DCMi) [1]. We recently showed that neutrophils play a critical role for cardiac inflammation in the experimental autoimmune myocarditis (EAM) mouse model which resembles the immunological and histopathological features of post-viral heart disease [2,3]. In this model, cardiac inflammation was induced by administration of a cardiac peptide together with complete Freund's adjuvant which triggered an autoimmune response with a peak of leukocyte infiltration at day 21 after immunization. We demonstrated that targeting neutrophil extracellular traps (NETs), a process by which neutrophils expel their nuclear DNA together with antibacterial proteins, thereby maintaining tissue inflammation, can substantially reduce cardiac inflammation in the EAM mouse model [2,4]. In order to undergo NET formation in the inflamed cardiac tissue, neutrophils must be recruited from the blood stream into the tissue by crossing the endothelial barrier of the cardiac vasculature. The neutrophil recruitment cascade consists of consecutive events including capturing of free-flowing neutrophils, rolling, adhesion, adhesion strengthening, intraluminal crawling, and transmigration [5]. The lymphocyte function-associated antigen 1 (LFA-1, CD11a/CD18) and the macrophage 1-antigen (Mac-1, CD11b/CD18), adhesion molecules of the beta2-integrin family, are of fundamental importance for neutrophil adhesion and subsequent recruitment into the inflamed tissue [6]. LFA-1 also acts as an important adhesion molecule for other leukocyte subsets during the recruitment process, e.g., for T cells. Beyond its role for leukocyte recruitment, the interaction of LFA-1 on T cells with its ligand (the intercellular molecule-1 (ICAM-1)) on antigen-presenting cells like dendritic cells can provide a co-stimulatory signal for T cell activation. It has been shown that LFA-1 signaling influences differentiation of T cells into specific effector subsets. Engagement of LFA-1 on T cells can activate the Notch pathway promoting Th1 differentiation and suppressing generation of Th17 cells and regulatory T cells [7]. LFA-1 has been investigated as target in different autoimmune diseases. In psoriasis, an antibody targeting LFA-1 resulted in significant improvement of plaque psoriasis in patients with moderate to severe disease [8]. In contrast, in the experimental autoimmune encephalomyelitis (EAE) model, infiltration of leukocytes into the spinal cord and the brain was substantially enhanced in LFA-1$^{-/-}$ mice which was accompanied by increased disease severity suggesting that LFA-1 was protective in this model [9]. The role of LFA-1 in myocarditis is unclear. In this study, we set out to investigate the role of LFA-1 in myocarditis using the EAM model.

2. Materials and Methods

2.1. Mice

Male wild-type mice (8 weeks) on a BALB/c background were obtained from Charles River (Sulzfeld, Germany). Animals were fed a standard chow diet ad libitum with free access to water. All animal experiments were approved by the Regierung von Oberbayern (55.2-1-54-2532-48-2014), Germany.

2.2. EAM Model

To induce EAM, a purified synthetic peptide of the cardiac myosin heavy chain alpha (Ac-RSLKLMATLFSTYASADR, Caslo, Kongens Lyngby, Denmark) emulsified in complete Freund's adjuvant (CFA, Sigma-Aldrich, St. Louis, MO) was applied subcutaneously (200 µg of cardiac peptide per mouse) at day 1 and day 7. Equal volumes of CFA + PBS were administered for sham controls. In order to block LFA-1, we used a chimeric rat-mouse IgG1 anti-mouse CD11a monoclonal antibody (muM17, Genentech, San Francisco, CA) which was previously described [10]. The antibody was applied subcutaneously (5 µg/g body weight) starting from day 1 once a week until day 21. Control animals were treated with a matching IgG1 isotype antibody (Genentech, San Francisco, CA) or PBS. Sham-immunized controls were treated with PBS. On day 21, mice were sacrificed and hearts were subsequently removed for analysis.

2.3. Histology and Heart Weight/Body Weight Ratio

To evaluate the cardiac tissue histologically, mouse hearts were rinsed with PBS and fixated with PFA 4% (Roth, Karlsruhe, Germany). Analysis of the heart weight/body weight ratio was conducted using a microbalance (CP64-0CE, Sartorius, Göttingen, Germany) after carefully removing the pericardium, connective tissue, and vascular remains. Thereafter, hearts were dehydrated in a graded series of ethanol concentrations and subsequently embedded in paraffin (Sigma-Aldrich,

St. Louis, MO, USA). To evaluate infiltration of leukocytes, sections were stained with hematoxylin (Roth) and eosin (Roth, H&E, day 21). The established EAM score (0: no inflammatory infiltrates; 1: small foci of <100 inflammatory cells between myocytes; 2: larger foci of >100 inflammatory cells; 3: >10% of a cross section shows infiltration of inflammatory cells; 4: >30% of a cross section shows infiltration of inflammatory cells) was used to evaluate leukocyte infiltration semi-quantitatively as previously described [2]. Analysis was performed in a blinded manner.

2.4. Flow Cytometry

To study infiltration of different leukocyte subsets into the cardiac tissue, flow cytometry was performed at day 21. The hearts were removed, perfused with PBS, subsequently cut in small pieces, and incubated with Liberase (Roche, Basel, Switzerland) for 45 min at 37 °C. Next, the suspension was mixed gently, filtered through a 40 μm cell strainer, and subsequently suspended in PBS. Cells were stained with a APC-conjugated rat anti-mouse CD11b antibody (clone M1-70, BD, Franklin Lakes, NJ), PE-conjugated rat anti-mouse Ly6G antibody (clone 1A8, Biolegend, San Diego, CA, USA), PerCP-conjugated rat anti-mouse CD45 antibody (clone 30F-11, BD), PB-conjugated rat anti-mouse CD4 antibody (clone RM4-5, BD), an APC rat anti-mouse F4/80 antibody (clone BM8, Biolegend) and FITC-conjugated rat anti-mouse CD133 antibody (clone EMK08, ThermoFisher, Waltham, MA, USA). Experiments were performed using a Gallios flow cytometer (Beckman Coulter, Krefeld, Germany). FlowJo software (TreeStar, Ashland, OR) was used to analyze data.

2.5. Statistical Analysis

Data shown represent the mean ± SEM. A Kolmogorov-Smirnoff test was conducted to test for normal distribution. As data were not normally distributed, a Kruskal Wallis test with pairwise comparison and Dunn-Bonferroni correction for multiple testing was applied for all statistical tests. An alpha level of 5% was considered as statistically significant. All data was analyzed using SPSS version 26.

3. Results

To investigate the role of LFA-1 for cardiac inflammation in myocarditis, we evaluated leukocyte infiltration and the heart weight/body weight ratio in the EAM model after blocking LFA-1. We targeted LFA-1 using a chimeric rat-mouse anti-mouse antibody from day 1 until day 21 during the course of EAM. Leukocyte infiltration into the inflamed cardiac tissue, which was analyzed histologically using a semi-quantitative score (referred to as EAM score), was significantly increased at day 21 in immunized mice without antibody treatment (PBS) or with isotype control antibody treatment compared to sham-immunized mice, as expected. Strikingly, leukocyte infiltration was substantially enhanced after blocking LFA-1 compared to mice treated with PBS or the matching isotype control antibody suggesting that LFA-1 suppressed leukocyte infiltration into the inflamed cardiac tissue in EAM (Figure 1a,b). The heart weight/body weight ratio was altered accordingly (Figure 1c).

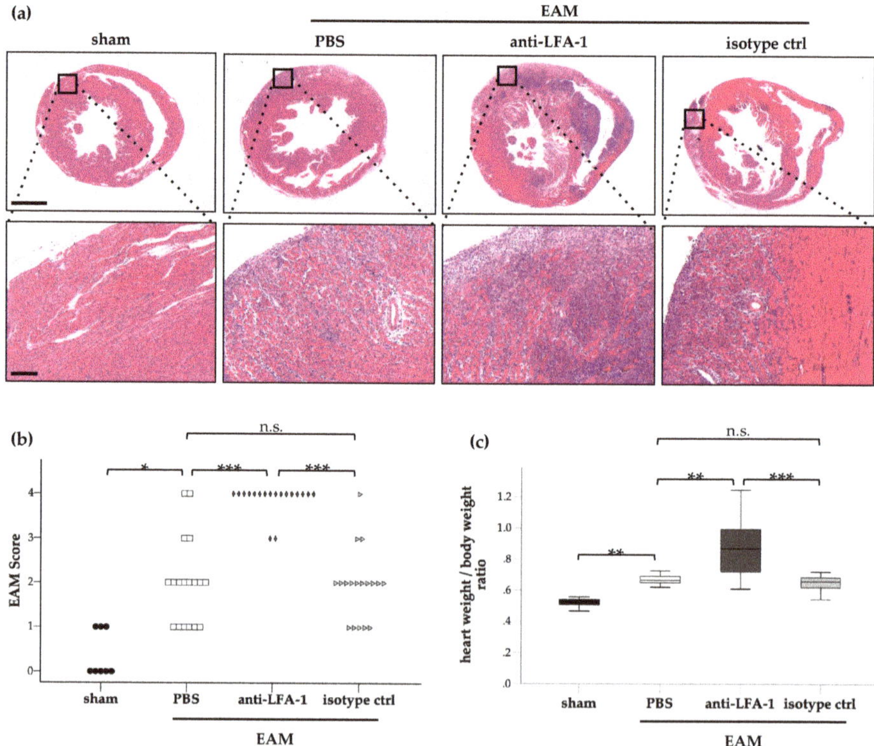

Figure 1. Blocking lymphocyte function-associated antigen 1 (LFA-1) aggravated cardiac inflammation in experimental autoimmune myocarditis (EAM): (**a**) Representative cross sections of heart tissue of sham-treated control mice, vehicle-treated EAM mice (PBS), anti-LFA-1-treated EAM mice as well as isotype control-treated mice on day 21. (**b**) EAM score (**c**) as well as evaluation of the heart weight/body weight ratio at day 21 after induction of EAM (EAM) or sham immunization (sham). EAM mice were treated with an anti-LFA-1 antibody (anti-LFA-1), a matching isotype control (isotype ctrl), or PBS as indicated, with $n = 8$ for sham group; $n = 18$ for EAM groups; * $p < 0.05$; ** $p < 0.01$; *** $p < 0.001$; n.s., not significant. Kruskal Wallis test followed by Dunn-Bonferroni post hoc test. Data are presented as (**b**) individual data points or (**c**) median with interquartile range, whiskers indicate 95% confidence interval.

Next, we determined whether infiltration of specific leukocyte subsets into the cardiac tissue in EAM would be affected by the blockade of LFA-1 in EAM. For this purpose, we administered the LFA-1 blocking antibody or the matching isotype or PBS for control from day 1 until day 21 and subsequently determined the percentage of infiltrated leukocyte subsets of all cells using flow cytometry. The percentage of infiltrated leukocytes determined by CD45+ cells was substantially enhanced by blockade of LFA-1 compared to PBS or the isotype control antibody (Figure 2a). To evaluate the specific leukocyte subsets affected by targeting LFA-1, we also stained for CD4, CD11b, F4-80, Ly6G, and CD133. Blockade of LFA-1 significantly increased the infiltration of all CD4+ T cells (Figure 2b). Moreover, infiltration of CD45/CD11b cells (mainly consisting of monocytes and neutrophils) was significantly elevated in LFA-1 antibody-treated mice compared to control mice (Figure 2c). Accordingly, an increased percentage of infiltrated F4-80+ cells resembling monocytes/macrophages was observed after blocking LFA-1 compared to control conditions (Figure 2d). Moreover, LFA-1 antibody treatment led to a higher number of neutrophils in the inflamed cardiac tissue (Figure 2e). Finally, the fraction of CD133+ progenitor cells, which represent the cellular source of TGF-β mediated fibrosis, was increased after LFA-1 blockade compared to PBS control albeit not reaching a statistical significant difference compared to isotype control (Figure 2f). These findings imply that blocking LFA-1 substantially promoted leukocyte infiltration into the inflamed cardiac tissue affecting CD4+ T cells, monocytes/macrophages, neutrophils, and possibly profibrotic CD133+ progenitor cells.

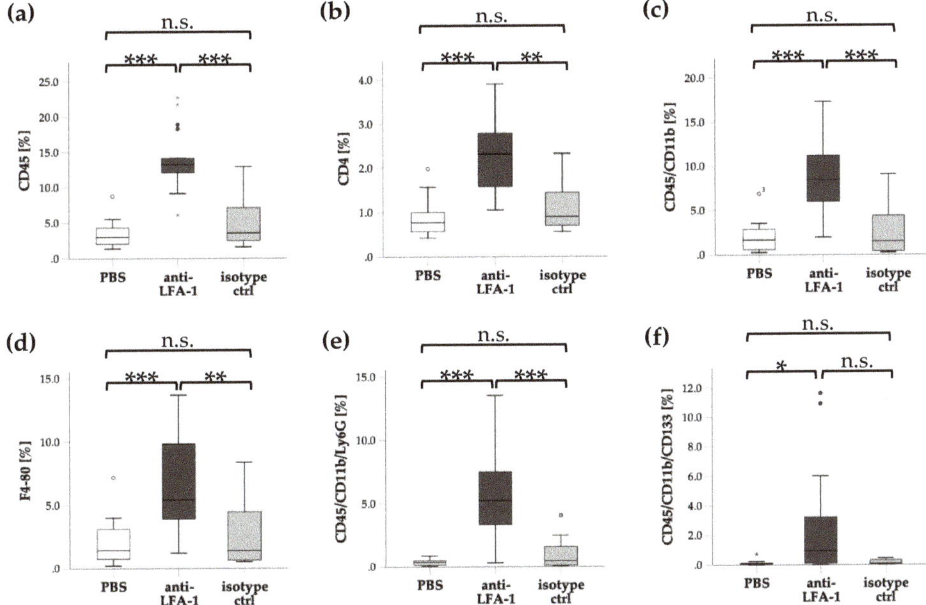

Figure 2. Infiltration of different leukocyte subsets is enhanced by blocking LFA-1: Flow cytometric analysis of leukocyte subpopulations in the inflamed cardiac tissue on day 21 after induction of EAM. Diagrams display the percentage of leukocytes (**a**, CD45) and leukocyte subpopulations (**b**, CD4; **c**, CD45/CD11b; **d**, F4-80; **e**, CD45/CD11b/Ly6G; **f**, CD45/CD11b/CD133) of all cells after blocking LFA-1 (anti-LFA-1) compared with the matching isotype control antibody (isotype ctrl) or PBS. n = 18 for PBS, n = 17 for anti-LFA-1 and isotype ctrl. * $p < 0.05$; ** $p < 0.01$; *** $p < 0.001$; n.s., not significant. Kruskal Wallis test followed by Dunn-Bonferroni post hoc test. Data are presented as median with interquartile range, whiskers indicate 95% confidence interval.

4. Discussion

Blocking LFA-1 or the absence of LFA-1 in different autoimmune diseases revealed a protective or detrimental role of this adhesion receptor depending on the autoimmune disease [8,9]. Moreover, the time-point of targeting LFA-1 may also have an enormous impact on the course of the particular disease. Our findings showed that biological blockade of LFA-1 from day 1 until day 21 substantially enhanced leukocyte infiltration into the inflamed cardiac tissue in the EAM model compared to immunized mice treated with an isotype antibody or PBS. Infiltration of inflammatory cells was accompanied by a dramatic increase of the heart weight/body weight ratio which is a very robust marker for cardiac inflammation indicating that blocking LFA-1 significantly promoted acute myocarditis in this model. Interestingly, all immunized mice treated with the anti-LFA-1 antibody showed the full phenotype of myocarditis (EAM score ≥ 3) whereas a significant percentage of mice generally develop only mild disease (EAM score ≤ 1). Analysis of different leukocyte subpopulations revealed that infiltration of CD4+ T cells, monocytes/macrophages, and neutrophils was enhanced after blocking LFA-1 compared to control conditions suggesting that all investigated leukocyte subsets were affected. In addition, we observed a numerical but not statistical significant increase of CD133+ progenitor cells after targeting LFA-1 which have been shown to promote fibrosis by expression and release of TGF-β [11]. These findings may point to a profibrotic phenotype after blocking LFA-1 in the EAM model.

The underlying mechanism of our preliminary findings remains unclear. As described above, LFA-1 serves as adhesion receptor in particular for neutrophils and T cells. In a peritonitis mouse model, blocking LFA-1 significantly reduced neutrophil recruitment into the peritoneal cavity [12]. The contribution of LFA-1 for neutrophil adhesion and subsequent recruitment depend on the tissue and the predominating inflammatory stimuli in the particular inflammatory setting [13]. The importance of LFA-1 for T cell adhesion and extravasation is also tissue-dependent as T cells also use the very late antigen-4 (VLA-4) as adhesion receptor [14]. Furthermore, the relevance of VLA-4 or LFA-1 may also depend on the T cell subset in specific inflammatory settings [15]. However, as we did not observe reduced but substantially enhanced CD4+ T cell infiltration, we did not assume reduced recruitment of a specific T cell subset in our model.

In line with our study, aggravated leukocyte infiltration was observed in LFA-1-deficient mice in the EAE model which was caused by an impairment of the generation of regulatory T cells and subsequent expansion in the absence of LFA-1 [9]. Enhanced CD4+ T cell infiltration into the spinal cord was also observed by targeting LFA-1 in the EAE model [15]. Whether blocking of LFA-1 also impacted expansion of regulatory T cells in the EAM model requires further investigation.

As described above, LFA-1 engagement can promote Th1 cell and suppress Th17 cell differentiation. Interferon γ-producing Th1 cells represent the predominant T cell subset in autoimmune myocarditis [16]. However, depletion of IL-17 reduced severity but did not prevent EAM, suggesting that both subsets contribute to the inflammatory milieu in the EAM model [17]. Blocking LFA-1 could potentially impact the fate of T cells towards Th17 effector cells thereby changing the inflammatory setting. However, the mechanism of enhanced cardiac inflammation in myocarditis after blocking LFA-1 still needs to be determined.

In summary, targeting LFA-1 substantially enhanced infiltration of neutrophils, monocytes/ macrophages, CD4+ T cells and potentially CD133+ progenitor cells and thereby promoted cardiac inflammation in the EAM model. These findings suggest that engagement of LFA-1 may prevent excessive inflammation in myocarditis.

Author Contributions: Conceptualization, L.T.W., S.B., and U.G.; methodology, L.T.W., A.U., V.S., and B.C.H.; formal analysis, L.T.W., A.U., V.S., and B.C.H.; investigation, L.T.W. and F.B.; resources, S.B. and U.G.; writing—original draft preparation, L.T.W.; writing—review and editing, G.K. and U.G.; visualization, L.T.W.; supervision, S.B. and U.G.; funding acquisition, U.G.

Funding: This research was funded by a grant of Deutsche Gesellschaft für Kardiologie. This work was supported by SFB 914 (Deutsche Forschungsgemeinschaft).

Acknowledgments: We thank Judith Arcifa for excellent technical assistance. We thank Genentech for providing the anti-LFA-1 as well as the isotype control antibody.

Conflicts of Interest: The authors declare no conflict of interest.

References

1. Maisch, B.; Richter, A.; Sandmoller, A.; Portig, I.; Pankuweit, S.; Network, B.M.-H.F. Inflammatory dilated cardiomyopathy (DCMI). *Herz* **2005**, *30*, 535–544. [CrossRef] [PubMed]
2. Weckbach, L.T.; Grabmaier, U.; Uhl, A.; Gess, S.; Boehm, F.; Zehrer, A.; Pick, R.; Salvermoser, M.; Czermak, T.; Pircher, J.; et al. Midkine drives cardiac inflammation by promoting neutrophil trafficking and NETosis in myocarditis. *J. Exp. Med.* **2019**, *216*, 350–368. [CrossRef] [PubMed]
3. Neu, N.; Klieber, R.; Fruhwirth, M.; Berger, P. Cardiac myosin-induced myocarditis as a model of postinfectious autoimmunity. *Eur. Heart J.* **1991**, *12*, 117–120. [CrossRef] [PubMed]
4. Brinkmann, V.; Reichard, U.; Goosmann, C.; Fauler, B.; Uhlemann, Y.; Weiss, D.S.; Weinrauch, Y.; Zychlinsky, A. Neutrophil extracellular traps kill bacteria. *Science* **2004**, *303*, 1532–1535. [CrossRef] [PubMed]
5. Weckbach, L.T.; Gola, A.; Winkelmann, M.; Jakob, S.M.; Groesser, L.; Borgolte, J.; Pogoda, F.; Pick, R.; Pruenster, M.; Muller-Hocker, J.; et al. The cytokine midkine supports neutrophil trafficking during acute

inflammation by promoting adhesion via beta2 integrins (CD11/CD18). *Blood* **2014**, *123*, 1887–1896. [CrossRef] [PubMed]

6. Ley, K.; Laudanna, C.; Cybulsky, M.I.; Nourshargh, S. Getting to the site of inflammation: the leukocyte adhesion cascade updated. *Nat. Rev. Immunol.* **2007**, *7*, 678–689. [CrossRef] [PubMed]

7. Verma, N.K.; Kelleher, D. An Introduction to LFA-1/ICAM-1 Interactions in T-Cell Motility. *Methods Mol. Biol.* **2019**, *1930*, 1–9. [PubMed]

8. Lebwohl, M.; Tyring, S.K.; Hamilton, T.K.; Toth, D.; Glazer, S.; Tawfik, N.H.; Walicke, P.; Dummer, W.; Wang, X.; Garovoy, M.R.; et al. A novel targeted T-cell modulator, efalizumab, for plaque psoriasis. *N. Engl. J. Med.* **2003**, *349*, 2004–2013. [CrossRef] [PubMed]

9. Gultner, S.; Kuhlmann, T.; Hesse, A.; Weber, J.P.; Riemer, C.; Baier, M.; Hutloff, A. Reduced Treg frequency in LFA-1-deficient mice allows enhanced T effector differentiation and pathology in EAE. *Eur. J. Immunol.* **2010**, *40*, 3403–3412. [CrossRef] [PubMed]

10. Clarke, J.; Leach, W.; Pippig, S.; Joshi, A.; Wu, B.; House, R.; Beyer, J. Evaluation of a surrogate antibody for preclinical safety testing of an anti-CD11a monoclonal antibody. *Regul Toxicol Pharm.* **2004**, *40*, 219–226. [CrossRef] [PubMed]

11. Kania, G.; Blyszczuk, P.; Valaperti, A.; Dieterle, T.; Leimenstoll, B.; Dirnhofer, S.; Zulewski, H.; Eriksson, U. Prominin-1+/CD133+ bone marrow-derived heart-resident cells suppress experimental autoimmune myocarditis. *Cardiovasc. Res.* **2008**, *80*, 236–245. [CrossRef] [PubMed]

12. Henderson, R.B.; Lim, L.H.; Tessier, P.A.; Gavins, F.N.; Mathies, M.; Perretti, M.; Hogg, N. The use of lymphocyte function-associated antigen (LFA)-1-deficient mice to determine the role of LFA-1, Mac-1, and alpha4 integrin in the inflammatory response of neutrophils. *J. Exp. Med.* **2001**, *194*, 219–226. [CrossRef] [PubMed]

13. Frommhold, D.; Kamphues, A.; Dannenberg, S.; Buschmann, K.; Zablotskaya, V.; Tschada, R.; Lange-Sperandio, B.; Nawroth, P.P.; Poeschl, J.; Bierhaus, A.; et al. RAGE and ICAM-1 differentially control leukocyte recruitment during acute inflammation in a stimulus-dependent manner. *BMC Immunol.* **2011**, *12*, 56. [CrossRef] [PubMed]

14. Walling, B.L.; Kim, M. LFA-1 in T Cell Migration and Differentiation. *Front. Immunol.* **2018**, *9*, 952. [CrossRef] [PubMed]

15. Rothhammer, V.; Heink, S.; Petermann, F.; Srivastava, R.; Claussen, M.C.; Hemmer, B.; Korn, T. Th17 lymphocytes traffic to the central nervous system independently of alpha4 integrin expression during EAE. *J. Exp. Med.* **2011**, *208*, 2465–2476. [CrossRef] [PubMed]

16. Heymans, S.; Eriksson, U.; Lehtonen, J.; Cooper, L.T., Jr. The Quest for New Approaches in Myocarditis and Inflammatory Cardiomyopathy. *J. Am. Coll. Cardiol.* **2016**, *68*, 2348–2364. [CrossRef] [PubMed]

17. Rangachari, M.; Mauermann, N.; Marty, R.R.; Dirnhofer, S.; Kurrer, M.O.; Komnenovic, V.; Penninger, J.M.; Eriksson, U. T-bet negatively regulates autoimmune myocarditis by suppressing local production of interleukin 17. *J. Exp. Med.* **2006**, *203*, 2009–2019. [CrossRef] [PubMed]

Article

18F-FDG PET-Based Imaging of Myocardial Inflammation Predicts a Functional Outcome Following Transplantation of mESC-Derived Cardiac Induced Cells in a Mouse Model of Myocardial Infarction

Praveen Vasudevan [1,2], Ralf Gaebel [1,2], Piet Doering [3,4], Paula Mueller [1,2], Heiko Lemcke [1,2], Jan Stenzel [5], Tobias Lindner [5], Jens Kurth [3], Gustav Steinhoff [1,2], Brigitte Vollmar [5], Bernd Joachim Krause [3], Hueseyin Ince [6], Robert David [1,2] and Cajetan Immanuel Lang [6,*]

[1] Department of Cardiac Surgery, Rostock University Medical Center, 18057 Rostock, Germany; praveen.vasudevan@med.uni-rostock.de (P.V.); Ralf.Gaebel@med.uni-rostock.de (R.G.); Paula.Mueller@uni-rostock.de (P.M.); Heiko.Lemcke@med.uni-rostock.de (H.L.); Gustav.Steinhoff@med.uni-rostock.de (G.S.); Robert.David@med.uni-rostock.de (R.D.)

[2] Department of Life, Light and Matter, University of Rostock, 18059 Rostock, Germany

[3] Department of Nuclear Medicine, Rostock University Medical Center, 18057 Rostock, Germany; Piet.doering@uni-rostock.de (P.D.); Jens.Kurth@med.uni-rostock.de (J.K.); Bernd.Krause@med.uni-rostock.de (B.J.K.)

[4] Rudolf-Zenker-Institute for Experimental Surgery, Rostock University Medical Center, 18057 Rostock, Germany

[5] Core Facility Multimodal Small Animal Imaging, Rostock University Medical Center, 18057 Rostock, Germany; Jan.Stenzel@med.uni-rostock.de (J.S.); Tobias.Lindner@med.uni-rostock.de (T.L.); Brigitte.Vollmar@med.uni-rostock.de (B.V.)

[6] Department of Cardiology, Rostock University Medical Center, 18057 Rostock, Germany; Hueseyin.Ince@med.uni-rostock.de

[*] Correspondence: Cajetan.Lang@med.uni-rostock.de; Tel.: +49-381-494-61-05

Received: 21 October 2019; Accepted: 9 December 2019; Published: 11 December 2019

Abstract: Cellular inflammation following acute myocardial infarction has gained increasing importance as a target mechanism for therapeutic approaches. We sought to investigate the effect of syngeneic cardiac induced cells (CiC) on myocardial inflammation using 18F-FDG PET (Positron emission tomography)-based imaging and the resulting effect on cardiac pump function using cardiac magnetic resonance (CMR) imaging in a mouse model of myocardial infarction. Mice underwent permanent left anterior descending coronary artery (LAD) ligation inducing an acute inflammatory response. The therapy group received an intramyocardial injection of 10^6 CiC into the border zone of the infarction. Five days after myocardial infarction, 18F-FDG PET was performed under anaesthesia with ketamine and xylazine (KX) to image the inflammatory response in the heart. Flow cytometry of the mononuclear cells in the heart was performed to analyze the inflammatory response. The effect of CiC therapy on cardiac function was determined after three weeks by CMR. The 18F-FDG PET imaging of the heart five days after myocardial infarction (MI) revealed high focal tracer accumulation in the border zone of the infarcted myocardium, whereas no difference was observed in the tracer uptake between infarct and remote myocardium. The CiC transplantation induced a shift in 18F-FDG uptake pattern, leading to significantly higher 18F-FDG uptake in the whole heart, as well as the remote area of the heart. Correspondingly, high numbers of $CD11^+$ cells could be measured by flow cytometry in this region. The CiC transplantation significantly improved the left ventricular ejection function (LVEF) three weeks after myocardial infarction. The CiC transplantation after myocardial infarction leads to an improvement in pump function through modulation of the cellular inflammatory response five days after myocardial infarction. By combining CiC transplantation and the cardiac glucose uptake suppression protocol with KX in a mouse model, we show for the first time, that imaging of cellular inflammation after myocardial infarction using 18F-FDG PET can be used as an early prognostic tool for assessing the efficacy of cardiac stem cell therapies.

Keywords: inflammation; 18F-FDG PET; cardiomyocytes; cardiac induced cells; cardiac function; non-invasive imaging

---

1. Introduction

Inflammatory activity of the innate immune system following myocardial infarction substantially influences remodeling after myocardial infarction (MI) and the evolution of heart failure [1]. Therefore, anti-inflammatory therapies have been under intense investigation since the early 1970s. Yet, none of the clinical trials testing pharmacological strategies aiming at unspecific reduction of myocardial inflammation has proven successful [2]. The main reason for these disappointing findings is the complexity of the well-orchestrated activity of different leukocyte populations. The amplitude and duration of inflammation and the timely and spatial resolution within the heart define the quality of the scar following MI and the amount of tissue loss rather than the mere extent of inflammation [1]. Hence, the concept of suppressing inflammation has changed to a further elaborated concept of "modulating cellular inflammation". The balance of pro- and anti-inflammatory immune cells and their recruiting mechanisms are discussed as both therapeutic and prognostic targets [1]. Further understanding of this intricate cellular immune response to myocardial ischemia in a translational setting requires non-invasive molecular imaging tools. The 18F-FDG PET (Positron emission tomography) using specific protocols for suppressing glucose uptake in cardiomyocytes has recently been introduced to detect cellular inflammation following MI in both patients and C57BL/6 mice [1,3–5]

Protocols to suppress physiological myocardial 18F-FDG uptake require low-carbohydrate diet the day before imaging followed by a 12 h fasting period and intravenous application of heparin before the actual scan [3]. In mice, myocardial glucose metabolism can be effectively suppressed by replacing the commonly used anaesthetic isofluorane for a ketamine/xylazine-based protocol [1].

However, to our knowledge, no studies have used these protocols for measuring therapeutic effects of therapies aiming at improving myocardial healing following MI. Therefore, we implanted syngeneic cardiac induced cells committed to the cardiac lineage in order to improve post-MI cardiac function in 129Sv mice as measured my cardiac magnetic resonance (CMR). Cellular inflammation was detected by 18F-FDG PET in vivo and by post-mortem flow cytometry in both infarcted and remote myocardium.

We hypothesize that PET-based imaging and quantification of cellular inflammation can be used as a molecular imaging tool for both quantification and spatial distribution of monocytes and as an early prognostic tool to predict the effect of cardiac stem cell therapies modulating post-MI inflammation.

2. Materials and Methods

2.1. Stem Cell Culture and Cardiovascular Differentiation

W4 murine embryonic stem cells (mESCs), originally isolated from the 129S6 mouse strain [6], were grown according to standard protocols as described previously [7]. In brief, cells were cultured in DMEM supplemented with 15% FBS Superior (Biochrom AG, Berlin, Germany), 1% Cell Shield® (Minerva Biolabs GmbH, Berlin, Germany), 100 µM non-essential amino acids, 1000 U/mL leukemia inhibitory factor (Phoenix Europe GmbH, Mannheim, Germany) and 100 µM β-mercaptoethanol (Sigma-Aldrich GmbH, Steinheim, Germany) at 37 °C, 5% CO_2, and 20% O_2. For cardiovascular differentiation we used cardiogenic differentiation medium, containing IBM (Iscove's Basal Medium, Biochrom AG) supplemented with 10% FBS Superior, 1% Cell Shield®, 100 µM non-essential amino acids, 450 μm 1-thioglycerol (Sigma-Aldrich GmbH), and 213 μg/mL ascorbic acid (Sigma-Aldrich GmbH), as described previously [7,8]. Cardiovascular differentiation was initiated by hanging-drop culture for two days at 37 °C, 5% CO_2, and 20% O_2. 400 cells per drop were plated on the cover of a square petri dish and grown for 2 days to start formation of embryoid bodies (EB). Afterwards, EB were grown additional four days in suspension culture [9] and then harvested for transplantation and PCR analysis. These cells will be referred to as cardiac induced cells (CiC). They were cultured till day 30 for beating foci analysis.

2.2. Animal Model

The present study was approved by the federal animal care committee of the Landesamt für Landwirtschaft, Lebensmittelsicherheit und Fischerei Mecklenburg-Vorpommern (LALLF, Germany) (registration no. LALLF M-V/TSD/7221.3-1.1-054/15). The 129S6/SvEvTac were bred in the animal facility of the Rostock University Medical Center and maintained in specified pathogen-free conditions. The mice had access to water and standard laboratory chow ad libitum and received humane care according to the German legislation on protection of animals and the Guide for the Care and Use of Laboratory Animals (NIH publication 86–23, revised 1985), and all efforts were made to minimize suffering. Mice were anaesthetized with pentobarbital (50 mg/kg, intraperitoneal). Following thoracotomy, the left anterior descending coronary artery (LAD) was ligated to induce acute myocardial infarction. The MI group received an intramyocardial injection of 10 μL PBS mixed with 10 μL Growth Factor Reduced Matrigel™ Matrix (Corning, Berlin, Germany). The MI induction plus cell transplantation (MIC) group received a suspension of 1×10^6 syngeneic in PBS (10μL) mixed with 10 μL Growth Factor Reduced Matrigel™ Matrix. Injections of 4×5 μL were given along the infarct border. The injection site was controlled visually at the time of transplantation [10].

Healthy mice (n = 33) were divided into 7 groups for 18F-FDG PET imaging 5 days following MI using different protocols for anaesthesia:

(1)   no intervention, isofluorane (n = 2)
(2)   no intervention, ketamine/xylazine (n = 2)
(3)   MI group, isofluorane (n = 4)
(4)   MI group, ketamine/xylazine (n = 6)
(5)   MIC group, ketamine/xylazine (n = 7)

Cardiac function was assessed in separate groups three weeks following MI by cardiac MRI:

(6)   MI only (n = 6)
(7)   MIC (n = 6)

Flow cytometric analysis was performed in separate groups (MI, n = 5 and MIC, n = 5).

2.3. qRT-PCR

RNA was isolated from the cells using the NucleoSpin® RNA isolation kit (Macherey-Nagel, Dueren, Germany). First strand cDNA was then synthesized using the cDNA synthesis kit (Thermo Fisher Scientific, Waltham, MA, USA) according to the manufacturer's instructions. The qPCR reaction was then carried out with the Taqman® Universal PCR Master Mix (Thermo Fisher Scientific) and performed on a StepOnePlus Real-Time PCR system (Applied Biosystems, Foster city, CA, USA). Primers of the following target genes: *Pou5f1* (Mm00658129_gH), *cTnnt2* (Mm01290256_m1), *MesP1* (Mm00801883_m1), and *Nkx2.5* (Mm01309813_s1) were purchased from Thermo Fisher Scientific. Gene expression values of the target genes at day 6 were then normalized to the housekeeping gene *Hprt* (Mm00446968_m1; Thermo Fisher Scientific) and compared relative to the expression values at day 0 using the ΔΔCt method for relative quantifications.

2.4. Beating Foci Analysis

The number of beating foci per EB was analyzed from day 7 to day 30 of differentiation. The EB were observed under a microscope (Carl Zeiss, Oberkochen, Germany) and the beating foci per each EB were then visually analyzed using the ZEN2011 software (Carl Zeiss).

2.5. Flow Cytometry

Single cell cardiac monocyte suspensions were prepared for flow cytometry, as previously described [11] Briefly, the remote and infarct tissue of the heart was dissected and enzymatically digested separately in HBSS with Ca^{2+} and Mg^{2+} (450 U/mL collagenase type I, 125 U/mL collagenase type XI, 120 U/mL DNase I, 60 U/mL hyaluronidase, all Sigma-Aldrich) for 30 min at 37 °C. The digested samples were then passed through a 100 µm filter and centrifuged to enrich for mononuclear cells. Red blood cells were then lysed using erythrocytes lysis buffer (eBioscience, San Diego, CA, USA) and the digest was then washed and suspended in MACS® buffer (PBS, 2 mM EDTA, 0.5% BSA). Samples were then labeled using Zombie Aqua dye (BioLegend, San Diego, CA, USA.), washed, resuspended in MACS buffer containing FCR Block (Miltenyi Biotec GmbH, Bergisch Gladbach, Germany), and stained (see Table 1 for antibody list). Stained samples were then analyzed on a BD FACS LSR II® running BD FACS Diva software (version 6.1.2, Franklin Lakes, NJ, USA). The various immune cell populations in the heart tissue were then assessed, as described in Figure 1.

Table 1. Antibodies used for flow cytometry.

Target	Clone	Source
CD45	30-F11	Biolegend
CD11b	M1/70	Biolegend
CD11c	N418	Biolegend
NK1.1	PK136	Biolegend
Ly6G	1A8	Biolegend
Ly6C	Hk1.4	Biolegend
CCR2	475301	R and D
MHC-II	AF6-120.1	Biolegend

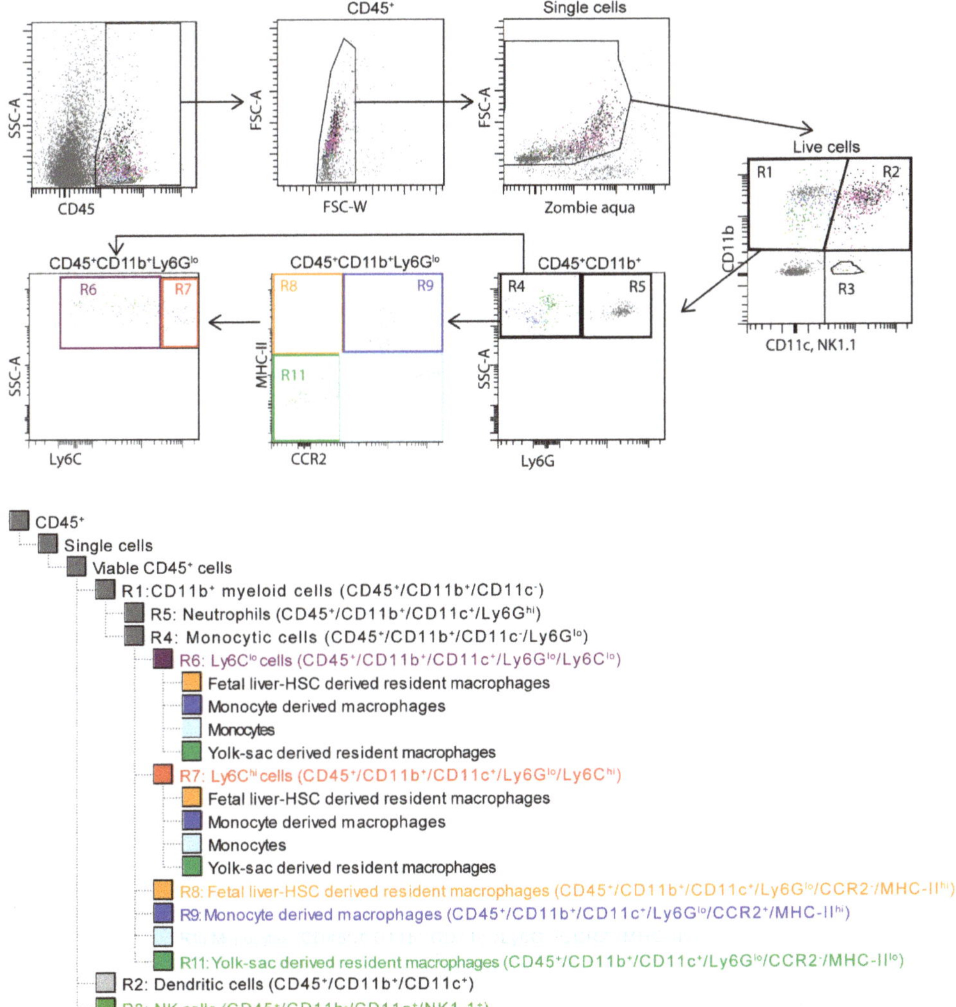

Figure 1. Gating strategy for identifying the different immune populations in the heart. Mononuclear cells expressing CD45 were gated and doublets (FSC-W vs. FSC-A) were excluded. Dead cells were excluded by Zombie aqua. The live single CD45$^+$ cells were then grouped into R1, CD11b+ myeloid cells (CD45$^+$/CD11b$^+$/CD11c$^-$); R2, dendritic cells (CD45$^+$/CD11b$^+$/CD11c$^+$); and R3, NK cells (CD45$^+$/CD11b$^-$/CD11c$^-$/NK1.1$^+$) based on their relative expression of CD11b and CD11c. R5, neutrophils (CD45$^+$/CD11b$^+$/CD11c$^-$/Ly6G^{hi}) were then excluded from R1 based on their Ly6G expression. The remaining R4 monocytic cells were then further characterized into R6, Ly6C^{hi} or commonly known as M1 cells (CD45$^+$/CD11b$^+$/CD11c$^-$/Ly6G^{lo}/Ly6C^{hi}); R7, Ly6C^{lo} or commonly known as M2 cells (CD45$^+$/CD11b$^+$/CD11c$^-$/Ly6G^{lo}/Ly6C^{lo}) based on their Ly6C expression; and into R8, fetal liver HSC-derived resident macrophages (CD45$^+$/CD11b$^+$/CD11c$^-$/Ly6G^{lo}/CCR2$^-$/MHC-II^{hi}); R9, monocyte derived macrophages (CD45$^+$/CD11b$^+$/CD11c$^-$/Ly6G^{lo}/CCR2$^+$/MHC-II^{hi}); R10, monocytes (CD45$^+$/CD11b$^+$/CD11c$^-$/Ly6G^{lo}/CCR2$^+$/MHC-II^{lo}); and R11, yolk sac-derived resident macrophages (CD45$^+$/CD11b$^+$/CD11c$^-$/Ly6G^{lo}/CCR2$^-$/MHC-II^{lo}) based on their CCR2 and MHC-II expression. These CCR2 and MHC-II gated populations were then back gated on R6 and R7 and their relative contribution to the M1 (Ly6C^{hi}) and M2 (Ly6C^{lo}) cells was assessed.

2.6. PET Imaging

For the PET study, mice of the groups 1,3, 6, and 7 were anaesthetized by inhalation of isoflurane (4% for induction and 1% to 2.5% maintenance during preparation and scanning), whereas mice of the groups 2,4, and 5 were anaesthetized by i.p. injection of ketamine/xylazine (ketamine 84 mg/kg and xylazine 11.2 mg/kg) 20 min before tracer application. All PET/CT scans were performed on a small animal PET/CT scanner (Inveon MM-PET/CT, Siemens Medical Solutions, Knoxville, TN, USA) [12] according to a standard protocol: Mice were injected intravenously with a dose of approximately 10 MBq 18F-FDG via a custom-made micro catheter placed in a tail vein. After an uptake period of 60 min, mice were imaged in prone position for 20 min. During the imaging session, respiration of the mice was controled and core body temperature was constantly kept at 38 °C via a heating pad. For attenuation correction and anatomical reference, whole body CT scans were acquired. CT images were reconstructed with a Feldkamp algorithm. The PET images were reconstructed with the three-dimensional (3D) iterative ordered subset expectation maximization reconstruction algorithm (3D-OSEM/OP-MAP) with the following parameters: 4 iterations (OSEM), 32 iterations (MAP), 1.7 mm target resolution, and 128 × 128 matrix size. Reconstruction included corrections for random coincidences, dead time, attenuation, scatter, and decay.

2.7. PET Image Analysis

Image analyses were performed using an Inveon Research Workplace (Siemens, Knoxville, TN, USA), as described previously [13]. PET (Positron emission tomography) and CT (Computerized tomography) images were fused by the use of an automated volumetric fusion algorithm and then verified by an experienced reader for perfect alignment. Consecutively, standardized representative volumes of interest (VOI) were manually placed in the remote area and in the infarcted region as well as the whole heart guided by anatomical landmarks, as described in detail in Figure 2. Correct VOI positioning was visually verified in axial, coronal, and sagittal projection.

Carimas 2 software (Turku PET Centre, Turku, Finland) was used for generating polar maps of the left ventricle according to the manual provided by the developer. Results are presented using 17-segmental standardized myocardial segmentation.

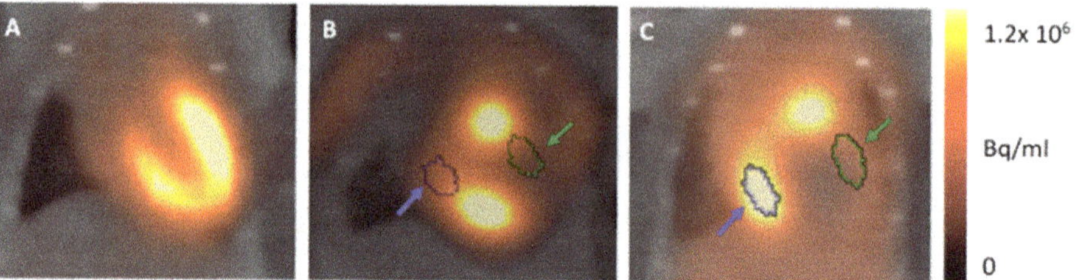

Figure 2. Representative images visualizing our volumes of interest (VOI) positioning strategy: (**A**) Myocardium of healthy mice anaesthetized with isofluorane can be clearly delineated and served as a reference for VOI positioning. (**B** and **C**) Anaesthesia with ketamine/xylazine. A VOI of 5 µL was positioned in both anterolateral wall (infarct area, green) and remote area (inferobasal, blue). Thereby both anatomical landmarks form the CT scan and the image of healthy myocardium (**A**) was used.

2.8. Cardiac Magnetic Resonance Imaging

Cardiac magnetic resonance (CMR) measurements were performed on a 7 Tesla small animal MRI system (BioSpec 70/30, maximum gradient strength 440 mT/m, Bruker BioSpin Gmbh, Ettlingen, Germany) equipped with a 1H transmit volume coil (86 mm, volume resonator) and a 2-by-2 receive-only surface coil array (both Bruker BioSpin GmbH). After induction of anaesthesia using 2% to 3.5% isoflurane in oxygen, animals were placed in a supine position on a dedicated mouse bed and surface coil was placed on the chest of the mice. Respiration rate and body temperature were monitored using an MR-compatible small animal monitoring and gating system (Model 1030, SA Instruments, Inc., Stony Brook, NY, USA), and stable body temperature was maintained by a warm water heating. Anaesthesia was maintained during the experiment with isoflurane oxygen (1.5% to 2%) to achieve a respiration rate of about 35 to 55 breaths.

After planning sequences, for the short axes view final images of the left ventricular ejection fraction (LVEF) measurements were acquired using a IntraGate gradient-echo cine sequences (IntraGate Cine-FLASH) in six short-axis planes completely covering the left ventricle. Acquisition parameters included: echo time (TE) 2.38 ms, repetition time (TR) 5.89 ms, flip angle 15°, 14 frames per cardiac cycle, oversampling 140, averages 1, field of view (FOV) 29.4 × 25.2 mm, matrix size 211 × 180, resolution in-plane 0.14 × 0.14 mm, slice thickness 1 mm, and scan time per slice 2 min.

2.9. Cardiac Magnetic Resonance Analysis

LVEF was assessed from the cine sequences using the freely available software Segment v2.0 R5165 (Medviso, Lund, Sweden) (http://segment.heiberg.se) [14]. The left ventricular (LV) endocardium in these slices was manually segmented to exclude the papillary muscles. The volumes of these segments were then integrated along the six planes of the LV and LVEF was then calculated from their summated end systolic and end diastolic volumes.

2.10. Statistics

All data are presented as mean values ± standard deviation (SD). Flow cytometric data and qPCR values are presented as mean values ± standard error of mean. Student's *t*-test was used for statistical analysis of parametric data and the Mann–Whitney test was used for nonparametric analysis of flow cytometric data. Values of $p < 0.05$ were considered statistically significant.

3. Results

3.1. Cardiac Induced Cells Show Increased Cardiac Markers and Beating Activity During Differentiation

In order to ascertain the differentiation status of the cells, we examined the expression of various markers at the beginning and at day six of differentiation (Figure 3A). We observed the expression of early cardiac markers such as *MesP1*, *Nkx2.5*, and cardiac troponin (*cTnnt2*). They were increased several fold at day six, whereas the expression of the pluripotency marker, *Pou5f1* (also known as *Oct4*), was strongly decreased. This indicates the cardiac lineage commitment of the cells. We also investigated the beating activity of the cells along the differentiation process to determine their functional capability (Figure 3B). In accordance with the strong induction with the induction of the cardiac markers, the cells were robustly beating and the number of beating foci per EB consistently increases over time till day 30 of differentiation. Therefore, these cardiac-induced cells have committed to the cardiac lineage with the potential to form beating cells in the host after transplantation.

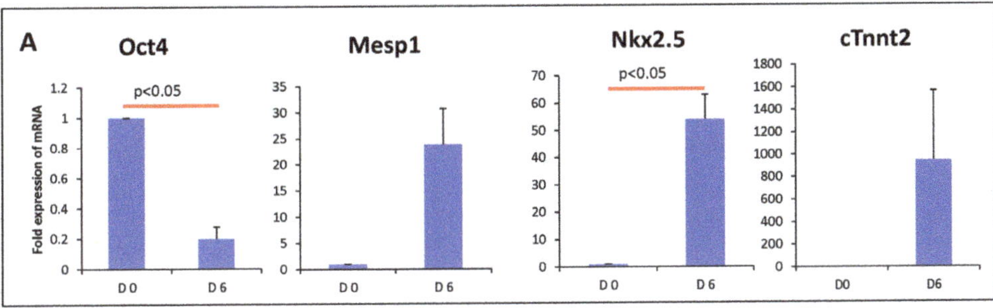

Figure 3. *Cont.*

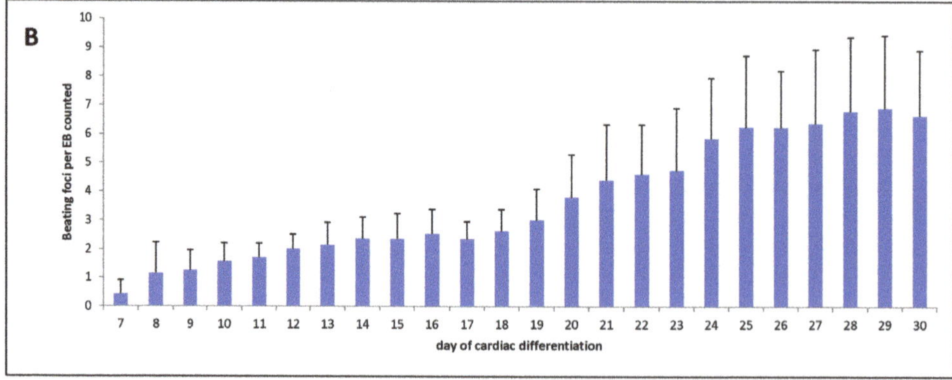

Figure 3. Characterization of cardiac-induced cells: (**A**) Analysis of mRNA expression after six days of cardiac differentiation shows a decrease of the pluripotency marker *Oct4*, whereas cardiac markers are upregulated. The mRNA expression was normalized to the expression of the house keeping gene *hprt*. Values are expressed as fold increase as compared with day 0. Values are presented as mean ± SEM. (**B**) Starting at day 7 of the cardiac differentiation protocol, beating foci per embryoid bodies (EB) were counted until the end of differentiation (n = 3). Values are presented as mean ± SD. *p*-value was calculated using the student *t*-test.

3.2. Impact of Anaesthesia on the FDG-Uptake Pattern in the Infarcted Heart

First, we examined if the myocardial glucose uptake suppression protocol established by Thackeray et al. [5] for C57BL/6 J mice can be applied in the mouse strain 129Sv used in our study on healthy mice (Figure 4).

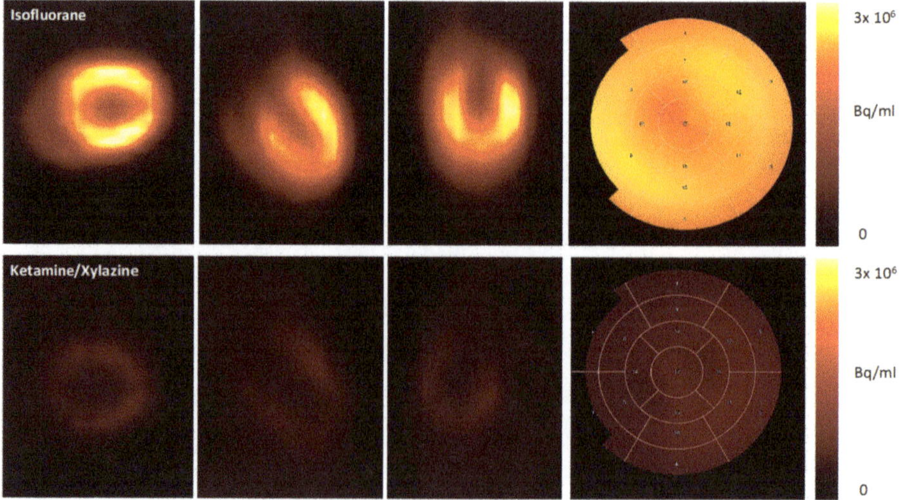

Figure 4. Representative images of healthy mice anaesthetized with isofluorane (upper row) and ketamine/xylazine (lower row). Uptake of 18F-FDG is effectively suppressed by the use of ketamine/xylazine as can be seen in axial, coronal, and sagittal planes, as well as the polar maps.

Secondly, 18F-FDG PET was performed five days after permanent LAD ligation. Under anaesthesia with isofluorane, the highest tracer accumulation was detected in the viable myocardium, whereas ketamine/xylazine (KX) led to accentuated tracer accumulation within the border zone (Figure 5).

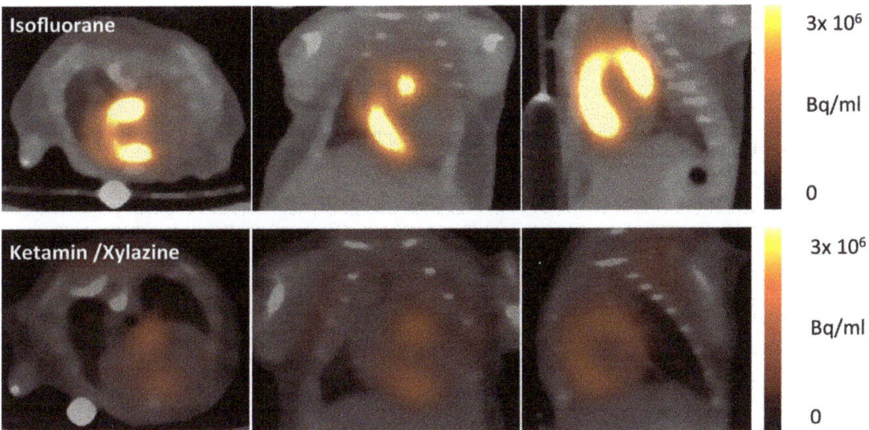

Figure 5. Sample axial, coronal, and sagittal myocardial 18F-FDG images at five days post-surgical myocardial infarction (MI) induction. Isofluorane leads to high tracer accumulation in the healthy myocardium, whereas the infarcted area can be clearly identified as an area of low glucose metabolism (upper row). When using ketamine/xylazine, the most intense tracer accumulation is detected in the area of the border zone, whereas 18F-FDG uptake in healthy myocardium remains suppressed.

The 18F-FDG uptake was significantly reduced by the use of ketamine/xylazine as compared with isofluorane for anaesthesia in the whole heart ($5.2 \pm 0.7\%$ ID/g vs. $46.1 \pm 11.2\%$ ID/g; $p = 0.02$) and both remote ($4.1 \pm 0.6\%$ ID/g vs. $79.3 \pm 22.7\%$ ID/g; $p < 0.0001$) and infarcted myocardium ($4.35 \pm 0.4\%$ ID/g vs. $11.6 \pm 6.0\%$ ID/g, $p = 0.002$) (Figure 6).

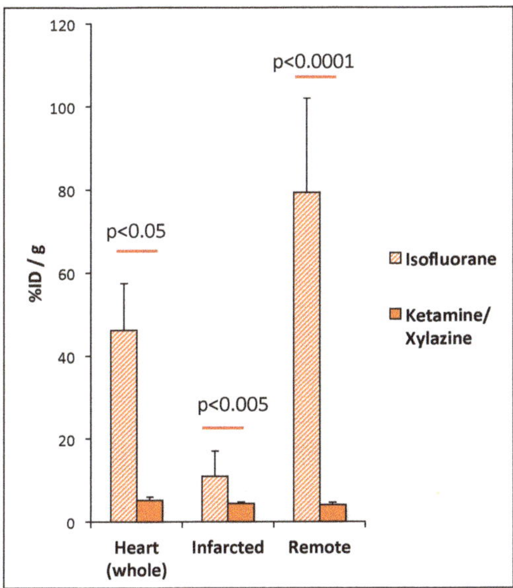

Figure 6. Regional quantitative analysis of 18F-FDG uptake five days after myocardial infarction in the whole heart, the infarct and the remote myocardium. Mice were anaesthetized with isofluorane or ketamine/xylazine, respectively. Values are presented as mean ± SD. *p*-value was calculated using the student *t*-test.

With KX, there was no difference in tracer uptake between remote and infarcted myocardium ($4.1 \pm 0.6\%$ ID/g vs. $4.4 \pm 0.4\%$ ID/g; $p = 0.5$). In contrast isoflurane lead to significantly higher 18F-FDG uptake in the remote as compared with the infarcted myocardium ($79.3 \pm 22.7\%$ ID/g vs. $11.6 \pm 6.0\%$ ID/g; $p = 0.001$).

3.3. FDG-Uptake Pattern is Fundamentally Changed by Cell Transplantation

We then compared 18F-FDG uptake patterns in untreated to animals treated with CiC therapy (MIC) using the ketamine/xylazine protocol (Figures 7 and 8). Transplantation of embryoid bodies containing 10^6 syngeneic cardiac induced cells following acute myocardial infarction led to an increase in 18F-FDG uptake in the remote myocardium as compared with the MI group ($10.7 \pm 4.3\%$ ID/g vs. $4.1 \pm 0.6\%$ ID/g; $p = 0.003$) (Figure 9). Interestingly, tracer accumulation in the center of the infarcted area was not altered by cell therapy ($4.3 \pm 1.4\%$ ID/g in MIC vs. $4.4 \pm 0.4\%$ ID/g in MI; $p = 0.9$). Furthermore 18F-FDG uptake in the whole heart was significantly increased in the MIC group ($8.0 \pm 2.9\%$ ID/g vs. $5.2 \pm 1.1\%$ ID/g; $p < 0.05$).

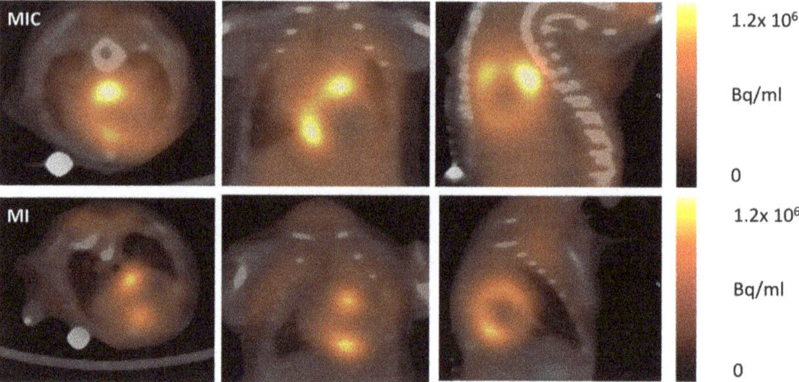

Figure 7. Sample axial, coronal, and sagittal myocardial 18F-FDG images at five days post-surgical MI induction and MI induction plus cell transplantation (MIC). Mice were anaesthetized with ketamine/xylazine. The MI group showed intense tracer accumulation in the border zone, whereas in cell-treated animals the highest tracer accumulation was found in the remote area.

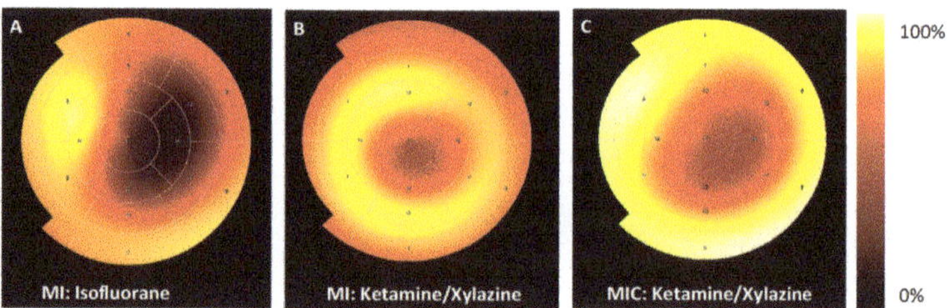

Figure 8. Representative 17-segment tomographic polar maps visualizing distinctive differences of FDG-distribution in the left ventricle between the groups. The apex is in the middle and the anterior wall at top, the inferior wall at bottom, the septum in left and the lateral wall in the right. High tracer uptake is visualized by the yellow colors, lower upake by red and black. (**A**) MI group anaesthetized with isofluorane, the infarct region shows low FDG uptake as compared with remote myocardium; (**B**) MI group anaesthetized with ketamine/xylazine, the most intense FDG uptake came in the border zone; and (**C**) MI + cardiac induced cells (CiC) group anaesthetized with ketamine/xylazine shows a change of the FDG uptake to a pattern, which is similar to (A) with low FDG accumulation within the infarct region and highest uptake in the remote area.

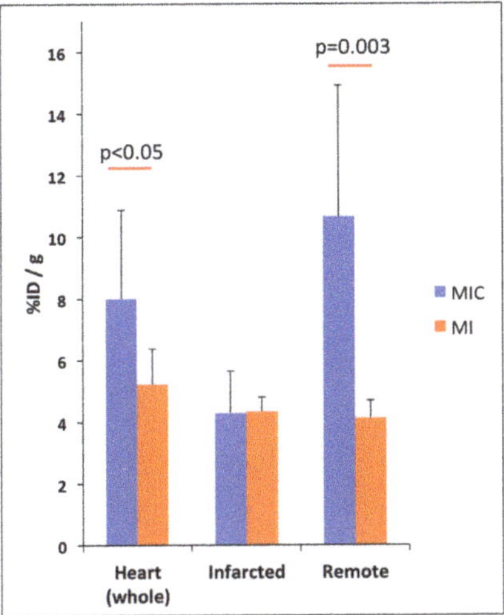

Figure 9. Regional quantitative analysis of 18F-FDG uptake five days after myocardial infarction in the whole heart, infarct, and remote myocardium. Both MI and MIC group were anaesthetized using ketamine/xylazine. Values are presented as mean ± SD. *p*-value was calculated using the student *t*-test.

3.4. Improvement of Cardiac Function Through Cell Therapy Assessed by CMR

In order to assess the value of 18F-FDG-based imaging of cellular inflammation post-MI as an early predictor of functional outcome following cardiac cell therapy, cine CMR was performed three weeks after LAD ligation in separate groups. LVEF was significantly reduced by LAD ligation as compared with healthy animals (24.2 ± 4.1% vs. 59.3 ± 3.7%; $p < 0.001$). Cell therapy led to a significant increase of LVEF (31.7 ± 3.5% vs. 24.2 ± 4.1%; $p = 0.007$) (Figure 10). From this we conclude, that the change in the 18F-FDG uptake pattern, as described above, is a valuable early predictor of therapeutic efficacy.

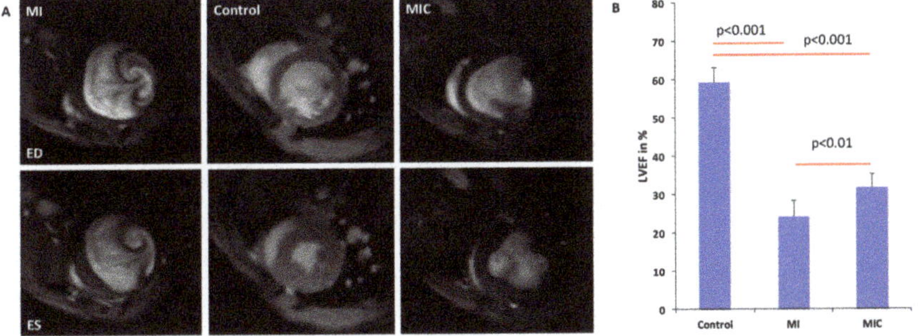

Figure 10. Improved cardiac function after transplantation of embryonic stem cells (ESC) derived CiC following acute myocardial infarction at 3 weeks following MI. (**A**) Magnetic resonance imaging analyses of infarcted animals receiving either CiC (MIC) or matrigel and healthy animals and (**B**) left ventricular ejection fraction was significantly higher in CiC animals vs. control animals (n = 6, *p*-value was calculated using the student *t*-test).

3.5. Modulation of the Immune Response Through Cell Therapy Assessed by Flow Cytometry

In order to identify the immune response within the myocardium, we analyzed the mononuclear cell (MNC) suspension isolated from the remote and infarct area of the hearts of 129Sv mice five days after MI or MIC (Figure 1).

We observed a significantly higher percentage of NK cells (CD45$^+$/CD11b$^-$/CD11c$^+$/NK1.1$^+$) in the infarct area (0.82 ± 0.04 vs. 0.56 ± 0.03; p = 0.01) as compared with the remote area in MI mice. Cell therapy influences the various immune subpopulations differently in infarct and remote areas of the heart. On the one hand, cell therapy led to a significantly higher percentage of CD11b$^+$ myeloid cells (Figure 11A) in the remote area (49.28 ± 3.92 vs. 31.17 ± 2.72; p = 0.01) of MIC hearts as compared with MI hearts. This agrees well with the increased 18F-FDG uptake observed in the remote area of MIC mice (Figure 9), which has been attributed mainly to CD11b$^+$ cells [1]. On the other hand, cell therapy led to a decrease in the percentage of NK cells (Figure 11A) in the infarct area (0.41 ± 0.05 vs. 0.82 ± 0.04; p = 0.007) when compared to MI hearts.

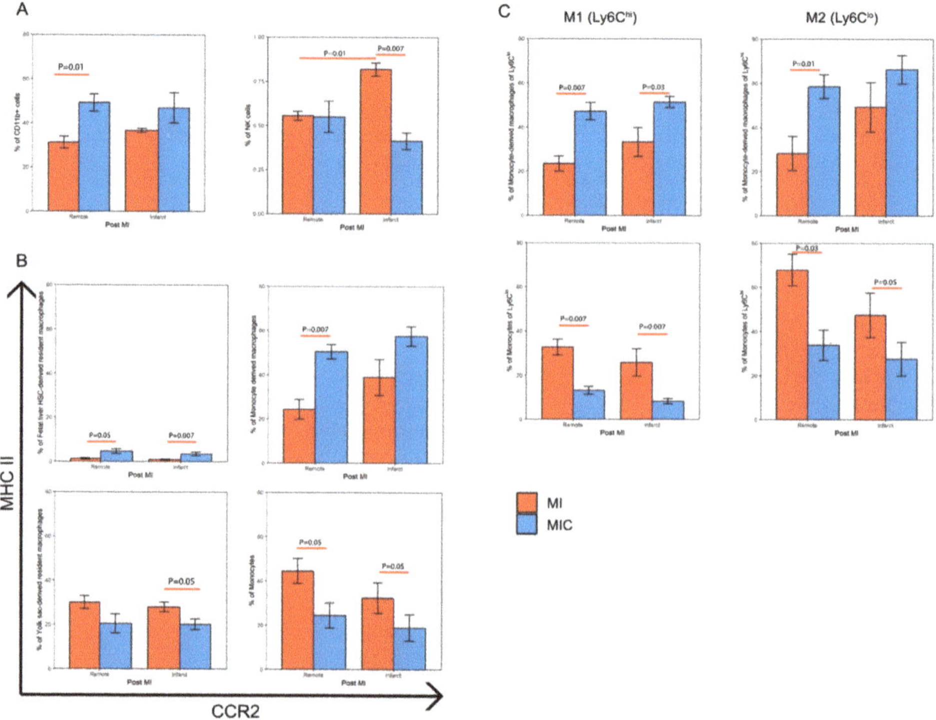

Figure 11. Effect of CiC transplantation on the percentage of various immune cell subpopulations in the heart based on flow cytometric analysis. Mice were subjected to MI/MI+ cells and the remote and infarct area of the hearts were dissected, digested, and the isolated mononuclear cells were stained using various antibodies. These immune cell populations were then characterized using the gating strategy described before and represented as three major groups. (**A**) CD11b$^+$ and NK cells, (**B**) populations based on their CCR2/MHC-II expression, and (**C**) contribution of the populations towards the M1 (Ly6C^{hi}) and M2 (Ly6C^{lo}) cells, n = 5. Values are represented as mean ± SEM. Significance was calculated using the Mann–Whitney test.

Interestingly, we also observed various differences in various CD11b$^+$ subpopulations based on their relative CCR2 and MHC-II expression between the MI and MIC groups (Figure 11B,C).

In the remote area, cell therapy led to an increase in the percentage of fetal liver HSC-derived resident macrophages (4.68 ± 1.03 vs. 1.29 ± 0.32; p = 0.05) and an increase in the percentage of monocyte-derived macrophages (50.48 ± 3.24 vs. 24.35 ± 4.54; p = 0.007), and their relative contribution to the percentage of M1 (Ly6C^{hi}) (58.48 ± 5.41 vs. 28.11 ± 7.72; p = 0.01) and M2 (Ly6C^{lo}) (47.28 ± 3.95 vs. 23.39 ± 3.38%; p = 0.007) cells along with a subsequent decrease in the percentage of monocytes

(24.33 ± 5.61 vs. 44.36 ± 5.62; $p = 0.05$) and their relative contribution to the percentage of M1 (Ly6C^{hi}) (33.76 ± 6.9 vs. 67.85 ± 7.22; $p = 0.03$) and M2 (Ly6C^{lo}) (13.25 ± 1.74 vs. 32.71 ± 3.56; $p = 0.007$) cells as compared with MI.

In the infarct area, we observed an increase in the percentage of fetal liver HSC-derived resident macrophages (4.68 ± 1.03 vs. 1.29 ± 0.32; $p = 0.05$) as well as monocyte-derived macrophages (57.46 ± 4.37 vs. 38.89 ± 8.13) and their relative contribution to the percentage of M2 (Ly6C^{lo}) (51.44 ± 2.47 vs. 33.25 ± 6.6%; $p = 0.03$) cells with a subsequent decrease in the percentage of monocytes (18.88 ± 5.99 vs. 32.25 ± 6.88; $p = 0.05$) and their relative contribution to the percentage of M1 (Ly6C^{hi}) (27.6 ± 7.59 vs. 47.4 ± 10.04; $p = 0.05$) and M2 (Ly6C^{lo}) (8.39 ± 1.17 vs. 25.85 ± 6.15; $p = 0.007$) along with a decrease in the percentage of yolk sac-derived resident macrophages (20.16 ± 2.43 vs. 27.95 ± 2.15; $p = 0.05$) as compared with the MI group.

Therefore, CiC transplantation in the heart increases the CD11b$^+$ cells in the remote area of the heart while favoring an increase in the MHC-II^{hi} subset of CD11b$^+$ cells (monocyte derived macrophages and fetal liver HSC-derived resident macrophages) in both the infarct and remote areas.

4. Discussion

Translational cardiovascular research has evolved at an incredible pace within the last decade and resulted in significantly higher survival rates after acute myocardial infarction. In contrast to highly efficient approaches based on early revascularization and pharmacological prevention of adverse ventricular remodeling, replacement of irreversibly lost cardiomyocytes has not been achieved yet, despite huge efforts in the field of regenerative medicine.

Recently, cellular inflammation following ischemic myocardial injury has been identified as a key player in the process of myocardial healing. Thereby, the local distribution patterns of specific monocyte and macrophage subpopulations have been proposed to determine the quality of myocardial healing [15,16].The vast majority of data has been obtained from rodent studies because their hearts can easily be excised for in vitro experiments such as flow cytometric measurements of single cells suspensions. Most researchers focus on the invasion of two distinct macrophage subpopulations, usually referred to as M1 and M2 macrophages. The early M1 macrophage subset is attracted to the site of myocardial injury via CCL2, expressing pro-inflammatory mediators and proteases for degradation of infarcted tissue. Subsequently, M2 macrophages are recruited via CX3CL1 for mediating synthesis of extracellular matrix and angiogenesis [3]. However, the M1/M2 classification does not adequately explain the complete spectrum of macrophage phenotypes. Recently, macrophages which do not express CCR2 in the neonatal heart have been shown to regenerate the infarcted tissue [15] while the adult heart involves CCR2$^+$ monocyte-derived macrophages also taking part in the remodeling process [17].

Preclinical studies have examined the effect of therapeutic applications, such as stem cell injections, to modify the qualitative and quantitative composition of the post infarct cellular immune response [18–20]. This modulation of the innate cellular immune response resulted in improved cardiac pump function, reduction of scar size, and adverse remodeling [18]. Findings from a meta-analysis by our group have ascribed high therapeutic potential to cardiovascular cell preparations [21]. Therefore, we sought to transplant ESC-derived cardiac induced cells to improve myocardial healing and also investigate whether they influence the ischemic cellular immune response.

The above-mentioned experiments are based on post-mortem analyses such as flow cytometry and immunohistochemistry of excised organs, and thus have been restricted to preclinical research. However, clinical translation requires methods that allow in-vivo visualization and quantification of inflammatory cells. Lee et al. established 18F-FDG PET for imaging the inflammatory cell activity in the heart, based on suppressing glucose metabolism in myocytes [1]. Interestingly, anaesthesia with ketamine/xylazine is both sufficient and highly effective in reducing glucose uptake in cardiomyocytes, hence, enabling visualization of inflammatory activity in the heart [1]. According to Lee et al.,

18F-FDG uptake at the site of myocardial inflammation related to the content of local CD11b[+] monocyte/macrophage concentration in C57BL6 mice [1], at day 5 after MI induction.

On the basis of these findings by other groups, we hypothesized the following:

(A) Distribution pattern of CD11b[+] myeloid cells at day five after MI induction can be modified by intramyocardial transplantation of CiC;

(B) This change can be visualized and quantified by 18F-FDG PET using ketamine/xylazine anaesthesia;

(C) The specific 18F-FDG uptake pattern correlates with functional outcome as measured by cardiac magnetic resonance imaging.

We were able to replicate the glucose uptake suppression protocol based on anaesthesia with KX [5] in 129sv mice and achieved an almost 88% reduction in the glucose uptake in the whole myocardium as compared with isoflurane anaesthesia. We did not find any significant difference in the 18F-FDG uptake levels between the remote and infarcted myocardium using KX anaesthesia, which is in line with findings from Thackeray et al. [4,5]. Furthermore, intense focal tracer uptake was localized in the border zone. Similarly, Lee et al. also reported accentuated FDG uptake in the border zone corresponding to high numbers of infiltrating CD11b[+] cells [1]. Despite visual assessment of this high focal tracer accumulation, quantitative analysis in this region was only performed by Thackeray et al. However, their VOI positioning strategy remains unclear from the manuscript. This might be due to a certain overlap of the border zone to the adjacent regions which makes demarcation of the border zone difficult. In order to produce comparable data in this aspect, VOI positioning strategies should be provided in future studies.

Interestingly, transplantation of 10^6 cardiac induced cells after MI significantly increased FDG uptake in both the whole heart and the remote myocardium, but not in the infarct region. Moreover, CiC transplantation led to a 7% improvement in LVEF three weeks after MI. The magnitude of LVEF improvement is well in line with previous reports [21]. Therefore, we conclude that the change in the myocardial 18F-FDG uptake pattern represents a valid tool for early PET-based in vivo prediction of myocardial healing post MI.

A deeper investigation of the immune response in the heart five days after MI using flow cytometry revealed an increase in the percentage of CD11b[+] cells and a shift towards increased monocyte-derived macrophages in the remote area of cell-treated mice. We also observed a reduction in the yolk sac-derived resident macrophages in the infarct area of cell-treated mice. We did not however observe a shift in the M1/M2 polarization phenotype, as observed in previous studies involving MSCs [18]. It is interesting to note that increased monocyte-derived macrophages and reduced yolk sac-derived macrophages in the heart have been previously attributed to adverse cardiac remodeling and scar formation [17]. However, these studies did not involve the transplantation of cells into the infarcted heart and it is well known that the gene expression profile and the characteristics of the immune cells are affected based on their location in the myocardium and the cells they interact with [22].

However, unbiased expression profiling of these cells over various time points and under the influence of CiC has yet to be carried out. The influence of CiC transplantation on the immune response in the heart has not been studied until now and our work suggests possible new mechanisms and targets for improving the efficiency of CiC.

The 18F-FDG-based imaging strategies for myocardial inflammation are highly attractive and have been applied in a clinical setting despite some limitations, which have yet to be overcome [3]. Both cardiomyocytes [4] and infiltrating inflammatory cells [23] in the setting of acute myocardial injury possess high levels of glucose metabolism. Metabolism of healthy cardiomyocytes is mainly based on fatty acid oxidation, whereas ischemia triggers increased anaerobic glycolysis, which requires much higher rates of glucose [24]. This might hamper direct correlation of focal 18F-FDG uptake and high concentration of CD11b[+] cells in the infarcted region. The use of KX for anaesthesia has been shown to reduce serum insulin levels in rodents, thus, preventing translocation of GLUT4 to membranes of cardiomyocytes and reducing 18F-FDG uptake [25]. In contrast, leucocyte glucose influx in a setting of acute inflammation depends more on GLUT1 and GLUT3, which are expressed and translocated independently of insulin [26]. Whereas, this KX protocol is well suited for preclinical research, different strategies are used for suppression of myocardial glucose uptake in patients including prolonged fasting, dietary modifications, and heparin loading before imaging [27]. This might hamper straightforward translation of imaging protocols established in rodents to clinical application.

Furthermore, the transplantation of CiC adds more complexity to its use in imaging the myocardium. It becomes difficult to attribute the observed 18F-FDG uptake pattern to the inflammatory cells alone since the ability of CiC to alter the cardiac metabolism is poorly understood. It is, therefore, difficult to delineate the relative contribution of 18F-FDG uptake between the host cardiomyocytes, immune cells, and the CiC. The heterogeneity of the inflammatory response and the relative 18F-FDG uptake between the different immune cell populations is also clearly not understood. Further research is necessary to understand metabolic changes in the different cells involved in healing myocardium.

Nevertheless, this is the first study to investigate 18F-FDG-based PET imaging of inflammation as an early indicator for assessing long term therapeutic efficiency in a rodent model of acute myocardial infarction. Furthermore, the current work illustrates the benefit of CiC transplantation to improved cardiac function after MI possibly through its beneficial modulation of the innate inflammatory response. Using such non-invasive techniques in the field of translational research will foster a better understanding of the inflammatory response and how its modulation could lead to altered cardiac function. It is also a valuable tool for monitoring various immune modulation and cell therapies and broadens the horizon for understanding the mechanisms of these therapeutic strategies.

Author Contributions: Conceptualization, C.I.L. and P.V.; methodology, T.L., J.S., and J.K.; formal analysis, C.I.L., P.V., and P.D.; investigation, C.I.L., P.V., R.G., P.M., and H.L.; writing—original draft preparation, P.V. and C.I.L.; writing—review and editing, J.K., R.D, G.S., and H.I.; visualization, P.V. and C.I.L; supervision, C.I.L. and R.D.; funding acquisition, C.I.L., R.D, B.J.K., and B.V.

Funding: This work was supported by the Federal Ministry of Education and Research Germany (FKZ 0312138A and FKZ 316159), the State Mecklenburg-Western Pomerania with EU Structural Funds (ESF/IVWM-B34-0030/10 and ESF/IVBM-B35-0010/12) and the DFG (DA1296/6-1), the German Heart Foundation (F/01/12) and the German Heart Research Foundation (F/34/15). In addition, R.D. is supported by the EU Structural Fund (ESF/14-BM-A55-0024/18), the FORUN Program of Rostock University Medical Centre (889001 and 889003), the Josef and Käthe Klinz Foundation (T319/29737/2017), the DAMP Foundation, and the BMBF (VIP+ 00240). The small animal PET/CT was funded by the Deutsche Forschungsgemeinschaft (INST 2268/6-1 FUGG). The small animal MRI system was funded by the European Regional Development Fund (EFRE UHROM 16). C.I.L. was supported by the Clinician Scientist Program of the Rostock University Medical Center.

Conflicts of Interest: The authors declare no conflict of interest.

Abbreviations

CiC	cardiac induced cells
mESCs	murine embryonic stem cells
MI	myocardial infarction
EB	embryoid bodies
LAD	left anterior descending coronary artery
FOV	field of view
VOI	volumes of interest
CMR	cardiac magnetic resonance
LV	left ventricle
LVEF	left ventricular ejection fraction
SD	standard deviation
SEM	standard error of mean
MNC	mononuclear cell
KX	ketamine/xylazine
PET	positron emission tomography
CT	computerized tomography

References

1. Lee, W.W.; Marinelli, B.; van der Laan, A.M.; Sena, B.F.; Gorbatov, R.; Leuschner, F.; Dutta, P.; Iwamoto, Y.; Ueno, T.; Begieneman, M.P.V.; et al. PET/MRI of inflammation in myocardial infarction. *J. Am. Coll. Cardiol.* **2012**, *59*, 153–163. [CrossRef] [PubMed]

2. Van Hout, G.P.; Jansen of Lorkeers, S.J.; Wever, K.E.; Sena, E.S.; Kouwenberg, L.H.; van Solinge, W.W.; Macleod, M.R.; Doevendans, P.A.; Pasterkamp, G.; Chamuleau, S.A.J.; et al. Translational failure of anti-inflammatory compounds for myocardial infarction: A meta-analysis of large animal models. *Cardiovasc. Res.* **2016**, *109*, 240–248. [CrossRef] [PubMed]

3. Rischpler, C.; Dirschinger, R.J.; Nekolla, S.G.; Kossmann, H.; Nicolosi, S.; Hanus, F.; van Marwick, S.; Kunze, K.P.; Meinicke, A.; Götze, K.; et al. Prospective Evaluation of 18F-Fluorodeoxyglucose Uptake in Postischemic Myocardium by Simultaneous Positron Emission Tomography/Magnetic Resonance Imaging as a Prognostic Marker of Functional Outcome. *Circ. Cardiovasc. Imaging* **2016**, *9*, e004316. [CrossRef] [PubMed]

4. Thackeray, J.T.; Bankstahl, J.P.; Wang, Y.; Wollert, K.C.; Bengel, F.M. Clinically relevant strategies for lowering cardiomyocyte glucose uptake for 18F-FDG imaging of myocardial inflammation in mice. *Eur. J. Nucl. Med. Mol. Imaging* **2015**, *42*, 771–780. [CrossRef]

5. Thackeray, J.T.; Bankstahl, J.P.; Wang, Y.; Korf-Klingebiel, M.; Walte, A.; Wittneben, A.; Wollert, K.C.; Bengel, F.M. Targeting post-infarct inflammation by PET imaging: Comparison of (68)Ga-citrate and (68)Ga-DOTATATE with (18)F-FDG in a mouse model. *Eur. J. Nucl. Med. Mol. Imaging* **2015**, *42*, 317–327. [CrossRef]

6. Auerbach, W.; Dunmore, J.H.; Fairchild-Huntress, V.; Fang, Q.; Auerbach, A.B.; Huszar, D.; Joyner, A.L. Establishment and chimera analysis of 129/SvEv- and C57BL/6-derived mouse embryonic stem cell lines. *Biotechniques* **2000**, *29*, 1024–1028. [CrossRef]

7. Thiele, F.; Voelkner, C.; Krebs, V.; Muller, P.; Jung, J.J.; Rimmbach, C.; Steinhoff, G.; Noack, T.; David, R.; Lemcke, H. Nkx2.5 Based Ventricular Programming of Murine ESC-Derived Cardiomyocytes. *Cell. Physiol. Biochem.* **2019**, *53*, 337–354.

8. Cao, N.; Liu, Z.; Chen, Z.; Wang, J.; Chen, T.; Zhao, X.; Ma, Y.; Qin, L.; Kang, J.; Wei, B.; et al. Ascorbic acid enhances the cardiac differentiation of induced pluripotent stem cells through promoting the proliferation of cardiac progenitor cells. *Cell Res.* **2012**, *22*, 219–236. [CrossRef]

9. Rimmbach, C.; Jung, J.J.; David, R. Generation of murine cardiac pacemaker cell aggregates based on ES-cell-programming in combination with Myh6-promoter-selection. *J. Vis. Exp.* **2015**, e52465. [CrossRef]

10. Muller, P.; Gaebel, R.; Lemcke, H.; Wiekhorst, F.; Hausburg, F.; Lang, C.; Zarniko, N.; Westphal, B.; Steinhoff, G.; David, R.; et al. Intramyocardial fate and effect of iron nanoparticles co-injected with MACS((R)) purified stem cell products. *Biomaterials* **2017**, *135*, 74–84. [CrossRef]

11. Klopsch, C.; Gaebel, R.; Lemcke, H.; Beyer, M.; Vasudevan, P.; Fang, H.Y.; Quante, M.; Vollmar, B.; Skorska, A.; David, R.; et al. Vimentin-Induced Cardiac Mesenchymal Stem Cells Proliferate in the Acute Ischemic Myocardium. *Cells Tissues Organs* **2018**, *206*, 35–45. [CrossRef] [PubMed]

12. Bao, Q.; Newport, D.; Chen, M.; Stout, D.B.; Chatziioannou, A.F. Performance evaluation of the inveon dedicated PET preclinical tomograph based on the NEMA NU-4 standards. *J. Nucl. Med.* **2009**, *50*, 401–408. [CrossRef] [PubMed]

13. Lang, C.; Lehner, S.; Todica, A.; Boening, G.; Franz, W.M.; Bartenstein, P.; Hacker, M.; David, R. Positron emission tomography based in-vivo imaging of early phase stem cell retention after intramyocardial delivery in the mouse model. *Eur. J. Nucl. Med. Mol. Imaging* **2013**, *40*, 1730–1738. [CrossRef] [PubMed]

14. Heiberg, E.; Sjogren, J.; Ugander, M.; Carlsson, M.; Engblom, H.; Arheden, H. Design and validation of Segment—Freely available software for cardiovascular image analysis. *BMC Med. Imaging* **2010**, *10*, 1. [CrossRef]

15. Aurora, A.B.; Olson, E.N. Immune modulation of stem cells and regeneration. *Cell Stem Cell* **2014**, *15*, 14–25. [CrossRef]

16. Ben-Mordechai, T.; Palevski, D.; Glucksam-Galnoy, Y.; Elron-Gross, I.; Margalit, R.; Leor, J. Targeting macrophage subsets for infarct repair. *J. Cardiovasc. Pharmacol. Ther.* **2015**, *20*, 36–51. [CrossRef]

17. Lavine, K.J.; Epelman, S.; Uchida, K.; Weber, K.J.; Nichols, C.G.; Schilling, J.D.; Ornitz, D.M.; Randolph, G.J.; Mann, D.L. Distinct macrophage lineages contribute to disparate patterns of cardiac recovery and remodeling in the neonatal and adult heart. *Proc. Natl. Acad. Sci. USA* **2014**, *111*, 16029–16034. [CrossRef]

18. Ben-Mordechai, T.; Holbova, R.; Landa-Rouben, N.; Harel-Adar, T.; Feinberg, M.S.; Abd Elrahman, I.; Blum, G.; Epstein, F.H.; Silman, Z.; Cohen, S.; et al. Macrophage subpopulations are essential for infarct repair with and without stem cell therapy. *J. Am. Coll. Cardiol.* **2013**, *62*, 1890–1901. [CrossRef]

19. Hasan, A.S.; Luo, L.; Yan, C.; Zhang, T.X.; Urata, Y.; Goto, S.; Mangoura, S.A.; Abdel-Raheem, M.H.; Zhang, S.; Li, T.-S. Cardiosphere-Derived Cells Facilitate Heart Repair by Modulating M1/M2 Macrophage Polarization and Neutrophil Recruitment. *PLoS ONE* **2016**, *11*, e0165255. [CrossRef]

20. Czapla, J.; Matuszczak, S.; Wisniewska, E.; Jarosz-Biej, M.; Smolarczyk, R.; Cichon, T.; Głowala-Kosińska, M.; Śliwka, J.; Garbacz, M.; Szczypior, M.; et al. Human Cardiac Mesenchymal Stromal Cells with CD105+CD34-Phenotype Enhance the Function of Post-Infarction Heart in Mice. *PLoS ONE* **2016**, *11*, e0158745. [CrossRef]

21. Lang, C.I.; Wolfien, M.; Langenbach, A.; Muller, P.; Wolkenhauer, O.; Yavari, A.; Ince, H.; Steinhoff, G.; Krause, B.; David, R.; et al. Cardiac Cell Therapies for the Treatment of Acute Myocardial Infarction: A Meta-Analysis from Mouse Studies. *Cell. Physiol. Biochem.* **2017**, *42*, 254–268. [CrossRef] [PubMed]

22. Sonnenberg, G.F.; Hepworth, M.R. Functional interactions between innate lymphoid cells and adaptive immunity. *Nat. Rev. Immunol.* **2019**, *19*, 599–613. [CrossRef] [PubMed]

23. Love, C.; Tomas, M.B.; Tronco, G.G.; Palestro, C.J. FDG PET of infection and inflammation. *Radiographics* **2005**, *25*, 1357–1368. [CrossRef] [PubMed]

24. Peterson, L.R.; Gropler, R.J. Radionuclide imaging of myocardial metabolism. *Circ. Cardiovasc. Imaging* **2010**, *3*, 211–222. [CrossRef]

25. Saha, J.K.; Xia, J.; Grondin, J.M.; Engle, S.K.; Jakubowski, J.A. Acute hyperglycemia induced by ketamine/xylazine anesthesia in rats: Mechanisms and implications for preclinical models. *Exp. Biol. Med. (Maywood)* **2005**, *230*, 777–784. [CrossRef]

26. Ahmed, N.; Kansara, M.; Berridge, M.V. Acute regulation of glucose transport in a monocyte-macrophage cell line: Glut-3 affinity for glucose is enhanced during the respiratory burst. *Biochem. J.* **1997**, *327 Pt 2*, 369–375. [CrossRef]

27. Schatka, I.; Bengel, F.M. Advanced imaging of cardiac sarcoidosis. *J. Nucl. Med.* **2014**, *55*, 99–106. [CrossRef]

Review

Recent Applications of Three Dimensional Printing in Cardiovascular Medicine

Chiara Gardin [1,2], **Letizia Ferroni** [1,2], **Christian Latremouille** [3], **Juan Carlos Chachques** [3], **Dinko Mitrečić** [4] **and Barbara Zavan** [1,2,*]

1. Maria Cecilia Hospital, GVM Care & Research, 48033 Cotignola (RA), Italy; chiara.gardin@unife.it (C.G.); letizia.ferroni@unife.it (L.F.)
2. Department of Medical Sciences, University of Ferrara, via Fossato di Mortara 70, 44121 Ferrara, Italy
3. Department of Cardiac Surgery Pompidou Hospital, Laboratory of Biosurgical Research, Carpentier Foundation, University Paris Descartes, 75105 Paris, France; Christian.Latremouillec@gmail.com (C.L.); j.chachques-ext@aphp.fr (J.C.C.)
4. Laboratory for Stem Cells, Croatian Institute for Brain Research, School of Medicine University of Zagreb, Šalata 12, 10 000 Zagreb, Croatia; dinko.mitrecic@mef.hr
* Correspondence: barbara.zavan@unife.it; Tel.: +39-0532-455-502

Received: 30 January 2020; Accepted: 9 March 2020; Published: 17 March 2020

Abstract: Three dimensional (3D) printing, which consists in the conversion of digital images into a 3D physical model, is a promising and versatile field that, over the last decade, has experienced a rapid development in medicine. Cardiovascular medicine, in particular, is one of the fastest growing area for medical 3D printing. In this review, we firstly describe the major steps and the most common technologies used in the 3D printing process, then we present current applications of 3D printing with relevance to the cardiovascular field. The technology is more frequently used for the creation of anatomical 3D models useful for teaching, training, and procedural planning of complex surgical cases, as well as for facilitating communication with patients and their families. However, the most attractive and novel application of 3D printing in the last years is bioprinting, which holds the great potential to solve the ever-increasing crisis of organ shortage. In this review, we then present some of the 3D bioprinting strategies used for fabricating fully functional cardiovascular tissues, including myocardium, heart tissue patches, and heart valves. The implications of 3D bioprinting in drug discovery, development, and delivery systems are also briefly discussed, in terms of in vitro cardiovascular drug toxicity. Finally, we describe some applications of 3D printing in the development and testing of cardiovascular medical devices, and the current regulatory frameworks that apply to manufacturing and commercialization of 3D printed products.

Keywords: 3D printing; 3D model; bioprinting; cardiovascular medicine; heart; myocardium; heart valves

---

1. Introduction

Three dimensional (3D) printing is a technique used to transform digital images in a physical 3D model by fusing or depositing material layers. The materials deposited can be powders, plastics, ceramics, metals, liquids, or even living cells, making the process extremely versatile [1,2]. The first technology for 3D printing, called stereolithography, was introduced in 1986 by Charles Hull [3]. From its invention, 3D printing has been largely developed, mostly in the last decades, and nowadays several techniques are available, with applications spanning from the industrial to the medical field [4].

In medicine, 3D printing is utilized for several purposes such as teaching, surgical planning, development of novel and/or personalized implantable devices, and also for creating scaffolds for tissue engineering and artificial functional tissue regeneration [5]. Since its first introduction, the application of 3D printing has greatly expanded mainly in the maxillofacial and orthopedic sectors [6]. With regard to the cardiovascular field, one of the most popular clinical uses of 3D printing is related to the possibility to create 3D printed heart models. These personalized models are proven to be particularly useful in pre-operative planning and pre-surgical simulation of complex cardiac interventions, intra-operative orientation for improving clinical decision-making, medical education and training, and communication in medical practice [7]. In this review, we firstly introduce the 3D printing process and technologies with relevance to cardiovascular medicine. Then, we present some cases of patient-specific 3D printing applications in cardiovascular pre-operative training and pre-surgical planning. Since 3D bioprinting currently represents the most attractive application of 3D printing in the healthcare sector, we then introduce methods for 3D bioprinting and the most commonly used bioinks. This review subsequently covers the applications of 3D bioprinting in the cardiovascular field through categories that include myocardium, heart valves, and cardiac patches for drug screening. In the last section, we describe current regulatory frameworks that USA and EU apply to 3D printed products. Finally, we summarize the major limitations of 3D printing and bioprinting, and the future directions that will enable the translation of these technologies to personalized therapeutic and pharmaceutical applications.

2. Process and Technologies of Cardiovascular 3D Printing

Generating a 3D model is a complex process comprising the sequential stages of diagnostic images acquisition, digital modeling, and 3D printing (Figure 1) [8]. Close collaboration between physicians, imagers, and engineers is therefore fundamental to obtain a functional and accurate 3D printed model.

The first step in the 3D printing process is the acquisition of accurate volumetric images formed by contiguous multiple slices that provide a dataset. Medical images suitable for 3D printing must have high contrast between adjacent structures, low noise, and high spatial resolution [9]. The methods usually employed to acquire cardiovascular imaging data are computed tomography (CT) and magnetic resonance imaging (MRI), but in some cases also 3D transthoracic echocardiography (TTE) or 3D transesophageal echocardiography (TEE) are utilized [10]. Since the quality of the imaging sourcing data is fundamental to obtain precise 3D models, it is essential to evaluate the advantages and limitations of each imaging modality prior to acquiring patient images for 3D modeling. CT represents the preferred imaging technique for 3D printing, because it can provide sub-millimetrical resolution of tissues. In the cardiovascular field, CT is an advantageous option for modeling both intracardiac (atria and ventricles) and extracardiac (great vessels) structures [11]. In addition, CT is able to clearly identify bone and pathologic calcium deposition, and to image patients with pacemakers, artificial valves and metal implants that are not compatible with MRI scanning [12]. However, the major limitation of CT is the exposure to radiation caused by the emission of X-rays, which has been correlated with increased risk of cancer [13]. On the other hand, MRI allows the acquisition of high-resolution images without the employment of ionizing radiation. This imaging modality is mostly employed for visualizing soft tissues; for example, it has been used for the creation of heart chambers and vasculature 3D models, and for the reconstruction of intracardiac tumors [14]. However, a limitation of MRI in comparison to CT includes a lower spatial resolution that limits its use for the evaluation of coronary arteries or the small morphological features within heart valve complexes [12]. 3D echocardiography is a widely available, relatively low-cost ultrasound imaging technique which, similarly to MRI, lacks ionizing radiation. 3D echocardiography is a valuable tool to 3D model ventricular chambers, cardiac valves, and the interatrial septum; however, it is subjected to artifacts and it is not suitable for visualizing extracardiac structures, such as the aorta and pulmonary arteries [15].

The second step in the 3D printing workflow is segmentation, namely, the delineation of appropriate regions of interest for creating patient-specific and highly accurate models of organs and tissues. In cardiovascular 3D printing, this process aims at discriminating the cardiovascular structures of interest, excluding irrelevant noncardiac structures, such as bone and lung [9]. Segmentation is a multi-step process. Prior to segmentation, the acquired diagnostic imaging dataset is exported into a digital imaging and communication in medicine (DICOM) format, which can be universally used in digital modeling softwares without the need for file type conversion [16]. Several medical image segmentation softwares are available; some of these are open-source and freely accessible, while others are commercial, licensed products. From the DICOM dataset, the target anatomic geometry is identified and segmented based on properties such as contrast or brightness [17]. As a result, segmentation masks are created such that pixels with the same intensity range are grouped and converted into 3D digital models using rendering techniques. These segmented digital models are then exported out in the standard tessellation language (STL) file type, which is the industry standard for 3D objects and 3D printing software [17]. Nevertheless, STL files do not contain color information and in order to print in multiple colors, a different file format will need to be used. For example, virtual reality modeling language (VRML) and additive manufacturing file format (AMF) provide multiple color and material options [18]. Once segmentation is completed, the final 3D digital model may be further modified within computer-aided design (CAD) software. The main post-processing adjustments of the 3D model are: Repairing errors and discontinuities that sometimes arise in the image segmentation and exporting processes, smoothing of the surface of the model due to scaling errors resulting from the resolution of the original medical image, and addition of other structures or elimination of unneeded parts from the segmented model [19].

The last step is the actual printing, and in this context it is important to identify the best type of technology and materials in relation to the final purpose, considering complexity, durability, and quality of the model to be obtained. Currently, several 3D printing technologies are available; the most frequently reported with potential applications in cardiovascular medicine are stereolithography (or vat polymerization), powder bed fusion (or selective laser sintering), material extrusion (or fused fiber filament/fused deposition modelling), and material jetting (or polyjet printing) [10,19]. Stereolithography and powder bed fusion technologies use a laser to fuse multiple layers of print material [17]. In particular, stereolithography uses ultraviolet laser light to harden the surface layer made of photopolymeric liquid resin, whereas powder bed fusion uses a laser to heat and fuse a bed of powdered printing material without requiring any support structure. In contrast, material extrusion and material jetting technologies use a nozzle or jet, respectively, to lay down a liquefied print material, which then solidifies before a new layer is built [17]. In detail, the extrusion technique creates 3D models by extruding thermoplastic materials or bioinks filaments layer by layer. Material jet printers, instead, produce 3D models by jetting thin layers of photopolymers that are instantly hardened by ultraviolet light, and allow the incorporation of multiple-materials and colors simultaneously. Once the print is completed, post-processing may also be required, which often includes cleaning to remove residual debris and support material and sanding if smooth surfaces are needed [14].

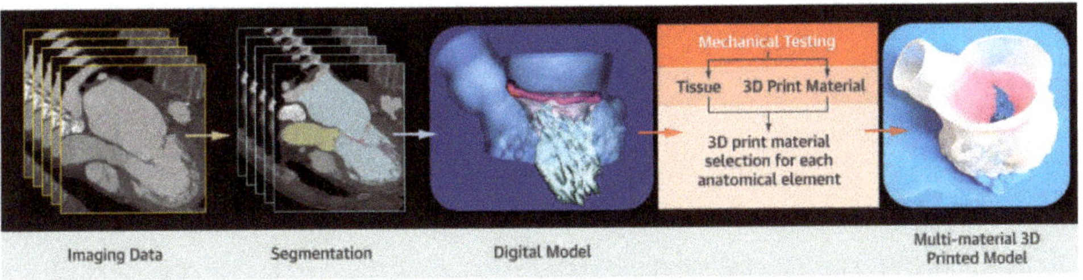

Figure 1. Cardiovascular 3D printing workflow includes acquisition of imaging data, segmentation, imaging modeling, and actual 3D printing. Reprinted with permission from Vukicevic et al. [12]. Copyright © 2020 American College of Cardiology Foundation.

3. 3D Printing for Teaching and Surgical Training in Cardiovascular Medicine

In modern medical schools, the use of 3D models is emerging as a successful novel tool to support and complement the classic methods for anatomy teaching and demonstration [20,21]. The creation of 3D models starting from a clinical dataset of images allows the realization of teaching models of normal and abnormal anatomy. Traditionally, cardiac morphology teaching is carried out using samples collected during cadaveric dissection or heart transplantation in pathological patients [18]. Nonetheless, the scarcity of these specimens, associated with the fact that these do not represent all possible heart diseases, makes 3D models more promising educational resources [18]. Interestingly, 3D models can be fabricated with materials that mimic the real consistency of tissues and they can be scaled to any size. In addition, these 3D models can be printed with different colors, variable material hardness, and even layered texturing if needed to reproduce sophisticated or unusual cardiovascular pathology (Figure 2). Another aspect not to be underestimated is the possibility for several generations of students to handle, manipulate and experience such anatomical variations because 3D printed models are reusable and safe. Consequently, there is no need of special laboratories, instruments, or equipments, such as cadaveric specimens, whose availability is becoming increasingly difficult due to logistical, financial and ethical issues [21–24].

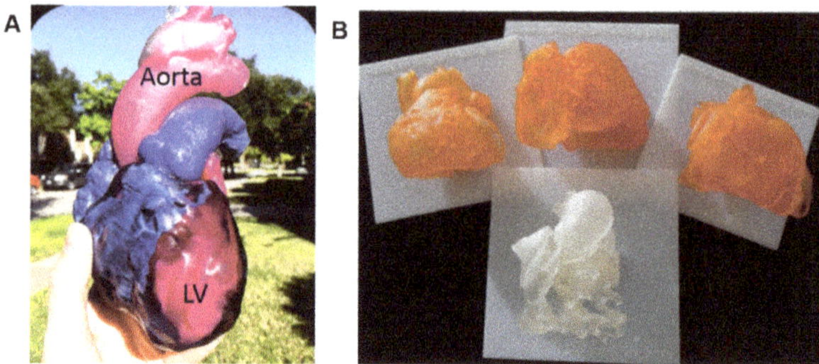

Figure 2. (**A**) Multi-material and multi-colored patient-specific 3D printed heart for educational purposes and communication with patients. LV, left ventricle. Reprinted with permission from Vukicevic et al. [12]. Copyright © 2020 American College of Cardiology Foundation. (**B**) Four models for surgical practice and training. Reprinted with permission from Yoo et al. [18]. Copyright © The Author(s) 2016.

3D printed anatomical models are great resources also for surgical and pre-operative training, thanks to the patient-specific information which can be obtained from the model. In particular, 3D printed models are considered more helpful than virtual reality (VR) and 3D digital images when visualizing complex cases. Indeed, VR and 3D digital imaging do not always demonstrate spatial relationships efficiently and may not always provide an accurate representation of the anatomy [24,25]. The use of such 3D printed models allows a realistic simulation of the surgical procedure to be performed before the actual surgery, thus acting as a surgical guide. If, on one hand, the trainees can learn, practice and repeat surgical procedures until they feel confident, the experienced surgeons, on the other hand, can use 3D printed models for developing new procedures or improving their surgical skills for rare diseases. This is especially true for certain rare congenital heart diseases (CHD), where each patient has a different intracardiac and extracardiac anatomy, such as the chest wall and the relation of the heart to the chest wall. In all these cases, 3D printed models can lead to better pre-operative planning [26,27]. Other advantages of training using 3D models include a reduction in operation time and the possibility to predict intra-operative complications and plan management as required [21]. The use of 3D printed models can ease the communication not only within the clinical staff but also with patients and their family [28]. For example, the patient's cardiovascular pathology and the intended or previously performed surgical procedure can be more easily understood when it is explained using 3D printed models [18].

Several cases of patient-specific 3D printing applications in cardiovascular pre-operative training have been reported. Recently, Bateman and coworkers at the University of Minnesota described three case studies in which heart 3D printed models resulted fundamental to evaluate the most appropriate surgical procedure [14]. The first case concerned a patient with dextrocardia that needed the placement of a left ventricular epicardial pacing. Because of the patient's anomalous anatomy, a 3D model was requested to highlight the paths of the coronary arteries across the myocardium of the left ventricle (LV), finally permitting to practice a minimal invasive surgical approach. The second case was about a patient with a very complex congenital heart disease accompanied by a complex vascular anatomy that requested heart transplantation. The creation of a detailed 3D printed model allowed to determine the best donor and recipient preparation to plan for the implantation of a normal anatomic heart into an anomalous cardiac vascular anatomy. In the last case study, two conjoined sisters attached from chest to navel had to be separated. 3D printed models of the two hearts revealed a previously unseen myocardial connection and led the surgeons to change the surgical procedure. These three case studies demonstrate that the close collaboration between the University of Minnesota's Academic Health Center and relevant departments of engineering has created an excellent environment for developing projects based on the clinical application of 3D printing in the cardiovascular field.

Patients with non-valvular atrial fibrillation (AF) undergoing left atrial appendage (LAA) occlusion can benefit from the advantages of 3D printing. LAA occlusion represents an alternative to anticoagulation for stroke prevention in patients with AF. In the retrospective and prospective analysis described by Fan and colleagues on 107 patients undergoing LAA occlusion, the creation of patient-specific 3D models from 3D TEE imaging data was associated with improved procedural safety and efficacy [29]. In particular, the use of such models offered outcome-relevant information anticipating the complexity of the surgical procedure, permitting to choose the best size device, thus improving intraprocedural performance and safety.

Motwani and colleagues reported a case where a 63-year-old man, several years after a cardiovascular intervention, developed a severe aortic regurgitation (AR) and needed to be re-operated for the closure of paravalvular leak (PVL) [30]. After evaluation with coronary angiography and echocardiography, the multi-disciplinary team decided to proceed with the transcatheter percutaneous closure of the PVL, which is a relatively novel procedure characterized by a small range of closure devices and limited sizing guidance. The creation of 3D models from CT images allowed to test different closure devices and to evaluate the best anatomical fit without interference with adjacent structures. After these bench tests, surgeons acquired high pre-procedural confidence with a particular closure PLD-W device. At 2 months from the intervention, the device remained well-seated with only trivial AR, and the patient returned to his active lifestyle without symptoms.

Over the last years, 3D printing has been demonstrated particularly helpful in the pre-surgical planning of minimally invasive procedures, such as mitral valve (MV) repair surgery [31–33]. Minimally invasive procedures are considered as very complicated, and reconstructive MV surgery is particularly challenging due to variation in valve pathologies. In addition, the possibility of surgeons to practice is limited and most of the pre-surgical training is conducted by assisting and performing real operations [34]. Various surgical repair techniques exist (e.g., annuloplasty, leaflet plasty, chordae plasty), which may address different parts of the valve [35]. Vukicevic and coworkers created patient-specific 3D MV models with different materials starting from TEE and CT imaging datasets. The authors stated that the material properties of such models facilitated functional evaluation of novel trans-catheter MV repair strategies, thus providing surgeons a realistic tool for surgical training [31]. More recently, Engelhardt and colleagues produced for the first time 3D printed MV silicone models containing the full MV apparatus of specific patients, that is the annulus, leaflets, chordae tendineae, and papillary muscles [33]. Then, twelve surgeons performed various surgical techniques, such as annuloplasty, neo-chordae implantation, and triangular leaflet resection, judging the realism of the valves very positively.

As mentioned above, 3D printed models result particularly helpful for a variety of complex CHD, including double-outlet right ventricle (DORV), atrial septal defect (ASD), and ventricular septal defect (VSD). Treatment of these conditions is generally surgical closure, although percutaneous approaches with closure devices are considered a safe alternative [10]. 3D printing has valuably aided in the spatial navigation of occluder devices during the operation and in optimizing patch sizing. In last years, a broad range of complex congenital heart anatomies have been reconstructed and 3D printed to enhance surgical planning (Figure 3) [36–39]. The Royal Brompton Hospital was the first to use images from cardiac MRI to replicate signs of scarring in 3D heart models for better understanding and improving treatments for CHD patients with arrhythmias [28]. In their models, scar tissues are printed with rigid, colored material for better visualization, allowing cardiologists to decrease invasive diagnostic cardiac catheterization procedures and to choose the best surgical procedures in relation to the patient conditions.

Despite the evidence that 3D printing could be very useful in teaching and clinical consultations, as well as in pre-operative planning and decision-making, additional studies are needed to determine whether 3D printed models are cost-effective and can reliably improve clinical outcomes before they become part of routine clinical practice.

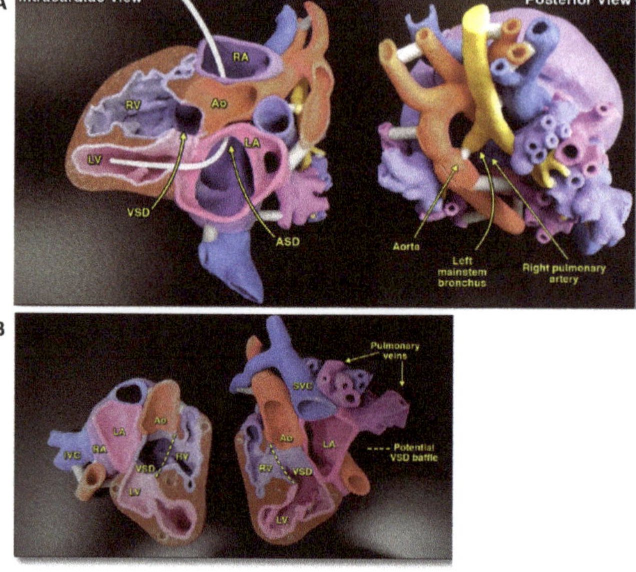

Figure 3. (**A**) 3D model printed to visualize the complex anatomy and aid surgical planning of a four-month-old patient with superoinferior ventricles, atrial septal defect (ASD), ventricular septal defect (VSD), left lung collapse, leftward shift of the heart, and compression of the left mainstem bronchus. (**B**) A 3D model obtained from cardiac MRI of an eight-month-old patient with situs inversus, dextrocardia, double-outlet right ventricle (DORV), and L-malposed great arteries was printed to guide next steps in surgical management. Reprinted with permission from Anwar et al. [39]. Copyright © 2020 Elsevier.

4. 3D Bioprinting for Cardiovascular Applications

The increase of organ requirement for transplantation and the ever-decreasing availability of donor has driven research to try to create biological substitutes mimicking native tissues to restore or replace multi-functioning organs. Therapies based on tissue engineering and regenerative medicine represent a potential solution for the organ donor shortage [40]. The traditional tissue engineering strategy is based on the incorporation of stem cells into natural or synthetic scaffolds that, in the presence of appropriate growth factors, direct cell proliferation and differentiation into 3D functioning tissues [41]. This approach has generated some significant successes in the past decades both in research and clinical applications, including myocardial tissue, vessel, and heart valve [42–45]. Nevertheless, complex 3D organs, such as kidney, liver, lungs, and heart, require precise multi-cellular structures with vascular network integration.

The new frontier of tissue engineering is to exploit 3D printing for functional organ reproduction. For this purpose, a special 3D printing method, named bioprinting, has been developed. 3D bioprinting was first introduced by Klebe in 1988, who used the term 'cytoscribing' for describing the technique of the precise placement of cells on a substrate using a computer-controlled inkjet printer or graphics plotter [46]. Currently, bioprinting refers to the 3D printing of biocompatible materials, cells, and biomolecules into complex 3D functional living tissues [47]. The bioprinting fabrication process is highly controllable, allowing precise positioning of various biomaterials and living cells simultaneously according to the natural compartments of the target tissue or organ. When compared with traditional tissue engineering techniques, 3D bioprinting offers additional important advantages, such as high precise cell placement and high digital control of speed, resolution, cell concentration, drop volume, and diameter of printed cells [48,49].

Based on the working principles and depending on the type of tissue to be fabricated, methods for 3D bioprinting include laser-assisted bioprinting, inkjet bioprinting, microextrusion bioprinting, and integrated approaches (Figure 4) [50]. Inkjet-based bioprinting, the most common method for 3D printing of live cells, is a non-contact technique in which living cells are printed in the form of droplets through cartridges instead of seeding them on scaffolds [49]. The mechanism used to generate the picolitre droplets can be thermal, piezoelectric, laser-induced, or based on pneumatic pressure. Thermal inkjet bioprinting has been shown to be more biocompatible to the living system comparing to the other inkjet bioprinting technologies, because cells are always maintained and protected in an aqueous environment during the printing process [40]. The presence of such a water-based environment also allows the printer to freely deliver from single cell to multiple cells, by simply regulating the bioink concentration and the printed patterns.

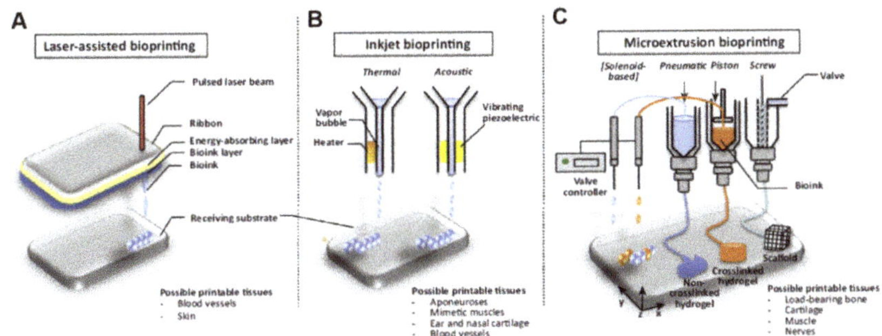

Figure 4. Schematic representation of 3D bioprinting technologies. (**A**) Laser-assisted bioprinting, (**B**) inkjet bioprinting, and (**C**) microextrusion bioprinting. Reprinted from Visscher et al. [50]. Copyright © 2020 Elsevier Ltd. All rights reserved.

Ink development is considered as one of the most challenging aspects in the bioprinting process. An ideal bioink should be highly biocompatible to hold live cells, mechanically stable after printing, and it should provide high resolution during printing [51]. Other important requirements for selecting a bioink for 3D bioprinting are permeability to oxygen and other gases, nutrients, and metabolic wastes, low cost, industrial scalability, and immunological compatibility [52].

The bioinks most commonly used in the 3D bioprinting are cell-laden hydrogels, decellularized extracellular matrix (ECM)-based solutions, and cell suspensions [53]. Hydrogels are hydrated networks of crosslinked polymers, in which cells can be encapsulated in 3D when the hydrogel undergoes gelation [54]. Over time, the scaffold biomaterial biodegrades and is replaced by cell-secreted ECM proteins; therefore, the encapsulated live cells grow occupying the space to form predesigned tissue structures [52]. These cell-laden hydrogel bioinks can be natural polymers, such as agarose, alginate, chitosan, collagen, gelatin, fibrin, and hyaluronic acid (HA); synthetic, such as pluronic and polyethylene glycol (PEG); or a combination of both. In any case, all hydrogel formulations require printing of a polymer solution followed by subsequent physical or chemical crosslinking [53]. Hydrogels are particularly attractive not only for their ability to accommodate live cells, but also for their adjustable chemical, mechanical and biodegradation properties, and the good resolution that can be achieved during printing [52]. Furthermore, hydrogels show low cytotoxicity and structural similarity to the native ECM. For cardiovascular bioprinting, soft bioinks can recapitulate the native elastic modulus, reliably mimicking the native physical properties of cardiac tissues and vasculature in the body.

Another bioink option is decellularization of allogenic or xenogenic tissues, which contain a variety of proteins, proteoglycans, and glycoproteins [48]. The use of decellularized ECM as bioinks offers the possibility to recreate more complex biologically and biochemically microenvironments, that mimic tissue specific ECM composition or resident cytokines [55]. Decellularized ECM-based bioinks are produced from the tissue of interest by removing the cells by sequential physical, chemical and enzymatic treatments while preserving an intact ECM [56]. The obtained ECM is then crushed into a powder and dissolved in a buffer solution to formulate the bioink for 3D printing. A polymer could be used as carrier to increase the solubility, to modify the viscosity, or to induce or increase post-crosslinking of the bioink [53]. Although decellularized ECM bioinks provide novel opportunities to fabricate tissue specific constructs, the decellularization process requires multiple steps, making it a costly approach. In addition, the use of decellularized ECM raises other concerns that must be addressed, such as immunogenicity and toxicity [57,58].

Other bioink materials recently explored in 3D bioprinting are cell aggregates. In cell aggregate configurations, the living cells are printed directly in a process resembling the normal embryonic growth. These approaches are emerging as an alternative to scaffold-based bioprinting [59]. One of the main advantages of the scaffold-free bioprinting is that problems related to the biomaterials' compatibility are avoided; on the other hand, cell–cell interactions play a crucial role in tissue formation. Another attractive property of scaffold-free bioprinting is its efficiency, as the speed can be comparable or even higher than other forms of bioprinting [60]. Nevertheless, the time of pre-printing preparations tend to be longer when compared with scaffolds-based approaches. For this reason, scaffold-free-methods are preferable for smaller, cell-heterogeneous, matrix-poor tissues where the continuous intercellular communication is important. Cell aggregates can be classified into tissue spheroids, cell pellets, and tissue strands [61]. Tissue-spheroids are 3D organized clusters of cells generally 200–400 μm in diameter that, once placed near each other, fuse into a living material thanks to surface tension forces [62]. These spheroids possess the ability to organize as organoids in native tissue, and 3D bioprinting can be used for their precise assembling [59]. Substantial body of evidence has indicated that preformed cell spheroids have been assembled in cartilage, bone, and cardiac muscle-like constructs [59].

Despite the complexity of the cardiac tissue, 3D bioprinting is emerging as one of the most advanced techniques for creating cardiovascular implants possessing biomimetic features that recapitulate both the native physiochemical and biomechanical characteristics of the cardiovascular system. In the following paragraphs, we present some of the 3D bioprinting strategies used for fabricating functional cardiovascular tissues, including myocardium, heart tissue patches, and heart valves.

4.1. 3D Bioprinting of Functional Myocardium

The myocardium is the thick intermediate muscular layer of the heart wall responsible for contraction and relaxation of the heart [63]. The myocardium contains 2–4 billion cardiomyocytes (CMs), that account for roughly 75% of the heart volume, although they represent only about 33% of the total cell number [64]. CMs associate with nonmyocyte cells, that is endothelial cells (ECs), smooth muscle cells (SMCs), and fibroblasts (FBs), generating an intricately organized 3D structure.

Most causes of heart failure are attributable to CMs deficiencies, and aging itself is associated with the loss of about 1 g of myocardium per year in the absence of specific heart disease [65]. CMs have a limited regenerative capacity, and during the progression of heart failure, also the cardiac ECM is modified and replaced by scar tissue [66]. Therefore, combining procedures aiming at regenerating both the CMs population and the ECM could improve the effectiveness of cellular therapy. In effect, over the years several tissue engineering strategies have been employed to create new myocardium that is electrically and mechanically integrated into the heart. The first successful contractile myocardium was fabricated in the late 1990s using chick embryonic CMs seeded onto a collagen scaffold [67]. Over the years, other myocardial tissue engineering procedures have been successfully explored to replace ECM in patients with ischemic heart disease or heart failure. For example, in the study of Chachques and colleagues, a biodegradable 3D collagen type I matrix was seeded with bone marrow cells (BMC) ($250 \pm 28 \times 10^6$ cells), then grafted on the infarcted LV of 10 patients with chronic ischemic heart disease following autologous implantation of BMC ($250 \pm 28 \times 10^6$ cells) on the scarred area (Figure 5A,B) [68]. Long-term results showed that a combined cell transplantation and matrix scaffolds approach offer further benefits with respect to cell therapy alone. Apart from being considered feasible and safe, this tissue-engineered approach improved the efficiency of cellular cardiomyoplasty. In addition, the cell-seeded collagen matrix allowed to normalize cardiac wall stress in injured regions, thus limiting ventricular remodeling and improving diastolic function. Subsequent studies combined the use of biodegradable cell-seeded hybrid scaffolds with synthetic mesh wrap devices for the creation of bioartificial myocardium and cardiowrap bioprostheses for ventricular support and myocardial repair (Figure 5C) [69,70]. In advanced heart failure patients, having large dilated ventricles, complete cardiac wrapping is associated with bioprosthetic helical myocardial bands, that follow the anatomical heart configuration of native muscular ventricular bands (Figure 5D). The role of such cardiopatches, complete wrap bioprostheses and helical myocardial bands is to reduce size and fibrosis of infarct scar, limit ventricular spherical dilatation, and recover elliptical LV shape. This positive chamber remodeling should contribute to improve diastolic filling and myocardial conditions by angiogenic, antiapoptotic and regenerative mechanisms. Preclinical studies in 18 sheep demonstrated that a cardiopatch manufactured using porous elastomeric polycaprolactone (PCL) 3D membrane filled with peptide hydrogel and stem cells improves myocardial infarct scars [71]. PCL elastomer was chosen for its good mechanical properties, namely flexibility and adaptation to curved surfaces, such as those of ventricles. The elastomeric membrane externally covered the heart surface, having advantageous restraining effects on ventricle dilation. In addition, a capillary network developed between the inserted scaffold and the heart.

These works provided the input for the development of scaffold-based 3D bioprinting works (Figure 6A). The bioinks mostly used to 3D bioprint myocardial tissue include alginate, collagen, gelatin, HA, and decellularized ECM [62]. Not only the choice of a suitable matrix, but also an appropriate cell source is critical for an effective cardiac regenerative therapy (Figure 6B). In this context, cardiac progenitor cells (CPCs) seem to be a promising cell population due to their natural differentiation potential to the cardiac lineages [72]. Nevertheless, during the years, other cell populations different from CMs have been used for brioprinting cardiac tissue models, including embryonic stem cells (ESCs), mesenchymal stem cells (MSCs), and induced pluripotent stem cells (iPSCs). The advantages in using these cell types is that most of them is produced pre-clinically and is safe and effective in clinical practice [48].

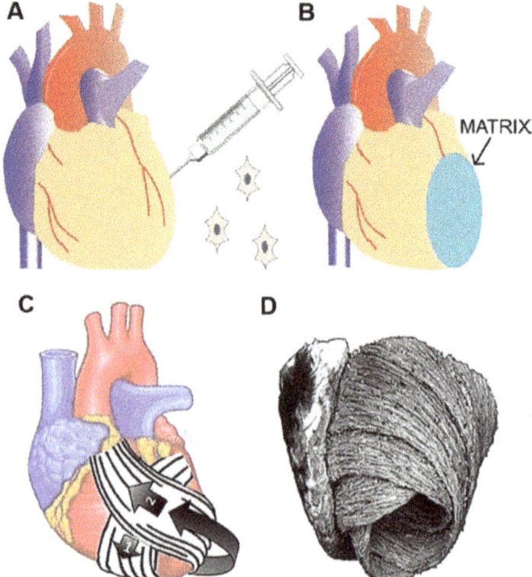

Figure 5. Schematic representation of the surgical procedure combining (**A**) the intra-infarct implantation of bone marrow cells (BMC) followed by (**B**) the fixation of a BMC-seeded collagen matrix onto the epicardial surface. Reprinted from Chachques et al. [68]. Copyright © 2020 The Society of Thoracic Surgeons. (**C**) Cardiowrap bioprostheses with helical loops, that follow (**D**) the anatomical heart configuration (native muscular ventricular bands). Modified from Chachques et al. [69]. Copyright © 2020 Informa UK Limited.

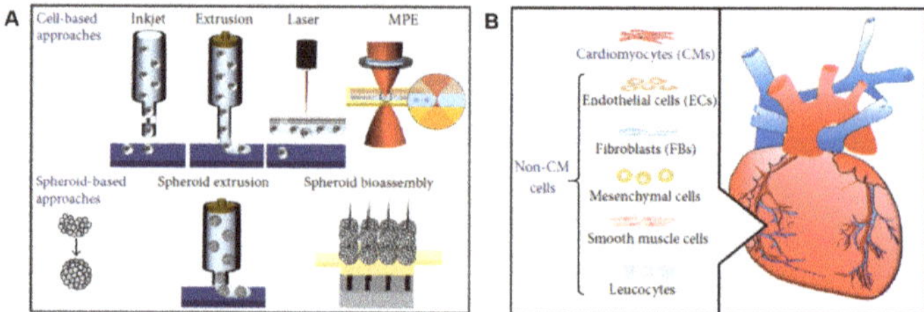

Figure 6. Schematic representation of 3D bioprinting of the myocardium, showing (**A**) the methods and (**B**) the cell types. Reprinted from Ong et al. [62]. Copyright © 2020 Chin Siang Ong et al.

Gaetani and coworkers used the method of pressure-based extrusion to bioprint a patch of human cardiac-derived cardiomyocyte progenitor cells (CMPCs) (30×10^6 cells/mL) on a scaffold made from alginate, which was characterized by precise pore size and microstructure [66]. The authors demonstrated the cardiogenic potential of this patch, determined by mRNA upregulation of the early cardiac transcription factors NK2 homeobox 5 (Nkx2.5), Gata-4, myocyte enhancement factor 2C (Mef-2c), and of the late cardiac marker Troponin T (TnT) by these cells 7 days post-printing. Furthermore, the printed cells were able to migrate from the alginate matrix to a matrigel layer, forming tubular like structures. Few years later, the same group bioprinted a myocardial patch consisting of human CMPCs (30×10^6 cells/mL) suspended in a scaffold made of HA and gelatin, then treated 10–12 weeks aged infarcted mice [73]. After 4 weeks from implantation, the scaffold proved to be an excellent vehicle to support cell survival, engraftment, and differentiation, finally improving mouse cardiac function (Figure 7A–C).

In order to bioprint 3D functional cardiac tissue constructs with improved conductive properties, Zhu and coworkers recently developed an attractive bioink containing an electrically conductive component [74]. To do this, the authors incorporated gold nanorods (GNRs) into a CMs- and cardiac FBs-laden hydrogel (1:1.5 × 10^6 cells/mL), consisting of gelatin methacrylate (GelMA) and alginate pre-polymer solutions. GelMA is a degradable hydrogel composed of gelatin, which is the denatured form of collagens, that has been tunably substituted with methacryloyl groups [53]. In addition, the presence of integrin-binding motifs and matrix metalloproteinase(MMPs)-sensitive groups make GelMA a biomaterial highly bioactive [75]. The advantage of using alginate, instead, is represented by its ability to maintain the viscosity of the bioink for a long period of time at a room temperature [76]. After 5 days of culture, both the CMs and cardiac FBs were successfully embedded within the GelMA/alginate bioprinted construct loaded with GNRs, and displayed good cell spreading [74]. One week later, the two cell populations entirely colonized the printed construct, forming a uniform and interconnected tissue layer. Compared to the constructs without GNRs, the presence of a conductive nanomaterial in the GelMA/alginate bioink further improved the electrical propagation between adjacent CMs, as evidenced by higher expression of the gap junction protein Cx-43 after 14 days and higher synchronized contraction of the bioprinted construct.

An important challenge in the 3D bioprinting technology remains the vascularization of constructs, that is the ability of building in vessels that are capable of anastomosing with host vessels following implantation [77]. With the aim to construct a stable vasculature, Gaebel and colleagues used a polyester urethane urea (PEUU) cardiac patch for bioprinting two types of human cells, ECs (4 × 10^6 cells) and MSCs (2 × 10^6 cells), in a defined arrangement mimicking the vasculature [78]. It has been reported that co-implantation of vascular ECs and MSCs could enhance the stability of the neovascularization of ECs [79]. The authors reported better cell viability and increased vessel formation after 8 days of culture when compared to randomly seeded patches. However, the most interesting finding of this study was that the 3D bioprinted cardiac patch was able to promote the creation of a vascular network 8 weeks after implantation in infarcted rat hearts, enhancing vessel formation also in the border zone of myocardial infarction. Furthermore, significant improvements of heart function were observed following implantation of the bioprinted constructs in the infarcted area.

In some other recent studies, bioinks derived from cardiac decellularized ECM have been developed for 3D bioprinting and appeared to be promising biomaterials in the repair of myocardial dysfunction and for the delivery of stem cells. For example, Pati and colleagues successfully decellularized the LV of porcine heart ECM with a combination of physical, chemical, and enzymatic processes [80,81]. The resulting heart decellularized ECM possessed levels of collagen and glycosaminoglycans similar to those of the native ECM, thus providing a viable environment for 3D bioprinting. By using a multi-head tissue/organ building system, the authors printed the obtained ECM with rat myoblast cells, producing a tissue that had a myocardial-like organization. The 3D bioprinted constructs further supported differentiation and maturation of encapsulated cells, as demonstrated by the expression level of the cardiac-specific genes fast myosin heavy chain (Myh6) and alpha-sarcomeric actinin (Actn1) during 14 days study-period. In agreement with gene expression results, immunofluorescence analysis confirmed the presence of Myh6 in the cell-printed constructs. Jang and collaborators produced a heart decellularized ECM from the LV of 6-month-old Korea domestic pigs [55]. They used three different bioink formulations to develop pre-vascularized stem cell patch. In detail, 2 layers of PCL were made up to provide mechanical support. Then, bioink I, containing 5 × 10^6 human CPCs/mL, and bioink II, containing 5 × 10^6 human MSCs/mL supplemented with 10 µg/mL vascular endothelial growth factor (VEGF), were alternatively printed on the PCL supporting layer. After printing, vitamin B2 was used for the crosslinking process by exposition to ultraviolet radiation. The result of this dual stem cell 3D printed structure was a tissue with stiffness similar to native myocardium, in which the cells were able to proliferate and differentiate. In addition, the developed stem cell patch promoted strong vascularization and tissue matrix formation when implanted for 28 days into 7-weeks-old balb/c nude mice. Another interesting example of 3D bioprinting involving two different stem cell populations comes from the recent work of Park and colleagues, who employed a multipronged approach for simultaneously restoring cardiac function and vessel formation in infarcted rat hearts [82]. To do this, they intramyocardially injected CMs derived from human iPSCs (iPSCs-CMs) (1 × 10^6 cells/rat), and epicardially implanted human MSCs-loaded patch (MSCs-PA) (1×10^6 cells/mL) generated from pig heart decellularized ECM. Epicardial patch carrying human MSCs improved vascular regeneration through prolonged secretion of angiogenic paracrine factors, but more importantly it promoted the engraftment, viability and maturation of the injected human iPSC-CMs, ultimately leading to restoration of cardiac function in infarcted-induced hearts.

Another group combined cardiac decellularized ECM hydrogel with GelMA and pediatric human CPCs (3×10^6 cells/mL) to print a patch to be used for pediatric patients suffering from right-ventricular (RV) failure or for adult myocardial dysfunction [83]. The presence of GelMA allowed for printability of the CPCs/cardiac ECM bioink through hydrogel polymerization via cooling to 10 °C, followed by white light radical polymerization and incubation at physiological temperatures. The authors demonstrated that the human CPCs cells in GelMA-cardiac ECM patches improved cardiogenic differentiation and angiogenic potential at 3 and 7 days when compared to cells in GelMA-only patches. In vivo, the printed GelMA-cardiac ECM patches remained attached to rat hearts epicardially, and vessels were formed after 14 days, indicating their integration with the native myocardium [83]. The importance of cardiac ECM as a component of bioinks was also underlined by Das and colleagues, who created bioinks with porcine heart tissue-derived ECM or collagen for encapsulating neonatal rat CMs (2×10^7 cells/mL) using an extrusion-based 3D bioprinter [84]. The patches were then cultured for a month in dynamic or static conditions with the aim to evaluate the structural arrangement of CMs and their subsequent gene expression. From a molecular point of view, heart tissue-derived ECM cultured under dynamic conditions promoted enhanced expression of cardiac specific genes, like cardiac TnT, Myh6, and Actn2, compared with the static condition during the 14-day study period. This was also true for genes encoding the proteins responsible for cell–cell adhesion, cell-matrix interaction (integrins), formation of basement proteins (laminins), and guidance of matrix remodeling events (MMPs). Furthermore, microscopic images revealed that CMs in decellularized ECM cultured dynamically exhibited a more aligned, uniform, and rod-like structural arrangement of the cardiac regulatory protein TnT with sarcomeric integrity compared to the respective static counterpart (Figure 7D–F). The results of this work would indicate that both matrix and microenvironment can be decisive factors for cell–cell and cell–matrix interactions, thus influencing engineered heart tissue maturation.

In a very recent work by Noor and colleagues, a biopsy of omental tissue was taken from patients, then the cells were reprogrammed to become iPSCs and differentiated to CMs and ECs, whereas the ECM was processed into a personalized hydrogel serving as a bioink for 3D printing [85]. The bioinks were then 3D printed for generating vascularized patches (2×10^7 iPSCs/mL and 2×10^7 ECs/mL) and complex cellularized structures. Structure and function of the patch was evaluated in vitro, while cardiac cell morphology was assessed after transplantation in between two layers of rat omentum, revealing elongated CMs with massive actinin striation. This work has demonstrated for the first time the possibility to engineer vascularized cardiac patches that fully match the immunological, biochemical, and anatomical properties of any individual.

As mentioned before, scaffold-free techniques are being increasingly developed because problems associated with using biomaterials, such as immunogenicity, fibrous tissue formation, biomaterial degradation, toxicity of degradation products, are avoided. One of the first work is the one of Atmanli and Domian, who addressed the need of a scaffold using a microtextured polydimethylsiloxane (PDMS) stamp to guide the microarchitecture of murine ventricular progenitor cells (CVPs) printed by microcontact printing [86]. Although this experiment did not test the in vivo therapeutic properties of the bioprinted myocardium, it nonetheless represents a pivotal early step in scaffold-free 3D bioprinting of myocardial tissue. Later, Ong and coworkers 3D bioprinted cardiac patches by mixing cell cardiospheres (33×10^5 cells/cardiosphere) composed of human iPSCs-CMs, FBs, and ECs at different ratios [87,88]. The assembling of cardiospheres resulted only when iPSCs or iPSC-CMs were co-cultured with at least 15% ECs or FBs. Patches of all cell ratios showed spontaneous beating, ventricular like action potential and uniform electrical conduction spontaneously within 3 days of bioprinting. By increasing FBs percentage, lower conduction velocities and longer action potential duration were recorded, suggesting some inhibition of electrical coupling of CMs by these cells. Nevertheless, after 1 week from implantation onto nude heart rats, all the bioprinted patches engrafted well on to the myocardium and were vascularized, indicating the regenerative potential of such a scaffold-free 3D bioprinting approach (Figure 7G–M). The applications of 3D bioprinting in the cardiac patches' fabrication discussed in this paragraph are listed in Table 1.

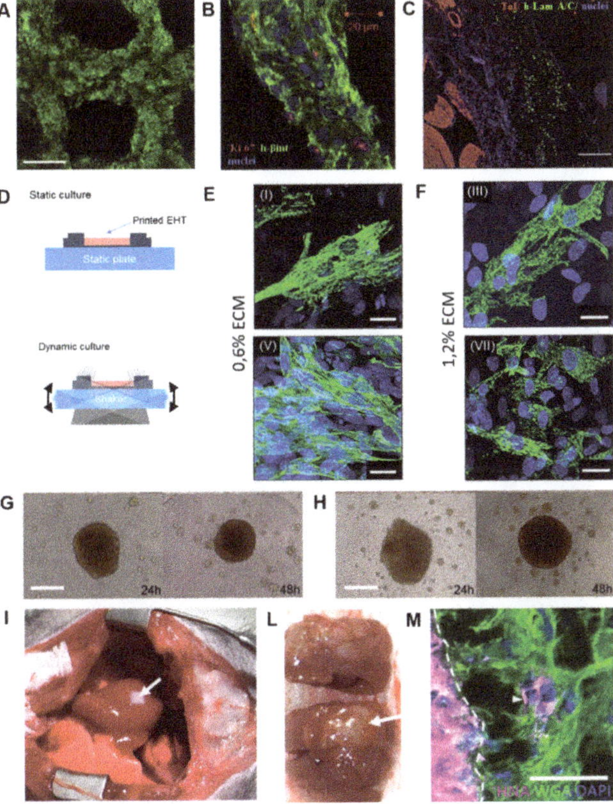

Figure 7. Examples of 3D bioprinting of myocardium using (**A–C**) cell-laden scaffolds, (**D–F**) decellularized heart ECM, and (**G–M**) cell-free scaffolds. (**A**) Human CMPCs 3D bioprinted in a scaffold made of HA and gelatin are alive (green) 2 hours after printing, (**B**) express the proliferation marker Ki-67 (red) after 7 days in culture, and (**C**) are visible in infarcted mice 4 weeks after transplantation. Reprinted from Gaetani et al. [73]. Copyright © 2020 Elsevier Ltd. (**D**) Schematic representation of 3D bioprinted cardiac decellularized ECM cultured under static and dynamic conditions. (**E**) confocal microscopy images of immunostaining for cardiac TnT (green) synthesized by CMs in 0.6%, and (**F**) 1.2% ECM cultured statically and dynamically for 14 days. Reprinted from Das et al. [84]. Copyright © 2020 Acta Materialia Inc. Cardiospheres form in 24 hours and start beating in 48 hours when iPSCs-CMs:FBs:ECs were co-cultured at (**G**) 70:15:15 or (**H**) 45:45:10 ratio. (**I**) Transplantation of 3D bioprinted cardiac patches (iPSCs-CMs:FBs:ECs 70:15:15) onto the anterior surface of the rat heart. (**L**) Anterior aspect of the heart explanted 1 week after implantation. (**M**) Confocal microscopy images of immunostaining for human nuclear antigen (HNA) (magenta), wheat germ agglutinin (WGA) (green), and DAPI (blue) showing the presence of human cells (white arrows) in native rat myocardium. White dotted line demarcates the cardiac patch (left) from the native rat myocardium (right). Reprinted from Ong et al. [88]. Copyright © 2020 Springer Nature.

Table 1. Applications of 3D bioprinting for production of cardiac patches.

Bioink	Cell (Concentration)	3D Bioprinting Technique	Significance	Ref.
alginate	human CMPCs [1] (30×10^6/mL)	extrusion-based bioprinting	cells were able to migrate out of the alginate matrix and fully colonize a matrigel layer, forming tubular-like structures in vitro	[66]
HA [2] and gelatin	human CMPCs (30×10^6/mL)	extrusion-based bioprinting	the scaffold was able to support cell survival, engraftment, and differentiation; in addition, it improved cardiac function after epicardial transplantation in a mouse model of myocardial infarction	[73]
GelMA [3] and alginate incorporating GNRs [4]	CMs [5] (1×10^6/mL) and cardiac FBs [6] (1.5×10^6/mL)	extrusion-based bioprinting	the presence of a conductive nanomaterial (GNRs) into the hydrogel improved the electrical propagation between adjacent CMs, that finally resulted in a synchronized contraction of the bioprinted construct in vitro	[74]
PEUU [7]	human ECs [8] (4×10^6) and human MSCs [9] (2×10^6)	laser-based bioprinting	co-implantation of ECs and MSCs in a defined printed pattern enhanced the vascularization of the construct and improved cardiac function after acute myocardial infarction in rats	[78]
decellularized ECM [10] from the LV [11] of porcine heart	rat myoblast cells (from 1 to 5×10^6/mL)	extrusion-based bioprinting	the construct possessed a microarchitecture having a native-like organization	[80,81]
decellularized ECM from the LV of 6-month-old Korea domestic pig	human CPCs [13] (5×10^6/mL) and human MSCs (5×10^6/mL)	extrusion-based bioprinting	the use of two different bioink formulations, one containing CPCs and the other made of MSCs supplemented with VEGF [14], allowed for the development of pre-vascularized cardiac patch	[55]
decellularized ECM from the LV of porcine heart	CMs derived from human iPSCs [15] (1×10^6/rat) and human MSCs (1×10^6/mL)	extrusion-based bioprinting	the strategy of intramyocardially applying CMs derived from human iPSCs-CM and epicardially implanting a cardiac patch containing human MSCs significantly improved cardiac function and vessel formation in a rat model of myocardial infarction	[82]
decellularized ECM from porcine ventricular tissue combined with GelMA	pediatric human CPCs (3×10^6/mL)	extrusion-based bioprinting	possibility of using the cardiac patch in pediatric patients suffering from RV [16] failure, or for treating adult myocardial dysfunction	[83]
decellularized ECM from the LV of porcine heart or collagen	neonatal rat CMs (2×10^7/mL)	extrusion-based bioprinting	the culture conditions (dynamic versus static) are decisive factors for the structural arrangement of CMs, and affect gene expression and the related signaling pathways	[84]
decellularized human omental tissue	human iPSCs-CMs [17] (2×10^7/mL) and ECs (2×10^7/mL)	extrusion-based bioprinting	possibility of generating vascularized patches that fully match the immunological, biochemical and anatomical properties of any individual	[85]
scaffold-free	cardiospheres (33×10^5 cells/cardiosphere) composed of human iPSCs-CMs, FBs and ECs at different ratios	3D bioprinting on a needle array	the biomaterial-free 3D printed cardiac patch produced from human iPSCs showed spontaneous beating, electrical integration of the cardiospheres, and in vivo engraftment and vascularization	[87,88]

[1] cardiomyocytes progenitor cells; [2] hyaluronic acid; [3] gelatin methacrylate; [4] gold nanorods; [5] cardiomyocytes; [6] fibroblasts; [7] polyester urethane urea; [8] endothelial cells; [9] mesenchymal stem cells; [10] extracellular matrix; [11] left ventricle; [13] cardiac progenitor cells; [14] vascular endothelial growth factor; [15] induced pluripotent stem cells; [16] right ventricle; [17] cardiomyocytes derived from human induced pluripotent stem cells.

4.2. 3D Bioprinting of Heart Valves

In addition to myocardial damage, heart valves disfunction represents another significant cause of heart failure [89]. Anatomically, heart valves consist of three leaflets and a root wall, mainly containing valve interstitial cells (VICs), SMCs, and valvular endothelial cells (VECs) [46]. These cells populate the valves in precise spatial locations having different flexure strength and stiffness [90]. The leaflets are indeed tri-layer structures, each characterized by unique ECM composition: the collagen-rich fibrosa layer, the intermediate proteoglycan-rich spongiosa, and the elastin-rich ventricularis [91]. These three layers continuously flaps, resulting in a closure and opening of the valve, and are subjected to shear stress and periodic loading and unloading. The roots, on the other hand, are annular structures providing support to the leaflet and serving as a base to be integrated with the major blood vessels of the heart [92]. In valve heart diseases, the valves become either too contracted to open-up entirely or incapable to close effectively. Currently, valve replacement surgeries are the only option for the vast majority of patients; these strategies usually employ mechanical or biological prosthetic valves [93]. Mechanical valves have high durability; however, the thrombogenicity of such valves is a major concern, forcing patients to take blood-thinning drugs for the rest of their lives. On the other hand, biological prosthetic valves suffer from other complications, including immune rejection and degeneration over time, thus requiring a possible reoperation after 10–20 years [93].

3D heart valve bioprinting has been explored as an alternative technology to traditional mechanical or biological prosthetic valve replacements, and also possesses further benefits over standard tissue engineering methods [94]. The advantages offered by 3D bioprinting include the possibility to generate a mechanically heterogeneous structure with spatial control of valve cells, therefore accurately replicating the complex biomimetic architecture of heart valves. Hydrogels are promising scaffold materials for bioprinting heart valves, due to their high physicochemical and mechanical tunability, and permeability to nutrients and waste for encapsulated cells. In addition, hydrogels can be designed to mimic biological properties of soft tissue heart valve scaffolds with high spatial precision [95,96]. Nonetheless, despite considerable efforts have been made in this field, it is still challenging to achieve completely controlled regional and spatial composition to bioprint the complex 3D structure of heart valves. Below, we report some examples of the most recent attempts of heart valves 3D bioprinting described in the literature.

Duan and colleagues experienced different types of hydrogel bioinks for creating 3D bioprinted valves. In one of their study, a 3D bioprinted aortic valve conduit was created by encapsulating human aortic root SMCs and porcine aortic VICs (1×10^7 cells/mL) in alginate/gelatin hydrogels (0.05 and 0.06 g/mL, respectively), using a dual syringe system to mimic the structure of valve root and leaflet, respectively [97]. Both cell types showed high cell viability after encapsulation within the hydrogel discs, being over 80% after 7 days from printing. Furthermore, encapsulated SMCs expressed elevated alpha-smooth muscle actin (αSMA), a contractile filament protein, whereas VICs expressed elevated vimentin, a protein typically present in FBs and other mesenchymal cells [98,99]. In a subsequent work, the same authors used human aortic VICs (HAVICs) (5×10^6 cells/mL) suspended in methacrylated-HA (Me-HA)/methacrylated-gelatin (Me-Gel) hybrid hydrogel solutions (4% w/v of Me-HA, and 6%, 10%, 12% w/v of Me-Gel) to bioprint a tri-leaflet valve [100]. In detail, the heart valve root was printed with acellular hydrogel, while the leaflet was printed with the Me-HA/Me-Gel hydrogels suspended with HAVICs. Cells in all the hydrogel formulations maintained a post-printing viability of over 90% after 7 days, without significant difference in proliferation rate. By increasing Me-Gel concentration resulted in lower stiffness and higher viscosity, facilitating cell spreading, and better maintaining HAVICs fibroblastic phenotype. Histological and biomolecular analysis showed significant expression of αSMA, vimentin, periostin, and collagen type I, demonstrating that the cells proliferated normally post-printing and secreted their own ECM (Figure 8). However, both the studies described here were short-term works (7 days), and the printed constructs did not completely fulfil the mechanical performance of native valve tissue.

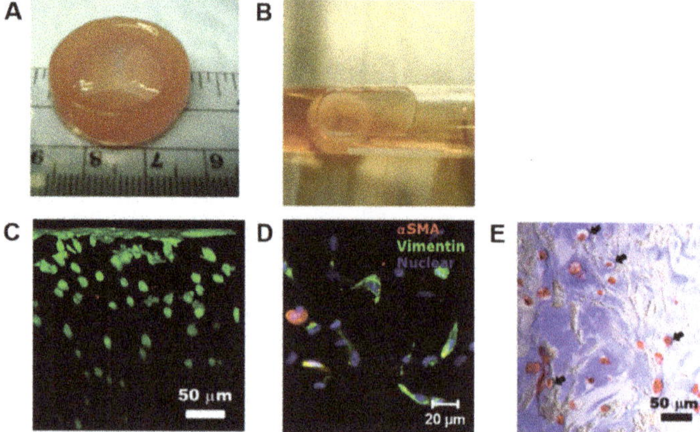

Figure 8. Example of 3D bioprinting of heart valve conduit with encapsulation of human aortic VICs (HAVICs) within the leaflets. (**A**) The valve conduit bioprinted using 4% w/v Me-HA/10% w/v Me-Gel hydrogels has inner diameter of 20 mm, outer diameter of 26 mm, height of 8 mm for valve root, and three leaflets with radius of 10 mm. (**B**) The bioprinted valve conduit shows an intact structure after photo-crosslinking and 7 days of static culture. (**C**) Cross-sectional view of Live/Dead image showing that nearly all encapsulated cells are alive from the surface to more than 300 μm below the surface. (**D**) Immunofluorescence images of the encapsulated HAVICs showing positivity for αSMA (red), vimentin (green), and nuclei (blue) after 7 days of culture. (**E**) Masson's Trichrome staining of bioprinted leaflets showing that more intense blue color is found around the encapsulated HAVICs, indicating the newly deposition of collagen. Reprinted from Duan et al. [100]. Copyright © 2020 Acta Materialia Inc.

In the 3D bioprinting process, a high degree of geometric control and shape fidelity of an hydrogel construct can be achieved by using photo-crosslinking [101]. Light exposure is generally sufficient to solidify the hydrogel structure, and it can be performed during printing or post-fabrication of the whole construct [102]. Kang and colleagues tested several parameter combinations of photoinitiator type and concentration, and light intensity for optimizing cell viability during 3D bioprinting in a mixture of methacrylated gelatin/poly-ethylene glycol diacrylate/alginate (MEGEL/PEGDA/alginate) [102]. The two compared photoinitiators were Irgacure 2959 and VA086: Irgacure 2959 was chosen because it is one of the most widely used in contact with cells, whereas VA086 has been reported to be less cytotoxic than Irgacure [103–105]. As cell source, the authors used HAVICs, human aortic valve sinus smooth muscle cells (HASSMCs), and human adipose derived mesenchymal stem cells (HADMSCs) (2.5×10^6 cells/mL). The encapsulated cell population that responded better to increased concentrations of photoinitiator was the HADMSCs. When using 0.25–1.0% w/v VA086, 0.025–0.1% w/v Irgacure 2959, and 365 nm light intensity 2–136 mW/cm^2 for encapsulating cells in the the MEGEL/PEGDA/alginate bioink, cell viability was 95% for HASSMCs, 93% for HAVICs, and 93% for HADMSCs. The major findings of this work were the identification of different parameters that can be combined for optimizing the viability of different cell populations within 3D bioprinted hydrogels. Some positive examples of other cell types are coming from applications in neuroscience and neurosurgery, where cells grown in either two-dimensional (2D) or 3D systems offer exciting opportunities in both basic [106] and in concrete regenerative tasks [107,108].

3D bioprinting has been successfully used also for generating 3D models of calcific aortic valve disease (CAVD) recapitulating leaflet layer biomechanics [109]. As mentioned above, the healthy aortic valve (AV) is composed of three semilunar leaflets, each comprising three stacked layers with its own ECM composition [91]. With the progression of CAVD, the leaflets of the AV become fibrotic and calcify as their constituent cell population of VICs undergo myofibrogenic and osteogenic differentiation [91].

In their study, Van der Valk and colleagues bioprinted a model of CAVD encapsulating human VICs (10×10^6 cells/mL) in GelMA/methacrylated HA (HAMA) hydrogels followed by UV crosslinking with the photoinitiator lithium phenyl-2,4,6-trimethylbenzoylphosphinate [109]. The 3D printed hydrogels with encapsulated VICs were then cultured with osteogenic factors for promoting the formation of microcalcifications. The exposition to osteogenic stimuli for 14 days effectively caused the formation of microcalcific nodules; nevertheless, negligible levels of apoptosis were measured in the encapsulated cells, thus suggesting that calcification was not related to cell death by apoptotic processes. In this study, the authors additionally quantified the compressive mechanical properties of each of the AV leaflet layers, finding that the ventricularis showed an intermediate Young's modulus (26.9 kPa) between the fibrosa and the spongiosa layers (37.1 kPa and 15.4 kPa, respectively). The findings of this work have established for the first time a novel 3D model for the study of valvular mechanobiology and could also facilitate high-throughput drug screening for CAVD in a biologically-relevant environment. 3D bioprinting applications for the generation of heart valve models discussed in this paragraph are summarized in Table 2.

Table 2. Applications of 3D bioprinting for production of heart valves.

Bioink	Cell (Concentration)	3D Bioprinting Technique	Significance	Ref.
alginate/gelatin	human aortic root smooth muscle cells (SMCs) [1] (1×10^7/mL) and porcine aortic valve interstitial cells (VICs) [2] (1×10^7/mL)	extrusion-based bioprinting	the use of a dual syringe system, each containing a defined cell population (SMCs or VICs), allowed for the creation of a 3D printed aortic valve conduit complete of valve root and leaflet	[97]
Me-HA [3] and Me-Gel [4]	HAVICs [5] (5×10^6/mL)	extrusion-based bioprinting	a heart valve conduit was bioprinted with acellular root and three leaflets encapsulating HAVICs; by varying the concentration of the hydrogel formulations it was possible to modulate the behavior of the encapsulated cells	[100]
MEGEL/PEGDA/ alginate [6]	HAVICs, human aortic valve sinus smooth muscle cells (HASSMCs) [7], and human adipose derived mesenchymal stem cells (HADMSCs) [8] (2.5×10^6/mL)	extrusion-based bioprinting	variable combinations of photoinitiator type (Irgagure 2959 versus VA086) and concentration, and light intensity (2–136 mW/cm^2) can be used to optimize cell viability during 3D printing for multiple cell types	[102]
GelMA/HAMA [9]	human VICs (10×10^6/mL)	extrusion-based bioprinting	a 3D model of calcific aortic valve disease (CAVD) [10] was created recapitulating leaflet layer-specific mechanical properties which is useful for studying the valvular mechanobiology and for high-throughput drug screening	[109]

[1] smooth muscle cells; [2] valve interstitial cells; [3] methacrylated-hyaluronic acid; [4] methacrylated-gelatin; [5] human aortic VICs; [6] methacrylated gelatin/poly-ethylene glycol diacrylate/alginate; [7] human aortic valve sinus smooth muscle cells; [8] human adipose derived mesenchymal stem cells; [9] gelatin methacrylated/methacrylated HA; [10] calcific aortic valve disease.

4.3. 3D Bioprinted Heart Tissue Patches for Drug Screening

An increasingly widespread opinion believes that 3D printing can offer great potential also in pharmaceutics for the discovery and development of new drugs, as well as for advances in drug delivery strategies [94]. Drug discovery represents the preclinical phase of the entire process, and it consists in testing thousands of compounds to select few candidates with possible beneficial effects against the clinical target diseases. The drug development phase involves further validation of the selected drug in terms of absorption, distribution, metabolism, excretion (ADME), toxicity, dosage, and treatment modalities, as well as interaction with other molecular compounds [110]. Drug development is a multistep process that involves phase I-III clinical trials, review and approval by the Food and Drug Administration (FDA), phase IV clinical trials, followed by launch in the market, and post-market surveillance by FDA.

About 94% of the drugs that passed the preclinical trials fail in the clinical phase [111]. It is believed that this is mainly attributable to inadequate screening in preclinical trials, most of which are still performed on traditional 2D monolayer culture systems that miss the native 3D extracellular microenvironment [112]. Furthermore, the efficacy and toxicity of drugs evaluated in animal studies do not always predict the response in human patients [113]. For example, during both the preclinical and clinical stages of drug development, cardiotoxicity remains a major cause of failure, thus representing the primary reason for the retraction of pharmaceuticals from the market [114]. The use of 3D engineered tissue platforms could overcome these problems, and 3D bioprinting, in particular, could play a key role in both the drug discovery and development phases. 3D bioprinted tissues could produce results similar to those obtained in in vivo tests, thus representing valuable platforms for conducting drug toxicity analysis in vitro.

In the cardiovascular field, most of the efforts toward developing drug discovery and screening platforms have focused on recreating microtissues of the left ventricular myocardium, the site of most cardiac pathologies and the primary pumping chamber of the heart [115]. For example, Zhang and colleagues developed a novel hybrid strategy based on 3D bioprinting to engineer endothelialized myocardial tissues [116]. In their work, the authors directly encapsulated ECs (1×10^7 cells/mL) within an alginate/GelMA bioink through a combination of extrusion and photocuring processes (Figure 9A). In approximatively 2 weeks of culture, the ECs gradually migrated toward the microfiber peripheries, organizing in a confluent layer of endothelium, and formed a pattern resembling the blood vessel walls (Figure 9B). This bioprinted microfiber scaffold was then seeded with rat-derived CMs (1×10^6 cells/mL) (Figure 9C). The CMs adhered and spread on the surface of the microfibers across the entire thickness of the scaffolds, and strongly expressed the contractile Actn1 protein and the intercellular conductive connexin-43, as demonstrated by immunostaining analysis. In addition, the CMs-populated constructs started beating spontaneously and synchronously after 48 hours of culture. In order to construct the endothelialized-heart-on-a-chip device and evaluate cardiovascular drug toxicity, a microfluidic perfusion bioreactor was generated. The endothelialized microtissue was tested with the common anti-cancer drug doxorubicin, which demonstrated dose-dependent effects towards both CMs and ECs. In particular, the beating rate of the CMs decreased to 70.5% and 1.62%, and the levels of von Willebrand factor (vWF) secreted by the ECs were reduced to 76.0% and 35.3% 6 days after exposure to 10 μM and 100 μM doxorubicin, respectively (Figure 9D,E). In the same study, the authors adopted a similar fabrication method and drug toxicity test on endothelialized human iPSCs-derived CMs microtissue, with results that were comparable to those observed in the rat-derived microtissue, thus suggesting potential translational for personalized drug screening.

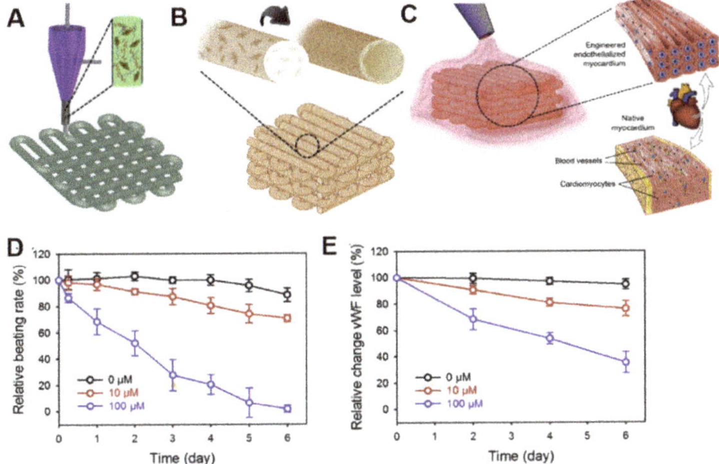

Figure 9. Example of 3D bioprinted heart tissue patches used for drug screening. (**A**) Schematic representation of the extrusion-based 3D bioprinting system used to generate microfibrous alginate/gelatin methacrylate (GelMA) scaffolds encapsulating endothelial cells (ECs), that (**B**) in approximatively 2 weeks form a vascular bed through migration of cells to the peripheries of the microfibers. (**C**) Cardiomyocytes (CMs) are then seeded into the interstitial space of the endothelialized scaffold. The doxorubicin dose-concentration response is evaluated as (**D**) relative beating rate of CMs and (**E**) relative expression levels of von Willebrand factor (vWF) in ECs. Reprinted from Zhang et al. [116]. Copyright © 2020 Elsevier Ltd.

More recently, Lind and coworkers designed cardiac microphysiological devices via multi-material 3D bioprinting by sequentially using six functional bioinks based on highly conductance, piezoresistive, and biocompatible soft materials [117]. After printing, the devices were seeded with human iPSCs-CMs (220 K/cm^2), which self-assembled into laminar tissues mimicking the structure of the native heart. Such a fabricated system allowed the authors to perform dose response studies of cardiac drugs that influence contraction strength or beat rate directly inside a cell incubator. In particular, the L-type calcium channel blocker verapamil and the β-adrenergic agonist isoproterenol were investigated. The engineered microtissue displayed inotropic responses to verapamil and isoproterenol, comparable to data obtained from engineered 3D neonatal rat ventricular myocardial tissues and isolated postnatal whole rat hearts, therefore demonstrating the potential of this model as a drug screening platform.

In another study, Wang and colleagues 3D bioprinted a functional cardiac tissue mimicking the structural, physiological, and functional features of native myocardium [118]. The authors used a fibrin-based composite hydrogel as bioink to print primary CMs (10×10^6 cells/mL) isolated from infant rat hearts. Bioprinted cardiac tissue constructs exhibited spontaneous synchronous contraction in culture, and positivity to Actn1 and connexin 43, indicating the generation of uniformly aligned and electromechanically coupled cardiac cells. The authors then evaluated physiologic responses of these bioprinted cardiac tissues to known cardiac drugs, the androgen agonist epinephrine and the androgen antagonist carbachol after 3 weeks of culture. Calcium imaging analysis was used to quantify spontaneous beating of bioprinted cardiac tissues, revealing that epinephrine increased the beating frequency from 80 to 110 beats per minute (BPM), whereas carbachol decreased the beating frequency to 40 BPM. Interestingly, the authors demonstrated the reversible effects of these drugs when these were removed from the bioprinted cardiac tissues, thus indicating their effectiveness in testing the physiological response of cardiac drugs.

As mentioned above, 3D bioprinting could be useful also for the development of drug delivery strategies [94]. Some current applications of 3D printing in drug delivery systems include implant surfaces (i.e., orthopedic implants) modified with drug-eluting solutions [119,120], medical devices (i.e., stent or catheter) containing and eluting drugs [121], and 3D printed delivery devices providing personalized drug release profiles [122]. 3D bioprinting applications for drug screening discussed in this paragraph are summarized in Table 3.

Table 3. Applications of 3D bioprinting for drug screening.

Bioink	Cell (Concentration)	Drug Tested	Significance	Ref.
alginate/GelMA [1]	human ECs [2] (1×10^7/mL) and neonatal rat CMs [3] (1×10^6/mL)	doxorubicin (anti-cancer drug)	the doxorubicin dose-concentration response was evaluated in the endothelialized-myocardium-on-a-chip both as beating rate in CMs and as relative expression levels of vWF [4] in ECs	[116]
dextran, TPU [5], CB [6]:TPU, Ag:PA [7], soft PDMS [8], rigid PDMS	iPSCs-CMs [9] (220 k/cm^2)	verapamil (cardiac drug), isoproterenol (cardiac drug)	the engineered microtissues displayed inotropic responses to verapamil and isoproterenol comparable to data obtained from engineered 3D neonatal rat ventricular myocardial tissues and isolated postnatal whole rat hearts	[117]
fibrin-based composite hydrogel (20 mg/mL fibrinogen, 30 mg/mL gelatin, 20 µg/mL aprotinin, 10% glycerol, and 3 mg/mL HA [10])	rat CMs (10×10^6/mL)	epinephrine (cardiac drug) and carbachol (cardiac drug)	the bioprinted cardiac tissues physiologically responded to the tested cardiac drugs by modulating the CMs beating frequency; reversible effects of the drugs were observed once these were removed from the bioprinted tissues, thus confirming the effectiveness of these constructs as in vitro 3D tissue models	[118]

[1] gelatin methacrylated; [2] endothelial cells; [3] cardiomyocytes; smooth muscle cells; [4] von Willebrand factor; [5] thermoplastic polyurethane; [6] carbon black nanoparticles; [7] silver particle-filled, polyamide; [8] polydimethylsiloxane; [9] cardiomyocytes derived from human induced pluripotent stem cells; [10] hyaluronic acid.

5. 3D Printing for Testing and Realizing New Heart Devices

Another important application of 3D printing is the development and testing of medical devices. With regard to the cardiovascular field, one of the first studies was the one of Kalejs and von Segesser, who created an aortic root model made of common house-hold silicone, then started to use this model in in-vitro valved stents testing, integrating the aortic root in an artificial circulatory loop [123]. Biglino and coworkers, instead, imaged a tract of descending aorta of a 29-year-old volunteer 50 mm in length with MRI, then used the polyjet technique for printing models having different wall thickness (0.6, 0.7, 0.8, 1.0, and 1.5 mm) but constant internal diameter (15.5 mm) [124]. These models of descending aorta were then subjected to compliance tests through gradual increase and decrease of the internal volume in order to track pressure variations. Two critical cases of vascular anatomies were then selected for evaluating the practicability of the method, after the identification of a range of material's distensibility. The first case concerned an adult patient affected by severe pulmonary regurgitation and dilated right ventricular outflow tract (RVOT), whereas the second case was about a pediatric patient with hypoplastic left heart syndrome (HLHS) and aortic coarctation (2.7 mm narrowing). Good anatomical finishing was qualitatively obtained for both models, suggesting their usefulness for device implantation and testing (Figure 10A). Later, Mashari and colleagues acquired TEE images of a patient subjected to a percutaneous MitraClip® operation, which is a minimally invasive procedure to reduce the mitral regurgitation, for creating a 3D MV model [125]. Hemodynamic tests were then carried out on this 3D model by insertion in a pulse duplicator chamber loaded with a fluid mimicking the blood (Figure 10B). Paulsen and colleagues used a 3D printed left heart simulator to compare two different conduits for valve-spearing aortic root replacement [126]. The simulator was efficient to show the comparability of straight tubular and Valsalva grafts in terms of gross hemodynamic and coronary blood flow, however it revealed that the two conduits differ considerably in terms of biomechanics. 3D printing also resulted effective in realizing personalized occluders for the treatment of CHD using new biodegradable materials. As an example, Sun and coworkers produced a 3D printed biodegradable occluder made of a copolimer composed by poly-L lactic acid, trimethylene carbonate and glycolide [127]. Such a 3D printed model was then tested both in vitro and in vivo in a rabbit model, demonstrating good results in terms of safety, reliability and biocompatibility.

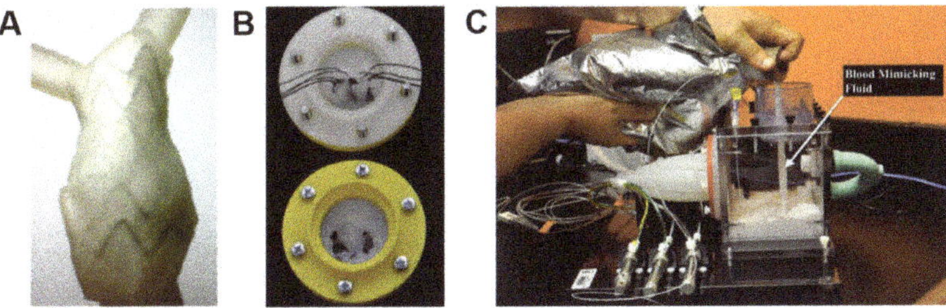

Figure 10. Examples of 3D printed models for device testing. (**A**) 3D printed right ventricular outflow tract (RVOT) model used for physical insertion of a stent-graft for assessing patient's suitability for the device. Reprinted from Biglino et al. [124]. Copyright © 2020, Springer Nature. (**B**) Silicone 3D printed mitral valve (MV) incorporated into (**C**) a pulse-duplicator chamber filled with a blood mimicking fluid for hemodynamic testing. Reprinted from Mashari et al. [125]. Copyright © 2020 Elsevier Inc.

6. Regulatory Considerations and Commercialization of 3D Printed and 3D Bioprinted Products

The 3D printing industry is a rapidly expanding field, and it has been estimated that the global market for 3D printed devices will reach around \$34.8 billion with a compound annual growth rate (CAGR) of 23.2% during the forecast period 2019–2024, while that of 3D bioprinting will be near \$1.647 million with a CAGR of 20.4% by the end of 2024 [128,129]. For this reason, it is necessary that regulatory agencies around the world will soon try to establish quality standards before devices manufactured with 3D printing technologies become part of common clinical practice.

Worldwide, the regulatory requirements for a given type of medicinal product depend strictly on its classification. In the USA, medicinal products are categorized as drugs, biologics, or medical devices [130]. Each type of product is regulated by a different center within the FDA: The center for drug evaluation and research (CDER), the center for biologics evaluation and research (CBER), or the center for devices and radiological health (CDRH), respectively. The FDA regulatory agency in the USA considers 3D printed devices in the same way to traditional medical devices, and therefore subject them to the same regulatory requirements and submission information [131]. The CDRH classifies medical devices into three classes, on the basis of the intended use of the device, the indications for use, and the risk level to patient or user (Figure 11) [131]. Medical devices of Class I have a minimal harm potential to the patient or user, require general controls without clinical trials and are exempt from the regulatory process, meaning there is no need for proof of safety or efficacy. Medical devices of Class II have a moderate risk to the patient or user, require both general and special controls, and most of them need also filing a premarket notification, known as 510 (k), for notifying the FDA the intent of marketing. A 510 (k) is obtained when the device proves to be safe, effective, and substantially equivalent to a device already legally marketed that is not subjected to premarket approval [131]. Finally, Class III includes those medical devices that support or sustain human life, are generally implanted, and thus require clinical trials or other evidences before the most rigorous premarket approval. If a Class III device presents only minor changes from an already approved device, named predicate device, it may not need the stringent premarket approval; on the contrary, all novel devices that do not have a predicate device are automatically classified as Class III devices [130,132].

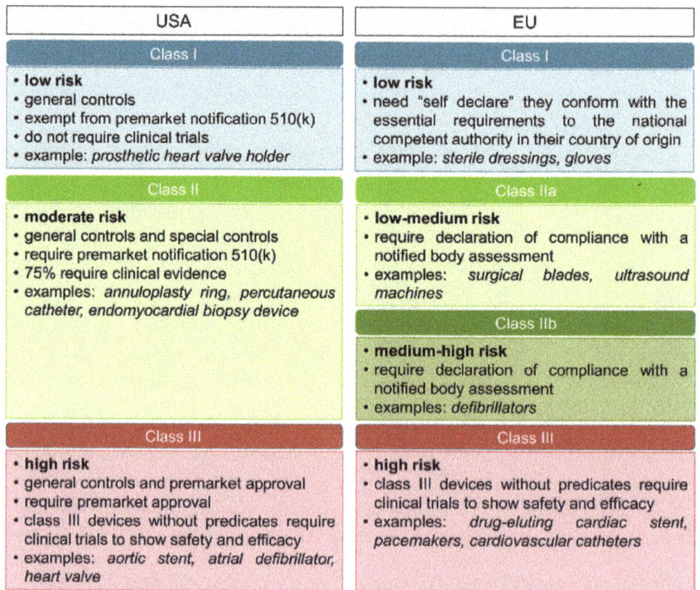

Figure 11. Classification of medical devices in the USA and in the EU.

In the EU, products are categorized as medicinal products (drugs or biologics) or medical devices, and they have their own unique regulatory processes [133]. With establishment of the EU, approval processes for medical devices followed a path of harmonization; nevertheless, medical device regulation does not fall solely to any one agency. Indeed, national competent authorities (NCAs) are responsible to designate and audit notified bodies (NBs), which are independent commercial organizations that contract with device manufacturers to supply these certifications for a fee [134]. NBs assess and assure conformity with requirements of the relevant European commission (EU) directives. Once NBs demonstrate the device meets requirements for conformity, the NBs issue a conformité européenne (CE) mark, and the device can be marketed in any EU member state [135]. Differently from the USA, medical devices in the EU are classified into four classes: class I, class IIa, class IIb, and class III (Figure 11) [136]. Class I are low risk medical devices, and only require a self-declaration of conformity by manufacturers without the involvement of NBs. In contrast, and similarly to what happens in the USA, medical devices of Classes IIa, IIb, and III require clinical and/or nonclinical evidence to support their approval.

In December 2017, the FDA released a guidance document entitled "Technical Considerations for Additive Manufactured Medical Devices", with the aim to outline technical considerations for manufacturing 3D printed devices, and containing recommendations for testing and characterizing such devices [137]. As already discussed in a dedicated section of this review, the technical workflow for manufacturing 3D medical devices involve several steps (Figure 12). The aspects of 3D printing addressed in the guidance document focused on device design, materials control, printing and post-printing validation; examination of printing parameters; assessment of physical and mechanical characteristics of final devices; and biological and safety considerations of final devices in terms of cleaning, sterility, and biocompatibility [137]. It has to be remembered that 3D printed medical devices, like those traditionally manufactured, must meet quality system requirements across all the production process.

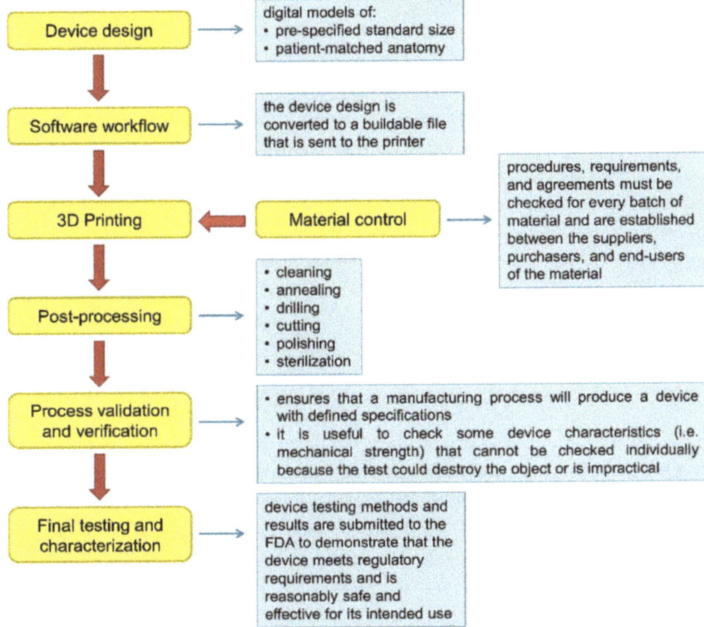

Figure 12. Technical workflow for manufacturing 3D printed medical devices.

Over the years, more than 80 3D printed medical devices have been approved by the FDA's center CDRH through 510(k) process [138]. Most of these devices are used in dental, orthopedic and cranio-maxillofacial fields (i.e., dental crowns, hearing aids, bone tether plates, skull plates, hip cups, spinal cages, facial implants, and screws), and are produced using powder bed fusion techniques (83%), stereolithography (12%), extrusion (3%), and jetting (2%) [138,139]. A postmarket analysis of the FDA's approved devices, conducted on the manufacturer and user facility device experience (MAUDE) database, evidenced that the principal reasons for product-related adverse events were improper size or incompatibility with other components of the device (46%), device fracture or shaving (45%), incorrect device size for the patient (5%), implant wear (2%) and incorrect device implanted (2%) [138]. To be fair, we have to say that many of these adverse events are common to medical devices manufactured with the traditional techniques.

As for 3D printed products containing biological components, no defined regulations currently exist, since these products cannot be simply classified as medical device, biologic, or drug, but can rather be considered both medical devices and biologics [140]. Indeed, combination products are products comprising two or more regulated components, that is drug-device, device-biologic, drug-biologic or drug-device-biologic [141]. A combination product is for example one in which living cells (biologics) are combined with a synthetic scaffold matrix (medical device) that provides physically support for the growth of new tissue [133]. The legislation applied to these products depends on their classification, which is determined by several factors, mainly by the way in which the finished product achieves its intended medical purpose and by its primary mode of action (PMOA) [133]. In the USA, the office of combination product (OPC) assigns combination products to the most appropriate FDA center [142]. If the PMOA of a cell-based medicinal product is provided by the biological cellular constituent, then the product is regulated as biologics by CBER in the USA, and as combined advanced-therapy medicinal products (ATMPs) in the EU [140,142]. On the other hand, if the therapeutic effect derives from the device component, the combination product is considered a Class III medical device in the EU. Regarding 3D bioprinted products, their classification mainly depends on whether these devices are mass-produced or custom-made. In current practice, custom-made devices, which are manufactured in accordance with a specific prescription for a specific individual, are exempt from normal quality system requirements or conformity assessment requirements by most regulatory bodies in both in the USA and in the EU [143,144].

The technical workflow for 3D bioprinting medical device is similar to that seen for 3D printing, although additional steps are included in the manufacturing process, such as selection of a design approach, cell type, and biological material [138]. Consequently, additional technical considerations need to be taken into account respect to 3D printed medical devices. These include, for example, printing parameters (printing temperature, resolution, and speed) or material selection. Mechanical and physicochemical properties, biocompatibility, as well as vascularization and biological function of the finished product are other parameters to be evaluated. Although 3D bioprinting technologies are largely still in the research and development stage, urgent regulatory and manufacturing process considerations are required to meet the growing global interest and needs of such 3D printed products.

7. Current Limitations, Future Perspectives, and Conclusions

3D printing attracted remarkable attention of the scientific community due to its high versatility, and in recent years the technology has considerably expanded also to the cardiovascular medicine. 3D printing allows the complex cardiac anatomy to be accurately replicated using materials resembling human heart tissues. For this reason, 3D printed heart models could serve as an excellent tool in facilitating medical education, pre-operative planning, and communication between doctor and patients or their family. Nevertheless, cardiac surgeons and cardiologists agree with the fact that these 3D printed heart models should only be used to complement the current diagnostic tools. This is particularly true when creating a model for cardiovascular pre-surgical planning, where an error made during segmentation of an image data set could have devastating consequences. In addition, there is still no statistical data currently available to demonstrate the real clinical value of these 3D printed models. Another obstacle that limits diffusion and routine application of 3D printing in clinical practice is related to its high cost. Further investigations are needed to analyze the cost-benefit of 3D printing for creating anatomical models. Interestingly, a recent work demonstrated that 3D printed heart models produced with low cost materials (A\$50) show efficiency and precision similar to models created with high cost materials (A\$300) [6]. Another limitation to overcome in a near future is the availability of the 3D printing technology in situ; most research hospitals, indeed, have access to facilities through academic collaborators. Building a 3D printing laboratory would require planning in hardware, software and dedicated staff [145].

Currently, the most advanced application of 3D printing is 3D bioprinting, which has the great potential to engineer highly organized functional tissues and/or organs with complex geometries and tailored components for widespread applications, including transplantation, drug discovery and development, and disease modelling. The most ambitious goal of the 3D bioprinting technology, however, is the realization of customized devices for clinical applications. In the cardiovascular field, fabrication of 3D bioprinted functional tissues greatly depends on the availability of biomimetic materials recapitulating native heart ECM and electrical conductivity properties. Although several specialized 3D bioprinters have been developed to fabricate various types of 3D heart tissues, such as myocardium and heart valves, bioprinting technology still faces many technical challenges. Some of the main efforts focus on obtaining high resolution distribution of cells, while minimizing their loss and maximizing cell–cell interactions. The ideal bioprinter should have resolution in the submicron range to allow bioprinting of a matrix with an orientation able to direct the alignment of cardiovascular cells [146]. Achieving the proper cellular composition is another issue in cardiovascular 3D bioprinting [62]. As emerged in different studies discussed in the manuscript, alteration of the composition of nonmyocyte cells, such as FBs and ECs, strongly influences the function, vascularization and vitality of the 3D bioprinted cardiac tissue.

Another crucial aspect to consider in 3D bioprinting relates the development of an ideal cardiac bioink, possessing appropriate stiffness and cell microenvironment, and controllable degradation rate. At the time of this writing, hydrogels are among the most used bioprintable materials for cardiovascular applications, as they provide good support for cells. Nevertheless, hydrogels do not contain specific ECM proteins for certain cell types, thus not fully reflecting the native environment of the heart. In addition, although hydrogels encapsulate and confine cells, they limit cell–cell interactions, and do not allow to obtain the same high cell density as native tissues. Another important shortcoming is the instability of hydrogels during the bioprinting process: Increasing hydrogel concentration improves its mechanical properties, but simultaneously limits biological activities [147]. Use of decellularized heart ECM as bioink material represents an alternative solution in 3D bioprinting, although further optimization is still needed. Besides the fact that decellularization is a tedious process, the native ECM loses its mechanical integrity and some of the biochemical properties when homogenized, and toxic residues can remain after the decellularization process. In addition, cells seeded on the decellularized ECM produce MMPs that rapid degrade the bioink [147]. Despite the advancements in scaffold-free techniques, limitations still exist also in the use of tissue spheroids for 3D bioprinting. Some are technical difficulties related to the printing process: The nozzle used to load tissue spheroids should be large enough to hold the spheroids inside and allow extrusion without clogging, since these structures do not have a homogeneous size and can be easily deformed or broken. Furthermore, necrosis can occur in the core of tissue spheroids, but this can be overcome by proper vascularization. Considering that each type of bioink has its own advantages and disadvantages over others, a future winning strategy could involve the combination of bioinks composed of multiple and complementary biomaterials. This in turn would require the development of hybrid systems integrating different but compatible bioprinting technologies. At present, the majority of the 3D bioprinting techniques require post-processing of the printed product through crosslinking. In most cases, this step complicates the workflow and leads to a potential deterioration in the quality of the final product [148]. For these reasons, there is an urgent need to improve material design or the printing technology itself for preventing the post-printing step.

As discussed in the manuscript, another challenge in 3D bioprinting is the achievement of early vascularization of tissue constructs, which is fundamental for exchange of materials and oxygen supply, and consequently for improving vascular integration with the host cardiovascular system. Supply of oxygen to 3D printed tissue constructs could be addressed by generation of oxygen biomaterials that, together with conductive elastomers, could be the future of 3D bioprinted cardiac tissues. In addition, the development of 3D printed bioreactors that can integrate electrical and mechanical stimulation could represent another potential area for generating functional heart tissues [149].Advancements in 3D bioprinting technologies will also be dependent on the capability to accelerate large-scale tissue constructs production in a rapid manner. At the moment, only extrusion-based bioprinting has the potential to produce volumetric human-scale tissues but this method lacks speed and resolution. A final challenge in this field is the conduction of long-term studies for evaluating cell viability after printing, cell phenotypic changes and their biological functionality, as many works published so far are short-term studies. Long-term studies could better assess the biological safety and fidelity of constructs before clinical trials. In this context, immunogenicity of the 3D bioprinted construct is an issue that has to be addressed as well.

A future area of research is undoubtedly 4D printing, which essentially combines a 3D printing technique with smart materials that can respond to external stimuli, thereby reshaping or changing their function over time [150]. To do this, 4D printing employs hydrogels responsive to, i.e., temperature, pH, light, electric or magnetic field, or shape-memory polymers [151]. The most intriguing aspect of 4D bioprinting is the possibility to create sophisticated high resolution dynamic and animated structures, otherwise inaccessible with the current static 3D bioprinting techniques. Nonetheless, an aspect that should be carefully considered in 4D printing is the presence of a stimulus that may have a negative effect on living cells. At the moment, 4D printing of cardiac tissue is still in its early stages, but we believe that it can give the opportunity to generate dynamic structures better mimicking the types of tissue structures in vivo. It is hoped that in a near future these important major challenges and regulatory issues could be addressed to enable the translation of the technology to personalized therapeutic and pharmaceutical applications.

Author Contributions: C.G., writing—original draft preparation and visualization; L.F., C.L., B.Z., J.C.C. and D.M. writing—review and editing. All authors have read and agreed to the published version of the manuscript.

Funding: This research received no external funding.

Acknowledgments: The authors would like to acknowledge networking support by the COST Action CA16122—Bioneca. D.M. work is co-financed by the Scientific Centre of Excellence for Basic, Clinical and Translational Neuroscience (project "Experimental and clinical research of hypoxic-ischemic damage in perinatal and adult brain"; GA KK01.1.1.01.0007 funded by the European Union through the European Regional Development Fund).

Conflicts of Interest: The authors declare no conflict of interest.

References

1. Paul, G.M.; Rezaienia, A.; Wen, P.; Condoor, S.; Parkar, N.; King, W.; Korakianitis, T. Medical Applications for 3D Printing: Recent Developments. *Mo Med.* **2018**, *115*, 75–81. [PubMed]
2. Ventola, C.L. Medical Applications for 3D Printing: Current and Projected Uses. *Pharm. Ther.* **2014**, *39*, 704–711.
3. Hull, C.W. Apparatus for Production of Three-Dimensional Objects by Stereolithography. U.S. Patent Application No. 4,575,330, 1986, 11 March 1986.
4. Zuniga, J.; Katsavelis, D.; Peck, J.; Stollberg, J.; Petrykowski, M.; Carson, A.; Fernandez, C. Cyborg beast: A low-cost 3d-printed prosthetic hand for children with upper-limb differences. *BMC Res. Notes* **2015**, *8*, 10. [CrossRef]
5. Vaccarezza, M.; Papa, V. 3D printing: A valuable resource in human anatomy education. *Anat. Sci. Int.* **2015**, *90*, 64–65. [CrossRef]
6. Lau, I.; Wong, Y.H.; Yeong, C.H.; Abdul Aziz, Y.F.; Md Sari, N.A.; Hashim, S.A.; Sun, Z. Quantitative and qualitative comparison of low- and high-cost 3D-printed heart models. *Quant. Imaging Med. Surg.* **2019**, *9*, 107–114. [CrossRef]
7. Lau, I.; Sun, Z. Three-dimensional printing in congenital heart disease: A systematic review. *J. Med. Radiat. Sci.* **2018**, *65*, 226–236. [CrossRef] [PubMed]
8. Garner, K.H.; Singla, D.K. 3D modeling: A future of cardiovascular medicine. *Can. J. Physiol. Pharmacol.* **2019**, *97*, 277–286. [CrossRef]
9. Valverde, I. Three-dimensional Printed Cardiac Models: Applications in the Field of Medical Education, Cardiovascular Surgery, and Structural Heart Interventions. *Rev. Esp. Cardiol. (Engl. Ed.)* **2017**, *70*, 282–291. [CrossRef] [PubMed]
10. Giannopoulos, A.A.; Mitsouras, D.; Yoo, S.J.; Liu, P.P.; Chatzizisis, Y.S.; Rybicki, F.J. Applications of 3D printing in cardiovascular diseases. *Nat. Rev. Cardiol.* **2016**, *13*, 701–718. [CrossRef]
11. Byrne, N.; Velasco Forte, M.; Tandon, A.; Valverde, I.; Hussain, T. A systematic review of image segmentation methodology, used in the additive manufacture of patient-specific 3D printed models of the cardiovascular system. *JRSM Cardiovasc. Dis.* **2016**, *5*, 2048004016645467. [CrossRef]
12. Vukicevic, M.; Mosadegh, B.; Min, J.K.; Little, S.H. Cardiac 3D Printing and its Future Directions. *JACC Cardiovasc. Imaging* **2017**, *10*, 171–184. [CrossRef] [PubMed]
13. Brenner, D.J.; Hall, E.J. Computed tomography—An increasing source of radiation exposure. *N. Engl. J. Med.* **2007**, *357*, 2277–2284. [CrossRef] [PubMed]
14. Bateman, M.G.; Durfee, W.K.; Iles, T.L.; Martin, C.M.; Liao, K.; Erdman, A.G.; Iaizzo, P.A. Cardiac patient-specific three-dimensional models as surgical planning tools. *Surgery* **2019**, *167*, 259–263. [CrossRef] [PubMed]
15. Shah, B.N. Echocardiography in the era of multimodality cardiovascular imaging. *BioMed Res. Int.* **2013**, *2013*, 310483. [CrossRef]
16. Wang, D.D.; Gheewala, N.; Shah, R.; Levin, D.; Myers, E.; Rollet, M.; O'Neill, W.W. Three-Dimensional Printing for Planning of Structural Heart Interventions. *Interv. Cardiol. Clin.* **2018**, *7*, 415–423. [CrossRef]

17. Otton, J.M.; Birbara, N.S.; Hussain, T.; Greil, G.; Foley, T.A.; Pather, N. 3D printing from cardiovascular CT: A practical guide and review. *Cardiovasc. Diagn. Ther.* **2017**, *7*, 507–526. [CrossRef]

18. Yoo, S.J.; Thabit, O.; Kim, E.K.; Ide, H.; Yim, D.; Dragulescu, A.; Seed, M.; Grosse-Wortmann, L.; van Arsdell, G. 3D printing in medicine of congenital heart diseases. *3D Print. Med.* **2015**, *2*, 3. [CrossRef]

19. Bücking, T.M.; Hill, E.R.; Robertson, J.L.; Maneas, E.; Plumb, A.A.; Nikitichev, D.I. From medical imaging data to 3D printed anatomical models. *PLoS ONE* **2017**, *12*, e0178540. [CrossRef]

20. Kerby, J.; Shukur, Z.N.; Shalhoub, J. The relationships between learning outcomes and methods of teaching anatomy as perceived by medical students. *Clin. Anat.* **2011**, *24*, 489–497. [CrossRef]

21. Li, K.H.C.; Kui, C.; Lee, E.K.M.; Ho, C.S.; Wong, S.H.; Wu, W.; Wong, W.T.; Voll, J.; Li, G.; Liu, T.; et al. *The Role of 3D Printing in Anatomy Education and Surgical Training: A Narrative Review*; MedEdPublisher: Dundee, UK, 2017.

22. Fasel, J.H.; Aguiar, D.; Kiss-Bodolay, D.; Montet, X.; Kalangos, A.; Stimec, B.V.; Ratib, O. Adapting anatomy teaching to surgical trends: A combination of classical dissection, medical imaging, and 3D-printing technologies. *Surg. Radiol. Anat.* **2016**, *38*, 361–367. [CrossRef]

23. Al-Ramahi, J.; Luo, H.; Fang, R.; Chou, A.; Jiang, J.; Kille, T. Development of an Innovative 3D Printed Rigid Bronchoscopy Training Model. *Ann. Otol. Rhinol. Laryngol.* **2016**, *125*, 965–969. [CrossRef] [PubMed]

24. Jonas, R.A. Training fellows in paediatric cardiac surgery. *Cardiol. Young* **2016**, *26*, 1474–1483. [CrossRef] [PubMed]

25. Shui, W.; Zhou, M.; Chen, S.; Pan, Z.; Deng, Q.; Yao, Y.; Pan, H.; He, T.; Wang, X. The production of digital and printed resources from multiple modalities using visualization and three-dimensional printing techniques. *Int. J. Comput. Assist. Radiol. Surg.* **2017**, *12*, 13–23. [CrossRef]

26. Ong, C.S.; Hibino, N. The use of 3D printing in cardiac surgery. *J. Thorac. Dis.* **2017**, *9*, 2301–2302. [CrossRef] [PubMed]

27. Lee, M.; Moharem-Elgamal, S.; Beckingham, R.; Hamilton, M.; Manghat, N.; Milano, E.G.; Bucciarelli-Ducci, C.; Caputo, M.; Biglino, G. Evaluating 3D-printed models of coronary anomalies: A survey among clinicians and researchers at a university hospital in the UK. *BMJ Open* **2019**, *9*, e025227. [CrossRef] [PubMed]

28. Nicholls, M. Three-dimensional imaging and printing in cardiology. *Eur. Heart J.* **2017**, *38*, 230–231. [CrossRef] [PubMed]

29. Fan, Y.; Yang, F.; Cheung, G.S.; Chan, A.K.; Wang, D.D.; Lam, Y.Y.; Chow, M.C.; Leong, M.C.; Kam, K.K.; So, K.C.; et al. Device Sizing Guided by Echocardiography-Based Three-Dimensional Printing Is Associated with Superior Outcome after Percutaneous Left Atrial Appendage Occlusion. *J. Am. Soc. Echocardiogr.* **2019**, *32*, 708–719.e701. [CrossRef]

30. Motwani, M.; Burley, O.; Luckie, M.; Cunnington, C.; Pisaniello, A.D.; Hasan, R.; Malik, I.; Fraser, D.G. 3D-printing assisted closure of paravalvular leak. *J. Cardiovasc. Comput. Tomogr.* **2019**. [CrossRef]

31. Vukicevic, M.; Puperi, D.S.; Jane Grande-Allen, K.; Little, S.H. 3D Printed Modeling of the Mitral Valve for Catheter-Based Structural Interventions. *Ann. Biomed. Eng.* **2017**, *45*, 508–519. [CrossRef]

32. Premyodhin, N.; Mandair, D.; Ferng, A.S.; Leach, T.S.; Palsma, R.P.; Albanna, M.Z.; Khalpey, Z.I. 3D printed mitral valve models: Affordable simulation for robotic mitral valve repair. *Interact. Cardiovasc. Thorac. Surg.* **2018**, *26*, 71–76. [CrossRef]

33. Engelhardt, S.; Sauerzapf, S.; Preim, B.; Karck, M.; Wolf, I.; De Simone, R. Flexible and comprehensive patient-specific mitral valve silicone models with chordae tendineae made from 3D-printable molds. *Int. J. Comput. Assist. Radiol. Surg.* **2019**, *14*, 1177–1186. [CrossRef]

34. Holzhey, D.M.; Seeburger, J.; Misfeld, M.; Borger, M.A.; Mohr, F.W. Learning minimally invasive mitral valve surgery: A cumulative sum sequential probability analysis of 3895 operations from a single high-volume center. *Circulation* **2013**, *128*, 483–491. [CrossRef]

35. Van Praet, K.M.; Stamm, C.; Sündermann, S.H.; Meyer, A.; Unbehaun, A.; Montagner, M.; Nazari Shafti, T.Z.; Jacobs, S.; Falk, V.; Kempfert, J. Minimally Invasive Surgical Mitral Valve Repair: State of the Art Review. *Interv. Cardiol.* **2018**, *13*, 14–19. [CrossRef]

36. Olivieri, L.J.; Krieger, A.; Loke, Y.H.; Nath, D.S.; Kim, P.C.; Sable, C.A. Three-dimensional printing of intracardiac defects from three-dimensional echocardiographic images: Feasibility and relative accuracy. *J. Am. Soc. Echocardiogr.* **2015**, *28*, 392–397. [CrossRef] [PubMed]

37. Farooqi, K.M.; Nielsen, J.C.; Uppu, S.C.; Srivastava, S.; Parness, I.A.; Sanz, J.; Nguyen, K. Use of 3-dimensional printing to demonstrate complex intracardiac relationships in double-outlet right ventricle for surgical planning. *Circ. Cardiovasc. Imaging* **2015**, *8*. [CrossRef] [PubMed]

38. Garekar, S.; Bharati, A.; Chokhandre, M.; Mali, S.; Trivedi, B.; Changela, V.P.; Solanki, N.; Gaikwad, S.; Agarwal, V. Clinical Application and Multidisciplinary Assessment of Three Dimensional Printing in Double Outlet Right Ventricle With Remote Ventricular Septal Defect. *World J. Pediatr. Congenit. Heart Surg.* **2016**, *7*, 344–350. [CrossRef]

39. Anwar, S.; Singh, G.K.; Varughese, J.; Nguyen, H.; Billadello, J.J.; Sheybani, E.F.; Woodard, P.K.; Manning, P.; Eghtesady, P. 3D Printing in Complex Congenital Heart Disease: Across a Spectrum of Age, Pathology, and Imaging Techniques. *JACC Cardiovasc. Imaging* **2017**, *10*, 953–956. [CrossRef] [PubMed]

40. Cui, X.; Boland, T.; D'Lima, D.D.; Lotz, M.K. Thermal inkjet printing in tissue engineering and regenerative medicine. *Recent Pat. Drug Deliv. Formul.* **2012**, *6*, 149–155. [CrossRef]

41. Langer, R.; Vacanti, J.P. Tissue engineering. *Science* **1993**, *260*, 920–926. [CrossRef] [PubMed]

42. Giraud, M.N.; Guex, A.G.; Tevaearai, H.T. Cell therapies for heart function recovery: Focus on myocardial tissue engineering and nanotechnologies. *Cardiol. Res. Pract.* **2012**, *2012*, 971614. [CrossRef]

43. Akhyari, P.; Kamiya, H.; Haverich, A.; Karck, M.; Lichtenberg, A. Myocardial tissue engineering: The extracellular matrix. *Eur. J. Cardiothorac. Surg.* **2008**, *34*, 229–241. [CrossRef] [PubMed]

44. Steinhoff, G.; Stock, U.; Karim, N.; Mertsching, H.; Timke, A.; Meliss, R.R.; Pethig, K.; Haverich, A.; Bader, A. Tissue engineering of pulmonary heart valves on allogenic acellular matrix conduits: In Vivo restoration of valve tissue. *Circulation* **2000**, *102*, iii-50–iii-55. [CrossRef] [PubMed]

45. Hamilton, N.; Bullock, A.J.; Macneil, S.; Janes, S.M.; Birchall, M. Tissue engineering airway mucosa: A systematic review. *Laryngoscope* **2014**, *124*, 961–968. [CrossRef]

46. Klebe, R.J. Cytoscribing: A method for micropositioning cells and the construction of two- and three-dimensional synthetic tissues. *Exp. Cell Res.* **1988**, *179*, 362–373. [CrossRef]

47. Murphy, S.V.; Atala, A. 3D bioprinting of tissues and organs. *Nat. Biotechnol.* **2014**, *32*, 773–785. [CrossRef] [PubMed]

48. Cui, H.; Miao, S.; Esworthy, T.; Zhou, X.; Lee, S.J.; Liu, C.; Yu, Z.X.; Fisher, J.P.; Mohiuddin, M.; Zhang, L.G. 3D bioprinting for cardiovascular regeneration and pharmacology. *Adv. Drug Deliv. Rev.* **2018**, *132*, 252–269. [CrossRef]

49. Ozbolat, I.T.; Yu, Y. Bioprinting toward organ fabrication: Challenges and future trends. *IEEE Trans. Biomed. Eng.* **2013**, *60*, 691–699. [CrossRef]

50. Visscher, D.O.; Farré-Guasch, E.; Helder, M.N.; Gibbs, S.; Forouzanfar, T.; van Zuijlen, P.P.; Wolff, J. Advances in Bioprinting Technologies for Craniofacial Reconstruction. *Trends Biotechnol.* **2016**, *34*, 700–710. [CrossRef]

51. Chung, J.H.Y.; Naficy, S.; Yue, Z.; Kapsa, R.; Quigley, A.; Moulton, S.E.; Wallace, G.G. Bio-ink properties and printability for extrusion printing living cells. *Biomater. Sci.* **2013**, *1*, 763–773. [CrossRef]

52. Gopinathan, J.; Noh, I. Recent trends in bioinks for 3D printing. *Biomater. Res.* **2018**, *22*, 11. [CrossRef]

53. Ji, S.; Guvendiren, M. Recent Advances in Bioink Design for 3D Bioprinting of Tissues and Organs. *Front. Bioeng. Biotechnol.* **2017**, *5*, 23. [CrossRef] [PubMed]

54. Donderwinkel, I.; van Hest, J.C.M.; Cameron, N.R. Bio-inks for 3D bioprinting: Recent advances and future prospects. *Polym. Chem.* **2017**, *8*, 4451–4471. [CrossRef]

55. Jang, J.; Park, H.J.; Kim, S.W.; Kim, H.; Park, J.Y.; Na, S.J.; Kim, H.J.; Park, M.N.; Choi, S.H.; Park, S.H.; et al. 3D printed complex tissue construct using stem cell-laden decellularized extracellular matrix bioinks for cardiac repair. *Biomaterials* **2017**, *112*, 264–274. [CrossRef]

56. Jung, J.P.; Bhuiyan, D.B.; Ogle, B.M. Solid organ fabrication: Comparison of decellularization to 3D bioprinting. *Biomater. Res.* **2016**, *20*, 27. [CrossRef]

57. Song, J.J.; Ott, H.C. Organ engineering based on decellularized matrix scaffolds. *Trends Mol. Med.* **2011**, *17*, 424–432. [CrossRef] [PubMed]

58. Moroni, F.; Mirabella, T. Decellularized matrices for cardiovascular tissue engineering. *Am. J. Stem Cells* **2014**, *3*, 1–20. [PubMed]

59. Moldovan, N.I.; Hibino, N.; Nakayama, K. Principles of the Kenzan Method for Robotic Cell Spheroid-Based Three-Dimensional Bioprinting. *Tissue Eng. Part B Rev.* **2017**, *23*, 237–244. [CrossRef] [PubMed]

60. Moldovan, N.I. Progress in scaffold-free bioprinting for cardiovascular medicine. *J. Cell. Mol. Med.* **2018**, *22*, 2964–2969. [CrossRef] [PubMed]

61. Peng, W.; Unutmaz, D.; Ozbolat, I.T. Bioprinting towards Physiologically Relevant Tissue Models for Pharmaceutics. *Trends Biotechnol.* **2016**, *34*, 722–732. [CrossRef] [PubMed]

62. Ong, C.S.; Nam, L.; Ong, K.; Krishnan, A.; Huang, C.Y.; Fukunishi, T.; Hibino, N. 3D and 4D Bioprinting of the Myocardium: Current Approaches, Challenges, and Future Prospects. *BioMed Res. Int.* **2018**, *2018*, 6497242. [CrossRef] [PubMed]

63. Laflamme, M.A.; Murry, C.E. Heart regeneration. *Nature* **2011**, *473*, 326–335. [CrossRef] [PubMed]

64. Mathur, A.; Ma, Z.; Loskill, P.; Jeeawoody, S.; Healy, K.E. In Vitro cardiac tissue models: Current status and future prospects. *Adv. Drug Deliv. Rev.* **2016**, *96*, 203–213. [CrossRef] [PubMed]

65. Olivetti, G.; Melissari, M.; Capasso, J.M.; Anversa, P. Cardiomyopathy of the aging human heart. Myocyte loss and reactive cellular hypertrophy. *Circ. Res.* **1991**, *68*, 1560–1568. [CrossRef] [PubMed]

66. Gaetani, R.; Doevendans, P.A.; Metz, C.H.; Alblas, J.; Messina, E.; Giacomello, A.; Sluijter, J.P. Cardiac tissue engineering using tissue printing technology and human cardiac progenitor cells. *Biomaterials* **2012**, *33*, 1782–1790. [CrossRef]

67. Eschenhagen, T.; Fink, C.; Remmers, U.; Scholz, H.; Wattchow, J.; Weil, J.; Zimmermann, W.; Dohmen, H.H.; Schäfer, H.; Bishopric, N.; et al. Three-dimensional reconstitution of embryonic cardiomyocytes in a collagen matrix: A new heart muscle model system. *FASEB J.* **1997**, *11*, 683–694. [CrossRef]

68. Chachques, J.C.; Trainini, J.C.; Lago, N.; Cortes-Morichetti, M.; Schussler, O.; Carpentier, A. Myocardial Assistance by Grafting a New Bioartificial Upgraded Myocardium (MAGNUM trial): Clinical feasibility study. *Ann. Thorac. Surg.* **2008**, *85*, 901–908. [CrossRef]

69. Chachques, J.C.; Pradas, M.M.; Bayes-Genis, A.; Semino, C. Creating the bioartificial myocardium for cardiac repair: Challenges and clinical targets. *Expert Rev. Cardiovasc. Ther.* **2013**, *11*, 1701–1711. [CrossRef]

70. Madonna, R.; Van Laake, L.W.; Botker, H.E.; Davidson, S.M.; De Caterina, R.; Engel, F.B.; Eschenhagen, T.; Fernandez-Aviles, F.; Hausenloy, D.J.; Hulot, J.S.; et al. ESC Working Group on Cellular Biology of the Heart: Position paper for Cardiovascular Research: Tissue engineering strategies combined with cell therapies for cardiac repair in ischaemic heart disease and heart failure. *Cardiovasc. Res.* **2019**, *115*, 488–500. [CrossRef]

71. Chachques, J.C.; Lila, N.; Soler-Botija, C.; Martinez-Ramos, C.; Valles, A.; Autret, G.; Perier, M.C.; Mirochnik, N.; Monleon-Pradas, M.; Bayes-Genis, A.; et al. Elastomeric cardiopatch scaffold for myocardial repair and ventricular support. *Eur. J. Cardiothorac. Surg.* **2019**. [CrossRef]

72. Gaetani, R.; Barile, L.; Forte, E.; Chimenti, I.; Ionta, V.; Di Consiglio, A.; Miraldi, F.; Frati, G.; Messina, E.; Giacomello, A. New perspectives to repair a broken heart. *Cardiovasc. Hematol. Agents Med. Chem.* **2009**, *7*, 91–107. [CrossRef]

73. Gaetani, R.; Feyen, D.A.; Verhage, V.; Slaats, R.; Messina, E.; Christman, K.L.; Giacomello, A.; Doevendans, P.A.; Sluijter, J.P. Epicardial application of cardiac progenitor cells in a 3D-printed gelatin/hyaluronic acid patch preserves cardiac function after myocardial infarction. *Biomaterials* **2015**, *61*, 339–348. [CrossRef] [PubMed]

74. Zhu, K.; Shin, S.R.; van Kempen, T.; Li, Y.C.; Ponraj, V.; Nasajpour, A.; Mandla, S.; Hu, N.; Liu, X.; Leijten, J.; et al. Gold Nanocomposite Bioink for Printing 3D Cardiac Constructs. *Adv. Funct. Mater.* **2017**, *27*. [CrossRef] [PubMed]

75. Yue, K.; Trujillo-de Santiago, G.; Alvarez, M.M.; Tamayol, A.; Annabi, N.; Khademhosseini, A. Synthesis, properties, and biomedical applications of gelatin methacryloyl (GelMA) hydrogels. *Biomaterials* **2015**, *73*, 254–271. [CrossRef] [PubMed]

76. Colosi, C.; Shin, S.R.; Manoharan, V.; Massa, S.; Costantini, M.; Barbetta, A.; Dokmeci, M.R.; Dentini, M.; Khademhosseini, A. Microfluidic Bioprinting of Heterogeneous 3D Tissue Constructs Using Low-Viscosity Bioink. *Adv. Mater.* **2016**, *28*, 677–684. [CrossRef] [PubMed]

77. Ashammakhi, N.; Ahadian, S.; Xu, C.; Montazerian, H.; Ko, H.; Nasiri, R.; Barros, N.; Khademhosseini, A. Bioinks and bioprinting technologies to make heterogeneous and biomimetic tissue constructs. *Mater. Today Bio* **2019**, *1*, 100008. [CrossRef] [PubMed]

78. Gaebel, R.; Ma, N.; Liu, J.; Guan, J.; Koch, L.; Klopsch, C.; Gruene, M.; Toelk, A.; Wang, W.; Mark, P.; et al. Patterning human stem cells and endothelial cells with laser printing for cardiac regeneration. *Biomaterials* **2011**, *32*, 9218–9230. [CrossRef]

79. Koike, N.; Fukumura, D.; Gralla, O.; Au, P.; Schechner, J.S.; Jain, R.K. Tissue engineering: Creation of long-lasting blood vessels. *Nature* **2004**, *428*, 138–139. [CrossRef]

80. Pati, F.; Jang, J.; Ha, D.H.; Won Kim, S.; Rhie, J.W.; Shim, J.H.; Kim, D.H.; Cho, D.W. Printing three-dimensional tissue analogues with decellularized extracellular matrix bioink. *Nat. Commun.* **2014**, *5*, 3935. [CrossRef]

81. Pati, F.; Cho, D.W. Bioprinting of 3D Tissue Models Using Decellularized Extracellular Matrix Bioink. *Methods Mol. Biol.* **2017**, *1612*, 381–390. [CrossRef]

82. Park, S.J.; Kim, R.Y.; Park, B.W.; Lee, S.; Choi, S.W.; Park, J.H.; Choi, J.J.; Kim, S.W.; Jang, J.; Cho, D.W.; et al. Dual stem cell therapy synergistically improves cardiac function and vascular regeneration following myocardial infarction. *Nat. Commun.* **2019**, *10*, 3123. [CrossRef] [PubMed]

83. Bejleri, D.; Streeter, B.W.; Nachlas, A.L.Y.; Brown, M.E.; Gaetani, R.; Christman, K.L.; Davis, M.E. A Bioprinted Cardiac Patch Composed of Cardiac-Specific Extracellular Matrix and Progenitor Cells for Heart Repair. *Adv. Healthc. Mater.* **2018**, *7*, e1800672. [CrossRef]

84. Das, S.; Kim, S.W.; Choi, Y.J.; Lee, S.; Lee, S.H.; Kong, J.S.; Park, H.J.; Cho, D.W.; Jang, J. Decellularized extracellular matrix bioinks and the external stimuli to enhance cardiac tissue development in vitro. *Acta Biomater.* **2019**, *95*, 188–200. [CrossRef]

85. Noor, N.; Shapira, A.; Edri, R.; Gal, I.; Wertheim, L.; Dvir, T. 3D Printing of Personalized Thick and Perfusable Cardiac Patches and Hearts. *Adv. Sci. (Weinh)* **2019**, *6*, 1900344. [CrossRef] [PubMed]

86. Atmanli, A.; Domian, I.J. Generation of aligned functional myocardial tissue through microcontact printing. *J. Vis. Exp.* **2013**, e50288. [CrossRef] [PubMed]

87. Ong, C.S.; Fukunishi, T.; Nashed, A.; Blazeski, A.; Zhang, H.; Hardy, S.; DiSilvestre, D.; Vricella, L.; Conte, J.; Tung, L.; et al. Creation of Cardiac Tissue Exhibiting Mechanical Integration of Spheroids Using 3D Bioprinting. *J. Vis. Exp.* **2017**. [CrossRef]

88. Ong, C.S.; Fukunishi, T.; Zhang, H.; Huang, C.Y.; Nashed, A.; Blazeski, A.; DiSilvestre, D.; Vricella, L.; Conte, J.; Tung, L.; et al. Biomaterial-Free Three-Dimensional Bioprinting of Cardiac Tissue using Human Induced Pluripotent Stem Cell Derived Cardiomyocytes. *Sci. Rep.* **2017**, *7*, 4566. [CrossRef] [PubMed]

89. Farrar, E.J.; Butcher, J.T. Valvular heart diseases in the developing world: Developmental biology takes center stage. *J. Heart Valve Dis.* **2012**, *21*, 234–240.

90. Vashistha, R.; Kumar, P.; Dangi, A.K.; Sharma, N.; Chhabra, D.; Shukla, P. Quest for cardiovascular interventions: Precise modeling and 3D printing of heart valves. *J. Biol. Eng.* **2019**, *13*, 12. [CrossRef]

91. Aikawa, E.; Libby, P. A Rock and a Hard Place: Chiseling Away at the Multiple Mechanisms of Aortic Stenosis. *Circulation* **2017**, *135*, 1951–1955. [CrossRef]

92. Butcher, J.T.; Mahler, G.J.; Hockaday, L.A. Aortic valve disease and treatment: The need for naturally engineered solutions. *Adv. Drug Deliv. Rev.* **2011**, *63*, 242–268. [CrossRef]

93. Harris, C.; Croce, B.; Cao, C. Tissue and mechanical heart valves. *Ann. Cardiothorac. Surg.* **2015**, *4*, 399. [CrossRef] [PubMed]

94. Vijayavenkataraman, S.; Yan, W.C.; Lu, W.F.; Wang, C.H.; Fuh, J.Y.H. 3D bioprinting of tissues and organs for regenerative medicine. *Adv. Drug Deliv. Rev.* **2018**, *132*, 296–332. [CrossRef] [PubMed]

95. Flanagan, T.C.; Cornelissen, C.; Koch, S.; Tschoeke, B.; Sachweh, J.S.; Schmitz-Rode, T.; Jockenhoevel, S. The In Vitro development of autologous fibrin-based tissue-engineered heart valves through optimised dynamic conditioning. *Biomaterials* **2007**, *28*, 3388–3397. [CrossRef] [PubMed]

96. Jia, X.; Kiick, K.L. Hybrid multicomponent hydrogels for tissue engineering. *Macromol. Biosci.* **2009**, *9*, 140–156. [CrossRef]

97. Duan, B.; Hockaday, L.A.; Kang, K.H.; Butcher, J.T. 3D bioprinting of heterogeneous aortic valve conduits with alginate/gelatin hydrogels. *J. Biomed. Mater. Res. Part A* **2013**, *101*, 1255–1264. [CrossRef]

98. Mendez, M.G.; Kojima, S.; Goldman, R.D. Vimentin induces changes in cell shape, motility, and adhesion during the epithelial to mesenchymal transition. *FASEB J.* **2010**, *24*, 1838–1851. [CrossRef]

99. Lu, S.H.; Lin, A.T.; Chen, K.K.; Chiang, H.S.; Chang, L.S. Characterization of smooth muscle differentiation of purified human skeletal muscle-derived cells. *J. Cell. Mol. Med.* **2011**, *15*, 587–592. [CrossRef]

100. Duan, B.; Kapetanovic, E.; Hockaday, L.A.; Butcher, J.T. Three-dimensional printed trileaflet valve conduits using biological hydrogels and human valve interstitial cells. *Acta Biomater.* **2014**, *10*, 1836–1846. [CrossRef] [PubMed]

101. Hockaday, L.A.; Kang, K.H.; Colangelo, N.W.; Cheung, P.Y.; Duan, B.; Malone, E.; Wu, J.; Girardi, L.N.; Bonassar, L.J.; Lipson, H.; et al. Rapid 3D printing of anatomically accurate and mechanically heterogeneous aortic valve hydrogel scaffolds. *Biofabrication* **2012**, *4*, 035005. [CrossRef]

102. Kang, L.H.; Armstrong, P.A.; Lee, L.J.; Duan, B.; Kang, K.H.; Butcher, J.T. Optimizing Photo-Encapsulation Viability of Heart Valve Cell Types in 3D Printable Composite Hydrogels. *Ann. Biomed. Eng.* **2017**, *45*, 360–377. [CrossRef]

103. Fairbanks, B.D.; Schwartz, M.P.; Bowman, C.N.; Anseth, K.S. Photoinitiated polymerization of PEG-diacrylate with lithium phenyl-2,4,6-trimethylbenzoylphosphinate: Polymerization rate and cytocompatibility. *Biomaterials* **2009**, *30*, 6702–6707. [CrossRef] [PubMed]

104. Duan, B.; Hockaday, L.A.; Kapetanovic, E.; Kang, K.H.; Butcher, J.T. Stiffness and adhesivity control aortic valve interstitial cell behavior within hyaluronic acid based hydrogels. *Acta Biomater.* **2013**, *9*, 7640–7650. [CrossRef] [PubMed]

105. Rouillard, A.D.; Berglund, C.M.; Lee, J.Y.; Polacheck, W.J.; Tsui, Y.; Bonassar, L.J.; Kirby, B.J. Methods for photocrosslinking alginate hydrogel scaffolds with high cell viability. *Tissue Eng. Part C Methods* **2011**, *17*, 173–179. [CrossRef] [PubMed]

106. Alić, I.; Kosi, N.; Kapuralin, K.; Gorup, D.; Gajović, S.; Pochet, R.; Mitrečić, D. Neural stem cells from mouse strain Thy1 YFP-16 are a valuable tool to monitor and evaluate neuronal differentiation and morphology. *Neurosci. Lett.* **2016**, *634*, 32–41. [CrossRef]

107. Konig, N.; Trolle, C.; Kapuralin, K.; Adameyko, I.; Mitrecic, D.; Aldskogius, H.; Shortland, P.J.; Kozlova, E.N. Murine neural crest stem cells and embryonic stem cell-derived neuron precursors survive and differentiate after transplantation in a model of dorsal root avulsion. *J. Tissue Eng. Regen Med.* **2017**, *11*, 129–137. [CrossRef]

108. Mitrecic, D.; Nicaise, C.; Klimaschewski, L.; Gajovic, S.; Bohl, D.; Pochet, R. Genetically modified stem cells for the treatment of neurological diseases. *Front. Biosci. (Elite Ed.)* **2012**, *4*, 1170–1181. [CrossRef]

109. Van der Valk, D.C.; van der Ven, C.F.T.; Blaser, M.C.; Grolman, J.M.; Wu, P.J.; Fenton, O.S.; Lee, L.H.; Tibbitt, M.W.; Andresen, J.L.; Wen, J.R.; et al. Engineering a 3D-Bioprinted Model of Human Heart Valve Disease Using Nanoindentation-Based Biomechanics. *Nanomaterials* **2018**, *8*, 296. [CrossRef]

110. Peng, W.; Datta, P.; Ayan, B.; Ozbolat, V.; Sosnoski, D.; Ozbolat, I.T. 3D bioprinting for drug discovery and development in pharmaceutics. *Acta Biomater.* **2017**, *57*, 26–46. [CrossRef]

111. Knowlton, S.; Tasoglu, S. A Bioprinted Liver-on-a-Chip for Drug Screening Applications. *Trends Biotechnol.* **2016**, *34*, 681–682. [CrossRef]

112. Wrzesinski, K.; Fey, S.J. From 2D to 3D—A New Dimension for Modelling the Effect of Natural Products on Human Tissue. *Curr. Pharm. Des.* **2015**, *21*, 5605–5616. [CrossRef]

113. Vanderburgh, J.; Sterling, J.A.; Guelcher, S.A. 3D Printing of Tissue Engineered Constructs for In Vitro Modeling of Disease Progression and Drug Screening. *Ann. Biomed. Eng.* **2017**, *45*, 164–179. [CrossRef] [PubMed]

114. Mozaffarian, D.; Benjamin, E.J.; Go, A.S.; Arnett, D.K.; Blaha, M.J.; Cushman, M.; de Ferranti, S.; Després, J.P.; Fullerton, H.J.; Howard, V.J.; et al. Heart disease and stroke statistics—2015 update: A report from the American Heart Association. *Circulation* **2015**, *131*, e29–e322. [CrossRef] [PubMed]

115. Ma, X.; Liu, J.; Zhu, W.; Tang, M.; Lawrence, N.; Yu, C.; Gou, M.; Chen, S. 3D bioprinting of functional tissue models for personalized drug screening and in vitro disease modeling. *Adv. Drug Deliv. Rev.* **2018**, *132*, 235–251. [CrossRef]

116. Zhang, Y.S.; Arneri, A.; Bersini, S.; Shin, S.R.; Zhu, K.; Goli-Malekabadi, Z.; Aleman, J.; Colosi, C.; Busignani, F.; Dell'Erba, V.; et al. Bioprinting 3D microfibrous scaffolds for engineering endothelialized myocardium and heart-on-a-chip. *Biomaterials* **2016**, *110*, 45–59. [CrossRef] [PubMed]

117. Lind, J.U.; Busbee, T.A.; Valentine, A.D.; Pasqualini, F.S.; Yuan, H.; Yadid, M.; Park, S.J.; Kotikian, A.; Nesmith, A.P.; Campbell, P.H.; et al. Instrumented cardiac microphysiological devices via multimaterial three-dimensional printing. *Nat. Mater.* **2017**, *16*, 303–308. [CrossRef] [PubMed]

118. Wang, Z.; Lee, S.J.; Cheng, H.J.; Yoo, J.J.; Atala, A. 3D bioprinted functional and contractile cardiac tissue constructs. *Acta Biomater.* **2018**, *70*, 48–56. [CrossRef]

119. Gu, Y.; Chen, X.; Lee, J.H.; Monteiro, D.A.; Wang, H.; Lee, W.Y. Inkjet printed antibiotic- and calcium-eluting bioresorbable nanocomposite micropatterns for orthopedic implants. *Acta Biomater.* **2012**, *8*, 424–431. [CrossRef] [PubMed]

120. Tarcha, P.J.; Verlee, D.; Hui, H.W.; Setesak, J.; Antohe, B.; Radulescu, D.; Wallace, D. The application of ink-jet technology for the coating and loading of drug-eluting stents. *Ann. Biomed. Eng.* **2007**, *35*, 1791–1799. [CrossRef]

121. Sandler, N.; Salmela, I.; Fallarero, A.; Rosling, A.; Khajeheian, M.; Kolakovic, R.; Genina, N.; Nyman, J.; Vuorela, P. Towards fabrication of 3D printed medical devices to prevent biofilm formation. *Int. J. Pharm.* **2014**, *459*, 62–64. [CrossRef]

122. Goyanes, A.; Wang, J.; Buanz, A.; Martínez-Pacheco, R.; Telford, R.; Gaisford, S.; Basit, A.W. 3D Printing of Medicines: Engineering Novel Oral Devices with Unique Design and Drug Release Characteristics. *Mol. Pharm.* **2015**, *12*, 4077–4084. [CrossRef]

123. Kalejs, M.; von Segesser, L.K. Rapid prototyping of compliant human aortic roots for assessment of valved stents. *Interact. Cardiovasc. Thorac. Surg.* **2009**, *8*, 182–186. [CrossRef] [PubMed]

124. Biglino, G.; Verschueren, P.; Zegels, R.; Taylor, A.M.; Schievano, S. Rapid prototyping compliant arterial phantoms for in-vitro studies and device testing. *J. Cardiovasc. Magn. Reson.* **2013**, *15*, 2. [CrossRef] [PubMed]

125. Mashari, A.; Knio, Z.; Jeganathan, J.; Montealegre-Gallegos, M.; Yeh, L.; Amador, Y.; Matyal, R.; Saraf, R.; Khabbaz, K.; Mahmood, F. Hemodynamic Testing of Patient-Specific Mitral Valves Using a Pulse Duplicator: A Clinical Application of Three-Dimensional Printing. *J. Cardiothorac. Vasc. Anesth.* **2016**, *30*, 1278–1285. [CrossRef] [PubMed]

126. Paulsen, M.J.; Kasinpila, P.; Imbrie-Moore, A.M.; Wang, H.; Hironaka, C.E.; Koyano, T.K.; Fong, R.; Chiu, P.; Goldstone, A.B.; Steele, A.N.; et al. Modeling conduit choice for valve-sparing aortic root replacement on biomechanics with a 3-dimensional-printed heart simulator. *J. Thorac. Cardiovasc. Surg.* **2019**, *158*, 392–403. [CrossRef] [PubMed]

127. Sun, Y.; Zhang, X.; Li, W.; Di, Y.; Xing, Q.; Cao, Q. 3D printing and biocompatibility study of a new biodegradable occluder for cardiac defect. *J. Cardiol.* **2019**, *74*, 182–188. [CrossRef] [PubMed]

128. 3D Printing Market by Offering (Printer, Material, Software, Service), Process (Binder Jetting, Direct Energy Deposition, Material Extrusion, Material Jetting, Powder Bed Fusion), Application, Vertical, Technology, and Geography—Global Forecast to 2024. Available online: https://www.marketsandmarkets.com/Market-Reports/3d-printing-market-1276.html (accessed on 25 February 2020).

129. 3D Bioprinting Market by Component (3D Bioprinters (Microextrusion, Inkjet, Laser), Bioink (Natural, Synthetic, Hybrid)), Material (Hydrogel, Living, Cells), Application (Skin, Drug Research), End User (Biopharma, Academia)—Global Forecast to 2024. Available online: https://www.marketsandmarkets.com/Market-Reports/3d-bioprinting-market-170201787 (accessed on 21 February 2020).

130. Sweet, B.V.; Schwemm, A.K.; Parsons, D.M. Review of the processes for FDA oversight of drugs, medical devices, and combination products. *J. Manag. Care Pharm.* **2011**, *17*, 40–50. [CrossRef]

131. U.S. Food and Drug Administration. Available online: https://www.fda.gov/medical-devices/overview-device-regulation/classify-your-medical-device (accessed on 20 February 2020).

132. Van Norman, G.A. Drugs, Devices, and the FDA: Part 2: An Overview of Approval Processes: FDA Approval of Medical Devices. *JACC Basic Transl. Sci.* **2016**, *1*, 277–287. [CrossRef]

133. Hourd, P.; Medcalf, N.; Segal, J.; Williams, D.J. A 3D bioprinting exemplar of the consequences of the regulatory requirements on customized processes. *Regen. Med.* **2015**, *10*, 863–883. [CrossRef] [PubMed]

134. Parvizi, N.; Woods, K. Regulation of medicines and medical devices: Contrasts and similarities. *Clin. Med.* **2014**, *14*, 6–12. [CrossRef] [PubMed]

135. Van Norman, G.A. Drugs and Devices: Comparison of European and U.S. Approval Processes. *JACC Basic Transl. Sci.* **2016**, *1*, 399–412. [CrossRef] [PubMed]

136. Classification of Medical Devices and Their Route to CE Markin. Available online: https://support.ce-check.eu/hc/en-us/articles/360008712879-Classification-Of-Medical-Devices-And-Their-Routes-To-CE-Marking (accessed on 25 February 2020).

137. Technical Considerations for Additive Manufactured Medical Devices—Guidance for Industry and Food and Drug Administration Staff. Available online: https://www.fda.gov/regulatory-information/search-fda-guidance-documents/technical-considerations-additive-manufactured-medical-devices-guidance-industry-and-food-and-drug (accessed on 20 February 2020).

138. Ricles, L.M.; Coburn, J.C.; Di Prima, M.; Oh, S.S. Regulating 3D-printed medical products. *Sci. Transl. Med.* **2018**, *10*. [CrossRef] [PubMed]

139. Rahman, Z.; Barakh Ali, S.F.; Ozkan, T.; Charoo, N.A.; Reddy, I.K.; Khan, M.A. Additive Manufacturing with 3D Printing: Progress from Bench to Bedside. *AAPS J.* **2018**, *20*, 101. [CrossRef] [PubMed]

140. Horst, A.; McDonald, F.; Hutmacher, D.W. A clarion call for understanding regulatory processes for additive manufacturing in the health sector. *Expert Rev. Med. Devices* **2019**, *16*, 405–412. [CrossRef] [PubMed]

141. Iglesias-López, C.; Agustí, A.; Obach, M.; Vallano, A. Regulatory Framework for Advanced Therapy Medicinal Products in Europe and United States. *Front. Pharmacol.* **2019**, *10*, 921. [CrossRef]

142. Early Development Considerations for Innovative Combination Products. Available online: https://www.fda.gov/regulatory-information/search-fda-guidance-documents/early-development-considerations-innovative-combination-products (accessed on 25 February 2020).

143. Gilbert, F.; O'Connell, C.D.; Mladenovska, T.; Dodds, S. Print Me an Organ? Ethical and Regulatory Issues Emerging from 3D Bioprinting in Medicine. *Sci. Eng. Ethics* **2018**, *24*, 73–91. [CrossRef]

144. Kritikos, M. *3D Bio-Printing for Medical and Enhancement Purposes: Legal and Ethical Aspects*; EPRS, Ed.; EPRS: Brussels, Belgium, 2018.

145. Meier, L.M.; Meineri, M.; Qua Hiansen, J.; Horlick, E.M. Structural and congenital heart disease interventions: The role of three-dimensional printing. *Neth. Heart J.* **2017**, *25*, 65–75. [CrossRef]

146. Duan, B. State-of-the-Art Review of 3D Bioprinting for Cardiovascular Tissue Engineering. *Ann. Biomed. Eng.* **2017**, *45*, 195–209. [CrossRef]

147. Hospodiuk, M.; Dey, M.; Sosnoski, D.; Ozbolat, I.T. The bioink: A comprehensive review on bioprintable materials. *Biotechnol. Adv.* **2017**, *35*, 217–239. [CrossRef]

148. Liaw, C.Y.; Guvendiren, M. Current and emerging applications of 3D printing in medicine. *Biofabrication* **2017**, *9*, 024102. [CrossRef]

149. Qasim, M.; Haq, F.; Kang, M.H.; Kim, J.H. 3D printing approaches for cardiac tissue engineering and role of immune modulation in tissue regeneration. *Int. J. Nanomed.* **2019**, *14*, 1311–1333. [CrossRef]

150. Yang, G.H.; Yeo, M.; Koo, Y.W.; Kim, G.H. 4D Bioprinting: Technological Advances in Biofabrication. *Macromol. Biosci.* **2019**, *19*, e1800441. [CrossRef] [PubMed]

151. Kirillova, A.; Maxson, R.; Stoychev, G.; Gomillion, C.T.; Ionov, L. 4D Biofabrication Using Shape-Morphing Hydrogels. *Adv. Mater.* **2017**, *29*. [CrossRef] [PubMed]

Review

Molecular Mechanisms of Cardiac Remodeling and Regeneration in Physical Exercise

Dominik Schüttler [1,2,3], **Sebastian Clauss** [1,2,3], **Ludwig T. Weckbach** [1,2,3,4] and **Stefan Brunner** [1,*]

[1]   Department of Medicine I, University Hospital Munich, Campus Grosshadern and Innenstadt, Ludwig-Maximilians University Munich (LMU), 81377 Munich, Germany; Dominik.Schuettler@med.uni-muenchen.de (D.S.); Sebastian.Clauss@med.uni-muenchen.de (S.C.); Ludwig.Wekcbach@med.uni-muenchen.de (L.T.W.)

[2]   DZHK (German Centre for Cardiovascular Research), Partner Site Munich, Munich Heart Alliance (MHA), 80336 Munich, Germany

[3]   Walter Brendel Centre of Experimental Medicine, Ludwig-Maximilians University Munich (LMU), 81377 Munich, Germany

[4]   Institute of Cardiovascular Physiology and Pathophysiology, Biomedical Center, Ludwig-Maximilians-University Munich, 82152 Planegg-Martinsried, Germany

*   Correspondence: stefan.brunner@med.uni-muenchen.de; Tel.: +49-89-4400-52248; Fax: +49-89-4400-52152

Received: 2 September 2019; Accepted: 19 September 2019; Published: 23 September 2019

Abstract: Regular physical activity with aerobic and muscle-strengthening training protects against the occurrence and progression of cardiovascular disease and can improve cardiac function in heart failure patients. In the past decade significant advances have been made in identifying mechanisms of cardiomyocyte re-programming and renewal including an enhanced exercise-induced proliferational capacity of cardiomyocytes and its progenitor cells. Various intracellular mechanisms mediating these positive effects on cardiac function have been found in animal models of exercise and will be highlighted in this review. 1) activation of extracellular and intracellular signaling pathways including phosphatidylinositol 3 phosphate kinase (PI3K)/protein kinase B (AKT)/mammalian target of rapamycin (mTOR), EGFR/JNK/SP-1, nitric oxide (NO)-signaling, and extracellular vesicles; 2) gene expression modulation via microRNAs (miR), in particular via miR-17-3p and miR-222; and 3) modulation of cardiac cellular metabolism and mitochondrial adaption. Understanding the cellular mechanisms, which generate an exercise-induced cardioprotective cellular phenotype with physiological hypertrophy and enhanced proliferational capacity may give rise to novel therapeutic targets. These may open up innovative strategies to preserve cardiac function after myocardial injury as well as in aged cardiac tissue.

Keywords: physical exercise; cardiac cellular regeneration; microRNA (miR); Akt signaling; cardiomyocyte proliferation; cardiac hypertrophy; cardioprotection

1. Introduction

Physical exercise has been shown to be protective against cardiovascular diseases (CVD), the leading cause of death worldwide [1]. Despite remarkable progress in medical, interventional, and surgical treatment options of CVD over the last years, prevention will be more and more vital for healthcare systems in an aging society. An alarming increase in the incidence of CVD-associated diseases such as insulin resistance, type II diabetes mellitus, and obesity demand altered lifestyle behaviors including dietary changes [2], cessation of smoking [3], and frequent physical exercise which all reduce risks for CVD clearly. The American Heart Association (AHA) therefore recommends at least 150 minutes of moderate-intensity aerobic activity or 75 minutes of vigorous aerobic activity or a combination of both per week as well as a moderate- to high-intensity muscle-strengthening activity on at least 2 days per week [4]. The beneficial effects on the cardiovascular system apply not only to young and healthy individuals [5] but also to patients with distinct cardiovascular risk factors or overt CVD and seem to decline after detraining [6,7]. Most importantly, exercise seems to be protective against myocardial ischemia-reperfusion injury [8].

Different acute and chronic changes in autonomic regulation, cardiac metabolism, signaling pathways, and protein expression in exercising hearts leading to cardiac growth and cellular reprogramming have been discovered over recent years. Especially, the beneficial effects of sports in heart failure and stable angina pectoris on patient outcomes, hospital admission, quality of life, and exercise capacity have been demonstrated [9–11]. However, the exact mechanisms of how physical activity delays the development of cardiovascular diseases remain unclear.

Physiologically, the heart can adapt to chronic exercise in order to meet the enhanced oxygen demand of the body, a process called 'remodeling'. Exercise above three hours per week results in a significantly lower resting heart rate as well as significantly higher maximum oxygen uptake (VO2) and left ventricular mass [12]. Aerobic training thus promotes physiological cardiac hypertrophy and can contribute to a preserved cardiac function. This has high clinical impact as training can counteract declined cardiac function to a certain extent in injured as well as in aging hearts.

Physiological hypertrophy is initiated via humoral factors and mechanical stress leading to changes in intracellular cardiac signaling to affect gene transcription, protein translation and modification, and metabolism [13]. These intracellular responses at a molecular level are different to those seen in pathological hypertrophy. In this context, exercise-modulated gene expression and cell signaling might protect the heart from further injuries and continuous maladaptive remodeling processes. Further understanding and identification of pathways responsible for physiologic exercise-induced adaption resulting in a cardioprotective phenotype with physiological hypertrophy and proliferation might help to identify triggers for physiologic/cardioprotective and pathologic/maladaptive remodeling that could potentially be used therapeutically to maintain cardiac function after ischemic or infectious injury of the heart as well as in aging hearts.

In the following review, we highlight these cellular mechanisms of cardiac remodeling in response to physical exercise with a focus on signaling pathways and microRNAs.

2. Cardiac Cellular Changes in Exercise

2.1. Cellular Regeneration and Physiological and Pathological Hypertrophy

Injuries to the heart such as biochemical stress, toxins, infections, or ischemia require regeneration in order to maintain proper cardiac function. In contrast to most other cells, however, adult mammalian cardiomyocytes lose the ability to proliferate resulting in a low cellular turnover rate of 0.3 to 1% per year in the heart [14,15]. Although pioneering reports lately showed that adult mammalian cardiomyocyte proliferation could potentially be targeted via highly conserved signaling cascades such as peroxisome proliferator-activated receptor delta (PPARδ) agonist carbacyclin (induction of PPARδ/PDK1/protein kinase B (Akt) pathway) [16], extracellular matrix (ECM) protein agrin [17], or Hippo-Yap pathways [18], this very low cellular turnover is insufficient to achieve sufficient cardiac regeneration after cardiac injury.

Instead of regeneration with full restoration of organ function, hypertrophy, and fibrotic healing with incomplete functional recovery occur in the heart in response to injury. Regarding hypertrophy, one has to distinguish physiologic hypertrophy (proportional growth of length and width with proportional chamber enlargement) in response to exercise and pathologic concentric (relatively greater increase in width with disturbed contractile elements) as well as eccentric hypertrophy (relatively greater increase in length) in response to injury [19,20].

In this context sports have been shown to induce physiologic cardiac hypertrophy. In mouse models exercise induced inhibition of the transcription factor C/EBP beta and increased expression of CBP/p300-interacting protein with ED-rich carboxy-terminal domain-4 (CITED4) resulting in cardiac hypertrophy and proliferation, and contributed substantially to resistance against adverse cardiac remodeling and subsequent heart failure [21]. CITED4 has been demonstrated to mediate its effect on hypertrophy and recovery after ischemic injury due to its regulation of mammalian target of rapamycin (mTOR) signaling [22]. Cardiac cell proliferation, per se, does not seem to be necessary for exercise-induced cardiac growth but is highly required as a protective mechanism counteracting ischemia/reperfusion injury [23]. In addition to an enhanced proliferation and division of differentiated cardiomyocytes exercise such as swimming activates C-kit and Sca1 positive cardiac progenitor cells. These adult cardiac progenitor cells provide a certain potential of self-renewal and support myocardial regeneration [24]. This positive effect of sports on stem cell recruiting has been detected in the heart and the vascular system: Exercise-induced activation of cardiac and endothelial progenitor cells protects the heart, the coronary, and vascular system and attenuates the decline in arterial elasticity mediating positive effects on hypertension [25–27]. Administration of exogenous stem cells has been linked myocardial repair and regeneration in cardiac diseases and could be a potential therapeutic option in the future [28].

Nevertheless, even though these recent reports show that cardiomyocytes are at least to some degree capable of proliferation, adult hearts mainly respond to exercise and stress with an increase in cell size. Athletes' hearts are consequently characterized by a benign increase in heart mass as an effect of regular training. This physiological transformation with increases in cardiac mitochondrial energy capacity has to be distinguished from pathological cardiac growth due to e.g., hypertension with diminished contractile function, ATP deficiency, and mitochondrial dysfunction [13,29,30]. Both physiological and pathological hypertrophy are associated with higher heart mass and size. Pathological hypertrophy, however, is associated with increased interstitial fibrosis, apoptosis, and loss of cardiomyocytes. It shows fetal gene expression, altered cell signaling, and a different metabolism with decreased fatty acid metabolism which results in cardiac dysfunction with increased risk of heart failure and sudden cardiac death compared to physiological cardiac hypertrophy in exercised hearts [19,20]. Activation of fetal genes is as mentioned above one of the prominent changes found in pathological cardiac hypertrophy. Alterations in gene expression patterns in hypertrophic hearts resemble patterns found during fetal cardiac development and involve regulations on transcriptional, posttranscriptional, and epigenetic level. This reactivation of a fetal gene program in failing hearts has been nicely reviewed by Dirkx et al. [31].

Physiological "benign" hypertrophy declines after long-term detraining with significant reduction in cavity size and normalization of wall thickness whereas pathological "malign" hypertrophy persists [32].

Reports of sudden cardiac deaths (SCD) among young athletes have gained pronounced attention in the media as well as in research over the past years and have initiated the debate whether there is a threshold between benign "healthy" hypertrophy and malignant "unhealthy" hypertrophy with pathological conditions due to regular high-intensity exercise. To date, there is no clear evidence that healthy athletes without an underlying cardiovascular disease or a genetic cardiomyopathy have an increased risk of SCD [33], although recent data reveals that the leading finding associated with SCD among athletes is actually a structurally normal heart (unexplained autopsy-negative sudden cardiac death) [34].

Nevertheless, there is strong evidence for beneficial effects of regular aerobic and muscle-strengthening activity despite the association with mild cardiac hypertrophy.

2.2. Animal Models of Exercise

To experimentally study exercise-induced cellular cardiac alterations and their effects on cardiovascular health, aging, and response to injury, various animal models of physical exercise have been developed over the years using zebrafish, rodents, and large animals (rabbits, dogs, pigs, goats, sheep, and horses). Exercise modalities mainly include swimming or treadmill and wheel running [35].

The zebrafish as an animal model has gained broad attention in the field of exercise and regeneration physiology of the heart in the past years [36,37]. Unlike mammals, adult zebrafish hearts are capable of proliferation in case of cardiac injury and therefore represent a valuable model to study cardioprotection, regeneration, and aging [38,39]. This plasticity of the zebrafish heart has been demonstrated in injured hearts [40,41] as well as in intensified swimming-trained hearts [36] and revealed useful insights in cardiac remodeling processes and their cellular basis.

Rodents, especially rats and mice, are the most frequently used species to investigate effects of sports on cardiovascular system due to several advantages: Short gestation periods, syngeneic strains, relatively low housing costs, and easily reproducible experiments [42]. Most importantly, however, exercise-induced cardiac hypertrophy in mice shows physiological cardiac responses similar to those seen in humans [43] which makes them a valuable pre-clinical model. Rodent exercise models therefore have been widely used to study effects on cardiac hypertrophy as well as regeneration and aging: Broad insights in aging processes including telomere shortening have been gathered in exercising mice and rats [44]. Physical activity has been highly useful to study exercise-induced cardiac hypertrophy and to distinguish it from its pathological form [35] as observed e.g., in mouse models of aortic constriction [45]. These models revealed alterations in intracellular signaling such as Akt/mTOR/S6K1/4EBP1 pathways [46–48]. In rodents, treadmill running is predominantly used as this modality allows to adjust different intensities including interval training as well as modulations in inclination, speed, and duration. These intensity-controlled treadmill workouts provide reproducible cardiac hypertrophy and increase heart weights by 12–29% and cardiomyocyte dimensions by 17–32% in mice [49]. Beneficial effects on cardiovascular function mediated by treadmill running with high intensity interval training (HIIT) or long-term aerobic exercise before myocardial infarction include for example significantly reduced infarct sizes as well as an increased induction of anti-apoptotic effects in cardiac cells [50,51].

In contrast, larger animals are rather infrequently used compared to rodent models to study exercise-induced effects on the heart due to higher costs and efforts in housing and experimental procedures. Nevertheless, especially pigs closely resemble human coronary and vascular anatomy, hemodynamic physiology, and electrophysiology [52]. Thus, pigs could be used to investigate exercise-induced cardiac effects. In this context, for example, improvements in myocardial contractile function and in collateral capacity after physical activity were detected in an ischemic porcine model [53].

2.3. Major Signaling Pathways in Exercise-Induced Cardiac Remodeling

Over the past years different signaling pathways have been identified mediating cardioprotective cardiac growth and adaption as well as damage repair and attenuating cellular aging in response to physical exercise. Figure 1 provides an overview of cellular reprograming in cardiomyocytes in response to physical exercise.

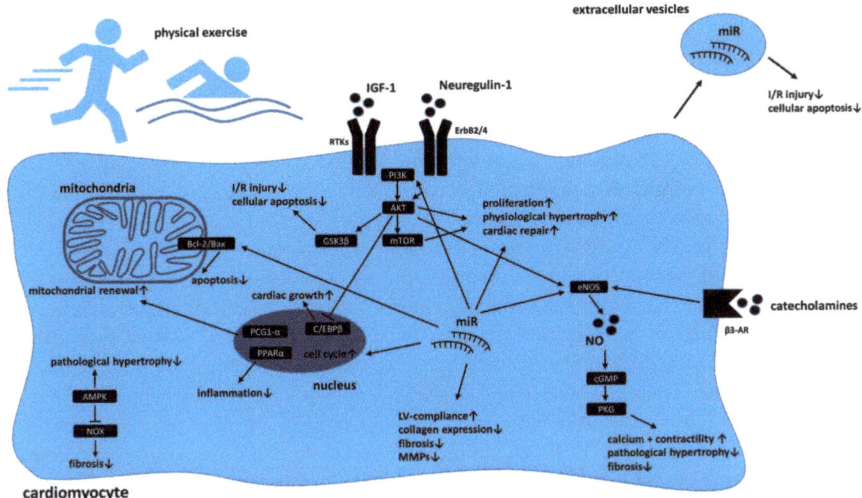

Figure 1. Schematic overview of cellular reprogramming in cardiomyocytes in response to physical exercise. Activation of receptor tyrosine kinases (RTKs) such as ErbB2/4 via growth factors (e.g., insulin-like growth factor 1 (IGF-1) or Neuregulin-1) enhance phosphatidylinositol 3 phosphate kinase (PI3K)/protein kinase B (Akt)/mammalian target of rapamycin (mTor)/ glycogen synthase kinase 3 beta (GSK3β) signaling which leads to proliferation, physiological hypertrophy, and cardiac repair mechanisms in response to injury. Beta3-adrenergic receptor (β3-AR) activation enhances endothelial nitric oxide synthase (eNOS) and subsequently intracellular nitric oxide (NO) levels which increases contractility and decreases fibrosis as well as pathological hypertrophy. Changes in miR expression influence intracellular signaling pathways (including Akt and eNOS), mediate apoptosis and cell cycle progression and influence cardiac compliance and fibrosis via alterations in collagen production and matrix metalloproteinase (MMP) expression. Sports induces mitochondrial renewal and decreases apoptosis via changes in B-cell lymphoma 2 (Bcl-2)/Bcl-2-associated X protein (Bax) ratio. Activation of adenosine monophosphate-activated protein kinase (AMPK) attenuates pathological hypertrophy and decrease profibrotic remodeling. Paracrine secretion of extracellular vesicles containing miR mediate I/R injury as well cellular apoptosis.

2.3.1. Akt-Signaling

The cellular key pathway in the regulation of physiological cardiac hypertrophy in response to exercise include phosphatidylinositol 3 phosphate kinase (PI3K) and Akt with their respective downstream signaling. These pathways are activated by extracellular growth factors such as insulin-like growth factor 1 (IGF-1) that has been linked to cardiac disease especially heart failure as it influences cardiac hypertrophy and contractile function [54]. This has been further demonstrated in rodents with heart failure where insulin-like growth factor-1 enhances ventricular hypertrophy and function [55] and suppresses apoptosis of cardiomyocytes [56]. Acute and chronic physical activity increase the local IGF-1 levels further confirming an important role in response to physical exercise but the precise role of IGF-1 is still not completely understood [57].

Mullen et al. identified PI3K and Akt signaling as crucial in the development of physiological (due to exercise) but not pathological (due to pressure overload) hypertrophy in mice [58]. These protective effects of enhanced PI3K signaling have been further demonstrated in dilated and hypertrophic cardiomyopathy with exercise and increase PI3K activity prolonging survival in a model of dilated cardiomyopathy up to 20% [59]. This implies that PI3K is essential for exercise-induced cardioprotection. Delivery of a constitutively active PI3K vector as a gene therapy can improve function of the failing heart in mice [60]. Akt acts as effector kinase downstream of PI3K and further activates mTOR signaling. Exercise training seems to be associated with activation of the Akt/mTOR pathway in contrast to pressure overload which is associated with its inactivation. This particular pathway may be one of the key regulators to distinguish between physiological and pathological cardiac hypertrophy. This cardiac growth is further mediated via downstream effectors of mTOR, including S6Kinase 1 and 4EBP1, as they play a crucial role in regulating protein biosynthesis, cell cycle progression, and hypertrophy [61,62].

Alterations of PI3K/Akt signaling is not only initiated via extracellular growth factors but also via changed intracellular microRNAs (miR) levels. miR-124 cluster among others suppress PI3K and are decreased in response to physical exercise leading to an enhanced activation of PI3K [63]. These changes mediated by miRNAs will be further elucidated in the next section.

Besides the mTOR pathway, Akt also mediates glycogen synthase kinase 3 beta (GSK3β) signaling. This key cellular pathway is involved in cellular apoptosis and alterations have been linked to various diseases including diabetes mellitus and carcinomas and was found to be altered in response to sports [64–67]. The role of GSK3 family in cardiac disease such as hypertrophy, aging, ischemic injury, or fibrogenesis has been shown in many rodent studies and has been comprehensively reviewed by Lal et al. [68]. Among this family: Akt; adenosine monophosphate-activated protein kinase (AMPK); beta3-adrenergic receptor (β3-AR); B-cell lymphoma 2 (Bcl-2); Bcl-2-associated X protein (Bax); CCAAT-enhancer-binding protein 1 (C/EBP1); cyclic guanosine monophosphate (cGMP); endothelial nitric oxide synthase (eNOS); GSK3β; IGF-1; ischemia/reperfusion (I/R); left ventricular (LV); miR; matrix metalloproteinase (MMP); mTOR; nitric oxide (NO); nicotinamide adenine dinucleotide phosphate (NADPH) oxidase (NOX); peroxisome proliferator-activated receptor gamma coactivator 1-alpha (PCG-1α); PI3K; PPARα; protein kinase G (PKG); receptor tyrosine kinase (RTK); and especially GSK3β seems to mediate positive effects in the context of ischemia/reperfusion injury which were found to be at least partly mediated via PI3K/Akt signaling [51].

2.3.2. Neuregulin-1/ErbB-Signaling

The neuregulin-1/ErbB2/ErbB4 pathway is another key pathway which was found to be altered in response to physical exercise and is connected with Akt/PI3K signaling underlining its cellular significance. Activation of the tyrosine kinases ErbB2 and ErbB4 by neuregulin-1 in response to exercise activates PI3K/Akt signaling and protects ventricular myocytes against apoptosis as demonstrated in a myocardial infarction model in rats [69,70]. ErbB2 and its downstream cascades have been shown vital in promoting mammalian heart regeneration, cardiomyocyte dedifferentiation, and proliferation [71]. In the past years this pathway has thus gained broad interest as potential therapeutic target in failing and ischemic hearts.

Neuregulin-1 is a small cell adhesion molecule that, among other tyrosine kinase receptors, is able to act on ErbB. It was shown to be a mediator of reverse cardiac remodeling in chronic heart failure [72]. In addition to its effects on ErbB receptors, neuregulin-1 induces paracrine secretion of cytokines in the heart including interleukin-1α and interferon-γ as well as pro-reparative factors (such as angiopoietin-2, brain-derived neurotrophic factor, and crypto-1), which have been demonstrated to contribute to cardiac repair mechanisms [73]. Rats showed exercise-induced upregulation of neuregulin-1 at the mRNA and protein level which was linked with physiological hypertrophy and cardiomyocyte proliferation [74]. Neuregulin-1 signaling can induce expression of transcription factor CAAT/enhancer-binding-protein beta (C/EBPβ) which is known to be involved in exercise-induced cardiac growth and protection in the context of pathological cardiac remodeling [21]. It has been further demonstrated that exercise activates neuregulin-1/ErbB signaling and promotes cardiac repair after myocardial infarction in rats [70] which shows its importance in cardiac regeneration in response to sports.

2.3.3. Nitric Oxide (NO) Signaling

Exercise has been shown to increase circulating catecholamines and consequently the expression of β3 adrenergic receptors (β3-AR) [8,75]. β3-AR stimulation in turn mediates endothelial nitric oxide synthase phosphorylation and increases cardiac NO metabolite levels (nitrite and nitrosothiols) which contribute to cardioprotective effects of exercise in ischemic hearts. Cessation of training in contrast extenuated NO levels and cardioprotective effects [8].

Interestingly, Akt signaling again is able to activate eNOS phosphorylating at Ser1176, an activation which has been shown to be critical for early ischemic preconditioning-induced cardioprotection [76]. NO has further been highlighted to inhibit ischemia/reperfusion injury, inflammation, and left ventricular remodeling in the absence of reactive oxygen species. Exercise inhibits ROS generation and promotes bioavailability of NO [77]. NO activates soluble guanylate cyclase, increases cGMP level, and activates PKG. Dysregulation in NO/PKG signaling is known to be involved in heart failure with enhanced calcium handling alterations, fibrosis, titin-based stiffness, pathological cellular hypertrophy, and microvascular dysfunction [78]. PKG in contrast inhibits pressure-induced cardiac remodeling in mice [79].

2.3.4. Other Pathways and Extracellular Vesicles

Physical exercise increases cardiac contractility to supply the raised demand of oxygenated blood. In this context training improves aging-induced downregulation of thyroid hormone receptor signaling mediated transcription of myosin heavy chain (MHC) and sarcoplasmatic reticulum Ca(2+)-ATPase and contributes to an improvement in cardiac function and contractility in aged rat hearts [80].

Signaling via extracellular vesicles (exosomes, EVs) has gained interest recently since they act as paracrine signaling mediators. They were found to be secreted by cardiac human progenitor cells containing microRNAs (e.g., miR-210, miR-132) and consequently inhibit cardiac apoptosis and improve cardiac function after acute myocardial infarction [81]. After three weeks of swimming mice showed increased levels of EVs and were more resistant against cardiac ischemia/reperfusion injury and demonstrated antiapoptotic effects by the activation of extracellular signal–regulated kinases 1/2 (ERK1/2) and heat shock protein 27 (HSP27) signaling [82]. Induced pluripotent stem cell (iPSC)-derived EVs seem to be more effective to induce cardiac repair mechanisms (including maintained left ventricular function and vascularization, amelioration of apoptosis, and hypertrophy) compared to iPSCs themselves [83].

3. MicroRNAs

MicroRNAs are small noncoding RNA molecules of approximately 22 nucleotides in length modulating gene expression post-transcriptionally by binding to its target messenger RNAs promoting their degradation [84]. Several animal and human studies have shown altered miR levels in cardiac diseases such as hypertrophy, ischemic, and dilated cardiomyopathy, aortic stenosis, or arrhythmias [85–90].

As described above cardiac cell regeneration was thought to be a rare event in adult mammalian hearts for a long time but recently different miRNA clusters have been linked to increased cardiomyocyte proliferation and cardiac regeneration whereas others have been related to reduced cardiomyocyte proliferation. Recently, however, animal studies revealed detailed insights in cardiac regeneration processes including cardiomyocyte proliferation in failing hearts [91,92] and inducible regeneration in adult cardiomyocytes [16–18]. Renewal of adult cardiomyocytes after acute injury of the heart is therefore present yet insufficient [93]. MicroRNAs are directly linked to control this cardiomyocyte regeneration, renewal, and proliferation. Many of these findings were obtained in exercise animal studies and may further contribute to develop novel innovative therapeutic strategies for patients with failing hearts in the future. Table 1 gives an overview of cardiac microRNAs changes in response to physical activity.

Table 1. Overview of microRNA levels altered in response to physical exercise and their contribution to cardioprotection.

MicroRNA	Cellular Target	Cardiac Function	Animal Model and Exercise Modality	References
miR-17-3p	TIMP3, PTEN	Cardiac hypertrophy Myocyte proliferation Cardiac apoptosis	Mice, swimming and wheel exercise	[94]
miR-222	P27, Hipk1, Hmbox1	Cell cycle Cardiac apoptosis Cardiac hypertrophy Myocyte proliferation	Mice, swimming and wheel exercise	[95]
miR-124	PI3K	Cardiac hypertrophy	Rats, swimming exercise	[63]
miR-21	PTEN	Cardiac hypertrophy	Rats, swimming exercise	[63,96]
miR-144	PTEN	Cardiac hypertrophy	Rats, swimming exercise	[63]
miR-145	TSC	Cardiac hypertrophy	Rats, swimming exercise	[63]
miR-126	Spred-1 Raf-1/ERK 1/2 signaling	Cardiac angiogenesis	Rats, swimming exercise	[97]
miR-133	Calcineurin PI3K/Akt signaling	Cardiac hypertrophy	Rats, swimming exercise	[96,98]
miR-29c	Collagen I und III TGFβ pathway	Left ventricular compliance	Rats, swimming exercise	[99]
miR-29b	MMP9	Fibrosis, matrix degradation	Mice, treadmill running	[100]
miR-455	MMP9	Fibrosis, matrix degradation	Mice, treadmill running	[100]
miR-199a	PGC1α	Cardiac hypertrophy	Mice, treadmill running	[101]
mi-R1	Bcl-2	Cardiac apoptosis	Mice, swimming exercise	[102]
miR-30	P53, Drp-1	Cardiac apoptosis	Mice, swimming exercise	[102]
miR-21	PDCD4	Cardiac apoptosis	Mice, swimming exercise	[102]

Extracellular signal–regulated kinases 1/2 (ERK1/2).

Several microRNAs have been discovered contributing to an improved cardiac repair after distinct cardiac injury [103]. Among those, the miR-17-92 cluster seems to play an important role in regulating cardiac growth, proliferation, hypertrophy, and survival in response to exercise [94]. This microRNA family is known for regulatory influences on cell cycle, apoptosis and proliferation and has been linked to enhanced proliferation and survival of colorectal cancer [104] as well as keratinocyte proliferation and metastasis [105]. Shi et al. confirmed the role of miR-17-3p in cardiomyocyte hypertrophy and proliferation in a rodent model of exercise: Inhibition of miR-17-3p in mice decreased exercise-induced cardiac growth, cardiomyocyte hypertrophy and expression of markers of myocyte proliferation. Furthermore, mice treated with a miR-17-3p agomir received protection against cardiac remodeling after cardiac ischemia. The authors identified that miR-17-3p suppresses TIMP3, an inhibitor of EGFR/JNK/SP-1, a pathway promoting cardiomyocyte proliferation [106], as well as PTEN which antagonizes PI3K/Akt pathway [107], a pathway vital for cardiac hypertrophy [58] as elucidated above. Samples of ventricular cardiomyocytes obtained in patients with dilated cardiomyopathy revealed a downregulation of miR-17-3p, which confirms its important role in cardiac remodeling, growth, and proliferation.

Exercise-induced activation of PI3K/AKT/mTOR signaling and subsequent left ventricular physiological hypertrophy is not only mediated via disrupted miR-17 levels but also via other miR-clusters as shown in a rat model [63]: Ma et al. found decreased miRNA-124 levels (targeting PI3K) and increased miRNAs-21, -144, and -145 (targeting PTEN and TSC-2), all leading to an induced activation of PI3K/Akt/mTOR signaling in cardiomyocytes. Furthermore swimming exercise and application of recombinant human growth hormone (r-hGH) in rats altered cardiac PI3K/Akt/mTOR signaling and miR-21 and miR-133 expression [96]. Reciprocal repression between miR-133 and calcineurin is involved in regulating cardiac hypertrophy [98]. These findings emphasize the significance of PI3K/Akt signaling in cardiac proliferation and physiological hypertrophy and support the beneficial effects of physical activity.

Importantly, miR-222 was identified as a key regulator of exercise-induced cardiomyocyte growth and proliferation in mice as induced miR-222 expression in cardiomyocytes led to resistance against adverse cardiac remodeling and ventricular dysfunction after ischemia [95]. These effects were mediated via inhibition of p27, Hipk1, and Hmbox1. Inhibition of endogenous miR-199a was shown to contribute to physiological cardiac hypertrophy probably due to the upregulation of PGC1α in treadmill-trained mice [101]. In another promising study treatment with either miR-199a or miR-590 elevated cardiomyocyte proliferation in postnatal mice promoting cell-cycle re-entry, recovered cardiac contractility, and decreased the levels of fibrosis after myocardial infarction [103].

In contrast miR-15 family attenuates heart regeneration through inhibition of postnatal cardiomyocyte proliferation and acute inhibition of miR-15 in adult mice is associated with improved contractile function after ischemic injury [108].

Physical activity also leads to altered miR clusters affecting increased cardiac angiogenesis in animal models. Among those miR-15 is involved in controlling angiogenesis and downregulated under hypoxic conditions [109]. Increased miRNA-126 expression is associated with exercise-induced cardiac angiogenesis in response to changes in vascular endothelial growth factor (VEGF) pathway as well as mitogen-activated protein kinase (MAPK) and PI3K/Akt/eNOS pathways in rats [97]. Decline in cardiac microvascularization is a finding often obtained in aging and diabetes mellitus. Exercise training in this context attenuated aging-induced downregulation of VEGF signaling cascades including phosphorylation of Akt and eNOS proteins contributing to an improvement of angiogenesis in old age rats [110].

Ventricular stiffness, increase in fibrosis and diastolic dysfunction often accompanies heart failure. MiR-29c was found to be involved in improving ventricular compliance: Exercising rats showed increased miRNA-29c expression correlating with a decrease in collagen I and III expression and improved LV compliance [99]. Exercise-induced release of exosomes from cardiomyocytes containing miR-29b and miR-455 downregulated matrix metalloproteinase 9 (MMP9) resulting in decreased fibrosis and matrix degradation [100].

Swimming-trained mice showed decreased cardiac apoptosis via increased Bcl-2/Bax ratio, an effect mediated via miR-1, miR-30b, and miR-21 [102].

Changes in miRNA expression found in animal studies seem to be at least in part transferable to humans though their significance is still incompletely understood: Increases in circulating miR-126 have been also shown in healthy adults after endurance exercise [111]. miR-1, miR-133a, and miR-206 levels were significantly elevated after exercise and correlated with performance parameters such as maximum oxygen uptake and anaerobic lactate threshold [112].

4. Metabolic and Mitochondrial Cardiac Changes

Exercising cardiomyocytes predominantly use glucose and fatty acids to generate energy. The systemic usage of substrates is measured via respiratory exchange ratio (RER). A RER of 1.0 represents total usage of carbohydrates while a ratio of 0.7 represents fatty acid usage (values in between represent a mixture). During exercise RER early approaches 1.0 but returns toward 0.7 in extended workouts indicating a shift towards fatty acid oxidation after longer duration [113].

Different metabolic and mitochondrial alterations related to training have been identified. This has high clinical implication for potential therapeutic targets as mitochondrial dysfunction is one of the key findings in heart failure and as exercise is protective for cardiac mitochondria against ischemia/reperfusion injury in mice [114]. Rather inconsistent data about changes in the proportion of glucose and fatty acid oxidation in cardiac cells have been gained in the past years from rodent models. Therefore, exact assignment of substrate utilization in response to chronic exercise remains uncertain [115]. Clear changes have been found in metabolic gene expression and mitochondrial pathways: Expression of genes involved in beta-oxidation of fatty acids and glucose metabolism in the heart differ between exercise-induced physiological cardiac growth and maladaptive pathological cardiac growth. Genes involved in β-oxidation are downregulated in maladaptive cardiac hypertrophy whereas they (including CD36, a fatty acid translocase and scavenger receptor) are found to be upregulated in exercise-induced cardiac hypertrophy [116]. CD36 deficiency is directly linked to insulin resistance and defective fatty acid metabolism in rats [117]. In contrast abnormal myocardial fatty acid uptake via redistribution of CD36 from intracellular stores to the plasma membrane was found in early stages of insulin resistance contributing to cardiac lipotoxicity [118]. The impact of CD36 on cardiac function is not entirely understood. CD36-deficient hearts were found not to be energetically or functionally compromised and were not more vulnerable to ischemia as energy generation through glucose oxidation was able to compensate for the loss of fatty acid-derived energy generation [119].

The transcriptional co-activators PGC-1α and PGC-1β regulate oxidative phosphorylation and fatty acid oxidation gene expression and control number and size of mitochondria. Heart failure is associated with repressed PGC-1α and PGC-1β gene expression [120]. In mice with diabetic cardiomyopathy running prevented cellular apoptosis and fibrosis, improved mitochondrial biogenesis, prevented diabetic cardiomyopathy-associated inhibition of PGC-1α, and activated Akt signaling [121].

Adenosine monophosphate-activated protein kinase is a serine/threonine kinase which participates in regulating cellular energy supply [122]. Long-term activation of AMPK blocks cardiac hypertrophy as well as NFAT, NF-kB, and MAPK signaling and thus preserves cardiac function in pressure-overload rats [123]. AMPK deficiency, on the other hand, exacerbates LV hypertrophy in mice [124]. Swimming-trained rats showed activated AMPK levels with reduced cardiac fibrosis due to inhibition of NADPH oxidase [125]. This finding has been confirmed as decreased AMPK activity by beta-adrenergic activation exacerbated cardiac fibrosis [126]. This is interesting as exercise activates AMPK [127] and consequently might be able to inhibit pathological hypertrophy and cardiac fibrosis.

The intrinsic mitochondrial apoptotic pathway is one of the most important mechanisms of myocyte degeneration in the progression of heart failure [128]. Exercise induces a cardiac mitochondrial phenotype that resists apoptotic stimuli increasing antioxidant enzymes [129]. The Bcl-2 pathway is one of those proapoptotic mitochondrial-mediated pathways and has been demonstrated to be critical in regulating apoptosis in aging patients [130]. Rat hearts showed 12 weeks after training attenuated age-induced elevation in Bax/Bcl-2 ratio and consequently lower apoptotic rates and remodeling [131].

Another key metabolic pathway induced by exercise is the peroxisome proliferator-activated receptors (PPAR)-pathway. PPAR are transcription factors mediating the development of cardiac hypertrophy and regulating fatty acid metabolism [132]. Exercise increases PPAR-alpha levels and decreases consequently inflammatory response including TNF-alpha and NF-kB levels [133]. This is important as PPAR-alpha stimulation downregulates inflammatory molecules and decreases infarct size [134]. Aging processes decrease PPAR-alpha levels, a finding that could be attenuated in swim-trained rat hearts which contributed to an improvement in fatty acid metabolic enzyme activity [135].

5. Conclusions

Alterations in cellular pathways including Akt, ErbB, and NO signaling as well as various miR clusters have been linked to cardiac disease such as heart failure. Physical exercise is known to be cardioprotective and can partly compensate cardiac damage as demonstrated in various animal models and patient studies. On a cellular level sports counteracts cardiac disease related alterations in these cellular pathways and is able to improve cardiac function. Insights in cellular cardioprotective pathways obtained from these exercise studies could contribute to the development of novel therapeutic strategies in failing hearts due to toxic, infectious or ischemic injury, or aging in the future.

Author Contributions: Conceptualization, D.S., S.C. and S.B.; methodology, D.S. and S.B.; software, D.S.; validation, D.S., S.C., L.T.W. and S.B.; formal analysis, D.S.; investigation, D.S., S.C., L.T.W. and S.B.; resources, D.S. and S.B.; data curation, D.S. and S.B.; writing—original draft preparation, D.S.; writing—review and editing, D.S., S.C., L.T.W. and S.B.; visualization, D.S.; supervision, S.B.; project administration, S.B.; funding acquisition, S.B.

Funding: This research received no external funding.

Conflicts of Interest: The authors declare there are no conflicts of interest.

References

1. Golbidi, S.; Laher, I. Exercise and the cardiovascular system. *Cardiol. Res. Pract.* **2012**, *2012*, 210852. [CrossRef] [PubMed]
2. Sacks, F.M.; Katan, M. Randomized clinical trials on the effects of dietary fat and carbohydrate on plasma lipoproteins and cardiovascular disease. *Am. J. Med.* **2002**, *113* (Suppl. 9B), 13S–24S. [CrossRef]
3. Wilson, K.; Gibson, N.; Willan, A.; Cook, D. Effect of smoking cessation on mortality after myocardial infarction: Meta-analysis of cohort studies. *Arch. Intern. Med.* **2000**, *160*, 939–944. [CrossRef] [PubMed]
4. Piercy, K.L.; Troiano, R.P. Physical Activity Guidelines for Americans from the US Department of Health and Human Services. *Circ. Cardiovasc. Qual. Outcomes* **2018**, *11*, e005263. [CrossRef] [PubMed]
5. Clarkson, P.; Montgomery, H.E.; Mullen, M.J.; Donald, A.E.; Powe, A.J.; Bull, T.; Jubb, M.; World, M.; Deanfield, J.E. Exercise training enhances endothelial function in young men. *J. Am. Coll. Cardiol.* **1999**, *33*, 1379–1385. [CrossRef]
6. Vona, M.; Codeluppi, G.M.; Iannino, T.; Ferrari, E.; Bogousslavsky, J.; von Segesser, L.K. Effects of different types of exercise training followed by detraining on endothelium-dependent dilation in patients with recent myocardial infarction. *Circulation* **2009**, *119*, 1601–1608. [CrossRef]
7. Pearson, M.J.; Smart, N.A. Aerobic Training Intensity for Improved Endothelial Function in Heart Failure Patients: A Systematic Review and Meta-Analysis. *Cardiol. Res. Pract.* **2017**, *2017*, 2450202. [CrossRef]
8. Calvert, J.W.; Condit, M.E.; Aragon, J.P.; Nicholson, C.K.; Moody, B.F.; Hood, R.L.; Sindler, A.L.; Gundewar, S.; Seals, D.R.; Barouch, L.A.; et al. Exercise protects against myocardial ischemia-reperfusion injury via stimulation of beta(3)-adrenergic receptors and increased nitric oxide signaling: Role of nitrite and nitrosothiols. *Circ. Res.* **2011**, *108*, 1448–1458. [CrossRef]
9. Taylor, R.S.; Long, L.; Mordi, I.R.; Madsen, M.T.; Davies, E.J.; Dalal, H.; Rees, K.; Singh, S.J.; Gluud, C.; Zwisler, A.D. Exercise-Based Rehabilitation for Heart Failure: Cochrane Systematic Review, Meta-Analysis, and Trial Sequential Analysis. *JACC Heart Fail.* **2019**, *7*, 691–705. [CrossRef]
10. Long, L.; Anderson, L.; He, J.; Gandhi, M.; Dewhirst, A.; Bridges, C.; Taylor, R. Exercise-based cardiac rehabilitation for stable angina: Systematic review and meta-analysis. *Open Heart* **2019**, *6*, e000989. [CrossRef]
11. Long, L.; Mordi, I.R.; Bridges, C.; Sagar, V.A.; Davies, E.J.; Coats, A.J.; Dalal, H.; Rees, K.; Singh, S.J.; Taylor, R.S. Exercise-based cardiac rehabilitation for adults with heart failure. *Cochrane Database Syst. Rev.* **2019**, *1*, CD003331. [CrossRef]
12. Fagard, R. Athlete's heart. *Heart* **2003**, *89*, 1455–1461. [CrossRef]
13. Maillet, M.; van Berlo, J.H.; Molkentin, J.D. Molecular basis of physiological heart growth: Fundamental concepts and new players. *Nat. Rev. Mol. Cell Biol.* **2013**, *14*, 38–48. [CrossRef] [PubMed]
14. He, L.; Zhou, B. Cardiomyocyte proliferation: Remove brakes and push accelerators. *Cell Res.* **2017**, *27*, 959–960. [CrossRef] [PubMed]
15. Bergmann, O.; Bhardwaj, R.D.; Bernard, S.; Zdunek, S.; Barnabe-Heider, F.; Walsh, S.; Zupicich, J.; Alkass, K.; Buchholz, B.A.; Druid, H.; et al. Evidence for cardiomyocyte renewal in humans. *Science* **2009**, *324*, 98–102. [CrossRef] [PubMed]
16. Magadum, A.; Ding, Y.; He, L.; Kim, T.; Vasudevarao, M.D.; Long, Q.; Yang, K.; Wickramasinghe, N.; Renikunta, H.V.; Dubois, N.; et al. Live cell screening platform identifies PPARdelta as a regulator of cardiomyocyte proliferation and cardiac repair. *Cell Res.* **2017**, *27*, 1002–1019. [CrossRef]
17. Bassat, E.; Mutlak, Y.E.; Genzelinakh, A.; Shadrin, I.Y.; Baruch Umansky, K.; Yifa, O.; Kain, D.; Rajchman, D.; Leach, J.; Riabov Bassat, D.; et al. The extracellular matrix protein agrin promotes heart regeneration in mice. *Nature* **2017**, *547*, 179–184. [CrossRef]
18. Morikawa, Y.; Heallen, T.; Leach, J.; Xiao, Y.; Martin, J.F. Dystrophin-glycoprotein complex sequesters Yap to inhibit cardiomyocyte proliferation. *Nature* **2017**, *547*, 227–231. [CrossRef]
19. Hunter, J.J.; Chien, K.R. Signaling pathways for cardiac hypertrophy and failure. *N. Engl. J. Med.* **1999**, *341*, 1276–1283. [CrossRef]

20. Bernardo, B.C.; Weeks, K.L.; Pretorius, L.; McMullen, J.R. Molecular distinction between physiological and pathological cardiac hypertrophy: Experimental findings and therapeutic strategies. *Pharmacol. Ther.* **2010**, *128*, 191–227. [CrossRef]

21. Bostrom, P.; Mann, N.; Wu, J.; Quintero, P.A.; Plovie, E.R.; Panakova, D.; Gupta, R.K.; Xiao, C.; MacRae, C.A.; Rosenzweig, A.; et al. C/EBPbeta controls exercise-induced cardiac growth and protects against pathological cardiac remodeling. *Cell* **2010**, *143*, 1072–1083. [CrossRef]

22. Bezzerides, V.J.; Platt, C.; Lerchenmuller, C.; Paruchuri, K.; Oh, N.L.; Xiao, C.; Cao, Y.; Mann, N.; Spiegelman, B.M.; Rosenzweig, A. CITED4 induces physiologic hypertrophy and promotes functional recovery after ischemic injury. *JCI Insight* **2016**, *1*, e85904. [CrossRef]

23. Bei, Y.; Fu, S.; Chen, X.; Chen, M.; Zhou, Q.; Yu, P.; Yao, J.; Wang, H.; Che, L.; Xu, J.; et al. Cardiac cell proliferation is not necessary for exercise-induced cardiac growth but required for its protection against ischaemia/reperfusion injury. *J. Cell. Mol. Med.* **2017**, *21*, 1648–1655. [CrossRef]

24. Beltrami, A.P.; Barlucchi, L.; Torella, D.; Baker, M.; Limana, F.; Chimenti, S.; Kasahara, H.; Rota, M.; Musso, E.; Urbanek, K.; et al. Adult cardiac stem cells are multipotent and support myocardial regeneration. *Cell* **2003**, *114*, 763–776. [CrossRef]

25. Yang, Z.; Xia, W.H.; Su, C.; Wu, F.; Zhang, Y.Y.; Xu, S.Y.; Liu, X.; Zhang, X.Y.; Ou, Z.J.; Lai, G.H.; et al. Regular exercise-induced increased number and activity of circulating endothelial progenitor cells attenuates age-related decline in arterial elasticity in healthy men. *Int. J. Cardiol.* **2013**, *165*, 247–254. [CrossRef]

26. Yang, Z.; Chen, L.; Su, C.; Xia, W.H.; Wang, Y.; Wang, J.M.; Chen, F.; Zhang, Y.Y.; Wu, F.; Xu, S.Y.; et al. Impaired endothelial progenitor cell activity is associated with reduced arterial elasticity in patients with essential hypertension. *Clin. Exp. Hypertens.* **2010**, *32*, 444–452. [CrossRef]

27. Vasa, M.; Fichtlscherer, S.; Aicher, A.; Adler, K.; Urbich, C.; Martin, H.; Zeiher, A.M.; Dimmeler, S. Number and migratory activity of circulating endothelial progenitor cells inversely correlate with risk factors for coronary artery disease. *Circ. Res.* **2001**, *89*, E1–E7. [CrossRef]

28. Buja, L.M. Cardiac repair and the putative role of stem cells. *J. Mol. Cell. Cardiol.* **2019**, *128*, 96–104. [CrossRef]

29. Doenst, T.; Nguyen, T.D.; Abel, E.D. Cardiac metabolism in heart failure: Implications beyond ATP production. *Circ. Res.* **2013**, *113*, 709–724. [CrossRef]

30. Ingwall, J.S. Energy metabolism in heart failure and remodelling. *Cardiovasc. Res.* **2009**, *81*, 412–419. [CrossRef]

31. Dirkx, E.; da Costa Martins, P.A.; De Windt, L.J. Regulation of fetal gene expression in heart failure. *Biochim. Biophys. Acta* **2013**, *1832*, 2414–2424. [CrossRef]

32. Pelliccia, A.; Maron, B.J.; De Luca, R.; Di Paolo, F.M.; Spataro, A.; Culasso, F. Remodeling of left ventricular hypertrophy in elite athletes after long-term deconditioning. *Circulation* **2002**, *105*, 944–949. [CrossRef]

33. Maron, B.J.; Pelliccia, A. The heart of trained athletes: Cardiac remodeling and the risks of sports, including sudden death. *Circulation* **2006**, *114*, 1633–1644. [CrossRef]

34. Asif, I.M.; Harmon, K.G. Incidence and Etiology of Sudden Cardiac Death: New Updates for Athletic Departments. *Sports Health* **2017**, *9*, 268–279. [CrossRef]

35. Thu, V.T.; Kim, H.K.; Han, J. Acute and Chronic Exercise in Animal Models. *Adv. Exp. Med. Biol.* **2017**, *999*, 55–71. [CrossRef]

36. Rovira, M.; Borras, D.M.; Marques, I.J.; Puig, C.; Planas, J.V. Physiological Responses to Swimming-Induced Exercise in the Adult Zebrafish Regenerating Heart. *Front. Physiol.* **2018**, *9*, 1362. [CrossRef]

37. Gonzalez-Rosa, J.M.; Burns, C.E.; Burns, C.G. Zebrafish heart regeneration: 15 years of discoveries. *Regeneration* **2017**, *4*, 105–123. [CrossRef]

38. Gilbert, M.J.; Zerulla, T.C.; Tierney, K.B. Zebrafish (Danio rerio) as a model for the study of aging and exercise: Physical ability and trainability decrease with age. *Exp. Gerontol.* **2014**, *50*, 106–113. [CrossRef]

39. Foglia, M.J.; Poss, K.D. Building and re-building the heart by cardiomyocyte proliferation. *Development* **2016**, *143*, 729–740. [CrossRef]

40. Gemberling, M.; Karra, R.; Dickson, A.L.; Poss, K.D. Nrg1 is an injury-induced cardiomyocyte mitogen for the endogenous heart regeneration program in zebrafish. *eLife* **2015**, *4*, e05871. [CrossRef]

41. Karra, R.; Knecht, A.K.; Kikuchi, K.; Poss, K.D. Myocardial NF-kappaB activation is essential for zebrafish heart regeneration. *Proc. Natl. Acad. Sci. USA* **2015**, *112*, 13255–13260. [CrossRef]

42. Vega, R.B.; Konhilas, J.P.; Kelly, D.P.; Leinwand, L.A. Molecular Mechanisms Underlying Cardiac Adaptation to Exercise. *Cell Metab.* **2017**, *25*, 1012–1026. [CrossRef]

43. Perrino, C.; Gargiulo, G.; Pironti, G.; Franzone, A.; Scudiero, L.; De Laurentis, M.; Magliulo, F.; Ilardi, F.; Carotenuto, G.; Schiattarella, G.G.; et al. Cardiovascular effects of treadmill exercise in physiological and pathological preclinical settings. *Am. J. Physiol. Heart Circ. Physiol.* **2011**, *300*, H1983–H1989. [CrossRef]

44. de Carvalho Cunha, V.N.; Dos Santos Rosa, T.; Sales, M.M.; Sousa, C.V.; da Silva Aguiar, S.; Deus, L.A.; Simoes, H.G.; de Andrade, R.V. Training Performed Above Lactate Threshold Decreases p53 and Shelterin Expression in Mice. *Int. J. Sports Med.* **2018**, *39*, 704–711. [CrossRef]

45. Merino, D.; Gil, A.; Gomez, J.; Ruiz, L.; Llano, M.; Garcia, R.; Hurle, M.A.; Nistal, J.F. Experimental modelling of cardiac pressure overload hypertrophy: Modified technique for precise, reproducible, safe and easy aortic arch banding-debanding in mice. *Sci. Rep.* **2018**, *8*, 3167. [CrossRef]

46. Kemi, O.J.; Ceci, M.; Wisloff, U.; Grimaldi, S.; Gallo, P.; Smith, G.L.; Condorelli, G.; Ellingsen, O. Activation or inactivation of cardiac Akt/mTOR signaling diverges physiological from pathological hypertrophy. *J. Cell. Physiol.* **2008**, *214*, 316–321. [CrossRef]

47. Shiojima, I.; Walsh, K. Regulation of cardiac growth and coronary angiogenesis by the Akt/PKB signaling pathway. *Genes Dev.* **2006**, *20*, 3347–3365. [CrossRef]

48. Heineke, J.; Molkentin, J.D. Regulation of cardiac hypertrophy by intracellular signalling pathways. *Nat. Rev. Mol. Cell Biol.* **2006**, *7*, 589–600. [CrossRef]

49. Kemi, O.J.; Loennechen, J.P.; Wisloff, U.; Ellingsen, O. Intensity-controlled treadmill running in mice: Cardiac and skeletal muscle hypertrophy. *J. Appl. Physiol.* **2002**, *93*, 1301–1309. [CrossRef]

50. Rahimi, M.; Shekarforoush, S.; Asgari, A.R.; Khoshbaten, A.; Rajabi, H.; Bazgir, B.; Mohammadi, M.T.; Sobhani, V.; Shakibaee, A. The effect of high intensity interval training on cardioprotection against ischemia-reperfusion injury in wistar rats. *EXCLI J.* **2015**, *14*, 237–246. [CrossRef]

51. Zhang, K.R.; Liu, H.T.; Zhang, H.F.; Zhang, Q.J.; Li, Q.X.; Yu, Q.J.; Guo, W.Y.; Wang, H.C.; Gao, F. Long-term aerobic exercise protects the heart against ischemia/reperfusion injury via PI3 kinase-dependent and Akt-mediated mechanism. *Apoptosis* **2007**, *12*, 1579–1588. [CrossRef] [PubMed]

52. Clauss, S.; Bleyer, C.; Schuttler, D.; Tomsits, P.; Renner, S.; Klymiuk, N.; Wakili, R.; Massberg, S.; Wolf, E.; Kaab, S. Animal models of arrhythmia: Classic electrophysiology to genetically modified large animals. *Nat. Rev. Cardiol.* **2019**, *16*, 457–475. [CrossRef] [PubMed]

53. Roth, D.M.; White, F.C.; Nichols, M.L.; Dobbs, S.L.; Longhurst, J.C.; Bloor, C.M. Effect of long-term exercise on regional myocardial function and coronary collateral development after gradual coronary artery occlusion in pigs. *Circulation* **1990**, *82*, 1778–1789. [CrossRef] [PubMed]

54. Castellano, G.; Affuso, F.; Conza, P.D.; Fazio, S. The GH/IGF-1 Axis and Heart Failure. *Curr. Cardiol. Rev.* **2009**, *5*, 203–215. [CrossRef] [PubMed]

55. Duerr, R.L.; Huang, S.; Miraliakbar, H.R.; Clark, R.; Chien, K.R.; Ross, J., Jr. Insulin-like growth factor-1 enhances ventricular hypertrophy and function during the onset of experimental cardiac failure. *J. Clin. Investig.* **1995**, *95*, 619–627. [CrossRef] [PubMed]

56. Lee, W.L.; Chen, J.W.; Ting, C.T.; Ishiwata, T.; Lin, S.J.; Korc, M.; Wang, P.H. Insulin-like growth factor I improves cardiovascular function and suppresses apoptosis of cardiomyocytes in dilated cardiomyopathy. *Endocrinology* **1999**, *140*, 4831–4840. [CrossRef] [PubMed]

57. Nindl, B.C.; Pierce, J.R. Insulin-like growth factor I as a biomarker of health, fitness, and training status. *Med. Sci. Sports Exerc.* **2010**, *42*, 39–49. [CrossRef] [PubMed]

58. McMullen, J.R.; Shioi, T.; Zhang, L.; Tarnavski, O.; Sherwood, M.C.; Kang, P.M.; Izumo, S. Phosphoinositide 3-kinase(p110alpha) plays a critical role for the induction of physiological, but not pathological, cardiac hypertrophy. *Proc. Natl. Acad. Sci. USA* **2003**, *100*, 12355–12360. [CrossRef] [PubMed]

59. McMullen, J.R.; Amirahmadi, F.; Woodcock, E.A.; Schinke-Braun, M.; Bouwman, R.D.; Hewitt, K.A.; Mollica, J.P.; Zhang, L.; Zhang, Y.; Shioi, T.; et al. Protective effects of exercise and phosphoinositide 3-kinase(p110alpha) signaling in dilated and hypertrophic cardiomyopathy. *Proc. Natl. Acad. Sci. USA* **2007**, *104*, 612–617. [CrossRef]

60. Weeks, K.L.; Gao, X.; Du, X.J.; Boey, E.J.; Matsumoto, A.; Bernardo, B.C.; Kiriazis, H.; Cemerlang, N.; Tan, J.W.; Tham, Y.K.; et al. Phosphoinositide 3-kinase p110alpha is a master regulator of exercise-induced cardioprotection and PI3K gene therapy rescues cardiac dysfunction. *Circ. Heart Fail.* **2012**, *5*, 523–534. [CrossRef] [PubMed]

61. Sciarretta, S.; Forte, M.; Frati, G.; Sadoshima, J. New Insights into the Role of mTOR Signaling in the Cardiovascular System. *Circ. Res.* **2018**, *122*, 489–505. [CrossRef] [PubMed]

62. Dorn, G.W., 2nd; Force, T. Protein kinase cascades in the regulation of cardiac hypertrophy. *J. Clin. Investig.* **2005**, *115*, 527–537. [CrossRef] [PubMed]

63. Ma, Z.; Qi, J.; Meng, S.; Wen, B.; Zhang, J. Swimming exercise training-induced left ventricular hypertrophy involves microRNAs and synergistic regulation of the PI3K/AKT/mTOR signaling pathway. *Eur. J. Appl. Physiol.* **2013**, *113*, 2473–2486. [CrossRef] [PubMed]

64. Schuettler, D.; Piontek, G.; Wirth, M.; Haller, B.; Reiter, R.; Brockhoff, G.; Pickhard, A. Selective inhibition of EGFR downstream signaling reverses the irradiation-enhanced migration of HNSCC cells. *Am. J. Cancer Res.* **2015**, *5*, 2660–2672. [PubMed]

65. Beurel, E.; Grieco, S.F.; Jope, R.S. Glycogen synthase kinase-3 (GSK3): Regulation, actions, and diseases. *Pharmacol. Ther.* **2015**, *148*, 114–131. [CrossRef] [PubMed]

66. Sugden, P.H.; Fuller, S.J.; Weiss, S.C.; Clerk, A. Glycogen synthase kinase 3 (GSK3) in the heart: A point of integration in hypertrophic signalling and a therapeutic target? A critical analysis. *Br. J. Pharmacol.* **2008**, *153* (Suppl. 1), S137–S153. [CrossRef]

67. Lee, Y.; Kwon, I.; Jang, Y.; Song, W.; Cosio-Lima, L.M.; Roltsch, M.H. Potential signaling pathways of acute endurance exercise-induced cardiac autophagy and mitophagy and its possible role in cardioprotection. *J. Physiol. Sci.* **2017**, *67*, 639–654. [CrossRef] [PubMed]

68. Lal, H.; Ahmad, F.; Woodgett, J.; Force, T. The GSK-3 family as therapeutic target for myocardial diseases. *Circ. Res.* **2015**, *116*, 138–149. [CrossRef] [PubMed]

69. Fukazawa, R.; Miller, T.A.; Kuramochi, Y.; Frantz, S.; Kim, Y.D.; Marchionni, M.A.; Kelly, R.A.; Sawyer, D.B. Neuregulin-1 protects ventricular myocytes from anthracycline-induced apoptosis via erbB4-dependent activation of PI3-kinase/Akt. *J. Mol. Cell. Cardiol.* **2003**, *35*, 1473–1479. [CrossRef] [PubMed]

70. Cai, M.X.; Shi, X.C.; Chen, T.; Tan, Z.N.; Lin, Q.Q.; Du, S.J.; Tian, Z.J. Exercise training activates neuregulin 1/ErbB signaling and promotes cardiac repair in a rat myocardial infarction model. *Life Sci.* **2016**, *149*, 1–9. [CrossRef]

71. D'Uva, G.; Aharonov, A.; Lauriola, M.; Kain, D.; Yahalom-Ronen, Y.; Carvalho, S.; Weisinger, K.; Bassat, E.; Rajchman, D.; Yifa, O.; et al. ERBB2 triggers mammalian heart regeneration by promoting cardiomyocyte dedifferentiation and proliferation. *Nat. Cell Biol.* **2015**, *17*, 627–638. [CrossRef] [PubMed]

72. Galindo, C.L.; Ryzhov, S.; Sawyer, D.B. Neuregulin as a heart failure therapy and mediator of reverse remodeling. *Curr. Heart Fail. Rep.* **2014**, *11*, 40–49. [CrossRef] [PubMed]

73. Kirabo, A.; Ryzhov, S.; Gupte, M.; Sengsayadeth, S.; Gumina, R.J.; Sawyer, D.B.; Galindo, C.L. Neuregulin-1beta induces proliferation, survival and paracrine signaling in normal human cardiac ventricular fibroblasts. *J. Mol. Cell. Cardiol.* **2017**, *105*, 59–69. [CrossRef] [PubMed]

74. Waring, C.D.; Vicinanza, C.; Papalamprou, A.; Smith, A.J.; Purushothaman, S.; Goldspink, D.F.; Nadal-Ginard, B.; Torella, D.; Ellison, G.M. The adult heart responds to increased workload with physiologic hypertrophy, cardiac stem cell activation, and new myocyte formation. *Eur. Heart J.* **2014**, *35*, 2722–2731. [CrossRef] [PubMed]

75. Barbier, J.; Rannou-Bekono, F.; Marchais, J.; Tanguy, S.; Carre, F. Alterations of beta3-adrenoceptors expression and their myocardial functional effects in physiological model of chronic exercise-induced cardiac hypertrophy. *Mol. Cell. Biochem.* **2007**, *300*, 69–75. [CrossRef] [PubMed]

76. Yang, C.; Talukder, M.A.; Varadharaj, S.; Velayutham, M.; Zweier, J.L. Early ischaemic preconditioning requires Akt- and PKA-mediated activation of eNOS via serine1176 phosphorylation. *Cardiovasc. Res.* **2013**, *97*, 33–43. [CrossRef] [PubMed]

77. Otani, H. The role of nitric oxide in myocardial repair and remodeling. *Antioxid. Redox Signal.* **2009**, *11*, 1913–1928. [CrossRef] [PubMed]

78. Kovacs, A.; Alogna, A.; Post, H.; Hamdani, N. Is enhancing cGMP-PKG signalling a promising therapeutic target for heart failure with preserved ejection fraction? *Neth. Heart J.* **2016**, *24*, 268–274. [CrossRef]

79. Blanton, R.M.; Takimoto, E.; Lane, A.M.; Aronovitz, M.; Piotrowski, R.; Karas, R.H.; Kass, D.A.; Mendelsohn, M.E. Protein kinase g ialpha inhibits pressure overload-induced cardiac remodeling and is required for the cardioprotective effect of sildenafil in vivo. *J. Am. Heart Assoc.* **2012**, *1*, e003731. [CrossRef]

80. Iemitsu, M.; Miyauchi, T.; Maeda, S.; Tanabe, T.; Takanashi, M.; Matsuda, M.; Yamaguchi, I. Exercise training improves cardiac function-related gene levels through thyroid hormone receptor signaling in aged rats. *Am. J. Physiol. Heart Circ. Physiol.* **2004**, *286*, H1696–H1705. [CrossRef]

81. Barile, L.; Lionetti, V.; Cervio, E.; Matteucci, M.; Gherghiceanu, M.; Popescu, L.M.; Torre, T.; Siclari, F.; Moccetti, T.; Vassalli, G. Extracellular vesicles from human cardiac progenitor cells inhibit cardiomyocyte apoptosis and improve cardiac function after myocardial infarction. *Cardiovasc. Res.* **2014**, *103*, 530–541. [CrossRef] [PubMed]

82. Bei, Y.; Xu, T.; Lv, D.; Yu, P.; Xu, J.; Che, L.; Das, A.; Tigges, J.; Toxavidis, V.; Ghiran, I.; et al. Exercise-induced circulating extracellular vesicles protect against cardiac ischemia-reperfusion injury. *Basic Res. Cardiol.* **2017**, *112*, 38. [CrossRef] [PubMed]

83. Adamiak, M.; Cheng, G.; Bobis-Wozowicz, S.; Zhao, L.; Kedracka-Krok, S.; Samanta, A.; Karnas, E.; Xuan, Y.T.; Skupien-Rabian, B.; Chen, X.; et al. Induced Pluripotent Stem Cell (iPSC)-Derived Extracellular Vesicles Are Safer and More Effective for Cardiac Repair Than iPSCs. *Circ. Res.* **2018**, *122*, 296–309. [CrossRef] [PubMed]

84. Huang, Y.; Shen, X.J.; Zou, Q.; Wang, S.P.; Tang, S.M.; Zhang, G.Z. Biological functions of microRNAs: A review. *J. Physiol. Biochem.* **2011**, *67*, 129–139. [CrossRef] [PubMed]

85. Verdoorn, K.S.; Matsuura, C.; Borges, J.P. Exercise for cardiac health and regeneration: Killing two birds with one stone. *Ann. Transl. Med.* **2017**, *5*, S13. [CrossRef] [PubMed]

86. Ultimo, S.; Zauli, G.; Martelli, A.M.; Vitale, M.; McCubrey, J.A.; Capitani, S.; Neri, L.M. Cardiovascular disease-related miRNAs expression: Potential role as biomarkers and effects of training exercise. *Oncotarget* **2018**, *9*, 17238–17254. [CrossRef]

87. Clauss, S.; Sinner, M.F.; Kaab, S.; Wakili, R. The Role of MicroRNAs in Antiarrhythmic Therapy for Atrial Fibrillation. *Arrhythmia Electrophysiol. Rev.* **2015**, *4*, 146–155. [CrossRef]

88. Clauss, S.; Wakili, R.; Hildebrand, B.; Kaab, S.; Hoster, E.; Klier, I.; Martens, E.; Hanley, A.; Hanssen, H.; Halle, M.; et al. MicroRNAs as Biomarkers for Acute Atrial Remodeling in Marathon Runners (The miRathon Study—A Sub-Study of the Munich Marathon Study). *PLoS ONE* **2016**, *11*, e0148599. [CrossRef]

89. Dawson, K.; Wakili, R.; Ordog, B.; Clauss, S.; Chen, Y.; Iwasaki, Y.; Voigt, N.; Qi, X.Y.; Sinner, M.F.; Dobrev, D.; et al. MicroRNA29: A mechanistic contributor and potential biomarker in atrial fibrillation. *Circulation* **2013**, *127*, 1466–1475. [CrossRef]

90. Chen, Y.; Wakili, R.; Xiao, J.; Wu, C.T.; Luo, X.; Clauss, S.; Dawson, K.; Qi, X.; Naud, P.; Shi, Y.F.; et al. Detailed characterization of microRNA changes in a canine heart failure model: Relationship to arrhythmogenic structural remodeling. *J. Mol. Cell. Cardiol.* **2014**, *77*, 113–124. [CrossRef]

91. Cahill, T.J.; Choudhury, R.P.; Riley, P.R. Heart regeneration and repair after myocardial infarction: Translational opportunities for novel therapeutics. *Nat. Rev. Drug Discov.* **2017**, *16*, 699–717. [CrossRef] [PubMed]

92. Uchida, S.; Dimmeler, S. Exercise controls non-coding RNAs. *Cell Metab.* **2015**, *21*, 511–512. [CrossRef] [PubMed]

93. Heallen, T.R.; Kadow, Z.A.; Kim, J.H.; Wang, J.; Martin, J.F. Stimulating Cardiogenesis as a Treatment for Heart Failure. *Circ. Res.* **2019**, *124*, 1647–1657. [CrossRef] [PubMed]

94. Shi, J.; Bei, Y.; Kong, X.; Liu, X.; Lei, Z.; Xu, T.; Wang, H.; Xuan, Q.; Chen, P.; Xu, J.; et al. miR-17-3p Contributes to Exercise-Induced Cardiac Growth and Protects against Myocardial Ischemia-Reperfusion Injury. *Theranostics* **2017**, *7*, 664–676. [CrossRef] [PubMed]

95. Liu, X.; Xiao, J.; Zhu, H.; Wei, X.; Platt, C.; Damilano, F.; Xiao, C.; Bezzerides, V.; Bostrom, P.; Che, L.; et al. miR-222 is necessary for exercise-induced cardiac growth and protects against pathological cardiac remodeling. *Cell Metab.* **2015**, *21*, 584–595. [CrossRef]

96. Palabiyik, O.; Tastekin, E.; Doganlar, Z.B.; Tayfur, P.; Dogan, A.; Vardar, S.A. Alteration in cardiac PI3K/Akt/mTOR and ERK signaling pathways with the use of growth hormone and swimming, and the roles of miR21 and miR133. *Biomed. Rep.* **2019**, *10*, 97–106. [CrossRef] [PubMed]

97. Da Silva, N.D., Jr.; Fernandes, T.; Soci, U.P.; Monteiro, A.W.; Phillips, M.I.; de Oliveira, E.M. Swimming training in rats increases cardiac MicroRNA-126 expression and angiogenesis. *Med. Sci. Sports Exerc.* **2012**, *44*, 1453–1462. [CrossRef]

98. Dong, D.L.; Chen, C.; Huo, R.; Wang, N.; Li, Z.; Tu, Y.J.; Hu, J.T.; Chu, X.; Huang, W.; Yang, B.F. Reciprocal repression between microRNA-133 and calcineurin regulates cardiac hypertrophy: A novel mechanism for progressive cardiac hypertrophy. *Hypertension* **2010**, *55*, 946–952. [CrossRef]

99. Soci, U.P.; Fernandes, T.; Hashimoto, N.Y.; Mota, G.F.; Amadeu, M.A.; Rosa, K.T.; Irigoyen, M.C.; Phillips, M.I.; Oliveira, E.M. MicroRNAs 29 are involved in the improvement of ventricular compliance promoted by aerobic exercise training in rats. *Physiol. Genom.* **2011**, *43*, 665–673. [CrossRef]

100. Chaturvedi, P.; Kalani, A.; Medina, I.; Familtseva, A.; Tyagi, S.C. Cardiosome mediated regulation of MMP9 in diabetic heart: Role of mir29b and mir455 in exercise. *J. Cell. Mol. Med.* **2015**, *19*, 2153–2161. [CrossRef]

101. Li, Z.; Liu, L.; Hou, N.; Song, Y.; An, X.; Zhang, Y.; Yang, X.; Wang, J. miR-199-sponge transgenic mice develop physiological cardiac hypertrophy. *Cardiovasc. Res.* **2016**, *110*, 258–267. [CrossRef] [PubMed]

102. Zhao, Y.; Ma, Z. Swimming training affects apoptosis-related microRNAs and reduces cardiac apoptosis in mice. *Gen. Physiol. Biophys.* **2016**, *35*, 443–450. [CrossRef] [PubMed]

103. Eulalio, A.; Mano, M.; Dal Ferro, M.; Zentilin, L.; Sinagra, G.; Zacchigna, S.; Giacca, M. Functional screening identifies miRNAs inducing cardiac regeneration. *Nature* **2012**, *492*, 376–381. [CrossRef] [PubMed]

104. Lu, D.; Tang, L.; Zhuang, Y.; Zhao, P. miR-17-3P regulates the proliferation and survival of colon cancer cells by targeting Par4. *Mol. Med. Rep.* **2018**, *17*, 618–623. [CrossRef] [PubMed]

105. Yan, H.; Song, K.; Zhang, G. MicroRNA-17-3p promotes keratinocyte cells growth and metastasis via targeting MYOT and regulating Notch1/NF-kappaB pathways. *Die Pharm.* **2017**, *72*, 543–549. [CrossRef]

106. Hammoud, L.; Burger, D.E.; Lu, X.; Feng, Q. Tissue inhibitor of metalloproteinase-3 inhibits neonatal mouse cardiomyocyte proliferation via EGFR/JNK/SP-1 signaling. *Am. J. Physiol. Cell Physiol.* **2009**, *296*, C735–C745. [CrossRef] [PubMed]

107. Worby, C.A.; Dixon, J.E. Pten. *Annu. Rev. Biochem.* **2014**, *83*, 641–669. [CrossRef] [PubMed]

108. Porrello, E.R.; Mahmoud, A.I.; Simpson, E.; Johnson, B.A.; Grinsfelder, D.; Canseco, D.; Mammen, P.P.; Rothermel, B.A.; Olson, E.N.; Sadek, H.A. Regulation of neonatal and adult mammalian heart regeneration by the miR-15 family. *Proc. Natl. Acad. Sci. USA* **2013**, *110*, 187–192. [CrossRef]

109. Hua, Z.; Lv, Q.; Ye, W.; Wong, C.K.; Cai, G.; Gu, D.; Ji, Y.; Zhao, C.; Wang, J.; Yang, B.B.; et al. MiRNA-directed regulation of VEGF and other angiogenic factors under hypoxia. *PLoS ONE* **2006**, *1*, e116. [CrossRef]

110. Iemitsu, M.; Maeda, S.; Jesmin, S.; Otsuki, T.; Miyauchi, T. Exercise training improves aging-induced downregulation of VEGF angiogenic signaling cascade in hearts. *Am. J. Physiol. Heart Circ. Physiol.* **2006**, *291*, H1290–H1298. [CrossRef]

111. Uhlemann, M.; Mobius-Winkler, S.; Fikenzer, S.; Adam, J.; Redlich, M.; Mohlenkamp, S.; Hilberg, T.; Schuler, G.C.; Adams, V. Circulating microRNA-126 increases after different forms of endurance exercise in healthy adults. *Eur. J. Prev. Cardiol.* **2014**, *21*, 484–491. [CrossRef] [PubMed]

112. Mooren, F.C.; Viereck, J.; Kruger, K.; Thum, T. Circulating microRNAs as potential biomarkers of aerobic exercise capacity. *Am. J. Physiol. Heart Circ. Physiol.* **2014**, *306*, H557–H563. [CrossRef] [PubMed]

113. Egan, B.; Zierath, J.R. Exercise metabolism and the molecular regulation of skeletal muscle adaptation. *Cell Metab.* **2013**, *17*, 162–184. [CrossRef] [PubMed]

114. Lee, Y.; Min, K.; Talbert, E.E.; Kavazis, A.N.; Smuder, A.J.; Willis, W.T.; Powers, S.K. Exercise protects cardiac mitochondria against ischemia-reperfusion injury. *Med. Sci. Sports Exerc.* **2012**, *44*, 397–405. [CrossRef] [PubMed]

115. Kolwicz, S.C., Jr. An "Exercise" in Cardiac Metabolism. *Front. Cardiovasc. Med.* **2018**, *5*, 66. [CrossRef]

116. Strom, C.C.; Aplin, M.; Ploug, T.; Christoffersen, T.E.; Langfort, J.; Viese, M.; Galbo, H.; Haunso, S.; Sheikh, S.P. Expression profiling reveals differences in metabolic gene expression between exercise-induced cardiac effects and maladaptive cardiac hypertrophy. *FEBS J.* **2005**, *272*, 2684–2695. [CrossRef]

117. Aitman, T.J.; Glazier, A.M.; Wallace, C.A.; Cooper, L.D.; Norsworthy, P.J.; Wahid, F.N.; Al-Majali, K.M.; Trembling, P.M.; Mann, C.J.; Shoulders, C.C.; et al. Identification of Cd36 (Fat) as an insulin-resistance gene causing defective fatty acid and glucose metabolism in hypertensive rats. *Nat. Genet.* **1999**, *21*, 76–83. [CrossRef]

118. Glatz, J.F.; Angin, Y.; Steinbusch, L.K.; Schwenk, R.W.; Luiken, J.J. CD36 as a target to prevent cardiac lipotoxicity and insulin resistance. *Prostaglandins Leukot. Essent. Fat. Acids* **2013**, *88*, 71–77. [CrossRef]

119. Kuang, M.; Febbraio, M.; Wagg, C.; Lopaschuk, G.D.; Dyck, J.R. Fatty acid translocase/CD36 deficiency does not energetically or functionally compromise hearts before or after ischemia. *Circulation* **2004**, *109*, 1550–1557. [CrossRef]

120. Riehle, C.; Abel, E.D. PGC-1 proteins and heart failure. *Trends Cardiovasc. Med.* **2012**, *22*, 98–105. [CrossRef]

121. Wang, H.; Bei, Y.; Lu, Y.; Sun, W.; Liu, Q.; Wang, Y.; Cao, Y.; Chen, P.; Xiao, J.; Kong, X. Exercise Prevents Cardiac Injury and Improves Mitochondrial Biogenesis in Advanced Diabetic Cardiomyopathy with PGC-1alpha and Akt Activation. *Cell. Physiol. Biochem.* **2015**, *35*, 2159–2168. [CrossRef] [PubMed]

122. Hardie, D.G.; Hawley, S.A.; Scott, J.W. AMP-activated protein kinase–development of the energy sensor concept. *J. Physiol.* **2006**, *574*, 7–15. [CrossRef] [PubMed]

123. Li, H.L.; Yin, R.; Chen, D.; Liu, D.; Wang, D.; Yang, Q.; Dong, Y.G. Long-term activation of adenosine monophosphate-activated protein kinase attenuates pressure-overload-induced cardiac hypertrophy. *J. Cell. Biochem.* **2007**, *100*, 1086–1099. [CrossRef] [PubMed]

124. Zhang, P.; Hu, X.; Xu, X.; Fassett, J.; Zhu, G.; Viollet, B.; Xu, W.; Wiczer, B.; Bernlohr, D.A.; Bache, R.J.; et al. AMP activated protein kinase-alpha2 deficiency exacerbates pressure-overload-induced left ventricular hypertrophy and dysfunction in mice. *Hypertension* **2008**, *52*, 918–924. [CrossRef] [PubMed]

125. Ma, X.; Fu, Y.; Xiao, H.; Song, Y.; Chen, R.; Shen, J.; An, X.; Shen, Q.; Li, Z.; Zhang, Y. Cardiac Fibrosis Alleviated by Exercise Training Is AMPK-Dependent. *PLoS ONE* **2015**, *10*, e0129971. [CrossRef]

126. Wang, J.; Song, Y.; Li, H.; Shen, Q.; Shen, J.; An, X.; Wu, J.; Zhang, J.; Wu, Y.; Xiao, H.; et al. Exacerbated cardiac fibrosis induced by beta-adrenergic activation in old mice due to decreased AMPK activity. *Clin. Exp. Pharmacol. Physiol.* **2016**, *43*, 1029–1037. [CrossRef]

127. Li, L.; Muhlfeld, C.; Niemann, B.; Pan, R.; Li, R.; Hilfiker-Kleiner, D.; Chen, Y.; Rohrbach, S. Mitochondrial biogenesis and PGC-1alpha deacetylation by chronic treadmill exercise: Differential response in cardiac and skeletal muscle. *Basic Res. Cardiol.* **2011**, *106*, 1221–1234. [CrossRef]

128. Chen, L.; Knowlton, A.A. Mitochondria and heart failure: New insights into an energetic problem. *Minerva Cardioangiol.* **2010**, *58*, 213–229.

129. Kavazis, A.N.; McClung, J.M.; Hood, D.A.; Powers, S.K. Exercise induces a cardiac mitochondrial phenotype that resists apoptotic stimuli. *Am. J. Physiol. Heart Circ. Physiol.* **2008**, *294*, H928–H935. [CrossRef]

130. Kwak, H.B. Effects of aging and exercise training on apoptosis in the heart. *J. Exerc. Rehabil.* **2013**, *9*, 212–219. [CrossRef]

131. Kwak, H.B.; Song, W.; Lawler, J.M. Exercise training attenuates age-induced elevation in Bax/Bcl-2 ratio, apoptosis, and remodeling in the rat heart. *FASEB J.* **2006**, *20*, 791–793. [CrossRef] [PubMed]

132. Robinson, E.; Grieve, D.J. Significance of peroxisome proliferator-activated receptors in the cardiovascular system in health and disease. *Pharmacol. Ther.* **2009**, *122*, 246–263. [CrossRef] [PubMed]

133. Santos, M.H.; Higuchi Mde, L.; Tucci, P.J.; Garavelo, S.M.; Reis, M.M.; Antonio, E.L.; Serra, A.J.; Maranhao, R.C. Previous exercise training increases levels of PPAR-alpha in long-term post-myocardial infarction in rats, which is correlated with better inflammatory response. *Clinics* **2016**, *71*, 163–168. [CrossRef]

134. Ibarra-Lara Mde, L.; Sanchez-Aguilar, M.; Soria, E.; Torres-Narvaez, J.C.; Del Valle-Mondragon, L.; Cervantes-Perez, L.G.; Perez-Severiano, F.; Ramirez-Ortega Mdel, C.; Pastelin-Hernandez, G.; Oidor-Chan, V.H.; et al. Peroxisome proliferator-activated receptors (PPAR) downregulate the expression of pro-inflammatory molecules in an experimental model of myocardial infarction. *Can. J. Physiol. Pharmacol.* **2016**, *94*, 634–642. [CrossRef] [PubMed]

135. Iemitsu, M.; Miyauchi, T.; Maeda, S.; Tanabe, T.; Takanashi, M.; Irukayama-Tomobe, Y.; Sakai, S.; Ohmori, H.; Matsuda, M.; Yamaguchi, I. Aging-induced decrease in the PPAR-alpha level in hearts is improved by exercise training. *Am. J. Physiol. Heart Circ. Physiol.* **2002**, *283*, H1750–H1760. [CrossRef] [PubMed]

MDPI

St. Alban-Anlage 66

4052 Basel

Switzerland

Tel. +41 61 683 77 34

Fax +41 61 302 89 18

www.mdpi.com

Cells Editorial Office

E-mail: cells@mdpi.com

www.mdpi.com/journal/cells